# STUDENT'S SOLUTIONS MANUAL

## JUDITH A. PENNA
*Indiana University Purdue University Indianapolis*

# INTERMEDIATE ALGEBRA:
## CONCEPTS AND APPLICATIONS

### SEVENTH EDITION

# Marvin L. Bittinger
*Indiana University Purdue University Indianapolis*

# David J. Ellenbogen
*Community College of Vermont*

PEARSON

Addison
Wesley

Boston  San Francisco  New York
London  Toronto  Sydney  Tokyo  Singapore  Madrid
Mexico City  Munich  Paris  Cape Town  Hong Kong  Montreal

Copyright © 2006 Pearson Education, Inc.
Publishing as Pearson Addison-Wesley, 75 Arlington Street, Boston, MA 02116.

ISBN    0-321-27822-4

6 BB 08 07

# Contents

# Chapter 1

# Algebra and Problem Solving

## Exercise Set 1.1

**1.** A letter representing a specific number that never changes is called a <u>constant</u>. (See page 2 in the text.)

**3.** In the expression $7y$, the multipliers 7 and $y$ are called <u>factors</u>. (See page 4 in the text.)

**5.** When no grouping symbols, exponents, division, or multiplication appear, we subtract before we add, provided the subtraction appears to the <u>left</u> of any addition. (See page 5 in the text.)

**7.** A number that can be written in the form $a/b$, where $a$ and $b$ are integers (with $b \neq 0$), is said to be a <u>rational</u> number. (See page 7 in the text.)

**9.** Division can be used to show that $\dfrac{7}{40}$ can be written as a <u>terminating</u> decimal. (See page 8 in the text.)

**11.** Six less than some number

Let $n$ represent the number. Then we have

$$n - 6.$$

**13.** Twelve times a number

Let $t$ represent the number. Then we have

$$12t.$$

**15.** Sixty-five percent of some number

Let $x$ represent the number. Then we have

$$0.65x, \text{ or } \frac{65}{100}x.$$

**17.** Nine more than twice a number

Let $y$ represent the number. Then we have

$$2y + 9.$$

**19.** Eight more than ten percent of some number

Let $s$ represent the number. Then we have

$$0.1s + 8, \text{ or } \frac{10}{100}s + 8$$

**21.** One less than the difference of two numbers

Let $m$ and $n$ represent the numbers. Then we have

$$m - n - 1.$$

**23.** Ninety miles per every four gallons of gas

We have

$$90 \div 4, \text{ or } \frac{90}{4}.$$

**25.** Substitute and carry out the operations indicated.

$$
\begin{aligned}
7x + y &= 7 \cdot 3 + 4 \\
&= 21 + 4 \\
&= 25
\end{aligned}
$$

**27.** Substitute and carry out the operations indicated.

$$
\begin{aligned}
2c \div 3b &= 2 \cdot 6 \div 3 \cdot 2 \\
&= 12 \div 3 \cdot 2 \\
&= 4 \cdot 2 \\
&= 8
\end{aligned}
$$

**29.** Substitute and carry out the operations indicated.

$$
\begin{aligned}
25 + r^2 - s &= 25 + 3^2 - 7 \\
&= 25 + 9 - 7 \\
&= 34 - 7 \\
&= 27
\end{aligned}
$$

**31.** $3n^2p - 3pn^2 = 3 \cdot 5^2 \cdot 9 - 3 \cdot 9 \cdot 5^2$

Observe that $3 \cdot 5^2 \cdot 9$ and $3 \cdot 9 \cdot 5^2$ represent the same number, so their difference is 0.

**33.** Substitute and carry out the operations indicated.

$$
\begin{aligned}
5x \div (2 + x - y) &= 5 \cdot 6 \div (2 + 6 - 2) \\
&= 5 \cdot 6 \div (8 - 2) \\
&= 5 \cdot 6 \div 6 \\
&= 30 \div 6 \\
&= 5
\end{aligned}
$$

**35.** Substitute and carry out the operations indicated.

$$
\begin{aligned}
[10 - (a - b)]^2 &= [10 - (7 - 2)]^2 \\
&= [10 - 5]^2 \\
&= 5^2 \\
&= 25
\end{aligned}
$$

**37.** Substitute and carry out the operations indicated.

$$
\begin{aligned}
[5(r + s)]^2 &= [5(1 + 2)]^2 \\
&= [5(3)]^2 \\
&= 15^2 \\
&= 225
\end{aligned}
$$

**39.** Substitute and carry out the operations indicated.

$$
\begin{aligned}
m^2 - [2(m - n)]^2 &= 7^2 - [2(7 - 5)]^2 \\
&= 7^2 - [2(2)]^2 \\
&= 7^2 - 4^2 \\
&= 49 - 16 \\
&= 33
\end{aligned}
$$

**41.** Substitute and carry out the operations indicated.

$$(r - s)^2 - 3(2r - s) = (11 - 3)^2 - 3(2 \cdot 11 - 3)$$
$$= 8^2 - 3(22 - 3)$$
$$= 8^2 - 3(19)$$
$$= 64 - 3(19)$$
$$= 64 - 57$$
$$= 7$$

**43.** We substitute 5 for $b$ and 7 for $h$ and multiply:

$$A = \frac{1}{2} \cdot b \cdot h = \frac{1}{2} \cdot 5 \cdot 7 = 17.5 \text{ sq ft}$$

**45.** We substitute 7 for $b$ and 3.2 for $h$ and multiply:

$$A = \frac{1}{2} \cdot b \cdot h = \frac{1}{2}(7)(3.2) = 11.2 \text{ sq m}$$

**47.** List the letters in the set: {a,e,i,o,u}, or {a,e,i,o,u,y}

**49.** List the numbers in the set: $\{1, 3, 5, 7, \ldots\}$

**51.** List the numbers in the set: $\{5, 10, 15, 20, \ldots\}$

**53.** Specify the conditions under which a number is in the set: $\{x | x \text{ is an odd number between 10 and 20}\}$

**55.** Specify the conditions under which a number is in the set: $\{x | x \text{ is a whole number less than 5}\}$

**57.** Specify the conditions under which a number is in the set: $\{n | n \text{ is a multiple of 5 between 7 and 79}\}$

**59.** (a) 0 and 6 are whole numbers.

(b) $-3$, 0, and 6 are integers.

(c) $-8.7$, $-3$, 0, $\frac{2}{3}$, and 6 are rational numbers.

(d) $\sqrt{7}$ is an irrational number.

(e) All of the given numbers are real numbers.

**61.** (a) 0 and 3 are whole numbers.

(b) $-17$, 0, and 3 are integers.

(c) $-17$, $-4.13$, 0, $\frac{5}{4}$, and 3 are rational numbers.

(d) $\sqrt{77}$ is an irrational number.

(e) All of the given numbers are real numbers.

**63.** Since 5.1 is not a natural numbers, the statement is false.

**65.** Since every whole number is an integer, the statement is true.

**67.** Since $\frac{2}{3}$ is a rational number, the statement is false.

**69.** Since $\sqrt{10}$ is an irrational number and every member of the set of irrational numbers is a member of the set of real numbers, the statement is false.

**71.** Since the set of integers includes some numbers that are not natural numbers, the statement is true.

**73.** Since the set of rational numbers includes some numbers that are not integers, he statement is false.

**75.** *Writing Exercise*

**77.** *Writing Exercise*

**79.** The quotient of the sum of two numbers and their difference

Let $a$ and $b$ represent the numbers. Then we have

$$\frac{a + b}{a - b}.$$

**81.** Half of the difference of the squares of two numbers

Let $r$ and $s$ represent the numbers. Then we have

$$\frac{1}{2}(r^2 - s^2), \text{ or } \frac{r^2 - s^2}{2}.$$

**83.** The only whole number that is not also a natural number is 0. Using roster notation to name the set, we have {0}.

**85.** List the numbers in the set:

$$\{5, 10, 15, 20, \ldots\}$$

**87.** List the numbers in the set:

$$\{1, 3, 5, 7, \ldots\}$$

**89.** Recall from geometry that when a right triangle has legs of length 2 and 3, the length of the hypotenuse is $\sqrt{2^2 + 3^2} = \sqrt{4 + 9} = \sqrt{13}$. We draw such a triangle:

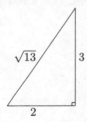

## Exercise Set 1.2

**1.** It is true that the sum of the two negative numbers is always negative. (See page 15 in the text.)

**3.** It is true that the product of a negative number and a positive number is always negative. (See page 17 in the text.)

**5.** The statement is false. Consider $-5 + 2 = -3$, for example.

**7.** The statement is false. Let $a = 2$ and $b = 6$. Then $2 < 6$ and $|2| < |6|$.

**9.** It is true that the associative law of multiplication states that for all real numbers $a$, $b$, and $c$, $(ab)c$ is equivalent to $a(bc)$. (See page 20 in the text.)

**11.** $|-9| = 9$     $-9$ is 9 units from 0.

**13.** $|6| = 6$     6 is 6 units from 0.

**15.** $|-6.2| = 6.2$     $-6.2$ is 6.2 units from 0.

**17.** $|0| = 0$     0 is 0 units from itself.

**19.** $\left|1\frac{7}{8}\right| = 1\frac{7}{8}$     $1\frac{7}{8}$ is $1\frac{7}{8}$ units from 0.

**21.** $|-4.21| = 4.21$      $-4.21$ is 4.21 units from 0.

**23.** $-6 \leq -2$

$-6$ is less than or equal to $-2$, a true statement since $-6$ is to the left of $-2$.

**25.** $-9 > 1$

$-9$ is greater than 1, a false statement since $-9$ is to the left of 1.

**27.** $3 \geq -5$

3 is greater than or equal to $-5$, a true statement since $-5$ is to the left of 3.

**29.** $-8 < -3$

$-8$ is less than $-3$, a true statement since $-8$ is to the left of $-3$.

**31.** $-4 \geq -4$

$-4$ is greater or equal to $-4$. Since $-4 = -4$ is true, $-4 \geq -4$ is true.

**33.** $-5 < -5$

$-5$ is less than $-5$, a false statement since $-5$ does not lie to the left of itself.

**35.** $3 + 9$

Two positive numbers: Add the numbers, getting 12. The answer is positive, 12.

**37.** $-4 + (-7)$

Two negative numbers: Add the absolute values, getting 11. The answer is negative, $-11$.

**39.** $-3.9 + 2.7$

A negative and a positive number: The absolute values are 3.9 and 2.7. Subtract 2.7 from 3.9 to get 1.2. The negative number is farther from 0, so the answer is negative, $-1.2$.

**41.** $\frac{2}{7} + \left(-\frac{3}{5}\right) = \frac{10}{35} + \left(-\frac{21}{35}\right)$

A positive and a negative number. The absolute values are $\frac{10}{35}$ and $\frac{21}{35}$. Subtract $\frac{10}{35}$ from $\frac{21}{35}$ to get $\frac{11}{35}$. The negative number is farther from 0, so the answer is negative, $-\frac{11}{35}$.

**43.** $-3.26 + (-5.8)$

Two negative numbers: Add the absolute values, getting 9.06. The answer is negative, $-9.06$.

**45.** $-\frac{1}{9} + \frac{2}{3} = -\frac{1}{9} + \frac{6}{9}$

A negative and a positive number. The absolute values are $\frac{1}{9}$ and $\frac{6}{9}$. Subtract $\frac{1}{9}$ from $\frac{6}{9}$ to get $\frac{5}{9}$. The positive number is farther from 0, so the answer is positive, $\frac{5}{9}$.

**47.** $0 + (-4.5)$

One number is zero: The sum is the other number, $-4.5$.

**49.** $-7.24 + 7.24$

A negative and a positive number: The numbers have the same absolute value, 7.24, so the answer is 0.

**51.** $15.9 + (-22.3)$

A positive and a negative number: The absolute values are 15.9 and 22.3. Subtract 15.9 from 22.3 to get 6.4. The negative number is farther from 0, so the answer is negative, $-6.4$.

**53.** The opposite of 3.14 is $-3.14$, because $3.14 + (-3.14) = 0$.

**55.** The opposite of $-4\frac{1}{3}$ is $4\frac{1}{3}$, because $-4\frac{1}{3} + 4\frac{1}{3} = 0$.

**57.** The opposite of 0 is 0, because $0 + 0 = 0$.

**59.** If $x = 9$, then $-x = -9$. (The opposite of 9 is $-9$.)

**61.** If $x = -2.7$, then $-x = -(-2.7) = 2.7$.

(The opposite of $-2.7$ is 2.7.)

**63.** If $x = 1.79$, then $-x = -1.79$. (The opposite of 1.79 is $-1.79$.)

**65.** If $x = 0$, then $-x = 0$. (The opposite of 0 is 0.)

**67.** $8 - 5 = 8 + (-5)$   Change the sign and add.

$\quad = 3$

**69.** $5 - 8 = 5 + (-8)$   Change the sign and add.

$\quad = -3$

**71.** $-5 - (-12) = -5 + 12$   Change the sign and add.

$\quad = 7$

**73.** $-5 - 14 = -5 + (-14) = -19$

**75.** $2.7 - 5.8 = 2.7 + (-5.8) = -3.1$

**77.** $-\frac{3}{5} - \frac{1}{2} = -\frac{3}{5} + \left(-\frac{1}{2}\right)$

$\quad = -\frac{6}{10} + \left(-\frac{5}{10}\right)$   Finding a common denominator

$\quad = -\frac{11}{10}$

**79.** $-3.9 - (-6.8) = -3.9 + 6.8 = 2.9$

**81.** $0 - (-7.9) = 0 + 7.9 = 7.9$

**83.** $(-5)6$

Two numbers with unlike signs: Multiply their absolute values, getting 30. The answer is negative, $-30$.

**85.** $(-4)(-9)$

Two numbers with the same sign: Multiply their absolute values, getting 36. The answer is positive, 36.

**87.** $(4.2)(-5)$

Two numbers with unlike signs: Multiply their absolute values, getting 21. The answer is negative, $-21$.

**89.** $\frac{3}{7}(-1)$

Two numbers with unlike signs: Multiply their absolute values, getting $\frac{3}{7}$. The answer is negative, $-\frac{3}{7}$.

**91.** $(-17.45)\cdot 0 = 0$

**93.** $(-3.2)\times(-1.7)$

Two numbers with the same sign: Multiply their absolute values, getting 5.44. The answer is positive, 5.44.

**95.** $\frac{-10}{-2}$

Two numbers with the same sign: Divide their absolute values, getting 5. The answer is positive, 5.

**97.** $\frac{-100}{20}$

Two numbers with unlike signs: Divide their absolute values, getting 5. The answer is negative, $-5$.

**99.** $\frac{73}{-1}$

Two numbers with unlike signs: Divide their absolute values, getting 73. The answer is negative, $-73$.

**101.** $\frac{0}{-7} = 0$

**103.** The reciprocal of 4 is $\frac{1}{4}$, because $4\cdot\frac{1}{4}=1$.

**105.** The reciprocal of $-9$ is $\frac{1}{-9}$, or $-\frac{1}{9}$, because $-9\left(-\frac{1}{9}\right)=1$.

**107.** The reciprocal of $\frac{2}{3}$ is $\frac{3}{2}$, because $\frac{2}{3}\cdot\frac{3}{2}=1$.

**109.** The reciprocal of $-\frac{3}{11}$ is $-\frac{11}{3}$, because $-\frac{3}{11}\left(-\frac{11}{3}\right)=1$.

**111.** $\frac{2}{3}\div\frac{4}{5}$

$=\frac{2}{3}\cdot\frac{5}{4}$  Multiplying by the reciprocal of 4/5

$=\frac{10}{12}$, or $\frac{5}{6}$

**113.** $\left(-\frac{3}{5}\right)\div\frac{1}{2}$

$=-\frac{3}{5}\cdot\frac{2}{1}$  Multiplying by the reciprocal of 1/2

$=-\frac{6}{5}$

**115.** $\left(-\frac{2}{9}\right)\div(-8)$

$=-\frac{2}{9}\cdot\left(-\frac{1}{8}\right)$  Multiplying by the reciprocal of $-8$

$=\frac{2}{72}$, or $\frac{1}{36}$

**117.** $-\frac{12}{7}\div\left(-\frac{12}{7}\right)$

This is a number divided by itself so the quotient is 1. We would also do this exercise as follows.

$-\frac{12}{7}\div\left(-\frac{12}{7}\right)=-\frac{12}{7}\cdot\left(-\frac{7}{12}\right)$  Multiplying by the reciprocal of $-\frac{12}{7}$

$=1$

**119.** $9-(8-3\cdot2^3)=9-(8-3\cdot8)$  Working within

$=9-(8-24)$  the parentheses

$=9-(-16)$  first

$=9+16$

$=25$

**121.** $\frac{5\cdot2-4^2}{27-2^4}=\frac{5\cdot2-16}{27-16}=\frac{10-16}{11}=\frac{-6}{11}$, or $-\frac{6}{11}$

**123.** $\frac{3^4-(5-3)^4}{8-2^3}=\frac{3^4-2^4}{8-8}=\frac{81-16}{0}$

Since division by 0 is undefined, this quotient is undefined.

**125.** $\frac{(2-3)^3-5|2-4|}{7-2\cdot5^2}=\frac{(-1)^3-5|-2|}{7-2\cdot25}=\frac{-1-5(2)}{7-50}=$ $\frac{-1-10}{-43}=\frac{-11}{-43}=\frac{11}{43}$

**127.** $|2^2-7|^3+4=|4-7|^3+4=|-3|^3+4=$ $3^3+4=27+4=31$

**129.** $32-(-5)^2+15\div(-3)\cdot2$

$=32-25+15\div(-3)\cdot2$  Evaluating the exponential expression

$=32-25-5\cdot2$  Dividing

$=32-25-10$  Multiplying

$=-3$  Subtracting

**131.** $17-\sqrt{11-(3+4)}\div[-5-(-6)]^2$

$=17-\sqrt{11-7}\div[-5+6]^2$

$=17-\sqrt{4}\div[1]^2$

$=17-2\div1$

$=17-2$

$=15$

**133.** Using the commutative law of addition, we have $4a+7b=7b+4a$.

Using the commutative law of multiplication, we have $4a+7b=a\cdot4+7b$

or $4a+7b=4a+b\cdot7$

or $4a+7b=a\cdot4+b\cdot7$.

**135.** Using the commutative law of multiplication, we have $(7x)y=y(7x)$

or $(7x)y=(x\cdot7)y$.

**137.** $(3x)y$

$=3(xy)$  Associative law of multiplication

**139.** $\quad x + (2y + 5)$

$= (x + 2y) + 5 \quad$ Associative law of addition

**141.** $\quad 2(a + 5)$

$= 2 \cdot a + 2 \cdot 5 \quad$ Using the distributive law

$= 2a + 10$

**143.** $\quad 4(x - y)$

$= 4 \cdot x - 4 \cdot y \quad$ Using the distributive law

$= 4x - 4y$

**145.** $\quad -5(2a + 3b)$

$= -5 \cdot 2a + (-5) \cdot 3b$

$= -10a - 15b$

**147.** $\quad 9a(b - c + d)$

$= 9a \cdot b - 9a \cdot c + 9a \cdot d$

$= 9ab - 9ac + 9ad$

**149.** $5x + 15 = 5 \cdot x + 5 \cdot 3 = 5(x + 3)$

**151.** $3p - 9 = 3 \cdot p - 3 \cdot 3 = 3(p - 3)$

**153.** $7x - 21y + 14z = 7 \cdot x - 7 \cdot 3y + 7 \cdot 2z = 7(x - 3y + 2z)$

**155.** $255 - 34b = 17 \cdot 15 - 17 \cdot 2b = 17(15 - 2b)$

**157.** *Writing Exercise*

**159.** Substitute and carry out the indicated operations.

$2(x + 5) = 2(3 + 5) = 2 \cdot 8 = 16$

$2x + 10 = 2 \cdot 3 + 10 = 6 + 10 = 16$

**161.** *Writing Exercise*

**163.** $(8 - 5)^3 + 9 = 36$

**165.** $5 \cdot 2^3 \div (3 - 4)^4 = 40$

**167.** Any value of $a$ such that $a \leq -6.2$ satisfies the given conditions. The largest of these values is $-6.2$.

**169.** *Writing Exercise*

---

# Exercise Set 1.3

**1.** By the distributive law, the expression $2(x - 5)$ is equivalent to the expression $2x - 10$, so they are equivalent expressions.

**3.** $3(t + 4) = 25$ and $3t + 12 = 25$ are equations and they have the same solution, $\dfrac{13}{3}$, so they are equivalent equations.

**5.** $4x - 9 = 7$ and $4x = 16$ are equations and they have the same solution, 4, so they are equivalent equations.

**7.** Combining like terms in the expression $8t + 5 - 2t + 1$, we get the expression $6t + 6$, so these are equivalent expressions.

**9.** $6x - 3 = 10x + 5$ and $-8 = 4x$ are equations and they have the same solution, $-2$, so they are equivalent equations.

**11.** $t + 5 = 11$ and $3t = 18$

Each equation has only one solution, the number 6. Thus the equations are equivalent.

**13.** $12 - x = 3$ and $2x = 20$

When $x$ is replaced by 9, the first equation is true, but the second equation is false. Thus the equations are not equivalent.

**15.** $5x = 2x$ and $\dfrac{4}{x} = 3$

When $x$ is replaced by 0, the first equation is true, but the second equation is not defined. Thus the equations are not equivalent.

**17.** $\quad x - 2.9 = 13.4$

$x - 2.9 + 2.9 = 13.4 + 2.9 \quad$ Addition principle; adding 2.9

$x + 0 = 13.4 + 2.9 \quad$ Law of opposites

$x = 16.3$

Check:

$$\begin{array}{c|c} x - 2.9 = 13.4 \\ \hline 16.3 - 2.9 & 13.4 \\ & \overset{?}{=} \\ 13.4 \overset{?}{=} 13.4 & \text{TRUE} \end{array}$$

The solution is 16.3.

**19.** $\quad 8t = 72$

$\dfrac{1}{8} \cdot 8t = \dfrac{1}{8} \cdot 72 \quad$ Multiplication principle; multiplying by $\dfrac{1}{8}$, the reciprocal of 8

$1t = 9$

$t = 9$

Check:

$$\begin{array}{c|c} 8t = 72 \\ \hline 8 \cdot 9 & 72 \\ & \overset{?}{=} \\ 72 \overset{?}{=} 72 & \text{TRUE} \end{array}$$

The solution is 9.

**21.** $\quad 4x - 12 = 60$

$4x - 12 + 12 = 60 + 12$

$4x = 72$

$\dfrac{1}{4} \cdot 4x = \dfrac{1}{4} \cdot 72$

$1x = \dfrac{72}{4}$

$x = 18$

Check:

$$\begin{array}{c|c} 4x - 12 = 60 \\ \hline 4 \cdot 18 - 12 & 60 \\ 72 - 12 & \\ & \overset{?}{=} \\ 60 \overset{?}{=} 60 & \text{TRUE} \end{array}$$

The solution is 18.

**23.** $\dfrac{3}{5}n + 2 = 7$

$\dfrac{3}{5}n + 2 - 2 = 17 - 2$

$\dfrac{3}{5}n = 15$

$\dfrac{5}{3} \cdot \dfrac{3}{5}n = \dfrac{5}{3} \cdot 15$

$1n = 25$

$n = 25$

Check:

$\dfrac{3}{5}n + 2 = 17$

$$\begin{array}{c|c} \dfrac{3}{5} \cdot 25 + 2 & 17 \\ 15 + 2 & \end{array}$$

$17 \overset{?}{=} 17$    TRUE

The solution is 25.

**25.** $2y - 11 = 37$

$2y - 11 + 11 = 37 + 11$

$2y = 48$

$\dfrac{1}{2} \cdot 2y = \dfrac{1}{2} \cdot 48$

$1y = \dfrac{48}{2}$

$y = 24$

Check:

$2y - 11 = 37$

$$\begin{array}{c|c} 2 \cdot 24 - 11 & 37 \\ 48 - 11 & \end{array}$$

$37 \overset{?}{=} 37$    TRUE

The solution is 24.

**27.** $3x + 7x = (3 + 7)x = 10x$

**29.** $7rt - 9rt = (7 - 9)rt = -2rt$

**31.** $9t^2 + t^2 = (9 + 1)t^2 = 10t^2$

**33.** $12a - a = (12 - 1)a = 11a$

**35.** $n - 8n = (1 - 8)n = -7n$

**37.** $5x - 3x + 8x = (5 - 3 + 8)x = 10x$

**39.** $4x - 2x^2 + 3x$

$= 4x + 3x - 2x^2$    Commutative law of
                          addition

$= (4 + 3)x - 2x^2$

$= 7x - 2x^2$

**41.** $4x - 7 + 18x + 25$

$= 4x + 18x - 7 + 25$

$= (4 + 18)x + (-7 + 25)$

$= 22x + 18$

**43.** $-7t^2 + 3t + 5t^3 - t^3 + 2t^2 - t$

$= (-7 + 2)t^2 + (3 - 1)t + (5 - 1)t^3$

$= -5t^2 + 2t + 4t^3$

**45.** $7a - (2a + 5)$

$= 7a - 2a - 5$

$= 5a - 5$

**47.** $m - (6m - 2)$

$= m - 6m + 2$

$= -5m + 2$

**49.** $3d - 7 - (5 - 2d)$

$= 3d - 7 - 5 + 2d$

$= 5d - 12$

**51.** $-2(x + 3) - 5(x - 4)$

$= -2x - 6 - 5x + 20$

$= -7x + 14$

**53.** $9a - [7 - 5(7a - 3)]$

$= 9a - [7 - 35a + 15]$

$= 9a - [22 - 35a]$

$= 9a - 22 + 35a$

$= 44a - 22$

**55.** $5\{-2a + 3[4 - 2(3a + 5)]\}$

$= 5\{-2a + 3[4 - 6a - 10]\}$

$= 5\{-2a + 3[-6 - 6a]\}$

$= 5\{-2a - 18 - 18a\}$

$= 5\{-20a - 18\}$

$= -100a - 90$

**57.** $2y + \{7[3(2y - 5) - (8y + 7)] + 9\}$

$= 2y + \{7[6y - 15 - 8y - 7] + 9\}$

$= 2y + \{7[-2y - 22] + 9\}$

$= 2y + \{-14y - 154 + 9\}$

$= 2y + \{-14y - 145\}$

$= 2y - 14y - 145$

$= -12y - 145$

**59.** $6x + 3x = 54$

$9x = 54$

$\dfrac{1}{9} \cdot 9x = \dfrac{1}{9} \cdot 54$

$x = 6$

Check:

$6x + 3x = 54$

$$\begin{array}{c|c} 6 \cdot 6 + 3 \cdot 6 & 54 \\ 36 + 18 & \end{array}$$

$54 \overset{?}{=} 54$    TRUE

The solution is 6.

**61.**  $\dfrac{2}{3}y - \dfrac{1}{4}y = 5$

$\dfrac{8}{12}y - \dfrac{3}{12}y = 5$

$\dfrac{5}{12}y = 5$

$\dfrac{12}{15} \cdot \dfrac{5}{12}y = \dfrac{12}{5} \cdot 5$

$1y = 12$

$y = 12$

Check:

$$\dfrac{2}{3}y - \dfrac{1}{4}y = 5$$

$$\begin{array}{c|c} \dfrac{2}{3} \cdot 12 - \dfrac{1}{4} \cdot 12 & 5 \\ 8 - 3 & \\ & \overset{?}{5 = 5} \quad \text{TRUE} \end{array}$$

The solution is 12.

**63.**  $5t - 13t = -32$

$-8t = -32$

$-\dfrac{1}{8} \cdot (-8t) = -\dfrac{1}{8} \cdot (-32)$

$t = 4$

Check:

$$5t - 13t = -32$$

$$\begin{array}{c|c} 5 \cdot 4 - 13 \cdot 4 & -32 \\ 20 - 52 & \\ & \overset{?}{-32 = -32} \quad \text{TRUE} \end{array}$$

The solution is 4.

**65.**  $3(x + 4) = 7x$

$3x + 12 = 7x$

$3x + 12 - 3x = 7x - 3x$

$12 = 4x$

$\dfrac{1}{4} \cdot 12 = \dfrac{1}{4} \cdot 4x$

$3 = x$

Check:

$$3(x + 4) = 7x$$

$$\begin{array}{c|c} 3(3 + 4) & 7 \cdot 3 \\ 3 \cdot 7 & \\ & \overset{?}{21 = 21} \quad \text{TRUE} \end{array}$$

The solution is 3.

**67.**  $70 = 10(3t - 2)$

$70 = 30t - 20$

$70 + 20 = 30t - 20 + 20$

$90 = 30t$

$\dfrac{1}{30} \cdot 90 = \dfrac{1}{30} \cdot 30t$

$3 = t$

Check:

$$70 = 10(3t - 2)$$

$$\begin{array}{c|c} 70 & 10(3 \cdot 3 - 2) \\ & 10(9 - 2) \\ & 10 \cdot 7 \\ \overset{?}{70 = 70} & \quad \text{TRUE} \end{array}$$

The solution is 3.

**69.**  $1.8(n - 2) = 9$

$1.8n - 3.6 = 9$

$1.8n - 3.6 + 3.6 = 9 + 3.6$

$1.8n = 12.6$

$\dfrac{1.8n}{1.8} = \dfrac{12.6}{1.8}$

$n = 7$

Check:

$$1.8(n - 2) = 9$$

$$\begin{array}{c|c} 1.8(7 - 2) & 9 \\ 1.8(5) & \\ & \overset{?}{9 = 9} \quad \text{TRUE} \end{array}$$

The solution is 7.

**71.**  $5y - (2y - 10) = 25$

$5y - 2y + 10 = 25$

$3y + 10 = 25$

$3y + 10 - 10 = 25 - 10$

$3y = 15$

$\dfrac{1}{3} \cdot 3y = \dfrac{1}{3} \cdot 15$

$y = 5$

Check:

$$5y - (2y - 10) = 25$$

$$\begin{array}{c|c} 5 \cdot 5 - (2 \cdot 5 - 10) & 25 \\ 25 - (10 - 10) & \\ 25 - 0 & \\ & \overset{?}{25 = 25} \quad \text{TRUE} \end{array}$$

The solution is 5.

**73.**
$$7y - 1 = 23 - 5y$$
$$7y - 1 + 5y = 23 - 5y + 5y$$
$$12y - 1 = 23$$
$$12y - 1 + 1 = 23 + 1$$
$$12y = 24$$
$$\frac{1}{12} \cdot 12y = \frac{1}{12} \cdot 24$$
$$y = 2$$

Check:

| $7y - 1 = 23 - 5y$ | |
|---|---|
| $7 \cdot 2 - 1$ | $23 - 5 \cdot 2$ |
| $14 - 1$ | $23 - 10$ |
| $13 \overset{?}{=} 13$ | TRUE |

The solution is 2.

**75.**
$$\frac{1}{5} + \frac{3}{10}x = \frac{4}{5}$$
$$\frac{1}{5} + \frac{3}{10}x - \frac{1}{5} = \frac{4}{5} - \frac{1}{5}$$
$$\frac{3}{10}x = \frac{3}{5}$$
$$\frac{10}{3} \cdot \frac{3}{10}x = \frac{10}{3} \cdot \frac{3}{5}$$
$$x = 2$$

Check:

| $\frac{1}{5} + \frac{3}{10}x = \frac{4}{5}$ | |
|---|---|
| $\frac{1}{5} + \frac{3}{10} \cdot 2$ | $\frac{4}{5}$ |
| $\frac{1}{5} + \frac{3}{5}$ | |
| $\frac{4}{5} \overset{?}{=} \frac{4}{5}$ | TRUE |

The solution is 2.

**77.**
$$\frac{9}{10}y - \frac{7}{10} = \frac{21}{5}$$
$$\frac{9}{10}y - \frac{7}{10} + \frac{7}{10} = \frac{21}{5} + \frac{7}{10}$$
$$\frac{9}{10}y = \frac{42}{10} + \frac{7}{10}$$
$$\frac{9}{10}y = \frac{49}{10}$$
$$\frac{10}{9} \cdot \frac{9}{10}y = \frac{10}{9} \cdot \frac{49}{10}$$
$$y = \frac{49}{9}$$

Check:

| $\frac{9}{10}y - \frac{7}{10} = \frac{21}{5}$ | |
|---|---|
| $\frac{9}{10} \cdot \frac{49}{9} - \frac{7}{10}$ | $\frac{21}{5}$ |
| $\frac{49}{10} - \frac{7}{10}$ | |
| $\frac{42}{10}$ | |
| $\frac{21}{5} \overset{?}{=} \frac{21}{5}$ | TRUE |

The solution is $\frac{49}{9}$.

**79.**
$$7r - 2 + 5r = 6r + 6 - 4r$$
$$12r - 2 = 2r + 6$$
$$12r - 2 - 2r = 2r + 6 - 2r$$
$$10r - 2 = 6$$
$$10r - 2 + 2 = 6 + 2$$
$$10r = 8$$
$$\frac{1}{10} \cdot 10r = \frac{1}{10}(8)$$
$$r = \frac{8}{10}$$
$$r = \frac{4}{5}$$

Check:

| $7r - 2 + 5r = 6r + 6 - 4r$ | |
|---|---|
| $7 \cdot \frac{4}{5} - 2 + 5 \cdot \frac{4}{5}$ | $6 \cdot \frac{4}{5} + 6 - 4 \cdot \frac{4}{5}$ |
| $\frac{28}{5} - 2 + 4$ | $\frac{24}{5} + 6 - \frac{16}{5}$ |
| $\frac{38}{5} \overset{?}{=} \frac{38}{5}$ | TRUE |

The solution is $\frac{4}{5}$.

**81.**
$$\frac{2}{3}(x - 2) - 1 = \frac{1}{4}(x - 3)$$
$$\frac{2}{3}x - \frac{4}{3} - 1 = \frac{1}{4}x - \frac{3}{4}$$
$$\frac{2}{3}x - \frac{7}{3} = \frac{1}{4}x - \frac{3}{4}$$
$$\frac{2}{3}x - \frac{7}{3} - \frac{1}{4}x = \frac{1}{4}x - \frac{3}{4} - \frac{1}{4}x$$
$$\frac{5}{12}x - \frac{7}{3} = -\frac{3}{4}$$
$$\frac{5}{12}x - \frac{7}{3} + \frac{7}{3} = -\frac{3}{4} + \frac{7}{3}$$
$$\frac{5}{12}x = \frac{19}{12}$$
$$\frac{12}{5} \cdot \frac{5}{12}x = \frac{12}{5} \cdot \frac{19}{12}$$
$$x = \frac{19}{5}$$

The check is left to the student. The solution is $\frac{19}{5}$.

**83.** $5 + 2(x - 3) = 2[5 - 4(x + 2)]$

$5 + 2x - 6 = 2[5 - 4x - 8]$

$2x - 1 = 2[-4x - 3]$

$2x - 1 = -8x - 6$

$2x - 1 + 1 = -8x - 6 + 1$

$2x = -8x - 5$

$2x + 8x = -8x - 5 + 8x$

$10x = -5$

$\dfrac{1}{10} \cdot 10x = \dfrac{1}{10}(-5)$

$x = -\dfrac{1}{2}$

Check:

$$\begin{array}{c|c} 5 + 2(x - 3) = 2[5 - 4(x + 2)] \\ \hline 5 + 2\left(-\dfrac{1}{2} - 3\right) & 2\left[5 - 4\left(-\dfrac{1}{2} + 2\right)\right] \\ 5 + 2\left(-\dfrac{7}{2}\right) & 2\left[5 - 4\left(\dfrac{3}{2}\right)\right] \\ 5 - 7 & 2[5 - 6] \\ -2 & 2[-1] \\ -2 \overset{?}{=} -2 & \text{TRUE} \end{array}$$

The solution is $-\dfrac{1}{2}$.

**85.** $7x - 2 - 3x = 4x$

$4x - 2 = 4x$

$4x - 2 - 4x = 4x - 4x$

$-2 = 0$

Since the original equation is equivalent to the false equation $-2 = 0$, there is no solution. The solution set is $\emptyset$. The equation is a contradiction.

**87.** $2 + 9x = 3(4x + 1) - 1$

$2 + 9x = 12x + 3 - 1$

$2 + 9x = 12x + 2$

$2 + 9x - 2 = 12x + 2 - 2$

$9x = 12x$

$9x - 9x = 12x - 9x$

$0 = 3x$

$\dfrac{1}{3} \cdot 0 = \dfrac{1}{3} \cdot 3x$

$0 = x$

The solution set is $\{0\}$. The equation is a conditional equation.

**89.** $-9t + 2 = -9t - 7(6 \div 2(49) + 8)$

Observe that $-7(6 \div 2(49) + 8)$ is a negative number. Then on the left side we have $-9t$ plus a positive number and on the right side we have $-9t$ plus a negative number. This is a contradiction, so the solution set is $\emptyset$.

**91.** $2\{9 - 3[-2x - 4]\} = 12x + 42$

$2\{9 + 6x + 12\} = 12x + 42$

$2\{21 + 6x\} = 12x + 42$

$42 + 12x = 12x + 42$

$42 + 12x - 12x = 12x + 42 - 12x$

$42 = 42$

The original equation is equivalent to the equation $42 = 42$, which is true for all real numbers. Thus the solution set is the set of all real numbers. The equation is an identity.

**93.** *Writing Exercise*

**95.** Let $n$ represent the number. Then we have

$2n + 9$, or $9 + 2n$.

**97.** *Writing Exercise*

**99.** $4.23x - 17.898 = -1.65x - 42.454$

$5.88x - 17.898 = -42.454$

$5.88x = -24.556$

$x = -\dfrac{24.556}{5.88}$

$x \approx -4.176190476$

The check is left to the student. The solution is approximately $-4.176190476$.

**101.** $4x - \{3x - [2x - (5x - (7x - 1))]\} = 4x + 7$

$4x - \{3x - [2x - (5x - 7x + 1)]\} = 4x + 7$

$4x - \{3x - [2x - (-2x + 1)]\} = 4x + 7$

$4x - \{3x - [2x + 2x - 1]\} = 4x + 7$

$4x - \{3x - [4x - 1]\} = 4x + 7$

$4x - \{3x - 4x + 1\} = 4x + 7$

$4x - \{-x + 1\} = 4x + 7$

$4x + x - 1 = 4x + 7$

$5x - 1 = 4x + 7$

$x - 1 = 7$

$x = 8$

The check is left to the student. The solution is 8.

**103.** $17 - 3\{5 + 2[x - 2]\} + 4\{x - 3(x + 7)\} =$
$\qquad\qquad 9\{x + 3[2 + 3(4 - x)]\}$

$17 - 3\{5 + 2x - 4\} + 4\{x - 3x - 21\} =$
$\qquad\qquad 9\{x + 3[2 + 12 - 3x]\}$

$17 - 3\{1 + 2x\} + 4\{-2x - 21\} = 9\{x + 3[14 - 3x]\}$

$17 - 3 - 6x - 8x - 84 = 9\{x + 42 - 9x\}$

$-14x - 70 = 9\{-8 + 42\}$

$-14x - 70 = -72x + 378$

$58x - 70 = 378$

$58x = 448$

$x = \dfrac{448}{58}$, or $\dfrac{224}{29}$

The check is left to the student. The solution is $\dfrac{224}{29}$.

**105.** *Writing Exercise*

## Exercise Set 1.4

**1. *Familiarize.*** There are two numbers involved, and we want to find both of them. We can let $x$ represent the first number and note that the second number is 7 more than the first. Also, the sum of the numbers is 65.

***Translate.*** The second number can be named $x + 7$. We translate to an equation:

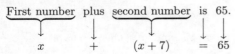

First number plus second number is 65.

$$x \quad + \quad (x+7) \quad = \quad 65$$

**3. *Familiarize.*** Let $t$ = the time, in hours, it will take thee swimmer to swim 1.8 km upstream. We will use the formula Distance = Speed × Time. The swimmer's speed upstream is $5 - 2.3$, or 2.7 km/h.

***Translate.*** We substitute in the formula.

$$1.8 = 2.7t$$

**5. *Familiarize.*** Since the sidewalk's speed is 5 ft/sec and Alida's walking speed is 4 ft/sec, Alida will move at a speed of $5+4$, or 9 ft/sec on the sidewalk. Let $t$ = the time, in seconds, it takes her to walk the length of the moving sidewalk, 300 ft.

***Translate.*** We will use the formula Distance = Speed × Time.

Distance = Speed × Time

$$300 \quad = \quad 9 \quad \times \quad t$$

**7. *Familiarize.*** There are three angle measures involved, and we want to find all three. We can let $x$ represent the smallest angle measure and note that the second is one more than $x$ and the third is one more than the second, or two more than $x$. We also note that the sum of the three angle measures must be 180°.

***Translate.*** The three angle measures are $x$, $x + 1$, and $x + 2$. We translate to an equation:

First plus second plus third is 180°.

$$x \quad + \quad (x+1) \quad + \quad (x+2) \quad = \quad 180$$

**9. *Familiarize.*** Let $w$ represent the wholesale price, in dollars. Then the wholesale price raised by 50% is $w + 0.5w$, or $1.5w$.

***Translate.***

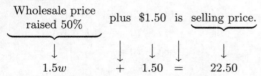

Wholesale price raised 50% plus $1.50 is selling price.

$$1.5w \quad + \quad 1.50 \quad = \quad 22.50$$

**11. *Familiarize.*** Let $t$ = the time, in hours, required for the plane to reach 29,000 ft. Since Distance = Speed × Time, the plane will travel $3500 \times t$ ft in $t$ hr. Note that the plane starts at an altitude of 8000 ft.

***Translate.***

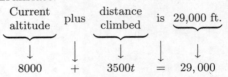

Current altitude plus distance climbed is 29,000 ft.

$$8000 \quad + \quad 3500t \quad = \quad 29{,}000$$

**13. *Familiarize.*** Let $x$ represent the measure of the second angle. Then the first angle is three times $x$, and the third is 12° less than twice $x$. The sum of the three angle measures is 180°.

***Translate.*** The first angle is $3x$, the second is $x$, and the third is $2x - 12$. Translate to an equation:

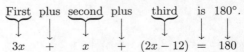

First plus second plus third is 180°.

$$3x \quad + \quad x \quad + \quad (2x-12) \quad = \quad 180$$

**15. *Familiarize.*** Note that each even integer is 2 more than the one preceding it. If we let $n$ represent the first even integer, then $n + 2$ represents the next even integer.

***Translate.***

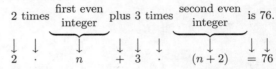

2 times first even integer plus 3 times second even integer is 76.

$$2 \quad \cdot \quad n \quad + \quad 3 \quad \cdot \quad (n+2) \quad = \quad 76$$

**17. *Familiarize.*** The perimeter of an equilateral triangle is 3 times the length of a side. Let $s$ = the length of a side of the smaller triangle. Then $2s$ = the length of a side of the larger triangle. The sum of the two perimeters is 90 cm.

***Translate.***

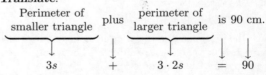

Perimeter of smaller triangle plus perimeter of larger triangle is 90 cm.

$$3s \quad + \quad 3 \cdot 2s \quad = \quad 90$$

**19. *Familiarize.*** Let $c$ represent the number of calls Brian will need on his next shift if he is to average 3 calls per shift. We find the average by adding the number of calls on each of the 5 shifts and then dividing by the number of addends.

***Translate.***

Average number of calls per shift is 3.

$$\frac{5+2+1+3+c}{5} \quad = \quad 3$$

**21. *Familiarize.*** Let $p$ represent the price Tony paid for his graphing calculator.

***Translate.***

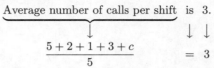

Price Ruth paid is price Tony paid less $13.

$$84 \quad = \quad p \quad - \quad 13$$

***Carry out.*** We solve the equation.

$$84 = p - 13$$
$$97 = p \qquad \text{Adding 13 to both sides}$$

**Check.** The price Ruth paid, $84, is $13 less than $97, so the answer checks.

**State.** Tony paid $97 for his graphing calculator.

23. **Familiarize.** Let $d$ represent the number of cases of diabetes diagnosed in the United States in 2000, in millions.

**Translate.**

$$\underbrace{\text{Number of cases in 2050}}_{29} \; \underbrace{\text{will be}}_{=} \; \underbrace{\tfrac{13}{5}}_{\tfrac{13}{5}} \; \underbrace{\text{of}}_{\cdot} \; \underbrace{\text{number of cases in 2000.}}_{d}$$

**Carry out.** We solve the equation.

$$29 = \frac{13}{5}d$$
$$\frac{5}{13} \cdot 29 = \frac{5}{13} \cdot \frac{13}{5}d$$
$$\frac{145}{13} = d$$
$$11.15 \approx d$$

**Check.** Since $\frac{13}{5}$ of 11.15 million is about 29 million, the answer checks.

**State.** About 11.15 million cases of diabetes had been diagnosed in the United States in 2000.

$c = \$1500$

25. **Familiarize.** Let $t$ represent the number of tickets Officer Schultz wrote up. Then $t + 9$ represents the number of tickets Office Reid wrote up.

**Translate.**

$$\underbrace{\text{Total number of tickets}}_{x + (x+9)} \; \underbrace{\text{is}}_{=} \; \underbrace{35.}_{35}$$

**Carry out.** We solve the equation.

$$x + x + 9 = 35$$
$$2x + 9 = 35$$
$$2x = 26 \quad \text{Subtracting 9 from both sides}$$
$$x = 13$$

If $x = 13$, then $x + 9 = 13 + 9$, or 22.

**Check.** 22 tickets is 9 more than 13 tickets. Also, $13 + 22 = 35$ tickets. The answer checks.

**State.** Officer Reid wrote 22 tickets.

27. **Familiarize.** Let $w$ represent the width of the mirror, in cm. Then $3w$ represents the length. Recall that the formula for the perimeter $P$ of a rectangle with length $l$ and width $w$ is $P = 2l + 2w$.

**Translate.**

$$\underbrace{\text{Perimeter}}_{2 \cdot 3w + 2 \cdot w} \; \underbrace{\text{is}}_{=} \; \underbrace{120 \text{ cm.}}_{120}$$

**Carry out.** We solve the equation.

$$2 \cdot 3w + 2 \cdot w = 120$$
$$6w + 2w = 120$$
$$8w = 120$$
$$w = 15$$

When $w = 15$, then $3w = 3 \cdot 15 = 45$.

**Check.** If the length is 45 cm and the width is 15 cm, then the length is three times the width. Also $P = 2 \cdot 45 + 2 \cdot 15 = 90 + 30 = 120$ cm. The answer checks.

**State.** The length of the mirror is 45 cm, and the width is 15 cm.

29. **Familiarize.** Let $l$ represent the length of the greenhouse, in meters. Then $\frac{1}{4}l$ represents the width. Recall that the formula for the perimeter $P$ of a rectangle with length $l$ and width $w$ is $P = 2l + 2w$.

**Translate.**

$$\underbrace{\text{Perimeter}}_{2l + 2 \cdot \frac{1}{4}l} \; \underbrace{\text{is}}_{=} \; \underbrace{130 \text{ m.}}_{130}$$

**Carry out.** We solve the equation.

$$2l + 2 \cdot \frac{1}{4}l = 130$$
$$2l + \frac{1}{2}l = 130$$
$$\frac{5}{2}l = 130$$
$$\frac{2}{5} \cdot \frac{5}{2}l = \frac{2}{5} \cdot 130$$

When $l = 52$, then $\frac{1}{4}l = 13$.

**Check.** If the length is 52 m and the width is 13 m, then the width is $\frac{1}{4}$ of the length. Also, $P = 2 \cdot 52 + 2 \cdot 13 = 104 + 26 = 130$ m. The answer checks.

**State.** The length of the greenhouse is 52 m, and the width is 13 m.

31. The Familiarize and Translate steps were done in Exercise 3.

**Carry out.** We solve the equation.

$$2.7t = 1.8$$
$$t = \frac{1}{2.7}(1.8)$$
$$t = \frac{2}{3}$$

**Check.** At a speed of 2.7 km/h, in $\frac{2}{3}$ hr Fran swims $\frac{2}{3}(2.7)$, or 1.8 km. Our answer checks.

**State.** It will take Fran $\frac{2}{3}$ hr to swim 1.8 km upriver.

33. The Familiarize and Translate steps were done in Exercise 14.

**Carry out**. We solve the equation.

$$4x + x + (2x + 5) = 180$$
$$7x + 5 = 180$$
$$7x = 180 - 5$$
$$7x = 175$$
$$\frac{1}{7} \cdot 7x = \frac{1}{7} \cdot 175$$
$$x = 25$$

When $x = 25$, then $4x = 4 \cdot 25 = 100$ and $2x + 5 = 2 \cdot 25 + 5 = 55$.

**Check**. One angle, 100°, is 4 times the 25°angle, and the third angle, 55°, is 5°more than twice the 25°angle. Also, $100° + 25° + 55° = 180°$. The answer checks.

**State**. The measures of the angles are 100°, 25°, and 55°.

**35.** The Familiarize and Translate steps were done in Exercise 10.

**Carry out**. We solve the equation.

$$c - 0.05c = 142.50$$
$$0.95c = 142.50$$
$$\frac{0.95c}{0.95} = \frac{142.50}{0.95}$$
$$c = 150$$

**Check**. 5% of $150 is $7.50 and $150 − $7.50 = $142.50. The answer checks.

**State**. The cost would have been $150 if the bill had not been paid promptly.

**37.** The Familiarize and Translate steps were done in Exercise 9.

**Carry out**. We solve the equation.

$$1.5w + 1.50 = 22.50$$
$$1.5w = 21$$
$$w = \frac{1}{1.5} \cdot 21$$
$$w = 14$$

**Check**. If a wholesale price of $14 is raised by 50%, we have $14 + 0.5($14) = $14 + $7 = $21. When $1.50 is added to this figure, we have $21 + $1.50 = $22.50. The answer checks.

**State**. The wholesale price is $14.

**39.** *Writing Exercise*

**41.**
$$7 = \frac{2}{3}(x + 6)$$
$$3 \cdot 7 = 3 \cdot \frac{2}{3}(x + 6)$$
$$21 = 2(x + 6)$$
$$21 = 2x + 12$$
$$21 - 12 = 2x$$
$$9 = 2x$$
$$\frac{1}{2} \cdot 9 = \frac{1}{2} \cdot 2x$$
$$\frac{9}{2} = x$$

The solution is $\frac{9}{2}$.

**43.**
$$8 = \frac{5 + t}{3}$$
$$3 \cdot 8 = 3\left(\frac{5 + t}{3}\right)$$
$$24 = 5 + t$$
$$24 - 5 = t$$
$$19 = t$$

The solution is 19.

**45.** *Writing Exercise*

**47.** **Familiarize**. The average score on the first four tests is $\frac{83 + 91 + 78 + 81}{4}$, or 83.25. Let $x =$ the number of points above this average that Tico scores on the next test. Then the score on the fifth test is $83.25 + x$.

**Translate**.

| Average score on 5 tests | is | 2 | more than | average score on 4 tests. |
|---|---|---|---|---|
| $\frac{83+91+78+81+(83.25+x)}{5}$ | = | 2 | + | 83.25 |

**Carry out**. Carry out some algebraic manipulation.

$$\frac{83 + 91 + 78 + 81 + (83.25 + x)}{5} = 2 + 83.25$$
$$\frac{416.25 + x}{5} = 85.25$$
$$416.25 + x = 426.25$$
$$x = 10$$

**Check**. If Tico scores 10 points more than the average of the first four tests on the fifth test, his score will be $83.25 + 10$, or 93.25. Then the five-test average will be $\frac{83 + 91 + 78 + 81 + 93.25}{5}$, or 85.25. This is 2 points above the four-test average, so the answer checks.

**State**. Tico must score 10 points above the four-test average in order to raise the average 2 points.

**49.** **Familiarize**. Let $p =$ the price of the house in 2001. From 2001 to 2002 real estate prices increased 6%, so the house was worth $p + 0.06p$, or $1.06p$. From 2002 to 2003 prices increased 2%, so the house was then worth $1.06p + 0.02(1.06p)$, or $1.02(1.06p)$. From 2003 to 2004 prices dropped 1%, so the value of the house became $1.02(1.06p) - 0.01(1.02)(1.06p)$, or $0.99(1.02)(1.06p)$.

**Translate**.

| The price of the house in 2004 | was | $117,743. |
|---|---|---|
| $0.99(1.02)(1.06p)$ | = | $117,743$ |

*Carry out*. We solve the equation.

$$0.99(1.02)(1.06p) = 117,743$$
$$p = \frac{117,743}{0.99(1.02)(1.06)}$$
$$p \approx 110,000$$

*Check*. If the price of the house in 2001 was $110,000, then in 2002 it was worth 1.06($110,000), or $116,600. In 2003 it was worth 1.02($116,600), or $118,932, and in 2004 it was worth 0.99($118,932), or $117,743. Our answer checks.

*State*. The house was worth $110,000 in 2001.

---

## Exercise Set 1.5

**1.** A formula is an <u>equation</u> that uses letters to represent a relationship between two or more quantities. (See page 42 in the text.)

**3.** For formula $C = \pi d$ is used to calculate the <u>circumference</u> of a circle. (See page 42 in the text.)

**5.** The formula <u>$A = bh$</u> is used to calculate the area of a parallelogram of height $h$ and base length $b$. (See page 42 in the text.)

**7.** In the formula for the area of a trapezoid, $A = \frac{h}{2}(b_1 + b_2)$, the numbers 1 and 2 are referred to as <u>subscripts</u>. (See Example 3.)

**9.**
$$d = rt$$
$$\frac{1}{t} \cdot d = \frac{1}{t} \cdot rt \quad \text{Multiplying both sides by } \frac{1}{t}$$
$$\frac{d}{t} = r \qquad \text{Simplifying}$$

**11.**
$$F = ma$$
$$\frac{1}{m} \cdot F = \frac{1}{m} \cdot ma \quad \text{Multiplying both sides by } \frac{1}{m}$$
$$\frac{F}{m} = a \qquad \text{Simplifying}$$

**13.**
$$W = EI$$
$$\frac{1}{E} \cdot W = \frac{1}{E} \cdot EI \quad \text{Multiplying both sides by } \frac{1}{E}$$
$$\frac{W}{E} = I \qquad \text{Simplifying}$$

**15.**
$$V = lwh$$
$$\frac{1}{lw} \cdot V = \frac{1}{lw} \cdot lwh \quad \text{Multiplying both sides by } \frac{1}{lw}$$
$$\frac{V}{lw} = h \qquad \text{Simplifying}$$

**17.**
$$L = \frac{k}{d^2}$$
$$d^2 \cdot L = d^2 \cdot \frac{k}{d^2} \quad \text{Multiplying both sides by } d^2$$
$$d^2 L = k \qquad \text{Simplifying}$$

**19.**
$$G = w + 150n$$
$$G - w = 150n \qquad \text{Subtracting } w \text{ from both sides}$$
$$\frac{1}{150}(G - w) = \frac{1}{150} \cdot 150n \quad \text{Multiplying both sides by } \frac{1}{150}$$
$$\frac{G - w}{150} = n \qquad \text{Simplifying}$$

**21.**
$$2w + 2h + l = p$$
$$l = p - 2w - 2h \quad \text{Adding } -2w - 2h \text{ to both sides}$$

**23.**
$$nl + nm = k$$
$$n(l + m) = k \qquad \text{Factoring}$$
$$n = \frac{k}{l + m} \quad \text{Dividing both sides by } l + m$$

**25.**
$$yx + zx = w$$
$$x(y + z) = w \qquad \text{Factoring}$$
$$x = \frac{w}{y + z} \quad \text{Dividing both sides by } y + z$$

**27.**
$$Ax + By = C$$
$$By = C - Ax \qquad \text{Subtracting } Ax \text{ from both sides}$$
$$\frac{1}{B} \cdot By = \frac{1}{B}(C - Ax) \quad \text{Multiplying both sides by } \frac{1}{B}$$
$$y = \frac{C - Ax}{B} \qquad \text{Simplifying}$$

**29.**
$$C = \frac{5}{9}(F - 32)$$
$$\frac{9}{5} \cdot C = \frac{9}{5} \cdot \frac{5}{9}(F - 32) \quad \text{Multiplying both sides by } \frac{9}{5}$$
$$\frac{9}{5}C = F - 32 \qquad \text{Simplifying}$$
$$\frac{9}{5}C + 32 = F \qquad \text{Adding 32 to both sides}$$

**31.**
$$V = \frac{4}{3}\pi r^3$$
$$\frac{3}{4\pi} \cdot V = \frac{3}{4\pi} \cdot \frac{4}{3}\pi r^3 \quad \text{Multiplying both sides by } \frac{3}{4\pi}$$
$$\frac{3V}{4\pi} = r^3 \qquad \text{Simplifying}$$

**33.**
$$ab = d - ac$$
$$ab + ac = d \qquad \text{Adding } ac \text{ to both sides}$$
$$a(b + c) = d \qquad \text{Factoring}$$
$$a = \frac{d}{b + c} \quad \text{Dividing both sides by } b + c$$

**35.**
$$xy = w + zy$$
$$xy - zy = w \qquad \text{Adding } -zy \text{ to both sides}$$
$$y(x - z) = w \qquad \text{Factoring}$$
$$y = \frac{w}{x - z} \quad \text{Dividing both sides by } x - z$$

**37.**
$$A = \frac{q_1 + q_2 + q_3}{n}$$

$$n \cdot A = n \cdot \frac{q_1 + q_2 + q_3}{n} \qquad \text{Clearing the fraction}$$

$$nA = q_1 + q_2 + q_3$$

$$nA \cdot \frac{1}{A} = (q_1 + q_2 + q_3) \cdot \frac{1}{A} \qquad \text{Multiplying both sides by } \frac{1}{A}$$

$$n = \frac{q_1 + q_2 + q_3}{A}$$

**39.**
$$v = \frac{d_2 - d_1}{t}$$

$$t \cdot v = t \cdot \frac{d_2 - d_1}{t} \qquad \text{Clearing the fraction}$$

$$tv = d_2 - d_1$$

$$tv \cdot \frac{1}{v} = (d_2 - d_1) \cdot \frac{1}{v} \qquad \text{Multiplying both sides by } \frac{1}{v}$$

$$t = \frac{d_2 - d_1}{v}$$

**41.**
$$v = \frac{d_2 - d_1}{t}$$

$$t \cdot v = t \cdot \frac{d_2 - d_1}{t} \qquad \text{Clearing the fraction}$$

$$tv = d_2 - d_1$$

$$tv - d_2 = -d_1 \qquad \text{Subtracting } d_2 \text{ from both sides}$$

$$-1 \cdot (tv - d_2) = -1 \cdot (-d_1) \qquad \text{Multiplying both sides by } -1$$

$$-tv + d_2 = d_1,$$
$$\text{or } d_2 - tv = d_1$$

**43.**
$$r - m = mnp$$
$$r = m + mnp$$
$$r = m(1 + np) \qquad \text{Factoring}$$

$$r \cdot \frac{1}{1 + np} = m(1 + np) \cdot \frac{1}{1 + np}$$

$$\frac{r}{1 + np} = m$$

**45.**
$$y - ab = ac^2$$
$$y = ab + ac^2$$
$$y = a(b + c^2) \qquad \text{Factoring}$$

$$y \cdot \frac{1}{b + c^2} = a(b + c^2) \cdot \frac{1}{b - c^2}$$

$$\frac{y}{b + c^2} = a$$

**47. Familiarize.** In Example 2 we find the formula for simple interest, $I = Prt$, when $I$ is the interest, $P$ is the principal, $r$ is the interest rate, and $t$ is the time, in years.

**Translate.** We want to find the interest rate, so we solve the formula for $r$.
$$I = Prt$$

$$\frac{1}{Pt} \cdot I = \frac{1}{Pt} \cdot Prt$$

$$\frac{I}{Pt} = r$$

**Carry out.** The model $r = \frac{I}{Pt}$ can be used to find the rate of interest at which an amount (the principal) must be invested in order to earn a given amount. We substitute \$2600 for $P$, $\frac{1}{2}$ for $t$ $\left(6 \text{ months} = \frac{1}{2} \text{ yr}\right)$, and \$156 for $I$.

$$\frac{I}{Pt} = r$$

$$\frac{\$156}{\$2600\left(\frac{1}{2}\right)} = r$$

$$\frac{156}{1300} = r$$

$$0.12 = r$$

$$12\% = r$$

**Check.** Since $\$2600(0.12)\left(\frac{1}{2}\right) = \$156$, the answer checks.

**State.** The interest rate must be 12%.

**49. Familiarize.** On page 42 of the text we find the formula for the area of a parallelogram, $A = bh$, where $b$ is the base and $h$ is the height.

**Translate.** We solve the formula for $h$.
$$A = bh$$

$$\frac{1}{b} \cdot A = \frac{1}{b} \cdot bh$$

$$\frac{A}{b} = h$$

**Carry out.** The model $h = \frac{A}{b}$ can be used to find the height of any parallelogram for which the area and base are known. We substitute 78 for $A$ and 13 for $b$.
$$h = \frac{A}{b}$$

$$h = \frac{78}{13}$$

$$h = 6$$

**Check.** We repeat the calculation. The answer checks.

**State.** The height is 6 cm.

**51. Familiarize and Translate.** In Example 5 the formula for body mass index is solved for $W$:
$$W = \frac{IH^2}{704.5}$$

**Carry out.** We substitute 30.2 for $I$ and 74 for $H$ (6 ft 2 in. is 74 in.) and calculate $W$.
$$W = \frac{30.2(74)^2}{704.5}$$

$$\approx 235$$

**Check.** We could repeat the calculations or substitute in the original formula and then solve for $W$. The answer checks.

**State.** Arnold Schwarzenegger weighs about 235 lb.

**53. Familiarize and Translate.** We will use the model developed in Example 6, $m = \pi r^2 h D$ to find the weight of the salt. Then we will add 28 g, the weight of the empty canister, to find the weight of the filled canister.

*Carry out*. We substitute 4 for $r$, 13.6 for $h$, and 2.16 for $D$ and calculate $m$.

$$m = \pi r^2 h D$$
$$= \pi(4)^2(13.6)(2.16)$$
$$\approx 1476.6$$

Add the weight of the empty canister:

$$1476.6 + 28 = 1504.6$$

*Check*. We repeat the calculations. The answer checks.

*State*. The filled canister weighs about 1504.6 g.

**55.** *Familiarize*. The formula for the area of a trapezoid is $A = \dfrac{1}{2}h(b_1 + b_2)$, where $A$ is the area, $h$ is the height, and $b_1$ and $b_2$ are the bases.

*Translate*. The unknown dimension is the height, so we solve the formula for $h$. This was done in Exercise 22. We have

$$h = \frac{2A}{b_1 + b_2}.$$

*Carry out*. We substitute.

$$h = \frac{2A}{b_1 + b_2}$$
$$h = \frac{2 \cdot 90}{8 + 12}$$
$$h = \frac{180}{20}$$
$$h = 9$$

*Check*. We repeat the calculation. The answer checks.

*State*. The unknown dimension, the height of the trapezoid, is 9 ft.

**57.** Observe that 9% of $1000 is $90, so $90 is the amount of simple interest that would be earned in 1 yr. Thus, it will take 1 yr for the investment to be worth $1090.

**59.** *Familiarize*. We will use the formula given in the text, $R = r + \dfrac{400(W - L)}{N}$.

*Translate*. We solve the formula for $r$.

$$R = r + \frac{400(W - L)}{N}$$
$$R - \frac{400 - (W - L)}{N} = r$$

*Carry out*. We substitute 1305 for $R$, 5 for $w$, 3 for $L$, and $5 + 3$, or 8, for $N$ and calculate $r$.

$$1305 - \frac{400(5 - 3)}{8} = r$$
$$1205 = r$$

*Check*. We can repeat the calculation or substitute in the original formula and then solve for $r$. The answer checks.

*State*. The average rating of Ulana's opponents was 1205.

**61.** *Familiarize*. We will use the formula given in the text, $K = 917 + 6(w + h - a)$.

*Translate*. We solve the formula for $h$.

$$K = 917 + 6(w + h - a)$$
$$K = 917 + 6w + 6h - 6a$$
$$K - 917 - 6w + 6a = 6h$$
$$\frac{K - 916 - 6w + 6a}{6} = h$$

*Carry out*. We substitute 1901 for $K$, 120 for $w$, and 23 for $a$ and calculate $h$.

$$\frac{1901 - 917 - 6 \cdot 120 + 6 \cdot 23}{6} = h$$
$$67 = h$$

*Check*. We can repeat the calculation or substitute in the original formula and then solve for $h$. The answer checks.

*State*. Julie is 67 in., or 5 ft 7 in., tall.

**63.** *Familiarize*. We will use the formula given in the text, $K = 19.18w + 7h - 9.52a + 92.4$.

*Translate*. We solve the formula for $a$.

$$K = 19.18w + 7h - 9.52a + 92.4$$
$$K - 19.18w - 7h - 92.4 = -9.52a$$
$$\frac{K - 19.18w - 7h - 92.4}{-9.52} = a$$

*Carry out*. We substitute 2476 for $K$, 87 for $w$, and 185 for $h$ and calculate $a$.

$$\frac{2476 - 19.18(87) - 7(185) - 92.4}{-9.52} = a$$
$$61 \approx a$$

*Check*. We can repeat the calculation or substitute in the original formula and solve for $a$. The answer checks.

*State*. Marv is 61 years old.

**65.** *Familiarize*. We will use Thurnau's model, $P = 9.337da - 299$.

*Translate*. Since we want to find the diameter of the fetus' head, we solve for $d$.

$$P = 9.337da - 299$$
$$P + 299 = 9.337da$$
$$\frac{P + 299}{9.337a} = d$$

*Carry out*. Substitute 1614 for $P$ and 24.1 for $a$ in the formula and calculate:

$$\frac{1614 + 299}{9.337(24.1)} = d$$
$$8.5 \approx d$$

*Check*. We repeat the calculation. The answer checks.

*State*. The diameter of the fetus' head at 29 weeks is about 8.5 cm.

**67.** *Familiarize*. We will use Goiten's model, $I = 1.08(T/N)$. Note that 8 hr $= 8 \times 1$ hr $= 8 \times 60$ min $= 480$ min.

**Translate**. We solve the formula for $N$.

$$I = 1.08\left(\frac{T}{N}\right)$$

$$N \cdot I = N(1.08)\left(\frac{T}{N}\right)$$

$$NI = 1.08T$$

$$N = \frac{1.08T}{I}$$

**Carry out**. We substitute 480 for $T$ and 15 for $I$.

$$N = \frac{1.08T}{I}$$

$$N = \frac{1.08(480)}{15}$$

$$N = 34.56$$

$$N \approx 34 \qquad \text{Rounding down}$$

**Check**. We repeat the calculations. The answer checks.

**State**. Dr. Cruz should schedule 34 appointments in one day.

**69.** *Writing Exercise*

**71.** $(7a)(3a) = (7 \cdot 3)(a \cdot a)$

$(7a)(3a) = a \cdot 3 \cdot a \cdot 7$

Answers may vary.

**73.** *Writing Exercise*

**75. Familiarize**. First we find the volume of the ring. Note that the inner diameter is 2 cm, so the inner radius is 2/2 or 1 cm. Then the volume of the ring is the volume of a right circular cylinder with height 0.5 cm and radius $1 + 0.15$, or 1.15 cm, less the volume of a right circular cylinder with height 0.5 cm and radius 1 cm. Recall that the formula for the volume of a right circular cylinder is $V = \pi r^2 h$. Then the volume of the ring is

$$\pi(1.15)^2(0.5) - \pi(1)^2(0.5) = 0.16125\pi \text{ cm}^3.$$

**Translate**. To find the weight of the ring we will use the formula $D = \dfrac{m}{V}$. Solving for $m$, we get

$$D = \frac{m}{V}$$

$$V \cdot D = V \cdot \frac{m}{V}$$

$$V \cdot D = m$$

**Carry out**. We substitute in the formula $m = V \cdot D$.

$$m = 0.16125\pi(21.5)$$

$$m \approx 10.9$$

**Check**. We repeat the calculations. The answer checks.

**State**. The ring will weigh about 10.9 g.

**77.** *Writing Exercise*

**79.**

$$A = 4lw + w^2$$

$$A - w^2 = 4lw$$

$$\frac{A - w^2}{4w} = l \qquad \text{Multiplying both sides by } \frac{1}{4w}$$

**81.**

$$\frac{P_1 V_1}{T_1} = \frac{P_2 V_2}{T_2}$$

$$P_1 V_1 T_2 = P_2 V_2 T_1 \quad \begin{array}{l}\text{Multiplying both sides}\\\text{by } T_1 T_2\end{array}$$

$$\frac{P_1 V_1 T_2}{P_2 V_2} = T_1 \qquad \begin{array}{l}\text{Multiplying both sides}\\[4pt]\text{by } \dfrac{1}{P_2 V_2}\end{array}$$

**83.**

$$m = \frac{\dfrac{d}{e}}{f}$$

$$\frac{me}{f} = \frac{d}{e} \qquad \text{Multiplying both sides by } \frac{e}{f}$$

$$\frac{me^2}{f} = d \qquad \text{Multiplying both sides by } e$$

**85.** $s + \dfrac{s+t}{s-t} = \dfrac{1}{t} + \dfrac{s+t}{s-t}$

Observe that if we subtract $\dfrac{s+t}{s-t}$ from both sides we are left with an equivalent equation, $s = \dfrac{1}{t}$. We solve this equation for $t$.

$$s = \frac{1}{t}$$

$$st = 1 \qquad \text{Multiplying both sides by } t$$

$$t = \frac{1}{s} \qquad \text{Multiplying both sides by } \frac{1}{s}$$

---

## Exercise Set 1.6

**1.** The power rule

**3.** Raising a product to a power

**5.** The product rule

**7.** Raising a quotient to a power

**9.** The quotient rule

**11.** $2^8 \cdot 2^4 = 2^{8+4} = 2^{12}$

**13.** $5^6 \cdot 5^4 = 5^{6+4} = 5^{10}$

**15.** $m^9 \cdot m^0 = m^{9+0} = m^9$

**17.** $6x^5 \cdot 3x^2 = 6 \cdot 3 \cdot x^5 \cdot x^2 = 18x^{5+2} = 18x^7$

**19.** $(-2m^4)(-8m^9) = (-2)(-8)m^4 \cdot m^9 = 16m^{4+9} = 16m^{13}$

**21.** $(x^3 y^4)(x^7 y^6 z^0) = (x^3 x^7)(y^4 y^6)(z^0) = x^{3+7} y^{4+6} \cdot 1 = x^{10} y^{10}$

**23.** $\dfrac{a^9}{a^3} = a^{9-3} = a^6$

**25.** $\dfrac{12t^7}{4t^2} = \dfrac{12}{4} \cdot t^{7-2} = 3t^5$

**27.** $\dfrac{m^7 n^9}{m^2 n^5} = m^{7-2} \cdot n^{9-5} = m^5 n^4$

**29.** $\dfrac{32x^8y^5}{8x^2y} = \dfrac{32}{8} \cdot x^{8-2} \cdot y^{5-1} = 4x^6y^4$

**31.** $\dfrac{28x^{10}y^9z^8}{-7x^2y^3z^2} = \dfrac{28}{-7} \cdot x^{10-2}y^{9-3}z^{8-2} = -4x^8y^6z^6$

**33.** $-x^0 = -(-2)^0 = -(1) = -1$

**35.** $(4x)^0 = (4(-2))^0 = (-8)^0 = 1$

**37.** $(-2)^4 = -2(-2)(-2)(-2) = 16$

**39.** $-2^4 = -2 \cdot 2 \cdot 2 \cdot 2 = -16$

**41.** $(-4)^{-2} = \dfrac{1}{(-4)^2} = \dfrac{1}{-4(-4)} = \dfrac{1}{16}$

**43.** $-4^{-2} = -\dfrac{1}{4^2} = -\dfrac{1}{16}$

**45.** $-2^{-4} = -\dfrac{1}{2^4} = -\dfrac{1}{16}$

**47.** $-2^{-6} = -\dfrac{1}{2^6} = -\dfrac{1}{64}$

**49.** $a^{-3} = \dfrac{1}{a^3}$

**51.** $\dfrac{1}{5^{-3}} = 5^3 = 125$

**53.** $8x^{-3} = 8 \cdot \dfrac{1}{x^3} = \dfrac{8}{x^3}$

**55.** $3a^8b^{-6} = 3a^8 \cdot \dfrac{1}{b^6} = \dfrac{3a^8}{b^6}$

**57.** We can move $z^{-4}$ to the other side of the fraction bar if we change the sign of the exponent.

$$\dfrac{z^{-4}}{3x^5} = \dfrac{1}{3x^5z^4}$$

**59.** We can move $x^{-2}$ and $z^{-4}$ to the other side of the fraction bar if we change the sign of the exponent.

$$\dfrac{x^{-2}y^7}{z^{-4}} = \dfrac{y^7z^4}{x^2}$$

**61.** $\dfrac{1}{8^4} = 8^{-4}$

**63.** $\dfrac{1}{(-12)^2} = (-12)^{-2}$

**65.** $x^5 = \dfrac{1}{x^{-5}}$

**67.** $4x^2 = 4 \cdot \dfrac{1}{x^{-2}} = \dfrac{4}{x^{-2}}$

**69.** $\dfrac{1}{(5y)^3} = (5y)^{-3}$

**71.** $\dfrac{1}{3y^4} = \dfrac{1}{3} \cdot \dfrac{1}{y^4} = \dfrac{1}{3} \cdot y^{-4} = \dfrac{y^{-4}}{3}$

**73.** $8^{-2} \cdot 8^{-4} = 8^{-2+(-4)} = 8^{-6}$, or $\dfrac{1}{8^6}$

**75.** $b^2 \cdot b^{-5} = b^{2+(-5)} = b^{-3}$, or $\dfrac{1}{b^3}$

**77.** $a^{-3} \cdot a^4 \cdot a^2 = a^{-3+4+2} = a^3$

**79.** $(4mn^3)(-2m^3n^2) = 4(-2) \cdot m \cdot m^3 \cdot n^3 \cdot n^2$
$$= -8n^{1+3}n^{3+2}$$
$$= -8m^4n^5$$

**81.** $(-2x^{-3})(7x^{-8}) = -2 \cdot 7 \cdot x^{-3} \cdot x^{-8} = -14x^{-3+(-8)}$
$$= -14x^{-11}, \text{ or } \dfrac{-14}{x^{11}}, \text{ or} -\dfrac{14}{x^{11}}$$

**83.** $(5a^{-2}b^{-3})(2a^{-4}b) = 5 \cdot 2 \cdot a^{-2} \cdot a^{-4} \cdot b^{-3} \cdot b$
$$= 10a^{-2+(-4)}b^{-3+1}$$
$$= 10a^{-6}b^{-2}, \text{ or } \dfrac{10}{a^6b^2}$$

**85.** $\dfrac{10^{-3}}{10^6} = 10^{-3-6} = 10^{-9}$, or $\dfrac{1}{10^9}$

**87.** $\dfrac{2^{-7}}{2^{-5}} = 2^{-7-(-5)} = 2^{-7+5} = 2^{-2}$, or $\dfrac{1}{2^2}$, or $\dfrac{1}{4}$

**89.** $\dfrac{y^4}{y^{-5}} = y^{4-(-5)} = y^{4+5} = y^9$

**91.** $\dfrac{24a^5b^3}{-8a^4b} = \dfrac{24}{-8}a^{5-4}b^{3-1} = -3ab^2$

**93.** $\dfrac{14a^4b^{-3}}{-8a^8b^{-5}} = \dfrac{14}{-8}a^{4-8}b^{-3-(-5)} = -\dfrac{7}{4}a^{-4}b^2$, or
$$-\dfrac{7b^2}{4a^4}$$

**95.** $\dfrac{-6x^{-2}y^4z^8}{24x^{-5}y^6z^{-3}} = \dfrac{-6}{24}x^{-2-(-5)}y^{4-6}z^{8-(-3)} =$
$$-\dfrac{1}{4}x^3y^{-2}z^{11}, \text{ or } -\dfrac{x^3z^{11}}{4y^2}$$

**97.** $(x^4)^3 = x^{4 \cdot 3} = x^{12}$

**99.** $(9^3)^{-4} = 9^{3(-4)} = 9^{-12}$, or $\dfrac{1}{9^{12}}$

**101.** $(t^{-8})^{-5} = t^{-8(-5)} = t^{40}$

**103.** $(5xy)^2 = 5^2x^2y^2 = 25x^2y^2$

**105.** $(a^3b)^4 = a^{3 \cdot 4}b^4 = a^{12}b^4$

**107.** $\dfrac{(x^5)^2(x^{-3})^4}{(x^2)^3} = \dfrac{x^{5 \cdot 2}x^{-3 \cdot 4}}{x^{2 \cdot 3}} = \dfrac{x^{10} \cdot x^{-12}}{x^6} = x^{10+(-12)-6} =$
$x^{-8}, \text{ or } \dfrac{1}{x^8}$

**109.** $\dfrac{(2a^3)^3 4a^{-3}}{(a^2)^5} = \dfrac{2^3a^{3 \cdot 3}4a^{-3}}{a^{2 \cdot 5}} = \dfrac{8a^9 4a^{-3}}{a^{10}} =$
$8 \cdot 4a^{9+(-3)-10} = 32a^{-4}, \text{ or } \dfrac{32}{a^4}$

**111.** $(8x^{-3}y^2)^{-4}(8x^{-3}y^2)^4 = (8x^{-3}y^2)^{-4+4} =$
$(8x^{-3}y^2)^0 = 1$

**113.** $\dfrac{(3x^3y^4)^3}{6xy^3} = \dfrac{3^3(x^3)^3(y^4)^3}{6xy^3} = \dfrac{27x^9y^{12}}{6xy^3} =$
$\dfrac{27}{6}x^{9-1}y^{12-3} = \dfrac{9}{2}x^8y^9, \text{ or } \dfrac{9x^8y^9}{2}$

**115.** $\left(\dfrac{-4x^4y^{-2}}{5x^{-1}y^4}\right)^{-4} = \left(\dfrac{-4}{5}x^{4-(-1)}y^{-2-4}\right)^{-4} =$

$\left(\dfrac{-4}{5}x^5y^{-6}\right)^{-4} = \dfrac{(-4)^{-4}(x^5)^{-4}(y^{-6})^{-4}}{5^{-4}} =$

$\dfrac{5^4x^{-20}y^{24}}{(-4)^4} = \dfrac{625x^{-20}y^{24}}{256}$, or $\dfrac{625y^{24}}{256x^{20}}$

**117.** $\left(\dfrac{6a^{-2}b^6}{8a^{-4}b^0}\right)^{-2} = \left(\dfrac{6}{8}a^{-2-(-4)}b^{6-0}\right)^{-2} = \left(\dfrac{3}{4}a^2b^6\right)^{-2} =$

$\left(\dfrac{3}{4}\right)^{-2}(a^2)^{-2}(b^6)^{-2} = \left(\dfrac{4}{3}\right)^2 a^{-4}b^{-12} = \dfrac{16}{9}a^{-4}b^{-12}$, or

$\dfrac{16}{9a^4b^{12}}$

**119.** $\left(\dfrac{4a^3b^{-9}}{6a^{-2}b^5}\right)^0 = 1$

(Any nonzero real number raised to the zero power is 1.)

**121.** *Writing Exercise*

**123.** $4.9t^2 + 3t = 4.9(-3)^2 + 3(-3)$

$\qquad\qquad = 4.9(9) + 3(-3)$

$\qquad\qquad = 44.1 - 9$

$\qquad\qquad = 35.1$

**125.** *Writing Exercise*

**127.** $\dfrac{8a^{x-2}}{2a^{2x+2}} = \dfrac{8}{2} \cdot a^{x-2-(2x+2)} = 4a^{x-2-2x-2} = 4a^{-x-4}$

**129.** $\dfrac{(2^{-2})^a \cdot (2^b)^{-a}}{(2^{-2})^{-b}(2^b)^{-2a}} = \dfrac{2^{-2a} \cdot 2^{-ab}}{2^{2b} \cdot 2^{-2ab}} = \dfrac{2^{-2a-ab}}{2^{2b-2ab}} =$

$2^{-2a-ab-(2b-2ab)} = 2^{-2a-ab-2b+2ab} = 2^{-2a-2b+ab}$

**131.** $(3^{a+2})^a = 3^{(a+2)(a)} = 3^{a^2+2a}$

**133.** $(7^{3-a})^{2b} = 7^{(3-a)2b} = 7^{6b-2ab}$

**135.** $\dfrac{3^{q+3} - 3^2(3^q)}{3(3^{q+4})} = \dfrac{3^{q+3} - 3^{q+2}}{3^{q+5}} = \dfrac{3^{q+2}(3-1)}{3^{q+2}(3^3)} =$

$\dfrac{2}{3^3} = \dfrac{2}{27}$

**137.** $\left[\left(\dfrac{a^{-2c}}{b^{7c}}\right)^{-3}\left(\dfrac{a^{4c}}{b^{-3c}}\right)^2\right]^{-a} = \left(\dfrac{a^{6c}}{b^{-21c}} \cdot \dfrac{a^{8c}}{b^{-6c}}\right)^{-a} =$

$\left(\dfrac{a^{14c}}{b^{-27c}}\right)^{-a} = \dfrac{a^{-14ac}}{b^{27ac}}$

## Exercise Set 1.7

**1.** The length of an Olympic marathon, in centimeters, is a large number so its representation in scientific notation would include a positive power of 10.

**3.** The mass of a hydrogen atom, in grams is a small number so its representation in scientific notation would include a negative power of 10.

**5.** The time between leap years, in seconds, is a large number so its representation in scientific notation would include a positive power of 10.

**7.** $5 \times 10^{-4} = 0.0005$  Moving the decimal point 4 places to the left

**9.** $9.73 \times 10^8 = 973,000,000$  Moving the decimal point 8 places to the right

**11.** $4.923 \times 10^{-10} = 0.0000000004923$  Moving the decimal point 10 places to the left

**13.** $9.03 \times 10^{10} = 90,300,000,000$  Moving the decimal point 10 places to the right

**15.** $4.037 \times 10^{-8} = 0.00000004037$  Moving the decimal point 8 places to the left

**17.** $7.01 \times 10^{12} = 7,010,000,000,000$  Moving the decimal point 12 places to the right

**19.** $\qquad 83,000,000,000$

$= \dfrac{83,000,000,000}{10^{10}} \cdot 10^{10}$  Multiplying by 1: $\dfrac{10^{10}}{10^{10}} = 1$

$= 8.3 \times 10^{10}$  This is scientific notation.

**21.** $\qquad 863,000,000,000,000,000$

$= \dfrac{863,000,000,000,000,000}{10^{17}} \cdot 10^{17}$

$= 8.63 \times 10^{17}$

**23.** $\qquad 0.000000016$

$= \dfrac{0.000000016}{10^8} \cdot 10^8$  Multiplying by 1: $\dfrac{10^8}{10^8} = 1$

$= \dfrac{1.6}{10^8}$

$= 1.6 \times 10^{-8}$  Writing scientific notation

**25.** $\qquad 0.00000000007$

$= \dfrac{0.00000000007}{10^{11}} \cdot 10^{11}$

$= \dfrac{7}{10^{11}}$

$= 7 \times 10^{-11}$

**27.** $\qquad 803,000,000,000$

$= \dfrac{803,000,000,000}{10^{11}} \cdot 10^{11}$

$= 8.03 \times 10^{11}$

**29.** $\qquad 0.000000904$

$= \dfrac{0.000000904}{10^7} \cdot 10^7$

$= \dfrac{9.04}{10^7}$

$= 9.04 \times 10^{-7}$

**31.** $431,700,000,000$

$= \dfrac{431,700,000,000}{10^{11}} \cdot 10^{11}$

$= 4.317 \times 10^{11}$

**33.** $(2.3 \times 10^6)(4.2 \times 10^{-11})$

$= (2.3 \times 4.2)(10^6 \times 10^{-11})$

$= 9.66 \times 10^{-5}$

$= 9.7 \times 10^{-5}$ Rounding to 2 significant digits

**35.** $(2.34 \times 10^{-8})(5.7 \times 10^{-4})$

$= (2.34 \times 5.7)(10^{-8} \times 10^{-4})$

$= 13.338 \times 10^{-12}$

$= (1.3338 \times 10^1) \times 10^{-12}$

$= 1.3338 \times (10^1 \times 10^{-12})$

$= 1.3338 \times 10^{-11}$

$= 1.3 \times 10^{-11}$ Rounding to 2 significant digits

**37.** $(5.2 \times 10^6)(2.6 \times 10^4) = (5.2 \times 2.6)(10^6 \times 10^4)$

$= 13.52 \times 10^{10}$

$= (1.352 \times 10) \times 10^{10}$

$= 1.352 \times (10 \times 10^{10})$

$= 1.352 \times 10^{11}$

$= 1.4 \times 10^{11}$

(2 significant digits)

**39.** $(7.01 \times 10^{-5})(6.5 \times 10^7)$

$= (7.01 \times 6.5)(10^{-5} \times 10^7)$

$= 45.565 \times 10^2$

$= (4.5565 \times 10^1) \times 10^2$

$= 4.5565 \times (10^1 \times 10^2)$

$= 4.5565 \times 10^3$

$= 4.6 \times 10^3$ Rounding to 2 significant digits

**41.** $(2.0 \times 10^6)(3.02 \times 10^{-6})$

Observe that $10^6 \times 10^{-6} = 1$, so the product is $2.0(3.02)$, or $6.04$, or $6.0$, rounded to two significant digits.

**43.** $\dfrac{5.1 \times 10^6}{3.4 \times 10^3} = \dfrac{5.1}{3.4} \times \dfrac{10^6}{10^3}$

$= 1.5 \times 10^3$

**45.** $\dfrac{7.5 \times 10^{-9}}{2.5 \times 10^{-4}} = \dfrac{7.5}{2.5} \times \dfrac{10^{-9}}{10^{-4}}$

$= 3.0 \times 10^{-5}$ (2 significant digits)

**47.** $\dfrac{3.2 \times 10^{-7}}{8.0 \times 10^8} = \dfrac{3.2}{8.0} \times \dfrac{10^{-7}}{10^8}$

$= 0.40 \times 10^{-15}$ (2 significant digits)

$= (4.0 \times 10^{-1}) \times 10^{-15}$

$= 4.0 \times 10^{-16}$

**49.** $\dfrac{9.36 \times 10^{-11}}{3.12 \times 10^{11}} = \dfrac{9.36}{3.12} \times \dfrac{10^{-11}}{10^{11}}$

$= 3.00 \times 10^{-22}$ (3 significant digits)

**51.** $\dfrac{6.12 \times 10^{19}}{3.06 \times 10^{-7}} = \dfrac{6.12}{3.06} \times \dfrac{10^{19}}{10^{-7}}$

$= 2.00 \times 10^{26}$ (3 significant digits)

**53.** $4.6 \times 10^{-9} + 3.2 \times 10^{-9} = (4.6 + 3.2) \times 10^{-9}$

$= 7.8 \times 10^{-9}$

**55.** $5.9 \times 10^{23} + 6.3 \times 10^{23}$

$= (5.9 + 6.3) \times 10^{23}$

$= 12.2 \times 10^{23}$

$= (1.22 \times 10) \times 10^{23}$

$= 1.22 \times 10^{24}$

$= 1.2 \times 10^{24}$ (2 significant digits)

**57.** *Familiarize*. Let $n =$ the smallest number of neutrinos that could have the same mass as an alpha particle of mass $3.62 \times 10^{-27}$ kg.

*Translate*. We divide.

$$n = \dfrac{3.62 \times 10^{-27} \text{ kg}}{1.8 \times 10^{-36} \text{ kg}}$$

*Carry out*. We perform the calculation and write scientific notation for the answer.

$$n = \dfrac{3.62 \times 10^{-27}}{1.8 \times 10^{-36}}$$

$$= \dfrac{3.62}{1.8} \times \dfrac{10^{-27}}{10^{-36}}$$

$$\approx 2.0 \times 10^9 \quad \text{(2 significant digits)}$$

*Check*. We multiply the answer by the maximum mass of a neutrino.

$(2.0 \times 10^9)(1.8 \times 10^{-36}) = 3.6 \times 10^{-27} \approx 3.62 \times 10^{-27}$

The answer checks.

*State*. The smallest umber of neutrinos that could have the same mass as an alpha particle of mass $3.62 \times 10^{-27}$ kg is about $2.0 \times 10^9$ neutrinos.

**59.** *Familiarize*. We have a cylinder with diameter $4.0 \times 10^{-10}$ in. and length 100 yd. We will use the formula for the volume of a cylinder $V = \pi r^2 h$. The radius is $\dfrac{4.0 \times 10^{-10}}{2}$, or $2.0 \times 10^{-10}$ in. We convert 100 yd to inches:

100 yd = $100 \times 1$ yd = $100 \times 36$ in. = 3600 in., or $3.6 \times 10^3$ in.

*Translate*. We substitute in the formula.

$V = \pi r^2 h$

$V = \pi(2.0 \times 10^{-10})^2(3.6 \times 10^3)$

*Carry out*. We do the calculation.

$V = \pi(2.0 \times 10^{-10})^2(3.6 \times 10^3)$

$= \pi \times 4.0 \times 10^{-20} \times 3.6 \times 10^3$

$= (\pi \times 4.0 \times 3.6) \times (10^{-20} \times 10^3)$

$\approx 45.2 \times 10^{-17}$

$\approx (4.52 \times 10) \times 10^{-17}$

$\approx 4.52 \times 10^{-16}$

$\approx 4.5 \times 10^{-16}$ (2 significant digits)

*Check*. Recheck the translation and the calculations. The answer checks.

*State*. The volume of a 100-yd carbon nanotube is about $4.5 \times 10^{-16}$ in$^3$.

**61.** *Familiarize*. We can think of the plastic as a rectangular solid whose length is the perimeter of the house and with width 8 mil, or $8 \times \dfrac{1}{1000}$ in. $= 8 \times 0.001$ in. $= 0.008$ in. $= 8 \times 10^{-3}$ in. and height 4 ft. The perimeter of the house is $2 \cdot 32$ ft $+ 2 \cdot 24$ ft, or 112 ft. We will convert the perimeter and the height to inches.

112 ft $= 112 \times 1$ ft $= 112 \times 12$ in. $= 1344$ in., or $1.344 \times 10^3$ in.

4 ft $= 4 \times 1$ ft $= 4 \times 12$ in. $= 48$ in., or $4.8 \times 10$ in.

Recall that the formula for the volume of a rectangular solid is $V = lwh$.

$$V = lwh$$
$$V = (1.344 \times 10^3) \times (4.8 \times 10) \times (8 \times 10^{-3})$$

*Carry out*. We do the calculation.

$$V = (1.344 \times 10^3) \times (4.8 \times 10) \times (8 \times 10^{-3})$$
$$V = (1.344 \times 4.8 \times 8) \times (10^3 \times 10 \times 10^{-3})$$
$$\approx 51.6 \times 10$$
$$\approx (5.16 \times 10) \times 10$$
$$\approx 5.16 \times 10^2$$
$$\approx 5 \times 10^2 \qquad \text{(1 significant digit)}$$

*Check*. Recheck the translation and the calculation. The answer checks.

*State*. The volume of the plastic is about $5 \times 10^2$ in$^3$. (If we had used feet as the unit of length, the result would be about $3 \times 10^{-1}$ ft$^3$.)

**63.** *Familiarize*. We know that 1 light year $= 5.88 \times 10^{12}$ mi. Let $y =$ the number of light years from the earth to Sirius.

*Translate*. The distance from the earth to Sirius is $y$ light years or $(5.88 \times 10^{12})y$ mi. It is also given by $4.704 \times 10^{13}$ mi. We write an equation:

$$(5.88 \times 10^{12})y = 4.704 \times 10^{13}$$

*Carry out*. We solve the equation.

$$(5.88 \times 10^{12})y = 4.704 \times 10^{13}$$
$$y = \frac{4.704 \times 10^{13}}{5.88 \times 10^{12}}$$
$$y = \frac{4.704}{5.88} \times \frac{10^{13}}{10^{12}}$$
$$y = 0.800 \times 10 \quad \text{The answer must}$$
$$\text{have 3 significant digits.}$$
$$y = 8.00$$

*Check*. Since light travels $5.88 \times 10^{12}$ mi in one year, in 8.00 yr it will travel $8.00 \times 5.88 \times 10^{12} = 4.704 \times 10^{13}$ mi, the distance from the earth to Sirius. The answer checks.

*State*. It is 8.00 light years from the earth to Sirius.

**65.** *Familiarize*. We are told that 1 Angstrom $= 10^{-10}$ m, 1 parsec $\approx 3.26$ light years, and 1 light year $= 9.46 \times 10^{15}$ m. Let $a$ represent the number of Angstroms in one parsec.

*Translate*. The length of one parsec is $a \times 10^{-10}$ m. It can also be expressed as 3.26 light years, or $3.26 \times 9.46 \times 10^{15}$ m. Since these quantities represent the same number, we can write the equation.

$$a \times 10^{-10} = 3.26 \times 9.46 \times 10^{15}.$$

*Carry out*. Solve the equation:

$$a \times 10^{-10} = 3.26 \times 9.46 \times 10^{15}$$
$$a \times 10^{-10} \times \frac{1}{10^{-10}} = 3.26 \times 9.46 \times 10^{15} \times \frac{1}{10^{-10}}$$
$$a = \frac{3.26 \times 9.46 \times 10^{15}}{10^{-10}}$$
$$= (3.26 \times 9.46) \times \frac{10^{15}}{10^{-10}}$$
$$= 30.8396 \times 10^{25}$$
$$= (3.08396 \times 10) \times 10^{25}$$
$$= 3.08396 \times (10 \times 10^{25})$$
$$= 3.08 \times 10^{26} \qquad \text{Rounding}$$
$$\text{to 3 significant digits}$$

*Check*. We recheck the translation and calculation.

*State*. There are about $3.08 \times 10^{26}$ Angstroms in one parsec.

**67.** *Familiarize*. We have a very long cylinder. Its length is the average distance from the earth to the sun, $1.5 \times 10^{11}$ m, and the diameter of its base is 3 Å. We will use the formula for the volume of a cylinder, $V = \pi r^2 h$. (See Example 8.)

*Translate*. We will express all distances in Angstroms.

Height (length) : $1.5 \times 10^{11}$ m $= \dfrac{1.5 \times 10^{11}}{10^{-10}}$ Å, or
$$1.5 \times 10^{21} \text{ Å}$$

Diameter: $3$ Å

The radius is half the diameter:

Radius: $\dfrac{1}{2} \times 3$ Å $= 1.5$ Å

Now substitute into the formula (using 3.14 for $\pi$):

$$V = \pi r^2 h$$
$$V = 3.14 \times 1.5^2 \times 1.5 \times 10^{21}$$

*Carry out*. Do the calculations.

$$V = 3.14 \times 1.5^2 \times 1.5 \times 10^{21}$$
$$= 10.5975 \times 10^{21}$$
$$= 1.05975 \times 10^{22}$$
$$= 1 \times 10^{22} \text{ Rounding to 1 significant digit}$$

We can convert this result to cubic meters, if desired.

$1$Å $= 10^{-10}$ m, so 1 cu Å $= (10^{-10})^3$ m$^3 = 10^{-30}$ m$^3$.

Then $1 \times 10^{22}$ cu Å $= 1 \times 10^{22}$ cu Å $\times \dfrac{10^{-30} \text{ m}^3}{1 \text{ cu Å}} =$

$1 \times 10^{22} \times 10^{-30} \times \dfrac{\text{cu Å}}{\text{cu Å}} \times \text{m}^3 = 1 \times 10^{-8}$ m$^3$.

*Check*. We recheck the translation and the calculations.

*State*. The volume of the sunbeam is about $1 \times 10^{22}$ cu Å, or $1 \times 10^{-8}$ m$^3$.

**69.** *Familiarize*. First we will find $d$, the number of drops in a pound. Then we will find $b$, the number of bacteria in a drop of U.S. mud.

*Translate*. To find $d$ we convert 1 pound to drops:
$$d = 1 \text{ lb} \cdot \frac{16 oz}{1 \text{ lb}} \cdot \frac{6.0 \text{ tsp}}{1 \text{ oz}} \cdot \frac{60.0 \text{ drops}}{1 \text{ tsp}}.$$

Then we divide to find $b$:
$$b = \frac{4.55 \times 10^{11}}{d}.$$

*Carry out*. We do the calculations.
$$d = 1 \text{ lb} \cdot \frac{16 oz}{1 \text{ lb}} \cdot \frac{6.0 \text{ tsp}}{1 \text{ oz}} \cdot \frac{60.0 \text{ drops}}{1 \text{ tsp}}$$
$$= 5760 \text{ drops}$$

Now we find $b$.
$$b = \frac{4.55 \times 10^{11}}{5760} = \frac{4.55 \times 10^{11}}{5.760 \times 10^3} \approx 0.790 \times 10^8 \approx$$
$$(7.90 \times 10^{-1}) \times 10^8 = 7.9 \times 10^7$$

(Our answer must have 2 significant digits.)

*Check*. If there are about $7.90 \times 10^7$ bacteria in a drop of U.S. mud, then in a pound there are about
$$\frac{7.9 \times 10^7}{1 \text{ drop}} \cdot \frac{60.0 \text{ drops}}{1 \text{ tsp}} \cdot \frac{6.0 \text{ tsp}}{1 \text{ oz}} \cdot \frac{16 \text{ oz}}{1 \text{ lb}} =$$
$$\frac{45,504 \times 10^7}{1 \text{ lb}} \approx 4.55 \times 10^{11} \text{ bacteria per pound. The an-}$$
swer checks.

*State*. About $7.9 \times 10^7$ bacteria live in a drop of U.S. mud.

**71.** *Familiarize*. First we will find the distance $C$ around Jupiter at the equator, in km. Then we will use the formula Speed × Time = Distance to find the speed $s$ at which Jupiter's equator is spinning.

*Translate*. We will use the formula for the circumference of a circle to find the distance around Jupiter at the equator:
$$C = \pi d = \pi(1.43 \times 10^5).$$

Then we find the speed $s$ at which Jupiter's equator is spinning:

$$\underbrace{\text{Speed}}_{s} \times \underbrace{\text{Time}}_{10} = \underbrace{\text{Distance}}_{C}$$

*Carry out*. First we find $C$.
$$C = \pi(1.43 \times 10^5) \approx 4.49 \times 10^5$$

Then we find $s$.
$$s \times 10 = C$$
$$s \times 10 = 4.49 \times 10^5$$
$$s = \frac{4.49 \times 10^5}{10}$$
$$s = 4.49 \times 10^4$$

*Check*. At $4.49 \times 10^4$ km/h, in 10 hr, Jupiter's equator travels $4.49 \times 10^4 \times 10$, or $4.49 \times 10^5$ km. A circle with circumference $4.49 \times 10^5$ km has a diameter of $\frac{4.49 \times 10^5}{\pi} \approx 1.43 \times 10^5$ km. The answer checks.

*State*. Jupiter's equator spins at a speed of about $4.49 \times 10^4$ km/h.

**73.** *Writing Exercise*

**75.** $3x - 7y = 3 \cdot 5 - 7 \cdot 1 = 15 - 7 = 8$

**77.** *Writing Exercise*

**79.** First we divide to find the weight of one-half of the compound:
$$\frac{1.2 \times 10^{-9}}{2} = 0.60 \times 10^{-9} = (6.0 \times (10^{-1}) \times 10^{-9}) =$$
$$6.0 \times 10^{-1} \times 10^{-9} = 6.0 \times 10^{-10}$$

Then we subtract $3.5 \times 10^{-10}$ from the weight to find the lower boundary of the weight range, and we add $3.5 \times 10^{-10}$ to find the upper boundary:
$$6.0 \times 10^{-10} - 3.5 \times 10^{-10} = (6.0 - 3.5) \times 10^{-10} =$$
$$2.5 \times 10^{-10}$$
$$6.0 \times 10^{-10} + 3.5 \times 10^{-10} = (6.0 + 3.5) \times 10^{-10} =$$
$$9.5 \times 10^{-10}$$

The actual weight of each half is between $2.5 \times 10^{-10}$ oz and $9.5 \times 10^{-10}$ oz.

**81.** The larger number is the one in which the power of ten has the larger exponent. Since $-90$ is larger than $-91$, $8 \cdot 10^{-90}$ is larger than $9 \cdot 10^{-91}$.
$$8 \cdot 10^{-90} - 9 \cdot 10^{-91} = 10^{-90}(8 - 9 \cdot 10^{-1})$$
$$= 10^{-90}(8 - 0.9)$$
$$= 7.1 \times 10^{-90}$$

Thus, $8 \cdot 10^{-90}$ is larger by $7.1 \cdot 10^{-90}$.

**83.**
$$(4096)^{0.05}(4096)^{0.2} = 4096^{0.25}$$
$$= (2^{12})^{0.25}$$
$$= 2^3$$
$$= 8$$

**85.** *Familiarize*. Observe that there are $2^{n-1}$ grains of sand on the $n$th square of the chessboard. Let $g$ represent this quantity. Recall that a chessboard has 64 squares. Note also that $2^{10} \approx 10^3$.

*Translate*. We write the equation
$$g = 2^{n-1}.$$

To find the number of grains of sand on the last (or 64th) square, substitute 64 for $n$: $g = 2^{64-1}$

*Carry out*. Do the calculations, expressing the result in scientific notation.
$$g = 2^{64-1} = 2^{63} = 2^3(2^{10})^6$$
$$\approx 2^3(10^3)^6 \approx 8 \times 10^{18}$$

*Check*. Recheck the translation and the calculations.

*State*. Approximately $8 \times 10^{18}$ grains of sand are required for the last square.

# Chapter 2

# Graphs, Functions, and Linear Equations

## Exercise Set 2.1

**1.** In the fourth quadrant, a point's first coordinate is always positive and its second coordinate is always <u>negative</u>.

**3.** The two perpendicular number lines that are used for graphing are called <u>axes</u>.

**5.** To graph an equation means to make a drawing that represents all the <u>solutions</u> of the equation.

**7.** $A$ is 5 units right of the origin and 3 units up, so its coordinates are $(5, 3)$.

$B$ is 4 units left of the origin and 3 units up, so its coordinates are $(-4, 3)$.

$C$ is 0 units right or left of the origin and 2 units up, so its coordinates are $(0, 2)$.

$D$ is 2 units left of the origin and 3 units down, so its coordinates are $(-2, -3)$.

$E$ is 4 units right of the origin and 2 units down, so its coordinates are $(4, -2)$.

$F$ is 5 units left of the origin and 0 units up or down, so its coordinates are $(-5, 0)$.

**9.**

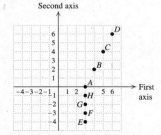

$A(3, 0)$ is 3 units right and 0 units up or down.

$B(4, 2)$ is 4 units right and 2 units up.

$C(5, 4)$ is 5 units right and 4 units up.

$D(6, 6)$ is 6 units right and 6 units up.

$E(3, -4)$ is 3 units right and 4 units down.

$F(3, -3)$ is 3 units right and 3 units down.

$G(3, -2)$ is 3 units right and 2 units down.

$H(3, -1)$ is 3 units right and 1 unit down.

**11.**

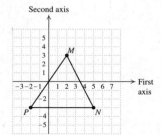

A triangle is formed. The area of a triangle is found by using the formula $A = \frac{1}{2}bh$. In this triangle the base and height are 7 units and 6 units, respectively.

$$A = \frac{1}{2}bh = \frac{1}{2} \cdot 7 \cdot 6 = \frac{42}{2} = 21 \text{ square units}$$

**13.** The first coordinate is negative and the second positive, so the point $(-4, 1)$ is in quadrant II.

**15.** Both coordinates are negative, so the point $(-6, -7)$ is in quadrant III.

**17.** Both coordinates are positive, so the point $\left(3, \frac{1}{2}\right)$ is in quadrant I.

**19.** The first coordinate is positive and the second negative, so the point $(6.9, -2)$ is in quadrant IV.

**21.**

$$\begin{array}{c|l} \multicolumn{2}{l}{y = 3x - 4} \\ \hline -1 & 3 \cdot 1 - 4 \\ & 3 - 4 \\ & \phantom{-}\overset{?}{} \\ \multicolumn{2}{l}{-1 \overset{?}{=} -1} \end{array}$$

Substituting 1 for $x$ and $-1$ for $y$ (alphabetical order of variables)

Since $-1 = -1$ is true, $(1, -1)$ is a solution of $y = 3x - 4$.

**23.**

$$\begin{array}{c|l} \multicolumn{2}{l}{5s - t = 8} \\ \hline 5 \cdot 2 - 4 & 8 \\ 10 - 4 & \\ \overset{?}{} & \\ 6 \overset{?}{=} 8 \end{array}$$

Substituting 2 for $s$ and 4 for $t$ (alphabetical order of variables)

Since $6 = 8$ is false, $(2, 4)$ is not a solution of $5s - t = 8$.

**25.**

$$\begin{array}{c|l} \multicolumn{2}{l}{4x - y = 7} \\ \hline 4 \cdot 3 - 5 & 7 \\ 12 - 5 & \\ \overset{?}{} & \\ 7 \overset{?}{=} 7 \end{array}$$

Substituting 3 for $x$ and 5 for $y$ (alphabetical order of variables)

Since $7 = 7$ is true, $(3, 5)$ is a solution of $4x - y = 7$.

**27.**   $6a + 5b = 3$

$$\frac{6 \cdot 0 + 5 \cdot \frac{3}{5} \;\bigm|\; 3}{\phantom{x}}$$   Substituting 0 for $a$ and $\frac{3}{5}$ for $b$

$\qquad 0 + 3 \;\bigm|$   (alphabetical order of variables)

$\qquad\qquad \overset{?}{3 = 3}$

Since $3 = 3$ is true, $\left(0, \frac{3}{5}\right)$ is a solution of $6a + 5b = 3$.

**29.**   $4r - 2s = 10$

$$\frac{4 \cdot 2 - 2 \cdot (-1) \;\bigm|\; 10}{\phantom{x}}$$   Substituting 2 for $r$ and

$\qquad\qquad\qquad\qquad$ $-1$ for $s$

$\qquad 8 + 2 \;\bigm|$   (alphabetical order of

$\qquad\qquad\qquad\qquad$ variables)

$\qquad\qquad \overset{?}{10 = 10}$

Since $10 = 10$ is true, $(2, -1)$ is a solution of $4r - 2s = 10$.

**31.**   $x - 3y = -4$

$$\frac{5 - 3 \cdot 3 \;\bigm|\; -4}{\phantom{x}}$$   Substituting 5 for $x$ and 3 for $y$

$\qquad 5 - 9 \;\bigm|$   (alphabetical order of variables)

$\qquad\qquad \overset{?}{-4 = -4}$

Since $-4 = -4$ is true, $(5, 3)$ is a solution of $x - 3y = -4$.

**33.**   $y = 3x^2$

$$\frac{-1 \;\bigm|\; 3(3)^2}{\phantom{x}}$$   Substituting 3 for $x$ and $-1$ for $y$

$\qquad\;\; \bigm|\; 3 \cdot 9$   (alphabetical order of variables)

$\qquad\qquad \overset{?}{-1 = 27}$

Since $-1 = 27$ is false, $(3, -1)$ is not a solution of $y = 3x^2$.

**35.**   $5s^2 - t = 7$

$$\frac{5(2)^2 - 3 \;\bigm|\; 7}{\phantom{x}}$$   Substituting 2 for $s$ and 3 for $t$

$\qquad 5 \cdot 4 - 3 \;\bigm|$   (alphabetical order of variables)

$\qquad 20 - 3 \;\bigm|$

$\qquad\qquad \overset{?}{17 = 7}$

Since $17 = 7$ is false, $(2, 3)$ is not a solution of $5s^2 - t = 7$.

**37.**   $y = x + 4$

To find an ordered pair, we choose any number for $x$ and then determine $y$ by substitution.

When $x = 0$, $y = 0 + 4 = 4$.

When $x = 1$, $y = 1 + 4 = 5$.

When $x = -2$, $y = -2 + 4 = 2$.

| $x$ | $y$ | $(x, y)$ |
|---|---|---|
| 0 | 4 | $(0, 4)$ |
| 1 | 5 | $(1, 5)$ |
| $-2$ | 2 | $(-2, 2)$ |

Plot these points, draw the line they determine, and label the graph $y = x + 1$.

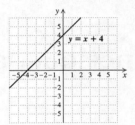

**39.**   $y = -x$

To find an ordered pair, we choose any number for $x$ and then determine $y$. For example, if we choose 1 for $x$, then $y = -1$. We find several ordered pairs, plot them, and draw the line.

| $x$ | $y$ | $(x, y)$ |
|---|---|---|
| 1 | $-1$ | $(1, -1)$ |
| 2 | $-2$ | $(2, -2)$ |
| $-1$ | 1 | $(-1, 1)$ |
| $-3$ | 3 | $(-3, 3)$ |

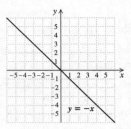

**41.**   $y = 3x - 1$

To find an ordered pair, we choose any number for $x$ and then determine $y$. For example, if $x = 2$, then $y = 3 \cdot 2 - 1 = 6 - 1 = 5$. We find several ordered pairs, plot them, and draw the line.

| $x$ | $y$ | $(x, y)$ |
|---|---|---|
| 2 | 5 | $(2, 5)$ |
| 0 | $-1$ | $(0, -1)$ |
| 1 | 2 | $(1, 2)$ |

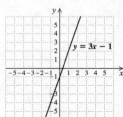

**43.**   $y = -2x + 3$

To find an ordered pair, we choose any number for $x$ and then determine $y$. For example, if $x = 1$, then $y = -2 \cdot 1 + 3 = -2 + 3 = 1$. We find several ordered pairs, plot them, and draw the line.

| $x$ | $y$ |
|---|---|
| 1 | 1 |
| 3 | $-3$ |
| $-1$ | 5 |
| 0 | 3 |

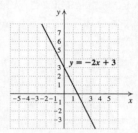

**45.**   $y + 2x = 3$

$\qquad y = -2x + 3$   Solving for $y$

Observe that this is the equation that was graphed in Exercise 43. The graph is shown above.

**47.** $y = -\dfrac{3}{2}x + 5$

To find an ordered pair, we choose any number for $x$ and then determine $y$. For example, if $x = 2$, then $y = -\dfrac{3}{2} \cdot 2 + 5 = -3 + 5 = 2$. We find several ordered pairs, plot them, and draw the line.

| $x$ | $y$ |
|-----|-----|
| 2 | 2 |
| 4 | $-1$ |
| 0 | 5 |
| 6 | $-4$ |

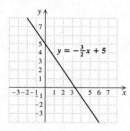

**49.** $y = \dfrac{3}{4}x + 1$

To find an ordered pair, we choose any number for $x$ and then determine $y$. For example, if $x = 4$, then $y = \dfrac{3}{4} \cdot 4 + 1 = 3 + 1 = 4$. We find several ordered pairs, plot them, and draw the line.

| $x$ | $y$ |
|-----|-----|
| 4 | 4 |
| 0 | 1 |
| $-4$ | $-2$ |

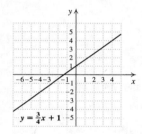

**51.** $y = |x| + 2$

We select $x$-values and find the corresponding $y$-values. The table lists some ordered pairs. We plot these points.

| $x$ | $y$ |
|-----|-----|
| 3 | 5 |
| 1 | 3 |
| 0 | 2 |
| $-1$ | 3 |
| $-3$ | 5 |

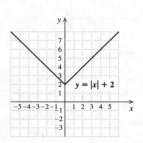

Note that the graph is V-shaped, centered at (0,2).

**53.** $y = |x| - 2$

We select $x$-values and find the corresponding $y$-values. The table lists some ordered pairs. We plot these points.

| $x$ | $y$ |
|-----|-----|
| 3 | 1 |
| 1 | $-1$ |
| 0 | $-2$ |
| $-1$ | $-1$ |
| $-4$ | 2 |

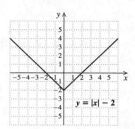

**55.** $y = x^2 + 2$

To find an ordered pair, we choose any number for $x$ and then determine $y$. For example, if $x = 2$, then $y = 2^2 + 2 = 6$. We find several ordered pairs, plot them, and connect them with a smooth curve.

| $x$ | $y$ |
|-----|-----|
| 2 | 6 |
| 1 | 3 |
| 0 | 2 |
| $-1$ | 3 |
| $-2$ | 6 |

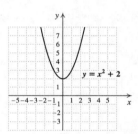

**57.** $y = x^2 - 2$

To find an ordered pair, we choose any number for $x$ and then determine $y$. For example, if $x = 2$, then $y = 2^2 - 2 = 4 - 2 = 2$. We find several ordered pairs, plot them, and connect them with a smooth curve.

| $x$ | $y$ |
|-----|-----|
| 2 | 2 |
| 1 | $-1$ |
| 0 | $-2$ |
| $-1$ | $-1$ |
| $-2$ | 2 |

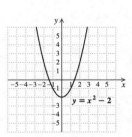

**59.** *Writing Exercise*

**61.** $5s - 3t = 5 \cdot 2 - 3 \cdot 4 = 10 - 12 = -2$

**63.** $(3x - y)^2 = (3 \cdot 4 - 2)^2 = (12 - 2)^2 = 10^2 = 100$

**65.** $(5 - x)^4(x + 2)^3 = (5 - (-2))^4(-2 + 2)^3$

Observe that $-2 + 2 = 0$, so the product is 0.

**67.** *Writing Exercise*

**69.** *Writing Exercise*

**71.** a) Graph IV seems most appropriate for this situation. It reflects driving speeds on local streets for the first 10 and last 5 minutes and freeway cruising speeds from 10 through 30 minutes.

b) Graph III seems most appropriate for this situation. It reflects driving speeds on local streets for the first 10 minutes, an express train speed for the next 20 minutes, and walking speeds for the final 5 minutes.

c) Graph I seems most appropriate for this situation. It reflects walking speeds for the first 10 and last 5 minutes and express bus speeds from 10 through 30 minutes.

d) Graph II seems most appropriate for this situation. It reflects that the speed was 0 mph for the first 10 minutes, the time spent waiting at the bus stop. Then it shows driving speeds that fall to 0 mph several times during the next 20 minutes, indicating that the school bus stops for other students during this period of time. Finally, it shows a walking speed for the last 5 minutes.

**73.** Substitute $-\frac{1}{3}$ for $x$ and $\frac{1}{4}$ for $y$ in each equation.

a)
$$-\frac{3}{2}x - 3y = -\frac{1}{4}$$
$$\begin{array}{c|c} -\frac{3}{2}\left(-\frac{1}{3}\right) - 3\left(\frac{1}{4}\right) & -\frac{1}{4} \\ \frac{1}{2} - \frac{3}{4} & \\ & \overset{?}{} \\ -\frac{1}{4} \overset{?}{=} -\frac{1}{4} \end{array}$$

Since $-\frac{1}{4} = -\frac{1}{4}$ is true, $\left(-\frac{1}{3}, \frac{1}{4}\right)$ is a solution.

b)
$$8y - 15x = \frac{7}{2}$$
$$\begin{array}{c|c} 8\left(\frac{1}{4}\right) - 15\left(-\frac{1}{3}\right) & \frac{7}{2} \\ 2 + 5 & \\ & \overset{?}{} \\ 7 \overset{?}{=} \frac{7}{2} \end{array}$$

Since $7 = \frac{7}{2}$ is false, $\left(-\frac{1}{3}, \frac{1}{4}\right)$ is not a solution.

c)
$$0.16y = -0.09x + 0.1$$
$$\begin{array}{c|c} 0.16\left(\frac{1}{4}\right) & -0.09\left(-\frac{1}{3}\right) + 0.1 \\ 0.04 & 0.03 + 0.1 \\ & \overset{?}{} \\ 0.04 \overset{?}{=} 0.13 \end{array}$$

Since $0.04 = 0.13$ is false, $\left(-\frac{1}{3}, \frac{1}{4}\right)$ is not a solution.

d)
$$2(-y+2) - \frac{1}{4}(3x-1) = 4$$
$$\begin{array}{c|c} 2\left(-\frac{1}{4}+2\right) - \frac{1}{4}\left[3\left(-\frac{1}{3}\right) - 1\right] & 4 \\ 2\left(\frac{7}{4}\right) - \frac{1}{4}(-2) & \\ \frac{7}{2} + \frac{1}{2} & \\ \frac{8}{2} & \\ & \overset{?}{} \\ 4 \overset{?}{=} 4 \end{array}$$

Since $4 = 4$ is true, $\left(-\frac{1}{3}, \frac{1}{4}\right)$ is a solution.

**75.** We make a drawing.

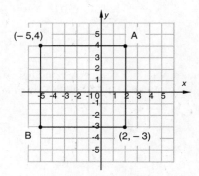

From the drawing we see that the vertex $A$ is 2 units right and up 4 units. Thus, its coordinates are $(2, 4)$. We also see that vertex $B$ is 5 units left and down 3 units, so its coordinates are $(-5, -3)$. We also see that the distance between any pair of adjacent vertices is 7 units. The area of a square whose side has length 7 units is $7 \cdot 7$, or 49 square units.

**77.** $y = \dfrac{1}{x-2}$

Choose $x$-values from $-2$ to $6$, and use a calculator to find the corresponding $y$-values. (Note that we cannot choose 2 as a first coordinate since $\dfrac{1}{2-2}$, or $\dfrac{1}{0}$, is not defined.) Plot the points, and draw the graph. Note that it has two branches, one on each side of a vertical line through $(2, 0)$.

| $x$ | $y$ |
|-----|-----|
| $-2$ | $-0.25$ |
| $0$ | $-0.5$ |
| $1$ | $-1$ |
| $1.9$ | $-10$ |
| $2.1$ | $10$ |
| $2.5$ | $2$ |
| $3$ | $1$ |
| $4$ | $0.5$ |
| $6$ | $0.25$ |

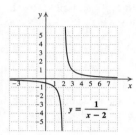

**79.** $y = \dfrac{1}{x^2}$

Choose $x$-values from $-4$ to $4$ and use a calculator to find the corresponding $y$-values. (Note that we cannot choose $0$ as a first coordinate since $1/0^2$, or $1/0$, is not defined.) Plot the points and draw the graph. Note that it has two branches, one on each side of the $y$-axis.

| $x$ | $y$ |
|-----|-----|
| $-4$ | $0.0625$ |
| $-3$ | $0.\overline{1}$ |
| $-2$ | $0.25$ |
| $-1$ | $1$ |
| $-0.5$ | $4$ |
| $0.5$ | $4$ |
| $1$ | $1$ |
| $2$ | $0.25$ |
| $3$ | $0.\overline{1}$ |
| $4$ | $0.0625$ |

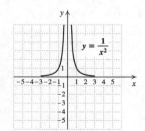

**81.** $y = \sqrt{x} + 1$

Choose $x$-values from $0$ to $10$ and use a calculator to find the corresponding $y$-values. Plot the points and draw the graph.

| $x$ | $y$ |
|-----|-----|
| $0$ | $1$ |
| $1$ | $2$ |
| $2$ | $2.414$ |
| $3$ | $2.732$ |
| $4$ | $3$ |
| $5$ | $3.236$ |
| $6$ | $3.449$ |
| $7$ | $3.646$ |
| $8$ | $3.828$ |
| $9$ | $4$ |
| $10$ | $4.162$ |

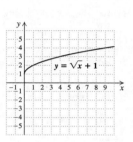

**83.** $y = x^3$

Choose $x$-values from $-2$ to $2$ and use a calculator to find the corresponding $y$-values. Plot the points and draw the graph.

| $x$ | $y$ |
|-----|-----|
| $-2$ | $-8$ |
| $-1.5$ | $-3.375$ |
| $-1$ | $-1$ |
| $-0.5$ | $-0.125$ |
| $-0.25$ | $-0.016$ |
| $0$ | $0$ |
| $0.5$ | $0.125$ |
| $1$ | $1$ |
| $1.5$ | $3.375$ |
| $2$ | $8$ |

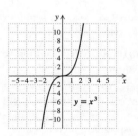

**85.** a) $y = 0.375x^3$

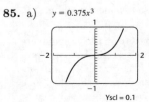

Yscl = 0.1

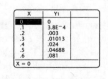

b) $y = -3.5x^2 + 6x - 8$

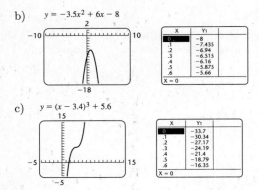

c) $y = (x - 3.4)^3 + 5.6$

## Exercise Set 2.2

1. For any function, the set of all inputs, or first values, is called the <u>domain</u>.

3. In any function, each member of the domain is paired with <u>exactly</u> one member of the range.

5. When a function is graphed, members of the domain are located on the <u>horizontal</u> axis.

7. The notation $f(3)$ is read "$f$ of 3," "$f$ at 3," or "the value of $f$ at 3."

9. The correspondence is a function, because each member of the domain corresponds to exactly one member of the range.

11. The correspondence is a function, because each member of the domain corresponds to exactly one member of the range.

13. The correspondence is a function, because each member of the domain corresponds to exactly one member of the range.

15. This correspondence is not a function because a member of the domain (July 24) corresponds to more than one member of the range. (July 18 also corresponds to more than one member of the range.)

17. This correspondence is a function, because each pumpkin has only one price.

19. This correspondence is a function, because each player has only one uniform number.

21. a) Locate 1 on the horizontal axis and then find the point on the graph for which 1 is the first coordinate. From that point, look to the vertical axis to find the corresponding $y$-coordinate, $-1$. Thus, $f(1) = -1$.

b) The domain is the set of all $x$-values in the graph. It is $\{x| -4 \le x \le 3\}$.

c) To determine which member(s) of the domain are paired with 2, locate 2 on the vertical axis. From there look left and right to the graph to find any points for which 2 is the second coordinate. One such point exists. Its first coordinate is $-3$. Thus, the $x$-value for which $f(x) = 2$ is $-3$.

d) The range is the set of all $y$-values in the graph. It is $\{y| -2 \le y \le 5\}$.

23. a) Locate 1 on the horizontal axis and then find the point on the graph for which 1 is the first coordinate. From that point, look to the vertical axis to find the corresponding $y$-coordinate, 3. Thus, $f(1) = 3$.

b) The set of all $x$-values in the graph extends from $-1$ to 4, so the domain is $\{x| -1 \le x \le 4\}$.

c) To determine which member(s) of the domain are paired with 2, locate 2 on the vertical axis. From there look left and right to the graph to find any points for which 2 is the second coordinate. One such point exists. Its first coordinate is 3. Thus, the $x$-value for which $f(x) = 2$ is 3.

d) The set of all $y$-values in the graph extends from 1 to 4, so the range is $\{y|1 \le y \le 4\}$.

25. a) Locate 1 on the horizontal axis and then find the point on the graph for which 1 is the first coordinate. From that point, look to the vertical axis to find the corresponding $y$-coordinate. It appears to be 3. Thus, $f(1) = 3$.

b) The set of all $x$-values in the graph extends from $-4$ to 3 so the domain is $\{x| -4 \le x \le 3\}$.

c) To determine which member(s) of the domain are paired with 2, locate 2 on the vertical axis. From there look left and right to the graph to find any points for which 2 is the second coordinate. One such point exists. Its first coordinate is 0, so the $x$-value for which $f(x) = 2$ is 0.

d) The set of all $y$-values in the graph extends from $-5$ to 4, so the range is $\{y| -5 \le y \le 4\}$.

27. a) Locate 1 on the horizontal axis and then find the point on the graph for which 1 is the first coordinate. From that point, look to the vertical axis to find the corresponding $y$-coordinate, 3. Thus, $f(1) = 3$.

b) The set of all $x$-values in the graph extends from $-4$ to 3, so the domain is $\{x| -4 \le x \le 3\}$.

c) To determine which member(s) of the domain are paired with 2, locate 2 on the vertical axis. From there look left and right to the graph to find any points for which 2 is the second coordinate. One such point exists. Its first coordinate is $-3$. Thus, the $x$-value for which $f(x) = 2$ is $-3$.

d) The set of all $y$-values in the graph extends from $-2$ to 5, so the range is $\{y| -2 \le y \le 5\}$.

29. a) Locate 1 on the horizontal axis and then find the point on the graph for which 1 is the first coordinate. From that point, look to the vertical axis to find the corresponding $y$-coordinate, 1. Thus, $f(1) = 1$.

b) The domain is the set of all $x$-values in the graph. It is $\{-3, -1, 1, 3, 5\}$.

c) To determine which member(s) of the domain are paired with 2, locate 2 on the vertical axis. From there look left and right to the graph to find any points for which 2 is the second coordinate. One such point exists. Its first coordinate is 3. Thus, the $x$-value for which $f(x) = 2$ is 3.

d) The range is the set of all $y$-values in the graph. It is $\{-1, 0, 1, 2, 3\}$.

**31.** a) Locate 1 on the horizontal axis and then find the point on the graph for which 1 is the first coordinate. From that point, look to the vertical axis to find the corresponding $y$-coordinate, 4. Thus, $f(1) = 4$.

b) The set of all $x$-values in the graph extends from $-3$ to 4, so the domain is $\{x| -3 \leq x \leq 4\}$.

c) To determine which member(s) of the domain are paired with 2, locate 2 on the vertical axis. From there look left and right to the graph to find any points for which 2 is the second coordinate. There are two such points, $(-1, 2)$ and $(3, 2)$. Thus, the $x$-values for which $f(x) = 2$ are $-1$ and 3.

d) The set of all $y$-values in the graph extends from $-4$ to 5, so the range is $\{y| -4 \leq y \leq 5\}$.

**33.** a) Locate 1 on the horizontal axis and then find the point on the graph for which 1 is the first coordinate. From that point, look to the vertical axis to find the corresponding $y$-coordinate, 2. Thus, $f(1) = 2$.

b) The set of all $x$-values in the graph extends from $-4$ to 4, so the domain is $\{x| -4 \leq x \leq 4\}$.

c) To determine which member(s) of the domain are paired with 2, locate 2 on the vertical axis. From there look left and right to the graph to find any points for which 2 is the second coordinate. All points in the set $\{x|0 < x \leq 2\}$ satisfy this condition. These are the $x$-values for which $f(x) = 2$.

d) The range is the set of all $y$-values in the graph. It is $\{1, 2, 3, 4\}$.

**35.** We can use the vertical line test:

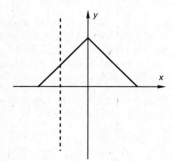

Visualize moving this vertical line across the graph. No vertical line will intersect the graph more than once. Thus, the graph is a graph of a function.

**37.** We can use the vertical line test:

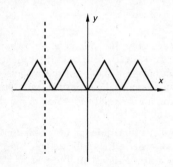

Visualize moving this vertical line across the graph. No vertical line will intersect the graph more than once. Thus, the graph is a graph of a function.

**39.** We can use the vertical line test.

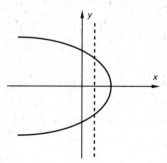

It is possible for a vertical line to intersect the graph more than once. Thus this is not the graph of a function.

**41.** We can use the vertical line test.

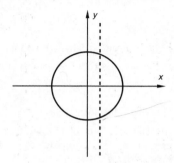

It is possible for a vertical line to intersect the graph more than once. Thus this is not a graph of a function.

**43.** $g(x) = x + 5$

a) $g(0) = 0 + 5 = 5$

b) $g(-4) = -4 + 5 = 1$

c) $g(-7) = -7 + 5 = -2$

d) $g(8) = 8 + 5 = 13$

e) $g(a + 2) = a + 2 + 5 = a + 7$

f) $g(a) + 2 = a + 5 + 2 = a + 7$

**45.** $f(n) = 5n^2 + 4n$

a) $f(0) = 5 \cdot 0^2 + 4 \cdot 0 = 0 + 0 = 0$

b) $f(-1) = 5(-1)^2 + 4(-1) = 5 - 4 = 1$

c) $f(3) = 5 \cdot 3^2 + 4 \cdot 3 = 45 + 12 = 57$

d) $f(t) = 5t^2 + 4t$

e) $f(2a) = 5(2a)^2 + 4 \cdot 2a = 5 \cdot 4a^2 + 8a = 20a^2 + 8a$

f) $f(3) - 9 = 5 \cdot 3^2 + 4 \cdot 3 - 9 = 5 \cdot 9 + 4 \cdot 3 - 9 = 45 + 12 - 9 = 48$

**47.** $f(x) = \dfrac{x - 3}{2x - 5}$

a) $f(0) = \dfrac{0 - 3}{2 \cdot 0 - 5} = \dfrac{-3}{0 - 5} = \dfrac{-3}{-5} = \dfrac{3}{5}$

b) $f(4) = \dfrac{4-3}{2 \cdot 4 - 5} = \dfrac{1}{8-5} = \dfrac{1}{3}$

c) $f(-1) = \dfrac{-1-3}{2(-1)-5} = \dfrac{-4}{-2-5} = \dfrac{-4}{-7} = \dfrac{4}{7}$

d) $f(3) = \dfrac{3-3}{2 \cdot 3 - 5} = \dfrac{0}{6-5} = \dfrac{0}{1} = 0$

e) $f(x+2) = \dfrac{x+2-3}{2(x+2)-5} = \dfrac{x-1}{2x+4-5} = \dfrac{x-1}{2x-1}$

**49.** a) $f(x) = \dfrac{5}{x-3}$

Since $\dfrac{5}{x-3}$ cannot be computed when the denominator is 0, we find the $x$-value that causes $x-3$ to be 0:

$$x - 3 = 0$$
$$x = 3 \quad \text{Adding 3 to both sides}$$

Thus, 3 is not in the domain of $f$, while all other real numbers are.   The domain of $f$ is $\{x | x \text{ is a real number and } x \neq 3\}$.

b) $f(x) = \dfrac{7}{6-x}$

Since $\dfrac{7}{6-x}$ cannot be computed when the denominator is 0, we find the $x$-value that causes $6-x$ to be 0:

$$6 - x = 0$$
$$6 = x \quad \text{Adding } x \text{ on both sides}$$

Thus, 6 is not in the domain of $f$, while all other real numbers are.   The domain of $f$ is $\{x | x \text{ is a real number and } x \neq 6\}$.

c) $f(x) = 2x + 1$

Since we can compute $2x+1$ for any real number $x$, the domain is the set of all real numbers.

d) $f(x) = x^2 + 3$

Since we can compute $x^2 + 3$ for any real number $x$, the domain is the set of all real numbers.

e) $f(x) = \dfrac{3}{2x-5}$

Since $\dfrac{3}{2x-5}$ cannot be computed when the denominator is 0, we find the $x$-value that causes $2x-5$ to be 0:

$$2x - 5 = 0$$
$$2x = 5$$
$$x = \dfrac{5}{2}$$

Thus, $\dfrac{5}{2}$ is not in the domain of $f$, while all other real numbers are.   The domain of $f$ is $\left\{ x | x \text{ is a real number and } x \neq \dfrac{5}{2} \right\}$.

f) $f(x) = |3x - 4|$

Since we can compute $|3x - 4|$ for any real number $x$, the domain is the set of all real numbers.

**51.** $A(s) = s^2 \dfrac{\sqrt{3}}{4}$

$A(4) = 4^2 \dfrac{\sqrt{3}}{4} = 4\sqrt{3} \approx 6.93$

The area is $4\sqrt{3}$ cm$^2 \approx 6.93$ cm$^2$.

**53.** $V(r) = 4\pi r^2$

$V(3) = 4\pi(3)^2 = 36\pi$

The area is $36\pi$ in$^2 \approx 113.10$ in$^2$.

**55.** $P(d) = 1 + \dfrac{d}{33}$

$P(20) = 1 + \dfrac{20}{33} = 1\dfrac{20}{33}$ atm

$P(30) = 1 + \dfrac{30}{33} = 1 + \dfrac{10}{11} = 1\dfrac{10}{11}$ atm

$P(100) = 1 + \dfrac{100}{33} = 1 + 3\dfrac{1}{33} = 4\dfrac{1}{33}$ atm

**57.** $H(x) = 2.75x + 71.48$

$H(32) = 2.75(32) + 71.48 = 159.48$

The predicted height is 159.48 cm.

**59.** $F(C) = \dfrac{9}{5}C + 32$

$F(-10) = \dfrac{9}{5}(-10) + 32 = -18 + 32 = 14$

The equivalent temperature is 14°F.

**61.** Locate the point that is directly above 225.   Then estimate its second coordinate by moving horizontally from the point to the vertical axis.   The rate is about 75 per 10,000 men.

**63.** Locate the point on the graph that is directly above '60. Then estimate its second coordinate by moving horizontally from the point to the vertical axis.   In 1960, about 56% of Americans were willing to vote for a woman for president.   That is, $P(1960) \approx 56\%$.

**65.** Plot and connect the points, using CFL wattage as the first coordinate and the wattage of the incandescent equivalent as the second coordinate.

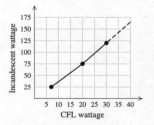

To estimate the wattage of an incandescent bulb that creates light equivalent to a 15-watt CFL bulb, first locate the point that is directly above 15.   Then estimate the second coordinate by moving horizontally from the point to the vertical axis.   Read the approximate function value there. The wattage is about 60 watts.

To predict the wattage of an incandescent bulb that creates light equivalent to a 35-watt CFL bulb, extend the graph and extrapolate.   The wattage is about 140 watts.

**67.** Plot and connect the points, using body weight as the first coordinate and the corresponding number of drinks as the second coordinate.

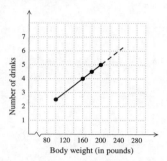

To estimate the number of drinks that a 140-lb person would have to drink to be considered intoxicated, first locate the point that is directly above 140. Then estimate its second coordinate by moving horizontally from the point to the vertical axis. Read the approximate function value there. The estimated number of drinks is 3.5.

To predict the number of drinks it would take for a 250-lb person to be considered intoxicated, extend the graph and extrapolate. It appears that it would take about 6 drinks.

**69.** Plot and connect the points, using the year as the first coordinate and the corresponding number of reported cases of AIDS as the second coordinate.

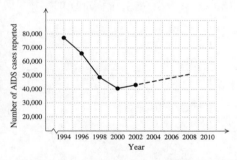

To estimate the number of cases of AIDS reported in 1997, first locate the point that is directly above 1997. Then move horizontally from the point to the vertical axis and read the approximate function value there. We estimate that about 57,000 cases of AIDS were reported in 1997.

To estimate the number of cases of AIDS that will be reported in 2007, extend the graph and extrapolate. It appears that about 50,000 cases of AIDS will be reported in 2007.

**71.** Plot and connect the points, using the year as the first coordinate and the total sales as the second coordinate.

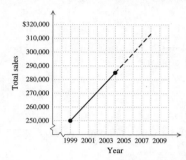

To estimate the total sales for 2000, first locate the point directly above 2000. Then estimate its second coordinate by moving horizontally to the vertical axis. Read the approximate function value there. The estimated 2000 sales total is about $257,000.

To estimate the sales for 2007, extend the graph and extrapolate. We estimate the sales for 2007 to be about $306,000.

**73.** *Writing Exercise*

**75.** $\dfrac{10 - 3^2}{9 - 2 \cdot 3} = \dfrac{10 - 9}{9 - 6} = \dfrac{1}{3}$

**77.**
$$S = 2lh + 2lw + 2wh$$
$$S - 2wh = 2lh + 2lw$$
$$S - 2wh = l(2h + 2w)$$
$$\dfrac{S - 2wh}{2h + 2w} = l$$

**79.**
$$2x + 3y = 6$$
$$3y = 6 - 2x$$
$$y = \dfrac{6 - 2x}{3}, \text{ or } 2 - \dfrac{2}{3}x, \text{ or } -\dfrac{2}{3}x + 2$$

**81.** *Writing Exercise*

**83.** To find $f(g(-4))$, we first find $g(-4)$:

$g(-4) = 2(-4) + 5 = -8 + 5 = -3$.

Then $f(g(-4)) = f(-3) = 3(-3)^2 - 1 = 3 \cdot 9 - 1 = 27 - 1 = 26$.

To find $g(f(-4))$, we first find $f(-4)$:

$f(-4) = 3(-4)^2 - 1 = 3 \cdot 16 - 1 = 48 - 1 = 47$.

Then $g(f(-4)) = g(47) = 2 \cdot 47 + 5 = 94 + 5 = 99$.

**85.** $f(\text{tiger}) = \text{dog}$

$f(\text{dog}) = f(f(\text{tiger})) = \text{cat}$

$f(\text{cat}) = f(f(f(\text{tiger}))) = \text{fish}$

$f(\text{fish}) = f(f(f(f(\text{tiger})))) = \text{worm}$

**87.** Locate the highest point on the graph. Then move vertically to the horizontal axis and read the corresponding time. It is about 2 min, 50 sec.

**89.** The two largest contractions occurred at about 2 minutes, 50 seconds and 5 minutes, 40 seconds. The difference in these times, is 2 minutes, 50 seconds, so the frequency is about 1 every 3 minutes.

**91.** We know that $(-1, -7)$ and $(3, 8)$ are both solutions of $g(x) = mx + b$. Substituting, we have

$$-7 = m(-1) + b, \text{ or } -7 = -m + b,$$

and $\quad 8 = m(3) + b, \quad$ or $\quad 8 = 3m + b.$

Solve the first equation for $b$ and substitute that expression into the second equation.

| | |
|---|---|
| $-7 = -m + b$ | First equation |
| $m - 7 = b$ | Solving for $b$ |
| $8 = 3m + b$ | Second equation |
| $8 = 3m + (m - 7)$ | Substituting |
| $8 = 3m + m - 7$ | |
| $8 = 4m - 7$ | |
| $15 = 4m$ | |
| $\dfrac{15}{4} = m$ | |

We know that $m - 7 = b$, so $\dfrac{15}{4} - 7 = b$, or $-\dfrac{13}{4} = b$.

We have $m = \dfrac{15}{4}$ and $b = -\dfrac{13}{4}$, so $g(x) = \dfrac{15}{4}x - \dfrac{13}{4}$.

## Exercise Set 2.3

**1.** Rise is (e), the difference in $y$.

**3.** Slope is (c), $\dfrac{\text{difference in } y}{\text{difference in } x}$.

**5.** Slope-intercept form is (a), $y = mx + b$.

**7.** Graph: $f(x) = 2x - 7$.

We make a table of values. Then we plot the corresponding points and connect them.

| $x$ | $f(x)$ |
|-----|--------|
| 1 | $-5$ |
| 2 | $-3$ |
| 3 | $-1$ |
| 5 | 3 |

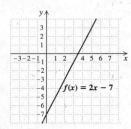

**9.** Graph: $g(x) = -\dfrac{1}{3}x + 2$.

We make a table of values. Then we plot the corresponding points and connect them.

| $x$ | $g(x)$ |
|-----|--------|
| $-3$ | 3 |
| 0 | 2 |
| 3 | 1 |
| 6 | 0 |

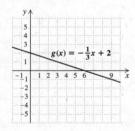

**11.** Graph: $h(x) = \dfrac{2}{5}x - 4.$

We make a table of values. Then we plot the corresponding points and connect them.

| $x$ | $h(x)$ |
|-----|--------|
| $-5$ | $-6$ |
| 0 | $-4$ |
| 5 | $-2$ |

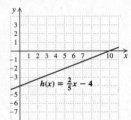

**13.** $y = 3x + 4$

The $y$-intercept is $(0, 4)$, or simply 4.

**15.** $g(x) = -4x - 3$

The $y$-intercept is $(0, -3)$, or simply $-3$.

**17.** $y = -\dfrac{3}{8}x - 4.5$

The $y$-intercept is $(0, -4.5)$, or simply $-4.5$.

**19.** $f(x) = 2.9x - 9$

The $y$-intercept is $(0, -9)$, or simply $-9$.

**21.** $y = 37x + 204$

The $y$-intercept is $(0, 204)$, or simply 204.

**23.** Slope $= \dfrac{\text{difference in } y}{\text{difference in } x} = \dfrac{5 - 9}{4 - 6} = \dfrac{-4}{-2} = 2$

**25.** Slope $= \dfrac{\text{difference in } y}{\text{difference in } x} = \dfrac{-4 - 8}{9 - 3} = \dfrac{-12}{6} = -2$

**27.** Slope $= \dfrac{\text{difference in } y}{\text{difference in } x} = \dfrac{\dfrac{1}{2} - \dfrac{4}{5}}{-\dfrac{2}{3} - \left(-\dfrac{4}{5}\right)} = \dfrac{-\dfrac{3}{10}}{\dfrac{2}{15}} =$

$-\dfrac{3}{10} \cdot \dfrac{15}{2} = -\dfrac{3 \cdot 3 \cdot \cancel{5}}{2 \cdot \cancel{5} \cdot 2} = -\dfrac{9}{4}$

**29.** Slope $= \dfrac{\text{difference in } y}{\text{difference in } x} = \dfrac{43.6 - 43.6}{4.5 - (-9.7)} = \dfrac{0}{14.2} = 0$

**31.** $y = \dfrac{5}{2}x + 3$

Slope is $\dfrac{5}{2}$; $y$-intercept is $(0, 3)$.

From the $y$-intercept, we go *up* 5 units and to the *right* 2 units. This gives us the point $(2, 8)$. We can now draw the graph.

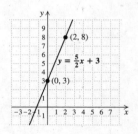

As a check, we can rename the slope and find another point.

$$\frac{5}{2} = \frac{5}{2} \cdot \frac{-1}{-1} = \frac{-5}{-2}$$

From the $y$-intercept, we go *down* 5 units and to the *left* 2 units. This gives us the point $(-2, -2)$. Since $(-2, -2)$ is on the line, we have a check.

**33.** $f(x) = -\dfrac{5}{2}x + 2$

Slope is $-\dfrac{5}{2}$, or $\dfrac{-5}{2}$; $y$-intercept is $(0, 2)$.

From the $y$-intercept, we go *down* 5 units and to the *right* 2 units. This gives us the point $(2, -3)$. We can now draw the graph.

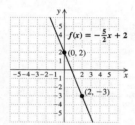

As a check, we can rename the slope and find another point.

$$\frac{-5}{2} = \frac{5}{-2}$$

From the $y$-intercept, we go *up* 5 units and to the *left* 2 units. This gives us the point $(-2, 7)$. Since $(-2, 7)$ is on the line, we have a check.

**35.** Convert to a slope-intercept equation.

$$2x - y = 5$$
$$-y = -2x + 5$$
$$y = 2x - 5$$

Slope is 2, or $\dfrac{2}{1}$; $y$-intercept is $(0, -5)$.

From the $y$-intercept, we go *up* 2 units and to the *right* 1 unit. This gives us the point $(1, -3)$. We can now draw the graph.

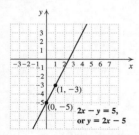

As a check, we can rename the slope and find another point.

$$2 = \frac{2}{1} \cdot \frac{3}{3} = \frac{6}{3}$$

From the $y$-intercept, we go *up* 6 units and to the *right* 3 units. This gives us the point $(3, 1)$. Since $(3, 1)$ is on the line, we have a check.

**37.** $F(x) = \dfrac{1}{3}x + 2$

Slope is $\dfrac{1}{3}$; $y$-intercept is $(0, 2)$.

From the $y$-intercept, we go *up* 1 unit and to the *right* 3 units. This gives us the point $(3, 3)$. We can now draw the graph.

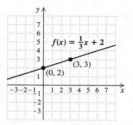

As a check, we can rename the slope and find another point.

$$\frac{1}{3} = \frac{1}{3} \cdot \frac{-1}{-1} = \frac{-1}{-3}$$

From the $y$-intercept, we go *down* 1 unit and to the *left* 3 units. This gives us the point $(-3, 1)$. Since $(-3, 1)$ is on the line, we have a check.

**39.** Convert to a slope intercept equation:

$$6y + 4x = 6$$
$$6y = -4x + 6$$
$$y = \frac{1}{6}(-4x + 6)$$
$$y = -\frac{2}{3}x + 1$$

Slope is $-\dfrac{2}{3}$ or $\dfrac{-2}{3}$; $y$-intercept is $(0, 1)$.

From the $y$-intercept we go *down* 2 units and to the *right* 3 units. This gives us the point $(3, -1)$. We can now draw the graph.

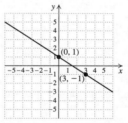

As a check we can rename the slope and find another point.

$$\frac{-2}{3} = \frac{-2}{3} \cdot \frac{-1}{-1} = \frac{2}{-3}$$

From the $y$-intercept, we go *up* 2 units and to the *left* 3 units. This gives us the point $(-3, 3)$. Since $(-3, 3)$ is on the line, we have a check.

**41.** $g(x) = -0.25x$

Slope is $-0.25$, or $\dfrac{-1}{4}$; $y$-intercept is $(0,0)$.

From the $y$-intercept, we go *down* 1 unit and to the *right* 4 units. This gives us the point $(4,-1)$. We can now draw the graph.

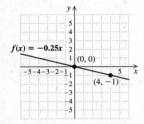

As a check, we can rename the slope and find another point.

$$\frac{-1}{4} = \frac{-1}{4} \cdot \frac{-1}{-1} = \frac{1}{-4}$$

From the $y$-intercept, we go *up* 1 unit and to the *left* 4 units. This gives us the point $(-4,1)$. Since $(-4,1)$ is on the line, we have a check.

**43.** Convert to a slope-intercept equation.

$$4x - 5y = 10$$
$$-5y = 4x + 10$$
$$y = \frac{4}{5}x - 2$$

Slope is $\dfrac{4}{5}$; $y$-intercept is $(0,-2)$.

From the $y$-intercept, we go *up* 4 units and to the *right* 5 units. This gives us the point $(5,2)$. We can now draw the graph.

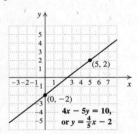

As a check, we choose some other value for $x$, say $-5$, and determine $y$:

$$y = \frac{4}{5}(-5) - 2 = -4 - 2 = -6$$

We plot the point $(-5,-6)$ and see that it *is* on the line.

**45.** $f(x) = \dfrac{5}{4}x - 2$

Slope is $\dfrac{5}{4}$; $y$-intercept is $(0,-2)$.

From the $y$-intercept, we go *up* 5 units and to the *right* 4 units. This gives us the point $(4,3)$. We can now draw the graph.

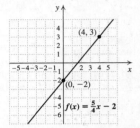

As a check, we choose some other value for $x$, say $-2$, and determine $f(x)$:

$$f(x) = \frac{5}{4}(-2) - 2 = -\frac{5}{2} - 2 = -\frac{9}{2}$$

We plot the point $\left(-2, -\dfrac{9}{2}\right)$ and see that it *is* on the line.

**47.** Convert to a slope-intercept equation:

$$12 - 4f(x) = 3x$$
$$-4f(x) = 3x - 12$$
$$f(x) = -\frac{1}{4}(3x - 12)$$
$$f(x) = -\frac{3}{4}x + 3$$

Slope is $-\dfrac{3}{4}$, or $\dfrac{-3}{4}$; $y$-intercept is $(0,3)$.

From the $y$-intercept, we go *down* 3 units and to the *right* 4 units. This gives us the point $(4,0)$. We can now draw the graph.

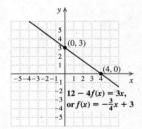

As a check, we choose some other value for $x$, say $-4$, and determine $f(x)$:

$$f(-4) = -\frac{3}{4}(-4) + 3 = 3 + 3 = 6$$

We plot the point $(-4,6)$ and see that it *is* on the line.

**49.** $g(x) = 4.5 = 0x + 4.5$

Slope is 0; $y$-intercept is $(0, 4.5)$.

From the $y$-intercept, we go up or down 0 units and any number of nonzero units to the left or right. Any point on the graph will lie on a horizontal line 4.5 units above the $x$-axis. We draw the graph.

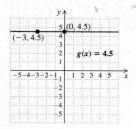

**51.** Use the slope-intercept equation, $f(x) = mx + b$, with $m = \dfrac{2}{3}$ and $b = -9$.

$$f(x) = mx + b$$
$$f(x) = \frac{2}{3}x + (-9)$$
$$f(x) = \frac{2}{3}x - 9$$

**53.** Use the slope-intercept equation, $f(x) = mx + b$, with $m = -6$ and $b = 2$.

$$f(x) = mx + b$$
$$f(x) = -6x + 2$$

**55.** Use the slope-intercept equation, $f(x) = mx + b$, with $m = -\dfrac{7}{9}$ and $b = 5$.

$$f(x) = mx + b$$
$$f(x) = -\frac{7}{9}x + 5$$

**57.** Use the slope-intercept equation, $f(x) = mx + b$, with $m = 5$ and $b = \dfrac{1}{2}$.

$$f(x) = mx + b$$
$$f(x) = 5x + \frac{1}{2}$$

**59.** We can use the coordinates of any two points on the line. We'll use $(0, 100)$ and $(9, 40)$.

Rate of change $= \dfrac{\text{change in } y}{\text{change in } x} = \dfrac{40 - 100}{9 - 0} = \dfrac{-60}{9} = -\dfrac{20}{3}$,

or $-6\dfrac{2}{3}$

The distance from the finish line is decreasing at a rate of $6\dfrac{2}{3}$ m per second.

**61.** We can use the coordinates of any two points on the line. We'll use $(0, 0)$ and $(2, 1)$.

Rate of change $= \dfrac{\text{change in } y}{\text{change in } x} = \dfrac{1 - 0}{2 - 0} = \dfrac{1}{2}$

The weight is increasing at a rate of $\dfrac{1}{2}$ pound per bag of feed.

**63.** We can use the coordinates of any two points on the line. We'll use $(0, 50)$ and $(2, 200)$.

Rate of change $= \dfrac{\text{change in } y}{\text{change in } x} = \dfrac{200 - 50}{2 - 0} = 75$

The number of pages read is increasing at a rate of 75 pages per day.

**65.** We can use the coordinates of any two points on the line. We'll use $(15, 470)$ and $(55, 510)$:

Rate of change $= \dfrac{\text{change in } y}{\text{change in } x} = \dfrac{510 - 470}{55 - 15} = \dfrac{40}{40} = 1$

The average SAT math score is increasing at a rate of 1 point per thousand dollars of family income.

**67.** The marathoner's speed is given by $\dfrac{\text{change in distance}}{\text{change in time}}$. Note that the runner reaches the 22-mi point 56 min after the 15-mi point was reached or after 2 hr, 56 min. We will express time in hours: 2 hr, 56 min $= 2\dfrac{14}{15}$ hr. Then

$$\frac{\text{change in distance}}{\text{change in time}} = \frac{22 - 15}{2\dfrac{14}{15} - 2} = \frac{7}{\dfrac{14}{15}} = 7 \cdot \frac{15}{14} = \frac{15}{2}, \text{ or}$$

7.5 mph.

The marathoner's speed is 7.5 mph.

**69.** The work rate is given by $\dfrac{\text{change in portion of house painted}}{\text{change in time}}$.

$$\frac{\text{change in portion of house painted}}{\text{change in time}} = \frac{\dfrac{2}{3} - \dfrac{1}{4}}{8 - 0} = \frac{\dfrac{5}{12}}{8} =$$
$$\frac{5}{12} \cdot \frac{1}{8} = \frac{5}{96}$$

The painter's work rate is $\dfrac{5}{96}$ of the house per hour.

**71.** The average rate of descent is given by $\dfrac{\text{change in altitude}}{\text{change in time}}$. We will express time in minutes:

$1\dfrac{1}{2}$ hr $= \dfrac{3}{2}$ hr $\cdot \dfrac{60 \text{ min}}{1 \text{ hr}} = 90$ min

2 hr, 10 min $=$ 2 hr $+$ 10 min $=$

2 hr $\cdot \dfrac{60 \text{ min}}{1 \text{ hr}} + 10$ min $= 120$ min $+ 10$ min $= 130$ min

Then

$$\frac{\text{change in altitude}}{\text{change in time}} = \frac{0 - 12{,}000}{130 - 90} = \frac{-12{,}000}{40} =$$
$$-300.$$

The average rate of descent is 300 ft/min.

**73.** a) Graph II indicated that 200 ml of fluid was dripped in the first 3 hr, a rate of $\dfrac{200}{3}$ ml/hr. It also indicates that 400 ml of fluid was dripped in the next 3 hr, a rate of $\dfrac{400}{3}$ ml/hr, and that this rate continues until the end of the time period shown. Since the rate of $\dfrac{400}{3}$ ml/hr is double the rate of $\dfrac{200}{3}$ ml/hr, this graph is appropriate for the given situation.

b) Graph IV indicates that 300 ml of fluid was dripped in the first 2 hr, a rate of 300/2, or 150 ml/hr. In the next 2 hr, 200 ml was dripped. This is a rate of 200/2, or 100 ml/hr. Then 100 ml was dripped in the next 3 hr, a rate of 100/3, or $33\dfrac{1}{3}$ ml/hr. Finally, in the remaining 2 hr, 0 ml of fluid was dripped, a rate of 0/2, or 0 ml/hr. Since the rate at which the fluid was given decreased as time progressed and eventually became 0, this graph is appropriate for the given situation.

c) Graph I is the only graph that shows a constant rate for 5 hours, in this case from 3 PM to 8 PM. Thus, it is appropriate for the given situation.

d) Graph III indicates that 100 ml of fluid was dripped in the first 4 hr, a rate of 100/4, or 25 ml/hr. In the next 3 hr, 200 ml was dripped. This is a rate of 200/3, or $66\frac{2}{3}$ ml/hr. Then 100 ml was dripped in the next hour, a rate of 100 ml/hr. In the last hour 200 ml was dripped, a rate of 200 ml/hr. Since the rate at which the fluid was given gradually increased, this graph is appropriate for the given situation.

**75.** $C(x) = 25x + 75$

25 signifies that the cost per person is $25; 75 signifies that the setup cost for the party is $75.

**77.** $L(t) = \frac{1}{2}t + 1$

$\frac{1}{2}$ signifies that Oscar's hair grows $\frac{1}{2}$ in. per month; 1 signifies that his hair is 1 in. long immediately after he gets it cut.

**79.** $A(t) = \frac{1}{7}t + 75.5$ is of the form $y = mx + b$ with $m = \frac{1}{7}$ and $b = 75.5$.

$\frac{1}{7}$ signifies that the life expectancy of American women increases $\frac{1}{7}$ yr per year for years after 1970; 75.5 signifies that the life expectancy of American women in 1970 was 75.5 years.

**81.** $P(t) = 0.227t + 4.29$

0.227 signifies that the price increases $0.227 per year, for years since 1995; 4.29 signifies that the average cost of a movie ticket in 1995 was $4.29.

**83.** $C(d) = 2d + 2.5$ is of the form $y = mx + b$ with $m = 2$ and $b = 2.5$.

2 signifies that the cost per mile of a taxi ride is $2; 2.5 signifies that the minimum cost of a taxi ride is $2.50.

**85.** $F(t) = -5000t + 90,000$

a) $-5000$ signifies that the truck's value depreciates $5000 per year; 90,000 signifies that the original value of the truck was $90,000.

b) We find the value of $t$ for which $F(t) = 0$.
$$0 = -5000t + 90,000$$
$$5000t = 90,000$$
$$t = 18$$

It will take 18 yr for the truck to depreciate completely.

c) The truck's value goes from $90,000 when $t = 0$ to $0 when $t = 18$, so the domain of $F$ is $\{x|0 \le t \le 18\}$.

**87.** $v(n) = -150n + 900$

a) $-150$ signifies that the snowblower's value depreciates $150 per winter of use; 900 signifies that the original value of the snowblower was $900.

b) We find the value of $n$ for which $v(n) = 300$.
$$300 = -150n + 900$$
$$-600 = -150n$$
$$4 = n$$

The snowblower's trade-in value will be $300 after 4 winters of use.

c) First we find the value of $n$ for which $v(n) = 0$.
$$0 = -150n + 900$$
$$-900 = -150n$$
$$6 = n$$

The value of the snowblower goes from $900 when $n = 0$ to $0 when $n = 6$, so the domain of $v$ is $\{n|0 \le n \le 6\}$.

**89.** *Writing Exercise*

**91.**
$$2x - 5 = 7x + 3$$
$$2x - 5 - 3 = 7x$$
$$2x - 8 = 7x$$
$$2x - 8 - 2x = 7x - 2x$$
$$-8 = 5x$$
$$\frac{1}{5}(-8) = \frac{1}{5} \cdot 5x$$
$$-\frac{8}{5} = x$$

The solution is $-\frac{8}{5}$.

**93.** $\frac{1}{5}x + 7 = 2$
$$\frac{1}{5}x = 2 - 7$$
$$\frac{1}{5}x = -5$$
$$5 \cdot \frac{1}{5}x = 5(-5)$$
$$x = -25$$

The solution is $-25$.

**95.**
$$3 \cdot 0 - 2y = 9$$
$$-2y = 9$$
$$-\frac{1}{2}(-2y) = -\frac{1}{2} \cdot 9$$
$$y = -\frac{9}{2}$$

The solution is $-\frac{9}{2}$.

**97.** *Writing Exercise*

**99.** $rx + py = s - ry$

$ry + py = -rx + s$

$y(r + p) = -rx + s$

$$y = -\frac{r}{r+p}x + \frac{s}{r+p}$$

The slope is $-\dfrac{r}{r+p}$, and the $y$-intercept is

$\left(0, \dfrac{s}{r+p}\right)$.

**101.** See the answer section in the text.

**103.** Let $c = 2$ and $d = 3$. Then $f(cd) = f(2 \cdot 3) = f(6) = m \cdot 6 + b = 6m + b$, but $f(c)f(d) = f(2)f(3) = (m \cdot 2 + b)(m \cdot 3 + b) = 6m^2 + 5mb + b^2$. Thus, the given statement is false.

**105.** Let $c = 5$ and $d = 2$. Then $f(c - d) = f(5 - 2) = f(3) = m \cdot 3 + b = 3m + b$, but $f(c) - f(d) = f(5) - f(2) = (m \cdot 5 + b) - (m \cdot 2 + b) = 5m + b - 2m - b = 3m$. Thus, the given statement is false.

**107.** a) Graph III indicates that the first 2 mi and the last 3 mi were traveled in approximately the same length of time and at a fairly rapid rate. The mile following the first two miles was traveled at a much slower rate. This could indicate that the first two miles were driven, the next mile was swum and the last three miles were driven, so this graph is most appropriate for the given situation.

b) The slope in Graph IV decreases at 2 mi and again at 3 mi. This could indicate that the first two miles were traveled by bicycle, the next mile was run, and the last 3 miles were walked, so this graph is most appropriate for the given situation.

c) The slope in Graph I decreases at 2 mi and then increases at 3 mi. This could indicate that the first two miles were traveled by bicycle, the next mile was hiked, and the last three miles were traveled by bus, so this graph is most appropriate for the given situation.

d) The slope in Graph II increases at 2 mi and again at 3 mi. This could indicate that the first two miles were hiked, the next mile was run, and the last three miles were traveled by bus, so this graph is most appropriate for the given situation.

**109.** a) $\dfrac{-c - (-6c)}{b - 5b} = \dfrac{5c}{-4b} = -\dfrac{5c}{4b}$

b) $\dfrac{(d + e) - d}{b - b} = \dfrac{e}{0}$

Since we cannot divide by 0, the slope is undefined.

c) $\dfrac{(-a - d) - (a + d)}{(c - f) - (c + f)} = \dfrac{-a - d - a - d}{c - f - c - f}$

$= \dfrac{-2a - 2d}{-2f}$

$= \dfrac{-2(a + d)}{-2f}$

$= \dfrac{a + d}{f}$

**111.** *Writing Exercise*

## Exercise Set 2.4

**1.** Every <u>horizontal</u> line has a slope of 0.

**3.** The graph of any equation of the form $x = a$ is a <u>vertical</u> line that crosses the $x$-axis at $(a, 0)$.

**5.** To find the $x$-intercept, we let $y = \underline{0}$ and solve the original equation for $\underline{x}$.

**7.** To solve $3x - 5 = 7$, we can graph $f(x) = 3x - 5$ and $g(x) = 7$ and find the $x$-value at the point of <u>intersection</u>.

**9.** Only <u>linear</u> equations have graphs that are straight lines.

**11.** $y - 9 = 3$

$y = 12$

The graph of $y = 12$ is a horizontal line. Since $y - 9 = 3$ is equivalent to $y = 12$, the slope of the line $y - 9 = 3$ is 0.

**13.** $8x = 6$

$x = \dfrac{3}{4}$

The graph of $x = \dfrac{3}{4}$ is a vertical line. Since $8x = 6$ is equivalent to $x = \dfrac{3}{4}$, the slope of the line $8x = 6$ is undefined.

**15.** $3y = 28$

$y = \dfrac{28}{3}$

The graph of $y = \dfrac{28}{3}$ is a horizontal line. Since $3y = 28$ is equivalent to $y = \dfrac{28}{3}$, the slope of the line $3y = 28$ is 0.

**17.** $9 + x = 12$

$x = 3$

The graph of $x = 3$ is a vertical line. Since $9 + x = 12$ is equivalent to $x = 3$, the slope of the line $9 + x = 12$ is undefined.

**19.** $2x - 4 = 3$

$2x = 7$

$x = \dfrac{7}{2}$

The graph of $x = \dfrac{7}{2}$ is a vertical line. Since $2x - 4 = 3$ is equivalent to $x = \dfrac{7}{2}$, the slope of the line $2x - 4 = 3$ is undefined.

**21.** $5y - 4 = 35$

$5y = 39$

$y = \dfrac{39}{5}$

The graph of $y = \dfrac{39}{5}$ is a horizontal line. Since $5y - 4 = 35$ is equivalent to $y = \dfrac{39}{5}$, the slope of the line $5y - 4 = 35$ is 0.

**23.**  $3y + x = 3y + 2$
$$x = 2$$

The graph of $x = 2$ is a vertical line. Since $3y + x = 3y + 2$ is equivalent to $x = 2$, the slope of the line $3y + x = 3y + 2$ is undefined.

**25.**  $5x - 2 = 2x - 7$
$$5x = 2x - 5$$
$$3x = -5$$
$$x = -\frac{5}{3}$$

The graph of $x = -\frac{5}{3}$ is a vertical line. Since $5x - 2 = 2x - 7$ is equivalent to $x = -\frac{5}{3}$, the slope of the line $5x - 2 = 2x - 7$ is undefined.

**27.**  $y = -\frac{2}{3}x + 5$

The equation is written in slope-intercept form. We see that the slope is $-\frac{2}{3}$.

**29.**  Graph $y = 5$.

This is a horizontal line that crosses the $y$-axis at $(0, 5)$. If we find some ordered pairs, note that, for any $x$-value chosen, $y$ must be 5.

| $x$ | $y$ |
|-----|-----|
| $-2$ | 5 |
| 0 | 5 |
| 3 | 5 |

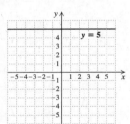

**31.**  Graph $x = 3$.

This is a vertical line that crosses the $x$-axis at $(3, 0)$. If we find some ordered pairs, note that, for any $y$-value chosen, $x$ must be 3.

| $x$ | $y$ |
|-----|-----|
| 3 | $-1$ |
| 3 | 0 |
| 3 | 2 |

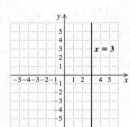

**33.**  Graph $4 \cdot f(x) = 20$.

First solve for $f(x)$.
$$4 \cdot f(x) = 20$$
$$f(x) = 5$$

This is a horizontal line that crosses the vertical axis at $(0, 5)$.

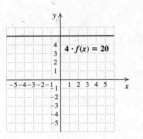

**35.**  Graph $3x = -15$.

Since $y$ does not appear, we solve for $x$.
$$3x = -15$$
$$x = -5$$

This is a vertical line that crosses the $x$-axis at $(-5, 0)$.

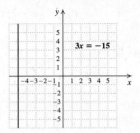

**37.**  Graph $4 \cdot g(x) + 3x = 12 + 3x$.

First solve for $g(x)$.
$$4 \cdot g(x) + 3x = 12 + 3x$$
$$4 \cdot g(x) = 12 \qquad \text{Subtracting } 3x \text{ on}$$
$$\qquad\qquad\qquad\text{both sides}$$
$$g(x) = 3$$

This is a horizontal line that crosses the vertical axis at $(0, 3)$.

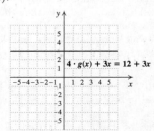

**39.**  Graph $x + y = 4$.

To find the $y$-intercept, let $x = 0$ and solve for $y$.
$$0 + y = 4$$
$$y = 4$$

The $y$-intercept is $(0, 4)$.

To find the $x$-intercept, let $y = 0$ and solve for $x$.
$$x + 0 = 4$$
$$x = 4$$

The $x$-intercept is $(4, 0)$.

Plot these points and draw the line. A third point could be used as a check.

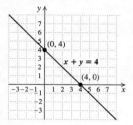

**41.** Graph $y = 2x + 6$.

To find the $y$-intercept, let $x = 0$ and solve for $y$.

$$y = 2 \cdot 0 + 6$$
$$y = 6$$

The $y$-intercept is $(0, 6)$.

To find the $x$-intercept, let $y = 0$ and solve for $x$.

$$0 = 2x + 6$$
$$-6 = 2x$$
$$-3 = x$$

The $x$-intercept is $(-3, 0)$.

Plot these points and draw the line. A third point could be used as a check.

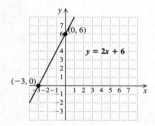

**43.** Graph $3x + 5y = -15$.

To find the $y$-intercept, let $x = 0$ and solve for $y$.

$$3 \cdot 0 + 5y = -15$$
$$5y = -15$$
$$y = -3$$

The $y$-intercept is $(0, -3)$.

To find the $x$-intercept, let $y = 0$ and solve for $x$.

$$3x + 5 \cdot 0 = -15$$
$$3x = -15$$
$$x = -5$$

The $x$-intercept is $(-5, 0)$.

Plot these points and draw the line. A third point could be used as a check.

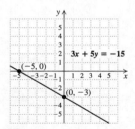

**45.** Graph $2x - 3y = 18$.

To find the $y$-intercept, let $x = 0$ and solve for $y$.

$$2 \cdot 0 - 3y = 18$$
$$-3y = 18$$
$$y = -6$$

The $y$-intercept is $(0, -6)$.

To find the $x$-intercept, let $y = 0$ and solve for $x$.

$$2x - 3 \cdot 0 = 18$$
$$2x = 18$$
$$x = 9$$

The $x$-intercept is $(9, 0)$.

Plot these points and draw the line. A third point could be used as a check.

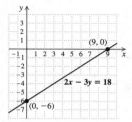

**47.** Graph $3y = 6x$.

To find the $y$-intercept, let $x = 0$ and solve for $y$.

$$3y = 6 \cdot 0$$
$$3y = 0$$
$$y = 0$$

The $y$-intercept is $(0, 0)$. This is also the $x$-intercept.

We find another point on the line. Let $x = 2$ and find the corresponding value of $y$.

$$3y = 6 \cdot 2$$
$$3y = 12$$
$$y = 4$$

The point $(2, 4)$ is on the graph.

Plot these points and draw the line. A third point could be used as a check.

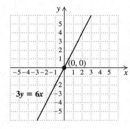

**49.** Graph $f(x) = 3x - 7$.

Because the function is in slope-intercept form, we know that the $y$-intercept is $(0, -7)$. To find the $x$-intercept, let $f(x) = 0$ and solve for $x$.

$$0 = 3x - 7$$
$$7 = 3x$$
$$\frac{7}{3} = x$$

The $x$-intercept is $\left(\frac{7}{3}, 0\right)$.

Plot these points and draw the line. A third point could be used as a check.

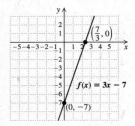

**51.** $1.4y - 3.5x = -9.8$

$\quad 14y - 35x = -98$   Multiplying by 10

$\quad 2y - 5x = -14$   Multiplying by $\frac{1}{7}$

Graph $2y - 5x = -14$.

To find the $y$-intercept, let $x = 0$.

$$2y - 5x = -14$$
$$2y - 5 \cdot 0 = -14$$
$$2y = -14$$
$$y = -7$$

$(0, -7)$ is the $y$-intercept.

To find the $x$-intercept, let $y = 0$.

$$2y - 5x = -14$$
$$2 \cdot 0 - 5x = -14$$
$$-5x = -14$$
$$x = 2.8$$

$(2.8, 0)$ is the $x$-intercept.

Plot these points and draw the line. A third point could be used as a check.

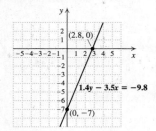

**53.** Graph $5x + 2g(x) = 7$

To find the $y$-intercept, let $x = 0$ and solve for $g(x)$.

$$5 \cdot 0 + 2g(x) = 7$$
$$2g(x) = 7$$
$$g(x) = \frac{7}{2}$$

$\left(0, \frac{7}{2}\right)$ is the $y$-intercept.

To find the $x$-intercept, let $g(x) = 0$ and solve for $x$.

$$5x + 2 \cdot 0 = 7$$
$$5x = 7$$
$$x = \frac{7}{5}$$

$\left(\frac{7}{5}, 0\right)$ is the $x$-intercept.

Plot these points and draw the line. A third point could be used as a check.

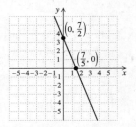

**55.** $x - 2 = 6$

Graph $f(x) = x - 2$ and $g(x) = 6$ on the same grid.

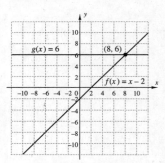

The lines appear to intersect at $(8, 6)$, so the solution is apparently 8.

Check:

$$\begin{array}{c|c} x - 2 = 6 \\ \hline 8 - 2 & 6 \\ & ? \\ 6 = 6 & \text{TRUE} \end{array}$$

The solution is 8.

**57.** $3x - 4 = -1$

Graph $f(x) = 3x - 4$ and $g(x) = -1$ on the same grid.

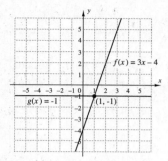

The lines appear to intersect at $(1, -1)$, so the solution is apparently 1.

Check:

$$
\begin{array}{c|c}
3x - 4 = -1 \\
\hline
3 \cdot 1 - 4 & 4 \\
3 - 4 & \\
& ? \\
-1 = -1 & \text{TRUE}
\end{array}
$$

The solution is 1.

**59.** $\dfrac{1}{2}x + 3 = 5$

Graph $f(x) = \dfrac{1}{2}x + 3$ and $g(x) = 5$ on the same grid.

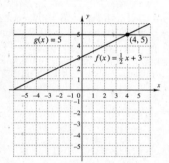

The lines appear to intersect at $(4, 5)$, so the solution is apparently 4.

Check:

$$
\begin{array}{c|c}
\dfrac{1}{2}x + 3 = 5 \\
\hline
\dfrac{1}{2}(4) + 3 & 5 \\
2 + 3 & \\
& ? \\
5 = 5 & \text{TRUE}
\end{array}
$$

The solution is 4.

**61.** $x - 8 = 3x - 5$

Graph $f(x) = x - 8$ and $g(x) = 3x - 5$ on the same grid.

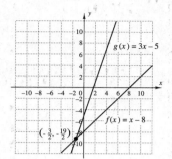

The lines appear to intersect at $\left(-\dfrac{3}{2}, -\dfrac{19}{2}\right)$, so the solution is apparently $-\dfrac{3}{2}$.

Check:

$$
\begin{array}{c|c}
x - 8 = 3x - 5 \\
\hline
-\dfrac{3}{2} - 8 & 3\left(-\dfrac{3}{2}\right) - 5 \\
-\dfrac{19}{2} & -\dfrac{9}{2} - 5 \\
& ? \\
-\dfrac{19}{2} = -\dfrac{19}{2} & \text{TRUE}
\end{array}
$$

The solution is $-\dfrac{3}{2}$.

**63.** $3 - x = \dfrac{1}{2}x - 3$

Graph $f(x) = 3 - x$ and $g(x) = \dfrac{1}{2}x - 3$ on the same grid.

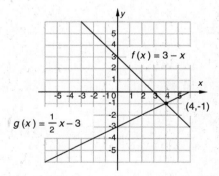

The lines appear to intersect at $(4, -1)$, so the solution is apparently 4.

Check:

$$
\begin{array}{c|c}
3 - x = \dfrac{1}{2}x - 3 \\
\hline
3 - 4 & \dfrac{1}{2}(4) - 3 \\
-1 & 2 - 3 \\
& ? \\
-1 = -1 & \text{TRUE}
\end{array}
$$

The solution is 4.

**65.** $2x + 1 = -x + 7$

Graph $f(x) = 2x + 1$ and $g(x) = -x + 7$ on the same grid.

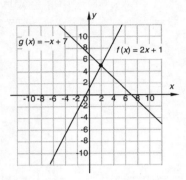

The lines appear to intersect at $(2, 5)$, so the solution is apparently 2.

Check:

$$\begin{array}{c|c} 2x + 1 = -x + 7 \\ \hline 2 \cdot 2 + 1 & -2 + 7 \\ 4 + 1 & 5 \\ & \overset{?}{} \\ 5 = 5 & \text{TRUE} \end{array}$$

The solution is 2.

**67. Familiarize.** After paying the first $100, the patient must pay $\frac{1}{5}$ of all additional charges. For a \$1000 bill, the patient pays $\$100 + \frac{1}{5}(\$1000 - \$100)$, or \$280. We can generalize this with a model, letting $C(b)$ represent the total cost to the patient, in dollars, for a hospital bill of $b$ dollars.

**Translate.**

$$\underbrace{\begin{array}{c}\text{Total cost}\\\text{to patient}\end{array}}_{C(b)} \ \underset{=}{\text{is}} \ \underset{100}{\$100} \ \underset{+}{\text{plus}} \ \underset{\frac{1}{5}}{\frac{1}{5}} \ \underset{\cdot}{\text{of}} \ \underbrace{\begin{array}{c}\text{cost in excess}\\\text{of \$100.}\end{array}}_{(b - 100)}$$

**Carry out.** First we rewrite the model in slope-intercept form.

$$C(b) = 100 + \frac{1}{5}(b - 100)$$

$$= 100 + \frac{1}{5}b - 20$$

$$= \frac{1}{5}b + 80$$

The vertical intercept is $(0, 80)$ and the slope is $\frac{1}{5}$. We plot $(0, 80)$ and, from there, go up 1 unit and right 5 units to $(5, 81)$. Draw a line through both points.

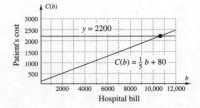

First we will find the value of $b$ for which $C(b) = 2200$. Then we will subtract 100 to find by how much Gerry's bill exceeded $100.

We find the solution of $\frac{1}{5}b + 80 = 2200$. We graph $y = 2200$ and find the point of intersection of the graphs. This point appears to be $(10, 600, 2200)$. Thus, Gerry's total hospital bill was \$10,600 and it exceeded \$100 by $\$10,600 - \$100$, or \$10,500.

**Check.** We evaluate:

$$C(10, 600) = \frac{1}{5}(10, 600) + 80 = 2120 + 80 = 2200.$$

The estimate is precise.

**State.** Gerry's hospital bill exceeded \$100 by \$10,500.

**69. Familiarize.** A monthly fee is charged after the purchase of the phone. After one month of service, the total cost will be $\$20 + \$25 = \$45$. After two months, the total cost will be $\$20 + 2 \cdot \$25 = \$90$. We can generalize this with a model, letting $C(t)$ represent the total cost, in dollars, for $t$ months of service.

**Translate.** We reword the problem and translate.

$$\underbrace{\begin{array}{c}\text{Total}\\\text{cost}\end{array}}_{C(t)} \ \underset{=}{\text{is}} \ \underbrace{\begin{array}{c}\text{cost of}\\\text{phone}\end{array}}_{20} \ \underset{+}{\text{plus}} \ \underbrace{\begin{array}{c}\$25 \text{ per}\\\text{month.}\end{array}}_{25 \cdot t}$$

**Carry out.** First write the model in slope-intercept form: $C(t) = 25t + 20$. The vertical intercept is $(0, 20)$ and the slope, or rate, is \$25 per month. Plot $(0, 20)$ and from there go *up* \$25 and to the *right* 1 month. This takes us to $(1, 45)$. Draw a line passing through both points.

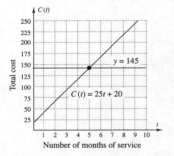

To estimate the time required for the total cost to reach \$145, we are estimating the solution of $25t + 20 = 145$. We do this by graphing $y = 145$ and finding the point of intersection of the graphs. This point appears to be $(5, 145)$. Thus, we estimate that it takes 5 months for the total cost to reach \$145.

**Check.** We evaluate.

$$C(5) = 25 \cdot 5 + 20$$

$$= 125 + 20$$

$$= 145$$

The estimate is precise.

**State.** It takes 5 months for the total cost to reach \$145.

**71. Familiarize.** After an initial \$3.00 parking fee, an additional 50¢ fee is charged for each 15-min unit of time.

After one 15-min unit of time the cost is $3.00 + $0.50, or $3.50. After two 15-min units, or 30 min, the cost is $3.00 + 2($0.50), or $4.00. We can generalize this with a model if we let $C(t)$ represent the total cost, in dollars, for $t$ 15-min units of time.

**Translate**. We reword the problem and translate.

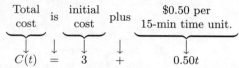

$$\begin{array}{c} \text{Total} \\ \text{cost} \end{array} \text{ is } \begin{array}{c} \text{initial} \\ \text{cost} \end{array} \text{ plus } \begin{array}{c} \$0.50 \text{ per} \\ \text{15-min time unit.} \end{array}$$

$$C(t) \quad = \quad 3 \quad + \quad 0.50t$$

**Carry out**. First write the model in slope-intercept form: $C(t) = 0.50t + 3$. The vertical intercept is $(0, 3)$ and the slope, or rate, is 0.50, or $\frac{1}{2}$. Plot $(0, 3)$ and from there go *up* $1 and to the *right* 2 15-min units of time. This takes us to $(2, 4)$. Draw a line passing through both points.

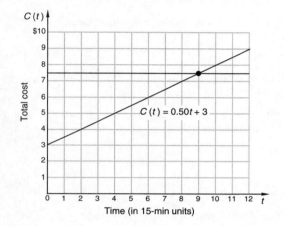

To estimate how long someone can park for $7.50, we are estimating the solution of $0.50t + 3 = 7.50$. We do this by graphing $y = 7.50$ and finding the point of intersection of the graphs.

The point appears to be $(9, 7.50)$. Thus, we estimate that someone can park for nine 15-min units of time, or 2 hr, 15 min, for $7.50.

**Check**. We evaluate:

$$C(9) = 3 + 0.50(9)$$
$$= 3 + 4.50$$
$$= 7.50$$

The estimate is precise.

**State**. Someone can park for 2 hr, 15 min for $7.50.

**73. Familiarize**. In addition to a charge of $109, FedEx charges $1.09 per pound. It costs $109 + $1.09(100), or $218, to ship a 100-lb package. It costs $109 + $1.09(125), or $245.25, to ship a 125-lb package. We can generalize this with a model if we let $C(w)$ represent the cost of shipping a package weighing $w$ lb, where $100 \le w \le 499$.

**Translate**.

$$\begin{array}{c} \$109 \text{ charge} \end{array} \text{ plus } \begin{array}{c} \$1.09 \text{ per} \\ \text{pound} \end{array} \text{ is } \begin{array}{c} \text{shipping cost.} \end{array}$$

$$109 \quad + \quad 1.09w \quad = \quad C(w)$$

**Carry out**. First write the model in slope-intercept form: $C(w) = 1.09w + 109$. The vertical intercept is $(0, 109)$ and the slope is 1.09, or $\frac{109}{100}$. Plot $(0, 109)$ and from there go up 109 units and right 100 units to $(100, 218)$. Draw a line passing through both points.

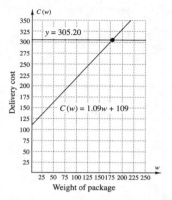

To estimate the weight of a package that costs $305.20 to ship, we are estimating the solution of $1.09w + 109 = 305.20$. We do this by graphing $y = 305.20$ and finding the point of intersection of the graphs. The point appears to be $(180, 305.20)$. Thus we estimate that a package that costs $305.20 to ship weighs 180 lb.

**Check**. We evaluate:

$$C(180) = 1.09(180) + 109$$
$$= 196.20 + 109$$
$$= 305.20$$

The estimate is precise.

**State**. A package that costs $305.20 to ship weighs 180 lb.

**75.** $5x - 3y = 15$

This equation is in the standard form for a linear equation, $Ax + By = C$, with $A = 5$, $B = -3$, and $C = 15$. Thus, it is a linear equation.

Solve for $y$ to find the slope.

$$5x - 3y = 15$$
$$-3y = -5x + 15$$
$$y = \frac{5}{3}x - 5$$

The slope is $\frac{5}{3}$.

**77.** $16 + 4y = 0$
$$4y = -16$$

This equation can be written in the standard form for a linear equation, $Ax + By = C$, with $A = 0$, $B = 4$, and $C = -16$. Thus, it is a linear equation.

Solve for $y$ to find the slope.

$$4y = -16$$
$$y = -4$$

This is a horizontal line, so the slope is 0. (We can think of this as $y = 0 \cdot x - 4$.)

**79.** $4g(x) = 6x^2$

Replace $g(x)$ with $y$ and attempt to write the equation in standard form.

$$4y = 6x^2$$
$$-6x^2 + 4y = 0$$

The equation is not linear, because it has an $x^2$-term.

**81.**
$$3y = 7(2x - 4)$$
$$3y = 14x - 28$$
$$-14x + 3y = -28$$

The equation can be written in the standard form for a linear equation, $Ax + By = C$, with $A = -14$, $B = 3$, and $C = -28$. Thus, it is a linear equation. Solve for $y$ to find the slope.

$$-14x + 3y = -28$$
$$3y = 14x - 28$$
$$y = \frac{14}{3}x - \frac{28}{3}$$

The slope is $\frac{14}{3}$.

**83.** $g(x) - \frac{1}{x} = 0$

Replace $g(x)$ with $y$ and attempt to write the equation in standard form.

$$y - \frac{1}{x} = 0$$
$$xy - 1 = 0 \quad \text{Multiplying by } x$$
$$xy = 1$$

The equation is not linear, because it has an $xy$-term.

**85.** $\frac{f(x)}{5} = x$

Replace $f(x)$ with $y$ and attempt to write the equation in standard form.

$$\frac{y}{5} = x$$
$$y = 5x$$
$$-5x + y = 0$$

The equation can be written in the standard form for a linear equation, $Ax + By = C$, with $A = -5$, $B = 1$, and $C = 0$. Thus, it is a linear equation. From our work above, we see that the equation can be written $y = 5x$, so the slope is 5.

**87.** *Writing Exercise*

**89.** $-\frac{3}{7} \cdot \frac{7}{3} = -\frac{3 \cdot 7}{7 \cdot 3} = -1$

**91.** $-5[x - (-3)] = -5[x + 3] = -5x - 15$

**93.** $\frac{2}{3}\left[x - \left(-\frac{1}{2}\right)\right] - 1 = \frac{2}{3}\left[x + \frac{1}{2}\right] - 1 =$

$\frac{2}{3}x + \frac{1}{3} - 1 = \frac{2}{3}x - \frac{2}{3}$

**95.** *Writing Exercise*

**97.** The line contains the points $(5, 0)$ and $(0, -4)$. We use the points to find the slope.

$$\text{Slope} = \frac{-4 - 0}{0 - 5} = \frac{-4}{-5} = \frac{4}{5}$$

Then the slope-intercept equation is $y = \frac{4}{5}x - 4$. We rewrite this equation in standard form.

$$y = \frac{4}{5}x - 4$$
$$5y = 4x - 20 \quad \text{Multiplying by 5 on both sides}$$
$$-4x + 5y = -20 \quad \text{Standard form}$$

This equation can also be written as $4x - 5y = 20$.

**99.** $rx + 3y = p^2 - s$

The equation is in standard form with $A = r$, $B = 3$, and $C = p^2 - s$. It is linear.

**101.** Try to put the equation in standard form.

$$r^2x = py + 5$$
$$r^2x - py = 5$$

The equation is in standard form with $A = r^2$, $B = -p$, and $C = 5$. It is linear.

**103.** Let equation $A$ have intercepts $(a, 0)$ and $(0, b)$. Then equation $B$ has intercepts $(2a, 0)$ and $(0, b)$.

$$\text{Slope of } A = \frac{b - 0}{0 - a} = -\frac{b}{a}$$

$$\text{Slope of } B = \frac{b - 0}{0 - 2a} = -\frac{b}{2a} = \frac{1}{2}\left(-\frac{b}{a}\right)$$

The slope of equation $B$ is $\frac{1}{2}$ the slope of equation $A$.

**105.** First write the equation in standard form.

$$ax + 3y = 5x - by + 8$$
$$ax - 5x + 3y + by = 8 \quad \text{Adding } -5x + by \text{ on both sides}$$
$$(a - 5)x + (3 + b)y = 8 \quad \text{Factoring}$$

If the graph is a vertical line, then the coefficient of $y$ is 0.

$$3 + b = 0$$
$$b = -3$$

Then we have $(a - 5)x = 8$.

If the line passes through $(4, 0)$, we have:

$$(a - 5)4 = 8 \quad \text{Substituting 4 for } x$$
$$a - 5 = 2$$
$$a = 7$$

**107.** We graph $C(t) = 0.50t + 3$, where $t$ represents the number of 15-min units of time, as a series of steps. The cost is constant within each 15-min unit of time. Thus,

for $0 < t \le 1$, $C(t) = 0.5(1) + 3 = \$3.50$;

for $1 < t \le 2$, $C(t) = 0.5(2) + 3 = \$4.00$;

for $2 < t \le 3$, $C(t) = 0.5(3) + 3 = \$4.50$;

and so on. We draw the graph. An open circle at a point indicates that the point is not on the graph.

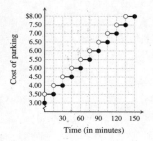

**109.** Graph $y_1 = 4x - 1$ and $y_2 = 3 - 2x$ in the same window and use the Intersect feature to find the first coordinate of the point of intersection, $0.66666667$.

We check by solving the equation algebraically.

$$4x - 1 = 3 - 2x$$
$$6x - 1 = 3$$
$$6x = 4$$
$$x = \frac{2}{3} = 0.\overline{6} \approx 0.66666667$$

**111.** Graph $y_1 = 8 - 7x$ and $y_2 = -2x - 5$ in the same window and use the Intersect feature to find the first coordinate of the point of intersection, $2.6$.

We check by solving the equation algebraically.

$$8 - 7x = -2x - 5$$
$$8 - 5x = -5$$
$$-5x = -13$$
$$x = 2.6$$

**113.** Graph $y = 38 + 2.35x$. Use the Intersect feature to find the $x$-coordinate that corresponds to the $y$-coordinate $623.15$. We find that 249 shirts were printed.

## Exercise Set 2.5

**1.** Three; see page 125 in the text.

**3.** False; given just one point, there is an infinite number of lines that can be drawn through it.

**5.** False; see page 128 in the text.

**7.** True; see page 127 in the text.

**9.** True; see Example 2.

**11.** $y - y_1 = m(x - x_1)$   Point-slope equation
$y - 4 = -2(x - 1)$   Substituting $-2$ for $m$, 1 for $x_1$, and 4 for $y_1$

To graph the equation, we count off a slope of $\dfrac{-2}{1}$, starting at $(1, 4)$, and draw the line.

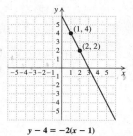

$$y - 4 = -2(x - 1)$$

**13.** $y - y_1 = m(x - x_1)$   Point-slope equation
$y - 2 = 3(x - 5)$   Substituting 3 for $m$, 5 for $x_1$, and 2 for $y_1$

To graph the equation, we count off a slope of $\dfrac{3}{1}$ or $\dfrac{-3}{-1}$, starting at $(5, 2)$ and draw the line.

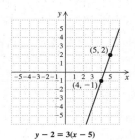

$$y - 2 = 3(x - 5)$$

**15.** $y - y_1 = m(x - x_1)$   Point-slope equation
$y - (-4) = \dfrac{1}{2}[x - (-2)]$   Substituting $\dfrac{1}{2}$ for $m$, $-2$ for $x_1$, and $-4$ for $y_1$

To graph the equation, we count off a slope of $\dfrac{1}{2}$, starting at $(-2, -4)$, and draw the line.

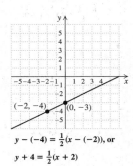

$$y - (-4) = \tfrac{1}{2}(x - (-2)), \text{ or}$$
$$y + 4 = \tfrac{1}{2}(x + 2)$$

**17.** $y - y_1 = m(x - x_1)$   Point-slope equation
$y - 0 = -1(x - 8)$   Substituting $-1$ for $m$, 8 for $x_1$, and 0 for $y_1$

To graph the equation, we count off a slope of $\dfrac{-1}{1}$, starting at $(8, 0)$, and draw the line.

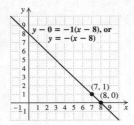

**19.**  $y - 9 = \frac{2}{7}(x - 8)$

$y - y_1 = m(x - x_1)$

$m = \frac{2}{7}$, $x_1 = 8$, and $y_1 = 9$, so the slope $m$ is $\frac{2}{7}$ and a point $(x_1, y_1)$ on the graph is $(8, 9)$.

**21.**  $y + 2 = -5(x - 7)$

$y - (-2) = -5(x - 7)$

$y - y_1 = m(x - x_1)$

$m = -5$, $x_1 = 7$, and $y_1 = -2$, so the slope $m$ is $-5$ and a point $(x_1, y_1)$ on the graph is $(7, -2)$.

**23.**  $y - 4 = -\frac{5}{3}(x + 2)$

$y - 4 = -\frac{5}{3}[x - (-2)]$

$y - y_1 = m(x - x_1)$

$m = -\frac{5}{3}$, $x_1 = -2$, and $y_1 = 4$, so the slope $m$ is $-\frac{5}{3}$ and a point $(x_1, y_1)$ on the graph is $(-2, 4)$.

**25.**  $y = \frac{4}{7}x$

The equation is of the form $y = mx$, so we know that its graph is a line through the origin with slope $m$. Thus, the slope is $\frac{4}{7}$ and a point on the graph is $(0, 0)$.

**27.**     $y - y_1 = m(x - x_1)$       Point-slope equation

$y - (-3) = 4(x - 2)$       Substituting 4 for $m$, 2 for $x_1$, and $-3$ for $y_1$

$y + 3 = 4x - 8$       Simplifying

$y = 4x - 11$       Subtracting 3 from both sides

$f(x) = 4x - 11$       Using function notation

To graph the equation, we count off a slope of $\frac{4}{1}$, starting at $(2, -3)$ or at $(0, -11)$, and draw the line.

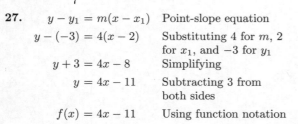

**29.**     $y - y_1 = m(x - x_1)$       Point-slope equation

$y - 8 = -\frac{3}{5}[x - (-4)]$       Substituting $-\frac{3}{5}$ for $m$, $-4$ for $x_1$, and 8 for $y_1$

$y - 8 = -\frac{3}{5}(x + 4)$

$y - 8 = -\frac{3}{5}x - \frac{12}{5}$       Simplifying

$y = -\frac{3}{5}x + \frac{28}{5}$       Adding 8 to both sides

$f(x) = -\frac{3}{5}x + \frac{28}{5}$       Using function notation

To graph the equation, we count off a slope of $\frac{-3}{5}$, starting at $(-4, 8)$, and draw the line.

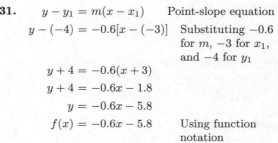

$$f(x) = -\tfrac{3}{5}x + \tfrac{28}{5}$$

**31.**     $y - y_1 = m(x - x_1)$       Point-slope equation

$y - (-4) = -0.6[x - (-3)]$       Substituting $-0.6$ for $m$, $-3$ for $x_1$, and $-4$ for $y_1$

$y + 4 = -0.6(x + 3)$

$y + 4 = -0.6x - 1.8$

$y = -0.6x - 5.8$

$f(x) = -0.6x - 5.8$       Using function notation

To graph the equation, we count off a slope of $-0.6$, or $\frac{-3}{5}$, starting at $(-3, -4)$, and draw the line.

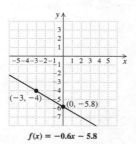

$$f(x) = -0.6x - 5.8$$

**33.**  $m = \frac{2}{7}$; $(0, -6)$

Observe that the slope is $\frac{2}{7}$ and the $y$-intercept is $(0, -6)$. Thus, we have $f(x) = \frac{2}{7}x - 6$.

To graph the equation, we count off a slope of $\frac{2}{7}$, starting at $(0, -6)$, and draw the line.

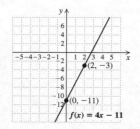

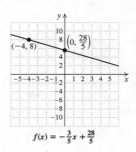

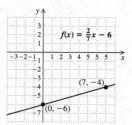

**35.** $y - y_1 = m(x - x_1)$  Point-slope equation

$y - 6 = \dfrac{3}{5}[x - (-4)]$  Substituting $\dfrac{3}{5}$ for $m$, $-4$

for $x_1$, and 6 for $y_1$

$y - 6 = \dfrac{3}{5}(x + 4)$

$y - 6 = \dfrac{3}{5}x + \dfrac{12}{5}$

$y = \dfrac{3}{5}x + \dfrac{42}{5}$

$f(x) = \dfrac{3}{5}x + \dfrac{42}{5}$     Using function notation

To graph the equation, we count off a slope of $\dfrac{3}{5}$, starting at $(-4, 6)$, and draw the line.

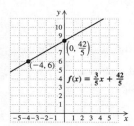

**37.** First find the slope of the line:

$$m = \frac{6 - 4}{5 - 1} = \frac{2}{4} = \frac{1}{2}$$

Use the point-slope equation with $m = \dfrac{1}{2}$ and

$(1, 4) = (x_1, y_1)$. (We could let $(5, 6) = (x_1, y_1)$ instead and obtain an equivalent equation.)

$y - 4 = \dfrac{1}{2}(x - 1)$

$y - 4 = \dfrac{1}{2}x - \dfrac{1}{2}$

$y = \dfrac{1}{2}x + \dfrac{7}{2}$

$f(x) = \dfrac{1}{2}x + \dfrac{7}{2}$   Using function notation

**39.** First find the slope of the line:

$$m = \frac{3 - (-3)}{6.5 - 2.5} = \frac{3 + 3}{4} = \frac{6}{4} = 1.5$$

Use the point-slope equation with $m = 1.5$ and $(6.5, 3) = (x_1, y_1)$.

$y - 3 = 1.5(x - 6.5)$

$y - 3 = 1.5x - 9.75$

$y = 1.5x - 6.75$

$f(x) = 1.5x - 6.75$   Using function notation

**41.** First find the slope of the line:

$$m = \frac{-2 - 3}{0 - 1} = \frac{-5}{-1} = 5$$

Use the point-slope equation with $m = 5$ and $(1, 3) = (x_1, y_1)$.

$y - 3 = 5(x - 1)$

$y - 3 = 5x - 5$

$y = 5x - 2$

$f(x) = 5x - 2$

**43.** First find the slope of the line:

$$m = \frac{-6 - (-3)}{-4 - (-2)} = \frac{-6 + 3}{-4 + 2} = \frac{-3}{-2} = \frac{3}{2}$$

Use the point-slope equation with $m = \dfrac{3}{2}$ and $(-2, -3) = (x_1, y_1)$.

$y - (-3) = \dfrac{3}{2}[x - (-2)]$

$y + 3 = \dfrac{3}{2}(x + 2)$

$y + 3 = \dfrac{3}{2}x + 3$

$y = \dfrac{3}{2}x$

$f(x) = \dfrac{3}{2}x$            Using function notation

**45.** a) Letting $t$ represent the number of years after 1971, we form the pairs $(0, 1542)$ and $(29, 1877)$. First we find the slope of the function that fits the data:

$$m = \frac{1877 - 1542}{29 - 0} = \frac{335}{29}.$$

We know that the $y$-intercept is $(0, 1542)$, so we write a function in slope-intercept form.

$$C(t) = \frac{335}{29}t + 1542.$$

b) In 2009, $t = 2009 - 1971 = 38$.

$$C(38) = \frac{335}{29} \cdot 38 + 1542 \approx 1981 \text{ calories}$$

c) We substitute 2000 for $C(t)$ and solve for $t$.

$$2000 = \frac{335}{29}t + 1542$$

$$458 = \frac{335}{29}t$$

$$\frac{29}{335} \cdot 458 = t$$

$$40 \approx t$$

The average number of calories an American woman will consume per day will reach 2000 approximately 40 yr after 1971, or in 2011.

**47.** a) Letting $t$ represent the number of years since 1990, we form the pairs $(0, 71.8)$ and $(10, 74.1)$. First we find the slope of the function that fits the data:

$$m = \frac{74.1 - 71.8}{10 - 0} = \frac{2.3}{10} = 0.23.$$

We know the $y$-intercept is $(0, 71.8)$, so we write a function in slope-intercept form.

$$E(t) = 0.23t + 71.8$$

b) In 2009, $t = 2009 - 1990 = 19$.

$$E(19) = 0.23(19) + 71.8 \approx 76.2 \text{ yr}$$

**49.** a) Letting $t$ represent the number of years since 1992, we form the pairs $(0, 178.6)$ and $(10, 282)$. First we find the slope of the function that fits the data:

$$m = \frac{282 - 178.6}{10 - 0} = \frac{103.4}{10} = 10.34$$

We know the $y$-intercept is $(0, 178.6)$, so we write a function in slope-intercept form.

$$A(t) = 10.34t + 178.6$$

b) In 2008, $t = 2008 - 1992 = 16$.

$$A(16) = 10.34(16) + 178.6 \approx \$344 \text{ million}$$

**51.** a) Letting $t$ represent the number of years since 1996, we form the pairs $(0, 57.3)$ and $(4, 69.9)$. First we find the slope of the function that fits the data:

$$m = \frac{69.9 - 57.3}{4 - 0} = \frac{12.6}{4} = 3.15$$

We know the $y$-intercept is $(0, 57.3)$, so we write a function in slope-intercept form.

$$N(t) = 3.15t + 57.3$$

b) In 2008, $t = 2008 - 1996 = 12$.

$$N(12) = 3.15(12) + 57.3 = 95.1 \text{ million tons}$$

**53.** a) Letting $t$ represent the number of years after 1999, we form the pairs $(0, 48)$ and $(4, 64)$. First we find the slope of the function that fits the data:

$$m = \frac{64 - 48}{4 - 0} = \frac{16}{4} = 4$$

We know the $y$-intercept is $(0, 48)$, so we write a function in slope-intercept form.

$$N(t) = 4t + 48$$

b) In 2009, $t = 2009 - 1999 = 10$.

$$N(10) = 4 \cdot 10 + 48 = 88 \text{ million Americans}$$

c) Substitute 104 for $N(t)$ and solve for $t$.

$$104 = 4t + 48$$
$$56 = 4t$$
$$14 = t$$

104 million Americans will find travel information on the Internet 14 yr after 1999, or in 2013.

**55.** a) Letting $t$ represent the number of years after 1994, we form the pairs $(0, 74.9)$ and $(6, 78.2)$. First we find the slope of the function that fits the data:

$$m = \frac{78.2 - 74.9}{6 - 0} = \frac{3.3}{6} = 0.55$$

We know the $y$-intercept is $(0, 74.9)$, so we write a function in slope-intercept form.

$$A(t) = 0.55t + 74.9$$

b) In 2009, $t = 2009 - 1994 = 15$.

$$A(15) = 0.55(15) + 74.9 = 83.15 \text{ million acres}$$

**57.** We first solve for $y$ and determine the slope of each line.

$$x + 8 = y$$
$$y = x + 8 \quad \text{Reversing the order}$$

The slope of $y = x + 8$ is 1.

$$y - x = -5$$
$$y = x - 5$$

The slope of $y = x - 5$ is 1.

The slopes are the same; the lines are parallel.

**59.** We first solve for $y$ and determine the slope of each line.

$$y + 9 = 3x$$
$$y = 3x - 9$$

The slope of $y = 3x - 9$ is 3.

$$3x - y = -2$$
$$3x + 2 = y$$
$$y = 3x + 2 \quad \text{Reversing the order}$$

The slope of $y = 3x + 2$ is 3.

The slopes are the same; the lines are parallel.

**61.** We determine the slope of each line.

The slope of $f(x) = 3x + 9$ is 3.

$$2y = 8x - 2$$
$$y = 4x - 1$$

The slope of $y = 4x - 1$ is 4.

The slopes are not the same; the lines are not parallel.

**63.** First solve the equation for $y$ and determine the slope of the given line.

$$x + 2y = 6 \qquad \text{Given line}$$
$$2y = -x + 6$$
$$y = -\frac{1}{2}x + 3$$

The slope of the given line is $-\frac{1}{2}$.

The slope of every line parallel to the given line must also be $-\frac{1}{2}$. We find the equation of the line with slope $-\frac{1}{2}$ and containing the point $(4, 7)$.

$$y - y_1 = m(x - x_1) \quad \text{Point-slope equation}$$
$$y - 7 = -\frac{1}{2}(x - 4) \quad \text{Substituting}$$
$$y - 7 = -\frac{1}{2}x + 2$$
$$y = -\frac{1}{2}x + 9$$

**65.** First solve the equation for $y$ and determine the slope of the given line.

$$5x - 3y = 8 \qquad \text{Given line}$$
$$-3y = -5x + 8$$
$$y = \frac{5}{3}x - \frac{8}{3}$$

The slope of the given line is $\frac{5}{3}$.

The slope of every line parallel to the given line must also be $\frac{5}{3}$. We find the equation of the line with slope $\frac{5}{3}$ and containing the point $(2, -6)$.

$$y - y_1 = m(x - x_1) \qquad \text{Point-slope equation}$$
$$y - (-6) = \frac{5}{3}(x - 2) \qquad \text{Substituting}$$
$$y + 6 = \frac{5}{3}x - \frac{10}{3}$$
$$y = \frac{5}{3}x - \frac{28}{3}$$

**67.** The slope of $y = 2x + 1$ is 2. The given point, $(0, -7)$, is the $y$-intercept, so we substitute in the slope-intercept equation.

$$y = 2x - 7$$

**69.** First solve the equation for $y$ and determine the slope of the given line.

$$2x + 3y = -7 \qquad \text{Given line}$$
$$3y = -2x - 7$$
$$y = -\frac{2}{3}x - \frac{7}{3}$$

The slope of the given line is $-\frac{2}{3}$.

The slope of every line parallel to the given line must also be $-\frac{2}{3}$. We find the equation of the line with slope $-\frac{2}{3}$ and containing the point $(-2, -3)$.

$$y - y_1 = m(x - x_1) \qquad \text{Point-slope equation}$$
$$y - (-3) = -\frac{2}{3}[x - (-2)] \qquad \text{Substituting}$$
$$y + 3 = -\frac{2}{3}(x + 2)$$
$$y + 3 = -\frac{2}{3}x - \frac{4}{3}$$
$$y = -\frac{2}{3}x - \frac{13}{3}$$

**71.** First solve the equation for $y$ and determine the slope of the given line.

$$3x - 9y = 2 \qquad \text{Given line}$$
$$3x - 2 = 9y$$
$$\frac{1}{3}x - \frac{2}{9} = y$$

The slope of the given line is $\frac{1}{3}$.

The slope of every line parallel to the given line must also be $\frac{1}{3}$. We find the equation of the line with slope $\frac{1}{3}$ and containing the point $(-6, 2)$.

$$y - y_1 = m(x - x_1) \qquad \text{Point-slope equation}$$
$$y - 2 = \frac{1}{3}[x - (-6)] \qquad \text{Substituting}$$
$$y - 2 = \frac{1}{3}(x + 6)$$
$$y - 2 = \frac{1}{3}x + 2$$
$$y = \frac{1}{3}x + 4$$

**73.** $x = 2$ is a vertical line. A line parallel to it that passes through $(5, -4)$ is the vertical line 5 units to the right of the $y$-axis, or $x = 5$.

**75.** We determine the slope of each line.

The slope of $f(x) = 4x - 3$ is 4.

$$4y = 7 - x$$
$$4y = -x + 7$$
$$y = -\frac{1}{4}x + \frac{7}{4}$$

The slope of $4y = 7 - x$ is $-\frac{1}{4}$.

The product of their slopes is $4\left(-\frac{1}{4}\right)$, or $-1$; the lines are perpendicular.

**77.** We determine the slope of each line.

$$x + 2y = 7$$
$$2y = -x + 7$$
$$y = -\frac{1}{2}x + \frac{7}{2}$$

The slope of $x + 2y = 7$ is $-\frac{1}{2}$.

$$2x + 4y = 4$$
$$4y = -2x + 4$$
$$y = -\frac{1}{2}x + 1.$$

The slope of $2x + 4y = 4$ is $-\frac{1}{2}$.

The product of their slopes is $\left(-\frac{1}{2}\right)\left(-\frac{1}{2}\right)$, or $\frac{1}{4}$; the lines are not perpendicular. For the lines to be perpendicular, the product must be $-1$.

**79.** First solve the equation for $y$ and determine the slope of the given line.

$$x + 2y = -7 \qquad \text{Given line}$$
$$2y = -x - 7$$
$$y = -\frac{1}{2}x - \frac{7}{2}$$

The slope of the given line is $-\frac{1}{2}$.

The slope of a perpendicular line is given by the opposite of the reciprocal of $-\frac{1}{2}$, 2.

We find the equation of the line with slope 2 containing the point $(2, 5)$.

$$y - y_1 = m(x - x_1) \quad \text{Point-slope equation}$$
$$y - 5 = 2(x - 2) \quad \text{Substituting}$$
$$y - 5 = 2x - 4$$
$$y = 2x + 1$$

**81.** First solve the equation for $y$ and determine the slope of the given line.

$$3x + 6y = 5 \qquad \text{Given line}$$
$$6y = -3x + 5$$
$$y = -\frac{1}{2}x + \frac{5}{6}$$

The slope of the given line is $-\frac{1}{2}$.

The slope of a perpendicular line is given by the opposite of the reciprocal of $-\frac{1}{2}$, 2.

We find the equation of the line with slope 2 and containing the point $(3, -2)$.

$$y - y_1 = m(x - x_1) \quad \text{Point-slope equation}$$
$$y - (-2) = 2(x - 3) \quad \text{Substituting}$$
$$y + 2 = 2x - 6$$
$$y = 2x - 8$$

**83.** First solve the equation for $y$ and determine the slope of the given line.

$$2x + 5y = 7 \qquad \text{Given line}$$
$$5y = -2x + 7$$
$$y = -\frac{2}{5}x + \frac{7}{5}$$

The slope of the given line is $-\frac{2}{5}$.

The slope of a perpendicular line is given by the opposite of the reciprocal of $-\frac{2}{5}$, $\frac{5}{2}$.

We find the equation of the line with slope $\frac{5}{2}$ and containing the point $(0, 9)$.

$$y - y_1 = m(x - x_1) \quad \text{Point-slope equation}$$
$$y - 9 = \frac{5}{2}(x - 0) \quad \text{Substituting}$$
$$y - 9 = \frac{5}{2}x$$
$$y = \frac{5}{2}x + 9$$

**85.** First solve the equation for $y$ and find the slope of the given line.

$$3x - 5y = 6$$
$$-5y = -3x + 6$$
$$y = \frac{3}{5}x - \frac{6}{5}$$

The slope of the given line is $\frac{3}{5}$. The slope of a perpendicular line is given by the opposite of the reciprocal of $\frac{3}{5}$, $-\frac{5}{3}$.

We find the equation of the line with slope $-\frac{5}{3}$ and containing the point $(-4, -7)$.

$$y - y_1 = m(x - x_1) \qquad \text{Point-slope equation}$$
$$y - (-7) = -\frac{5}{3}[x - (-4)]$$
$$y + 7 = -\frac{5}{3}(x + 4)$$
$$y + 7 = -\frac{5}{3}x - \frac{20}{3}$$
$$y = -\frac{5}{3}x - \frac{41}{3}$$

**87.** The slope of a line perpendicular to $2x - 5 = y$ is $-\frac{1}{2}$ and we are given the $y$-intercept of the desired line, $(0, 6)$. Then we have $y = -\frac{1}{2}x + 6$.

**89.** $y = 5$ is a horizontal line, so a line perpendicular to it must be vertical. The equation of the vertical line containing $(-3, 7)$ is $x = -3$.

**91.** *Writing Exercise*

**93.** $(3x^2 + 5x) + (2x - 4) = 3x^2 + (5 + 2)x - 4 = 3x^2 + 7x - 4$

**95.** $\dfrac{2t - 6}{4t + 1} = \dfrac{2 \cdot 3 - 6}{4 \cdot 3 + 1} = \dfrac{6 - 6}{12 + 1} = \dfrac{0}{13} = 0$

**97.** $2x - 5y = 2 \cdot 3 - 5(-1) = 6 + 5 = 11$

**99.** *Writing Exercise*

**101.** *Familiarize*. Celsius temperature $C$ corresponding to a Fahrenheit temperature $F$ can be modeled by a line that contains the points $(32, 0)$ and $(212, 100)$.

*Translate*. We find an equation relating $C$ and $F$.

$$m = \frac{100 - 0}{212 - 32} = \frac{100}{180} = \frac{5}{9}$$
$$C - 0 = \frac{5}{9}(F - 32)$$
$$C = \frac{5}{9}(F - 32)$$

*Carry out*. Using function notation we have $C(F) = \frac{5}{9}(F - 32)$. Now we find $C(70)$:

$$C(70) = \frac{5}{9}(70 - 32) = \frac{5}{9}(38) \approx 21.1.$$

*Check*. We can repeat the calculations. We could also graph the function and determine that $(70, 21.1)$ is on the graph.

*State*. A temperature of about $21.1°$ C corresponds to a temperature of $70°$ F.

**103.** *Familiarize*. The total cost $C$ of the phone, in dollars, after $t$ months, can be modeled by a line that contains the points $(5, 230)$ and $(9, 390)$.

*Translate*. We find an equation relating $C$ and $t$.

$$m = \frac{390 - 230}{9 - 5} = \frac{160}{4} = 40$$

$$C - 230 = 40(t - 5)$$
$$C - 230 = 40t - 200$$
$$C = 40t + 30$$

**Carry out**. Using function notation we have $C(t) = 40t + 30$. To find the costs already incurred when the service began we find $C(0)$:

$$C(0) = 40 \cdot 0 + 30 = 30$$

**Check**. We can repeat the calculations. We could also graph the function and determine that $(0, 30)$ is on the graph.

**State**. Mel had already incurred $30 in costs when his service just began.

**105.** We find the value of $p$ for which $A(p) = -2.5p + 26.5$ and $A(p) = 2p - 11$ are the same.

$$-2.5p + 26.5 = 2p - 11$$
$$26.5 = 4.5p - 11$$
$$37.5 = 4.5p$$
$$8.\overline{3} = p$$

Supply will equal demand at a price of about $8.33 per pound.

**107.** The price must be a positive number, so we have $p > 0$. Furthermore, the amount of coffee supplied must be a positive number. We have:

$$2p - 11 > 0$$
$$2p > 11$$
$$p > 5.5$$

Thus, the domain is $\{p | p > 5.5\}$.

**109.** Find the slope of $5y - kx = 7$:

$$5y - kx = 7$$
$$5y = kx + 7$$
$$y = \frac{k}{5}x + \frac{7}{5}$$

The slope is $\frac{k}{5}$.

Find the slope of the line containing $(7, -3)$ and $(-2, 5)$:

$$m = \frac{5 - (-3)}{-2 - 7} = \frac{5 + 3}{-9} = -\frac{8}{9}$$

If the lines are parallel, their slopes must be equal:

$$\frac{k}{5} = -\frac{8}{9}$$
$$k = -\frac{40}{9}$$

**111.** a) Following the instructions for entering data and using the linear regression option on a graphing calculator, we find the following function:

$$f(x) = 3.492366412x + 6.354961832$$

b) $f(35) = 3.492366412(35) + 6.354961832 \approx 129$ W

The corresponding answer to Exercise 65 in Section 2.2 is 140 W. The answer found using linear regression seems more reliable because it was found using a function that is based on more data points than the function in Section 2.2.

**113.**

## Exercise Set 2.6

**1.** domain; see page 135 in the text.

**3.** evaluate; see page 135 in the text.

**5.** excluding; see page 138 in the text.

**7.** Since $f(2) = -3 \cdot 2 + 1 = -5$, and $g(2) = 2^2 + 2 = 6$, we have $f(2) + g(2) = -5 + 6 = 1$.

**9.** Since $f(5) = -3 \cdot 5 + 1 = -14$ and $g(5) = 5^2 + 2 = 27$, we have $f(5) - g(5) = -14 - 27 = -41$.

**11.** Since $f(-1) = -3(-1) + 1 = 4$ and
$$g(-1) = (-1)^2 + 2 = 3, \text{ we have}$$
$$f(-1) \cdot g(-1) = 4 \cdot 3 = 12.$$

**13.** Since $f(-4) = -3(-4) + 1 = 13$ and
$$g(-4) = (-4)^2 + 2 = 18, \text{ we have}$$
$$f(-4)/g(-4) = 13/18.$$

**15.** Since $g(1) = 1^2 + 2 = 3$ and
$$f(1) = -3 \cdot 1 + 1 = -2, \text{ we have}$$
$$g(1) - f(1) = 3 - (-2) = 3 + 2 = 5.$$

**17.** $(f+g)(x) = f(x) + g(x) = (-3x+1) + (x^2+2) = x^2 - 3x + 3$

**19.** $(F + G)(x) = F(x) + G(x)$
$$= x^2 - 2 + 5 - x$$
$$= x^2 - x + 3$$

**21.** Using our work in Exercise 19, we have
$$(F + G)(-4) = (-4)^2 - (-4) + 3$$
$$= 16 + 4 + 3$$
$$= 23.$$

**23.** $(F - G)(x) = F(x) - G(x)$
$$= x^2 - 2 - (5 - x)$$
$$= x^2 - 2 - 5 + x$$
$$= x^2 + x - 7$$

Then we have
$$(F - G)(3) = 3^2 + 3 - 7$$
$$= 9 + 3 - 7$$
$$= 5.$$

**25.** $(F \cdot G)(x) = F(x) \cdot G(x)$
$$= (x^2 - 2)(5 - x)$$
$$= 5x^2 - x^3 - 10 + 2x$$

Then we have
$$(F \cdot G)(-3) = 5(-3)^2 - (-3)^3 - 10 + 2(-3)$$
$$= 5 \cdot 9 - (-27) - 10 - 6$$
$$= 45 + 27 - 10 - 6$$
$$= 56.$$

**27.** $(F/G)(x) = F(x)/G(x)$

$$= \frac{x^2 - 2}{5 - x}, \; x \neq 5$$

**29.** Using our work in Exercise 27, we have

$$(F/G)(-2) = \frac{(-2)^2 - 2}{5 - (-2)} = \frac{4 - 2}{5 + 2} = \frac{2}{7}.$$

**31.** $(P - L)(2) = P(2) - L(2) \approx 26.5\% - 22.5\% \approx 4\%$

**33.** $N(2000) = (R + W)(2000)$

$$= R(2000) + W(2000)$$
$$\approx 1.3 + 2.2$$
$$\approx 3.5$$

We estimate that 3.5 million U.S. women had children in 2000.

**35.** $(n + l)('98) = n('98) + l('98)$

From the middle line of the graph, we can see that $n('98) + l('98) \approx 50$ million.

This represents the total number of passengers serviced by Newark Liberty and LaGuardia airports in 1998.

**37.** From the top line of the graph, we see that $F('02) \approx 81$ million. This represents the number of passengers using the three airports in 2002.

**39.** $(F - k)('02) = F('02) - k('02)$

$$\approx 81 - 30$$
$$\approx 51 \text{ million}$$

This represents the number of passengers using LaGuardia and Newark Liberty in 2002.

**41.** The domain of $f$ and of $g$ is all real numbers. Thus, Domain of $f + g = $ Domain of $f - g = $ Domain of $f \cdot g = \{x | x$ is a real number$\}$.

**43.** Because division by 0 is undefined, we have

Domain of $f = \{x | x$ is a real number *and* $x \neq 3\}$,

and

Domain of $g = \{x | x$ is a real number$\}$.

Thus, Domain of $f + g = $ Domain of $f - g = $ Domain of $f \cdot g = \{x | x$ is a real number *and* $x \neq 3\}$.

**45.** Because division by 0 is undefined, we have

Domain of $f = \{x | x$ is a real number *and* $x \neq 0\}$,

and

Domain of $g = \{x | x$ is a real number$\}$.

Thus, Domain of $f + g = $ Domain of $f - g = $ Domain of $f \cdot g = \{x | x$ is a real number *and* $x \neq 0\}$.

**47.** Because division by 0 is undefined, we have

Domain of $f = \{x | x$ is a real number *and* $x \neq 1\}$,

and

Domain of $g = \{x | x$ is a real number$\}$.

Thus, Domain of $f + g = $ Domain of $f - g = $ Domain of $f \cdot g = \{x | x$ is a real number *and* $x \neq 1\}$.

**49.** Because division by 0 is undefined, we have

Domain of $f = \{x | x$ is a real number *and* $x \neq 2\}$,

and

Domain of $g = \{x | x$ is a real number *and* $x \neq 4\}$.

Thus, Domain of $f + g = $ Domain of $f - g = $ Domain of $f \cdot g = \{x | x$ is a real number *and* $x \neq 2$ *and* $x \neq 4\}$.

**51.** Domain of $f = $ Domain of $g = $

$\{x | x$ is a real number$\}$.

Since $g(x) = 0$ when $x - 3 = 0$, we have $g(x) = 0$ when $x = 3$. We conclude that Domain of $f/g = \{x | x$ is a real number *and* $x \neq 3\}$.

**53.** Domain of $f = $ Domain of $g = $

$\{x | x$ is a real number$\}$.

Since $g(x) = 0$ when $2x - 8 = 0$, we have $g(x) = 0$ when $x = 4$. We conclude that Domain of $f/g = \{x | x$ is a real number *and* $x \neq 4\}$.

**55.** Domain of $f = \{x | x$ is a real number *and* $x \neq 4\}$.

Domain of $g = \{x | x$ is a real number$\}$.

Since $g(x) = 0$ when $5 - x = 0$, we have $g(x) = 0$ when $x = 5$. We conclude that Domain of $f/g = \{x | x$ is a real number *and* $x \neq 4$ *and* $x \neq 5\}$.

**57.** Domain of $f = \{x | x$ is a real number *and* $x \neq -1\}$.

Domain of $g = \{x | x$ is a real number$\}$.

Since $g(x) = 0$ when $2x + 5 = 0$, we have $g(x) = 0$ when $x = -\frac{5}{2}$. We conclude that Domain of $f/g = \left\{x \middle| x$ is a real number *and* $x \neq -1$ *and* $x \neq -\frac{5}{2}\right\}$.

**59.** $(F + G)(5) = F(5) + G(5) = 1 + 3 = 4$

$(F + G)(7) = F(7) + G(7) = -1 + 4 = 3$

**61.** $(G - F)(7) = G(7) - F(7) = 4 - (-1) = 4 + 1 = 5$

$(G - F)(3) = G(3) - F(3) = 1 - 2 = -1$

**63.** From the graph we see that Domain of

$F = \{x | 0 \leq x \leq 9\}$ and Domain of

$G = \{x | 3 \leq x \leq 10\}$. Then Domain of

$F + G = \{x | 3 \leq x \leq 9\}$. Since $G(x)$ is never 0, Domain of $F/G = \{x | 3 \leq x \leq 9\}$.

**65.** We use $(F + G)(x) = F(x) + G(x)$.

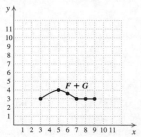

**67.** *Writing Exercise*

**69.** $4x - 7y = 8$

$\qquad 4x = 7y + 8 \qquad$ Adding $7y$

$\qquad \dfrac{1}{4} \cdot 4x = \dfrac{1}{4}(7y + 8) \quad$ Multiplying by $\dfrac{1}{4}$

$\qquad x = \dfrac{7}{4}y + 2$

**71.** $5x + 2y = -3$

$\qquad 2y = -5x - 3 \qquad$ Subtracting $5x$

$\qquad \dfrac{1}{2} \cdot 2y = \dfrac{1}{2}(-5x - 3) \quad$ Multiplying by $\dfrac{1}{2}$

$\qquad y = -\dfrac{5}{2}x - \dfrac{3}{2}$

**73.** Let $n$ represent the number; $2n + 5 = 49$.

**75.** Let $n$ represent the first integer; $x + (x + 1) = 145$.

**77.** *Writing Exercise*

**79.** Domain of $f = \left\{ x \Big| x \text{ is a real number } and\ x \neq -\dfrac{5}{2} \right\}$;

domain of $g = \{x | x \text{ is a real number } and\ x \neq -3\}$;

$g(x) = 0$ when $x^4 - 1 = 0$, or when $x = 1$ or $x = -1$.

Then domain of $f/g = \Big\{ x \Big| x \text{ is a real number } and$

$x \neq -\dfrac{5}{2} \text{ and } x \neq -3 \text{ and } x \neq 1 \text{ and } x \neq -1 \Big\}$.

**81.** Answers may vary.

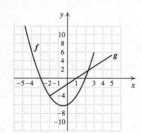

**83.** The problem states that Domain of $m = \{x | -1 < x < 5\}$. Since $n(x) = 0$ when $2x - 3 = 0$, we have $n(x) = 0$ when $x = \dfrac{3}{2}$. We conclude that Domain of $m/n = \Big\{ x | x \text{ is a real number } and\ -1 < x < 5 \text{ and } x \neq \dfrac{3}{2} \Big\}$.

**85.** Answers may vary. $f(x) = \dfrac{1}{x + 2}$, $g(x) = \dfrac{1}{x - 5}$

**87.**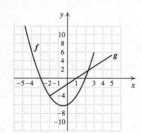

# Chapter 3

# Systems of Linear Equations and Problem Solving

**1.** True; see page 152 in the text.

**3.** True; see Example 4(c).

**5.** True; see Example 4(b).

**7.** False; see page 155 in the text.

**9.** We use alphabetical order for the variables. We replace $x$ by 1 and $y$ by 2.

$$
\begin{array}{c|c}
4x - y = 2 & \\
\hline
4 \cdot 1 - 2 & 2 \\
4 - 2 & \\
& \overset{?}{2 = 2} \text{ TRUE}
\end{array}
\qquad
\begin{array}{c|c}
10x - 3y = 4 & \\
\hline
10 \cdot 1 - 3 \cdot 2 & 4 \\
10 - 6 & \\
& \overset{?}{4 = 4} \text{ TRUE}
\end{array}
$$

The pair $(1, 2)$ makes both equations true, so it is a solution of the system.

**11.** We use alphabetical order for the variables. We replace $x$ by 2 and $y$ by 5.

$$
\begin{array}{c|c}
y = 3x - 1 & \\
\hline
5 & 3 \cdot 2 - 1 \\
& 6 - 1 \\
\overset{?}{5 = 5} & \text{TRUE}
\end{array}
\qquad
\begin{array}{c|c}
2x + y = 4 & \\
\hline
2 \cdot 2 + 5 & 4 \\
4 + 5 & \\
\overset{?}{9 = 4} & \text{FALSE}
\end{array}
$$

The pair $(2, 5)$ is not a solution of $2x + y = 4$. Therefore, it is not a solution of the system of equations.

**13.** We replace $x$ by 1 and $y$ by 5.

$$
\begin{array}{c|c}
x + y = 6 & \\
\hline
1 + 5 & 6 \\
\overset{?}{} & \text{TRUE} \\
6 = 6 &
\end{array}
\qquad
\begin{array}{c|c}
y = 2x + 3 & \\
\hline
5 & 2 \cdot 1 + 3 \\
& 2 + 3 \\
\overset{?}{5 = 5} & \text{TRUE}
\end{array}
$$

The pair $(1, 5)$ makes both equations true, so it is a solution of the system.

**15.** Observe that if we multiply both sides of the first equation by 2, we get the second equation. Thus, if we find that the given point makes the one equation true, we will also know that it makes the other equation true. We replace $x$ by 3 and $y$ by 1 in the first equation.

$$
\begin{array}{c|c}
3x + 4y = 13 & \\
\hline
3 \cdot 3 + 4 \cdot 1 & 13 \\
9 + 4 & \\
& \overset{?}{13 = 13} \text{ TRUE}
\end{array}
$$

The pair $(3, 1)$ makes both equations true, so it is a solution of the system.

**17.** Graph both equations.

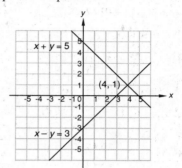

The solution (point of intersection) is apparently $(4, 1)$. Check:

$$
\begin{array}{c|c}
x - y = 3 & \\
\hline
4 - 1 & 3 \\
& \overset{?}{3 = 3} \text{ TRUE}
\end{array}
\qquad
\begin{array}{c|c}
x + y = 5 & \\
\hline
4 + 1 & 5 \\
& \overset{?}{5 = 5} \text{ TRUE}
\end{array}
$$

The solution is $(4, 1)$.

**19.** Graph the equations.

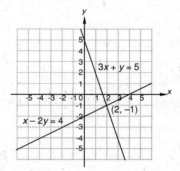

The solution (point of intersection) is apparently $(2, -1)$. Check:

$$
\begin{array}{c|c}
3x + y = 5 & \\
\hline
3 \cdot 2 + (-1) & 5 \\
6 - 1 & \\
& \overset{?}{5 = 5} \text{ TRUE}
\end{array}
\qquad
\begin{array}{c|c}
x - 2y = 4 & \\
\hline
2 - 2(-1) & 4 \\
2 + 2 & \\
& \overset{?}{4 = 4} \text{ TRUE}
\end{array}
$$

The solution is $(2, -1)$.

**21.** Graph both equations.

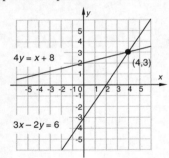

The solution (point of intersection) is apparently $(4, 3)$.

Check:

| $4y = x + 8$ | |
|---|---|
| $4 \cdot 3$ | $4 + 8$ |
| ? | TRUE |
| $12 = 12$ | |

| $3x - 2y = 6$ | |
|---|---|
| $3 \cdot 4 - 2 \cdot 3$ | $6$ |
| $12 - 6$ | |
| ? | |
| $6 = 6$ TRUE | |

The solution is $(4, 3)$.

**23.** Graph both equations.

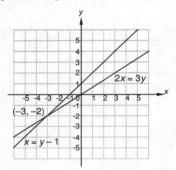

The solution (point of intersection) is apparently $(-3, -2)$.

Check:

| $x = y - 1$ | |
|---|---|
| $-3$ | $-2 - 1$ |
| ? | |
| $-3 = -3$ | TRUE |

| $2x = 3y$ | |
|---|---|
| $2(-3)$ | $3(-2)$ |
| ? | |
| $-6 = -6$ | TRUE |

The solution is $(-3, -2)$.

**25.** Graph both equations.

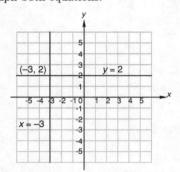

The ordered pair $(-3, 2)$ checks in both equations. It is the solution.

**27.** Graph both equations.

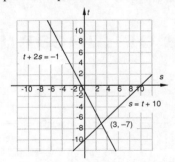

The solution (point of intersection) is apparently $(3, -7)$.

Check:

| $t + 2s = -1$ | |
|---|---|
| $-7 + 2 \cdot 3$ | $-1$ |
| $-7 + 6$ | |
| ? | |
| $-1 = -1$ TRUE | |

| $s = t + 10$ | |
|---|---|
| $3$ | $-7 + 10$ |
| ? | |
| $3 = 3$ | TRUE |

The solution is $(3, -7)$.

**29.** Graph both equations.

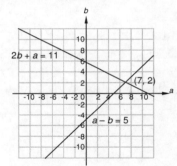

The solution (point of intersection) is apparently $(7, 2)$.

Check:

| $2b + a = 11$ | |
|---|---|
| $2 \cdot 2 + 7$ | $11$ |
| $4 + 7$ | |
| ? | |
| $11 = 11$ TRUE | |

| $a - b = 5$ | |
|---|---|
| $7 - 2$ | $5$ |
| ? | |
| $5 = 5$ | TRUE |

The solution is $(7, 2)$.

**31.** Graph both equations.

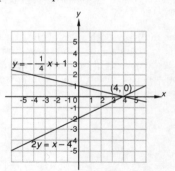

The solution (point of intersection) is apparently $(4, 0)$.

Check:

$$y = -\frac{1}{4}x + 1$$

| $0$ | $-\frac{1}{4} \cdot 4 + 1$ |
|---|---|
| | $-1 + 1$ |
| $?$ | |
| $0 = 0$ | TRUE |

$$2y = x - 4$$

| $2 \cdot 0$ | $4 - 4$ |
|---|---|
| $?$ | |
| $0 = 0$ | TRUE |

The solution is $(4, 0)$.

**33.** Graph both equations.

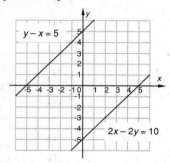

The lines are parallel. The system has no solution.

**35.** Graph both equations.

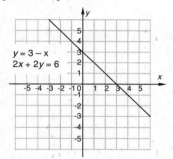

The graphs are the same. Any solution of one equation is a solution of the other. Each equation has infinitely many solutions. The solution set is the set of all pairs $(x, y)$ for which $y = 3 - x$, or $\{(x, y) | y = 3 - x\}$. (In place of $y = 3 - x$ we could have used $2x + 2y = 6$ since the two equations are equivalent.)

**37.** A system of equations is consistent if it has at least one solution. Of the systems under consideration, only the one in Exercise 33 has no solution. Therefore, all except the system in Exercise 33 are consistent.

**39.** A system of two equations in two variables is dependent if it has infinitely many solutions. Only the system in Exercise 35 is dependent.

**41. *Familiarize*.** Let $x$ = the first number and $y$ = the second number.

***Translate*.**

The sum of the numbers is 50.

$$x + y = 50$$

The first number is 25% of the second number.

$$x = 0.25 \cdot y$$

We have a system of equations:

$$x + y = 50,$$
$$x = 0.25y$$

**43. *Familiarize*.** Let $m$ = the number of ounces of mineral oil to be used and $v$ = the number of ounces of vinegar.

***Translate*.**

Number of ounces of mineral oil plus number of ounces of vinegar is 16 oz.

$$m + v = 16$$

Amount of mineral oil is two times the amount of vinegar plus 4 oz.

$$m = 2 \cdot v + 4$$

We have a system of equations:

$$m + v = 16,$$
$$m = 2v + 4$$

**45. *Familiarize*.** Let $x$ = the measure of one angle and $y$ = the measure of the other angle.

***Translate*.**

Two angles are supplementary.

Rewording: The sum of the measures is 180°.

$$x + y = 180$$

One angle is 3° less than twice the other.

Rewording: One angle is twice the other angle minus 3°.

$$x = 2y - 3$$

We have a system of equations:

$$x + y = 180,$$
$$x = 2y - 3$$

**47. *Familiarize*.** Let $g$ = the number of field goals and $t$ = the number of foul shots made.

***Translate*.** We organize the information in a table.

| Kind of shot | Field goal | Foul shot | Total |
|---|---|---|---|
| Number scored | $g$ | $t$ | 64 |
| Points per score | 2 | 1 | |
| Points scored | $2g$ | $t$ | 100 |

From the "Number scored" row of the table we get one equation:

$$g + t = 64$$

The "Points scored" row gives us another equation:

$$2g + t = 100$$

We have a system of equations:

$$g + t = 64,$$

$$2g + t = 100$$

**49.** *Familiarize*.  Let $x$ = the number of less expensive brushes sold and $y$ = the number of more expensive brushes sold.

*Translate*.  We organize the information in a table.

| Kind of brush | Less expen-sive | More expen-sive | Total |
|---|---|---|---|
| Number sold | $x$ | $y$ | 45 |
| Price | \$8.50 | \$9.75 | |
| Amount taken in | $8.50x$ | $9.75y$ | 398.75 |

The "Number sold" row of the table gives us one equation:

$$x + y = 45$$

The "Amount taken in" row gives us a second equation:

$$8.50x + 9.75y = 398.75$$

We have a system of equations:

$$x + y = 45,$$

$$8.50x + 9.75y = 398.75$$

We can multiply both sides of the second equation by 100 to clear the decimals:

$$x + y = 45,$$

$$850x + 975y = 39,875$$

**51.** *Familiarize*.  Let $h$ = the number of vials of Humulin Insulin sold and $n$ = the number of vials of Novolin Velosulin Insulin sold.

*Translate*.  We organize the information in a table.

| Brand | Humulin | Novolin | Total |
|---|---|---|---|
| Number sold | $h$ | $n$ | 50 |
| Price | \$27.06 | \$34.39 | |
| Amount taken in | $27.06h$ | $34.39n$ | 1565.57 |

The "Number sold" row of the table gives us one equation:

$$h + n = 50$$

The "Amount taken in" row gives us a second equation:

$$27.06h + 34.39n = 1565.57$$

We have a system of equations:

$$h + n = 50,$$

$$27.06h + 34.39n = 1565.57$$

We can multiply both sides of the second equation by 100 to clear the decimals:

$$h + n = 50,$$

$$2706h + 3439n = 156,557$$

**53.** *Familiarize*.  The basketball court is a rectangle with perimeter 288 ft.  Let $l$ = the length, in feet, and $w$ = width, in feet.  Recall that for a rectangle with length $l$ and width $w$, the perimeter $P$ is given by $P = 2l + 2w$.

*Translate*.  The formula for perimeter gives us one equation:

$$2l + 2w = 288$$

The statement relating length and width gives us another equation:

The length is 44 ft longer than the width.

$$l = w + 44$$

We have a system of equations:

$$2l + 2w = 288,$$

$$l = w + 44$$

**55.** *Writing Exercise*

**57.**   $2(4x - 3) - 7x = 9$

$$\begin{array}{ll} 8x - 6 - 7x = 9 & \text{Removing parentheses} \\ x - 6 = 9 & \text{Collecting like terms} \\ x = 15 & \text{Adding 6 to both sides} \end{array}$$

The solution is 15.

**59.**   $4x - 5x = 8x - 9 + 11x$

$$\begin{array}{ll} -x = 19x - 9 & \text{Collecting like terms} \\ -20x = -9 & \text{Adding } -19x \text{ to both sides} \\ x = \dfrac{9}{20} & \text{Multiplying both sides by } -\dfrac{1}{20} \end{array}$$

The solution is $\dfrac{9}{20}$.

**61.**   $3x + 4y = 7$

$$\begin{array}{ll} 4y = -3y + 7 & \text{Adding } -3x \text{ to both sides} \\ y = \dfrac{1}{4}(-3x + 7) & \text{Multiplying both sides by } \dfrac{1}{4} \\ y = -\dfrac{3}{4}x + \dfrac{7}{4} \end{array}$$

**63.** *Writing Exercise*

**65.** a) There are many correct answers.  One can be found by expressing the sum and difference of the two numbers:

$$x + y = 6,$$

$$x - y = 4$$

b) There are many correct answers.  For example, write an equation in two variables.  Then write a second equation by multiplying the left side of the first equation by one nonzero constant and multiplying the right side by another nonzero constant.

$$x + y = 1,$$

$$2x + 2y = 3$$

c) There are many correct answers.  One can be found by writing an equation in two variables and then writing a nonzero constant multiple of that equation:

$$x + y = 1,$$

$$2x + 2y = 2$$

**67.** Substitute 4 for $x$ and $-5$ for $y$ in the first equation:

$$A(4) - 6(-5) = 13$$
$$4A + 30 = 13$$
$$4A = -17$$
$$A = -\frac{17}{4}$$

Substitute 4 for $x$ and $-5$ for $y$ in the second equation:

$$4 - B(-5) = -8$$
$$4 + 5B = -8$$
$$5B = -12$$
$$B = -\frac{12}{5}$$

We have $A = -\dfrac{17}{4}$, $B = -\dfrac{12}{5}$.

**69. *Familiarize*.** Let $x =$ the number of years Lou has taught and $y =$ the number of years Juanita has taught. Two years ago, Lou and Juanita had taught $x - 2$ and $y - 2$ years, respectively.

***Translate*.**

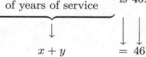

Together, the number of years of service $\quad$ is 46.

$$\underbrace{x + y}_{} \quad \underbrace{= 46}_{}$$

Two years ago
Lou had taught 2.5 times as many years as Juanita.

$$x - 2 = 2.5(y - 2)$$

We have a system of equations:

$$x + y = 46,$$
$$x - 2 = 2.5(y - 2)$$

**71. *Familiarize*.** Let $b =$ the number of ounces of baking soda and $v =$ the number of ounces of vinegar to be used. The amount of baking soda in the mixture will be four times the amount of vinegar.

***Translate*.**

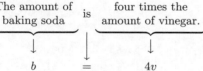

The amount of baking soda $\quad$ is $\quad$ four times the amount of vinegar.

$$\underbrace{b}_{} \quad \underbrace{=}_{} \quad \underbrace{4v}_{}$$

The total amount is 16 oz.

$$\underbrace{b + v}_{} \quad \underbrace{= \quad 16}_{}$$

We have a system of equations.

$$b = 4v,$$
$$b + v = 16$$

**73.** Graph both equations.

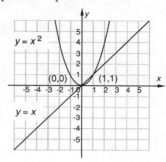

The solutions are apparently $(0,0)$ and $(1,1)$. Both pairs check.

**75.** $(0.07, -7.95)$

**77.** $(0.02, 1.25)$

## Exercise Set 3.2

**1.** Adding the equations, we get $8y = -1$, so choice (d) is correct.

**3.** Multiplying the second equation by 2 gives us the system of equations in (a), so choice (a) is correct.

**5.** Substituting $4x - 7$ for $y$ in the second equation gives us $6x + 3(4x - 7) = 19$, so choice (c) is correct.

**7.** $$y = 5 - 4x, \quad (1)$$
$$2x - 3y = 13 \quad (2)$$

We substitute $5 - 4x$ for $y$ in the second equation and solve for $x$.

$$2x - 3y = 13 \quad (2)$$
$$2x - 3(5 - 4x) = 13 \quad \text{Substituting}$$
$$2x - 15 + 12x = 13$$
$$14x - 15 = 13$$
$$14x = 28$$
$$x = 2$$

Next we substitute 2 for $x$ in either equation of the original system and solve for $y$.

$$y = 5 - 4x \quad (1)$$
$$y = 5 - 4 \cdot 2 \quad \text{Substituting}$$
$$y = 5 - 8$$
$$y = -3$$

We check the ordered pair $(2, -3)$.

$$\begin{array}{c|c} y = 5 - 4x \\ \hline -3 & 5 - 4 \cdot 2 \\ & 5 - 8 \\ & \phantom{5}? \\ -3 = -3 & \text{TRUE} \end{array}$$

$$2x - 3y = 13$$

$$\begin{array}{c|c} 2 \cdot 2 - 3(-3) & 13 \\ 4 + 9 & \\ & ? \\ 13 = 13 & \text{TRUE} \end{array}$$

Since $(2, -3)$ checks, it is the solution.

**9.**  $x = 8 - 4y,$    (1)

$3x + 5y = 3$    (2)

We substitute $8 - 4y$ for $x$ in the second equation and solve for $y$.

$$\begin{array}{ll} 3x + 5y = 3 & (2) \\ 3(8 - 4y) + 5y = 3 & \text{Substituting} \\ 24 - 12y + 5y = 3 & \\ 24 - 7y = 3 & \\ -7y = -21 & \\ y = 3 & \end{array}$$

Next we substitute 3 for $y$ in either equation of the original system and solve for $x$.

$$\begin{array}{ll} x = 8 - 4y & (1) \\ x = 8 - 4 \cdot 3 = 8 - 12 = -4 & \end{array}$$

We check the ordered pair $(-4, 3)$.

$$x = 8 - 4y$$

$$\begin{array}{c|c} -4 & 8 - 4 \cdot 3 \\ & 8 - 12 \\ & ? \\ -4 = -4 & \text{TRUE} \end{array}$$

$$3x + 5y = 3$$

$$\begin{array}{c|c} 3(-4) + 5 \cdot 3 & 3 \\ -12 + 15 & \\ & ? \\ 3 = 3 & \text{TRUE} \end{array}$$

Since $(-4, 3)$ checks, it is the solution.

**11.**  $3s - 4t = 14,$    (1)

$5s + t = 8$      (2)

We solve the second equation for $t$.

$$\begin{array}{ll} 5s + t = 8 & (2) \\ t = 8 - 5s & (3) \end{array}$$

We substitute $8 - 5s$ for $t$ in the first equation and solve for $s$.

$$\begin{array}{ll} 3s - 4t = 14 & (1) \\ 3s - 4(8 - 5s) = 14 & \text{Substituting} \\ 3s - 32 + 20s = 14 & \\ 23s - 32 = 14 & \\ 23s = 46 & \\ s = 2 & \end{array}$$

Next we substitute 2 for $s$ in Equation (1), (2), or (3). It is easiest to use Equation (3) since it is already solved for $t$.

$$t = 8 - 5 \cdot 2 = 8 - 10 = -2$$

We check the ordered pair $(2, -2)$.

$$3s - 4t = 14$$

$$\begin{array}{c|c} 3 \cdot 2 - 4(-2) & 14 \\ 6 + 8 & \\ & ? \\ 14 = 14 & \text{TRUE} \end{array}$$

$$5s + t = 8$$

$$\begin{array}{c|c} 5 \cdot 2 + (-2) & 8 \\ 10 - 2 & \\ & ? \\ 8 = 8 & \text{TRUE} \end{array}$$

Since $(2, -2)$ checks, it is the solution.

**13.**  $4x - 2y = 6,$    (1)

$2x - 3 = y$    (2)

We substitute $2x - 3$ for $y$ in the first equation and solve for $x$.

$$\begin{array}{ll} 4x - 2y = 6 & (1) \\ 4x - 2(2x - 3) = 6 & \\ 4x - 4x + 6 = 6 & \\ 6 = 6 & \end{array}$$

We have an identity, or an equation that is always true. The equations are dependent and the solution set is infinite: $\{(x, y) | 2x - 3 = y\}$.

**15.**  $-5s + t = 11,$    (1)

$4s + 12t = 4$    (2)

We solve the first equation for $t$.

$$\begin{array}{ll} -5s + t = 11 & (1) \\ t = 5s + 11 & (3) \end{array}$$

We substitute $5s + 11$ for $t$ in the second equation and solve for $s$.

$$\begin{array}{ll} 4s + 12t = 4 & (2) \\ 4s + 12(5s + 11) = 4 & \\ 4s + 60s + 132 = 4 & \\ 64s + 132 = 4 & \\ 64s = -128 & \\ s = -2 & \end{array}$$

Next we substitute $-2$ for $s$ in Equation (3).

$$t = 5s + 11 = 5(-2) + 11 = -10 + 11 = 1$$

We check the ordered pair $(-2, 1)$.

$$-5s + t = 11$$

$$\begin{array}{c|c} -5(-2) + 1 & 11 \\ 10 + 1 & \\ & ? \\ 11 = 11 & \text{TRUE} \end{array}$$

$$4s + 12t = 4$$

$$\begin{array}{c|c} 4(-2) + 12 \cdot 1 & 4 \\ -8 + 12 & \\ & ? \\ 4 = 4 & \text{TRUE} \end{array}$$

Since $(-2, 1)$ checks, it is the solution.

**17.** $2x + 2y = 2$, (1)

$3x - y = 1$ (2)

We solve the second equation for $y$.

$$3x - y = 1 \qquad (2)$$
$$-y = -3x + 1$$
$$y = 3x - 1 \qquad (3)$$

We substitute $3x - 1$ for $y$ in the first equation and solve for $x$.

$$2x + 2y = 2 \quad (1)$$
$$2x + 2(3x - 1) = 2$$
$$2x + 6x - 2 = 2$$
$$8x - 2 = 2$$
$$8x = 4$$
$$x = \frac{1}{2}$$

Next we substitute $\frac{1}{2}$ for $x$ in Equation (3).

$$y = 3x - 1 = 3 \cdot \frac{1}{2} - 1 = \frac{3}{2} - 1 = \frac{1}{2}$$

The ordered pair $\left(\frac{1}{2}, \frac{1}{2}\right)$ checks in both equations. It is the solution.

**19.** $x - 4y = 3$, (1)

$5x + 3y = 4$ (2)

We solve the first equation for $x$.

$$x - 4y = 3 \qquad (1)$$
$$x = 4y + 3 \quad (3)$$

Substitute $4y + 3$ for $x$ in the second equation and solve for $y$.

$$5x + 3y = 4 \qquad (2)$$
$$5(4y + 3) + 3y = 4$$
$$20y + 15 + 3y = 4$$
$$23y + 15 = 4$$
$$23y = -11$$
$$y = -\frac{11}{23}$$

Substitute $-\frac{11}{23}$ for $y$ in Equation (3).

$$x = 4\left(-\frac{11}{23}\right) + 3 = -\frac{44}{23} + \frac{69}{25} = \frac{25}{23}$$

The ordered pair is $\left(\frac{25}{23}, -\frac{11}{23}\right)$ checks in both equations. It is the solution.

**21.** $2x - 3 = y$ (1)

$y - 2x = 1$, (2)

We substitute $2x - 3$ for $y$ in the second equation and solve for $x$.

$$y - 2x = 1 \quad (2)$$
$$2x - 3 - 2x = 1 \quad \text{Substituting}$$
$$-3 = 1 \quad \text{Collecting like terms}$$

We have a contradiction, or an equation that is always false. Therefore, there is no solution.

**23.** $x + 3y = 7$ (1)

$\underline{-x + 4y = 7}$ (2)

$0 + 7y = 14$ Adding

$7y = 14$

$y = 2$

Substitute 2 for $y$ in one of the original equations and solve for $x$.

$$x + 3y = 7 \quad (1)$$
$$x + 3 \cdot 2 = 7 \quad \text{Substituting}$$
$$x + 6 = 7$$
$$x = 1$$

Check:

| $x + 3y = 7$ | | $-x + 4y = 7$ | |
|---|---|---|---|
| $1 + 3 \cdot 2$ | 7 | $-1 + 4 \cdot 2$ | 7 |
| $1 + 6$ | | $-1 + 8$ | |
| ? | | ? | |
| $7 = 7$ | TRUE | $7 = 7$ | TRUE |

Since $(1, 2)$ checks, it is the solution.

**25.** $x + y = 9$ (1)

$\underline{2x - y = -3}$ (2)

$3x = 6$ Adding

$x = 2$

Substitute 2 for $x$ in Equation (1) and solve for $y$.

$$x + y = 9 \quad (1)$$
$$2 + y = 9 \quad \text{Substituting}$$
$$y = 7$$

We obtain $(2, 7)$. This checks, so it is the solution.

**27.** $9x + 3y = -3$ (1)

$\underline{2x - 3y = -8}$ (2)

$11x + 0 = -11$ Adding

$11x = -11$

$x = -1$

Substitute $-1$ for $x$ in Equation (1) and solve for $y$.

$$9x + 3y = -3$$
$$9(-1) + 3y = -3 \quad \text{Substituting}$$
$$-9 + 3y = -3$$
$$3y = 6$$
$$y = 2$$

We obtain $(-1, 2)$. This checks, so it is the solution.

**29.** $5x + 3y = 19$,   (1)

$2x - 5y = 11$   (2)

We multiply twice to make two terms become opposites.

From (1):   $25x + 15y = \phantom{0}95$   Multiplying by 5

From (2):   $\phantom{0}6x - 15y = \phantom{0}33$   Multiplying by 3

$\phantom{00}31x + \phantom{00}0 = 128$   Adding

$$x = \frac{128}{31}$$

Substitute $\frac{128}{31}$ for $x$ in Equation (1) and solve for $y$.

$5x + 3y = 19$

$5 \cdot \dfrac{128}{31} + 3y = 19$      Substituting

$\dfrac{640}{31} + 3y = \dfrac{589}{31}$

$3y = -\dfrac{51}{31}$

$\dfrac{1}{3} \cdot 3y = \dfrac{1}{3} \cdot \left(-\dfrac{51}{31}\right)$

$y = -\dfrac{17}{31}$

We obtain $\left(\dfrac{128}{31}, -\dfrac{17}{31}\right)$. This checks, so it is the solution.

**31.** $5r - 3s = 24$,   (1)

$3r + 5s = 28$   (2)

We multiply twice to make two terms become additive inverses.

From (1):   $25r - 15s = 120$   Multiplying by 5

From (2):   $\phantom{0}9r + 15s = \phantom{0}84$   Multiplying by 3

$\phantom{00}34r + \phantom{000}0 = 204$   Adding

$r = 6$

Substitute 6 for $r$ in Equation (2) and solve for $s$.

$3r + 5s = 28$

$3 \cdot 6 + 5s = 28$   Substituting

$18 + 5s = 28$

$5s = 10$

$s = 2$

We obtain $(6, 2)$. This checks, so it is the solution.

**33.** $6s + 9t = 12$,   (1)

$4s + 6t = 5$   (2)

We multiply twice to make two terms become opposites.

From (1):   $12s + 18t = 24$   Multiplying by 2

From (2):   $-12s - 18t = -15$   Multiplying by $-3$

$0 = 9$

We get a contradiction, or an equation that is always false. The system has no solution.

**35.** $\dfrac{1}{2}x - \dfrac{1}{6}y = 3$   (1)

$\dfrac{2}{5}x + \dfrac{1}{2}y = 2$,   (2)

We first multiply each equation by the LCM of the denominators to clear fractions.

$3x - \phantom{0}y = 18$   (3)   Multiplying (1) by 6

$4x + 5y = 20$   (4)   Multiplying (2) by 10

We multiply by 5 on both sides of Equation (3) and then add.

$15x - 5y = \phantom{0}90$   Multiplying (3) by 5

$\phantom{0}4x + 5y = \phantom{0}20$   (4)

$\phantom{0}19x + \phantom{0}0 = 110$   Adding

$$x = \frac{110}{19}$$

Substitute $\frac{110}{19}$ for $x$ in one of the equations in which the fractions were cleared and solve for $y$.

$3x - y = 18$   (3)

$3\left(\dfrac{110}{19}\right) - y = 18$   Substituting

$\dfrac{330}{19} - y = \dfrac{342}{19}$

$-y = \dfrac{12}{19}$

$y = -\dfrac{12}{19}$

We obtain $\left(\dfrac{110}{19}, -\dfrac{12}{19}\right)$. This checks, so it is the solution.

**37.** $\dfrac{x}{2} + \dfrac{y}{3} = \dfrac{7}{6}$,   (1)

$\dfrac{2x}{3} + \dfrac{3y}{4} = \dfrac{5}{4}$   (2)

We first multiply each equation by the LCM of the denominators to clear fractions.

$3x + 2y = 7$   (3)   Multiplying (1) by 6

$8x + 9y = 15$   (4)   Multiplying (2) by 12

We multiply twice to make two terms become opposites.

From (3):   $27x + 18y = 63$   Multiplying by 9

From (4):   $-16x - 18y = -30$   Multiplying by $-2$

$11x \phantom{000000} = 33$   Adding

$x = 3$

Substitute 3 for $x$ in one of the equations in which the fractions were cleared and solve for $y$.

$3x + 2y = 7$   (3)

$3 \cdot 3 + 2y = 7$   Substituting

$9 + 2y = 7$

$2y = -2$

$y = -1$

We obtain $(3, -1)$. This checks, so it is the solution.

**39.** $12x - 6y = -15$,   (1)

$-4x + 2y = 5$   (2)

Observe that, if we multiply Equation (1) by $-\dfrac{1}{3}$, we obtain Equation (2). Thus, any pair that is a solution of Equation (1) is also a solution of Equation (2). The

equations are dependent and the solution set is infinite:
$\{(x,y)| -4x + 2y = 5\}$.

**41.**    $0.2a + 0.3b = 1,$

      $0.3a - 0.2b = 4,$

We first multiply each equation by 10 to clear decimals.

$2a + 3b = 10$    (1)

$3a - 2b = 40$    (2)

We multiply so that the $b$-terms can be eliminated.

From (1):    $4a + 6b = 20$      Multiplying by 2

From (2):    $\underline{9a - 6b = 120}$      Multiplying by 3

           $13a + \;\; 0 = 140$      Adding

$$a = \frac{140}{13}$$

Substitute $\frac{140}{13}$ for $a$ in Equation (1) and solve for $b$.

$$2a + 3b = 10$$

$$2 \cdot \frac{140}{13} + 3b = 10 \qquad \text{Substituting}$$

$$\frac{280}{13} + 3b = \frac{130}{13}$$

$$3b = -\frac{150}{13}$$

$$b = -\frac{50}{13}$$

We obtain $\left(\frac{140}{13}, -\frac{50}{13}\right)$. This checks, so it is the solution.

**43.**    $a - 2b = 16,$    (1)

      $b + 3 = 3a$    (2)

We will use the substitution method. First solve Equation (1) for $a$.

$$a - 2b = 16$$

$$a = 2b + 16 \quad (3)$$

Now substitute $2b + 16$ for $a$ in Equation (2) and solve for $b$.

$b + 3 = 3a$           (2)

$b + 3 = 3(2b + 16)$    Substituting

$b + 3 = 6b + 48$

$-45 = 5b$

$-9 = b$

Substitute $-9$ for $b$ in Equation (3).

$$a = 2(-9) + 16 = -2$$

We obtain $(-2, -9)$. This checks, so it is the solution.

**45.**    $10x + y = 306,$    (1)

      $10y + x = 90$      (2)

We will use the substitution method. First solve Equation (1) for $y$.

$$10x + y = 306$$

$$y = -10x + 306 \quad (3)$$

Now substitute $-10x + 306$ for $y$ in Equation (2) and solve for $y$.

$10y + x = 90$      (2)

$10(-10x + 306) + x = 90$      Substituting

$-100x + 3060 + x = 90$

$-99x + 3060 = 90$

$-99x = -2970$

$x = 30$

Substitute 30 for $x$ in Equation (3).

$$y = -10 \cdot 30 + 306 = 6$$

We obtain $(30, 6)$. This checks, so it is the solution.

**47.**    $3y = x - 2,$    (1)

      $x = 2 + 3y$    (2)

We will use the substitution method. Substitute $2 + 3y$ for $x$ in the first equation and solve for $y$.

$3y = x - 2$            (1)

$3y = 2 + 3y - 2$    Substituting

$3y = 3y$            Collecting like terms

We get an identity. The system is dependent and the solution set is infinite: $\{(x,y)|x = 2 + 3y\}$.

**49.**    $3s - 7t = 5,$

      $7t - 3s = 8,$

First we rewrite the second equation with the variables in a different order. Then we use the elimination method.

     $3s - 7t = 5,$    (1)

     $\underline{-3s + 7t = 8}$    (2)

            $0 = 13$

We get a contradiction, so the system has no solution.

**51.**    $0.05x + 0.25y = 22,$    (1)

      $0.15x + 0.05y = 24$    (2)

We first multiply each equation by 100 to clear decimals.

$5x + 25y = 2200$

$15x + 5y = 2400$

We multiply by $-5$ on both sides of the second equation and add.

     $5x + 25y = \;\;\;\;\;\; 2200$

   $\underline{-75x - 25y = -12,000}$    Multiplying (2) by $-5$

   $-70x \;\;\;\;\;\;\;\;\;\; = -9800$    Adding

$$x = \frac{-9800}{-70}$$

$$x = 140$$

Substitute 140 for $x$ in one of the equations in which the decimals were cleared and solve for $y$.

$$5x + 25y = 2200 \quad (1)$$
$$5 \cdot 140 + 25y = 2200 \quad \text{Substituting}$$
$$700 + 25y = 2200$$
$$25y = 1500$$
$$y = 60$$

We obtain $(140, 60)$. This checks, so it is the solution.

**53.**  $13a - 7b = 9, \quad (1)$

$\qquad 2a - 8b = 6 \quad (2)$

We will use the elimination method. First we multiply so that the $b$-terms can be eliminated.

From (1):  $\quad 104a - 56b = 72 \quad$ Multiplying by 8

From (2):  $\quad \underline{-14a + 56b = -42} \quad$ Multiplying by $-7$

$\qquad\qquad\quad 90a \qquad\quad = 30 \quad$ Adding

$$a = \frac{1}{3}$$

Substitute $\frac{1}{3}$ for $a$ in one of the equations and solve for $b$.

$$2a - 8b = 6 \qquad (2)$$
$$2 \cdot \frac{1}{3} - 8b = 6$$
$$\frac{2}{3} - 8b = 6$$
$$-8b = \frac{16}{3}$$
$$b = -\frac{2}{3}$$

We obtain $\left(\frac{1}{3}, -\frac{2}{3}\right)$. This checks, so it is the solution.

**55.**  *Writing Exercise*

**57.**  *Familiarize*. Let $m$ = the number of $\frac{1}{4}$-mi units traveled after the first $\frac{1}{2}$ mi. The total distance traveled will be $\frac{1}{2}$ mi $+ m \cdot \frac{1}{4}$ mi.

*Translate*.

| Fare for first $\frac{1}{2}$ mi | plus | fare for additional $\frac{1}{4}$-mi units | is | \$5.20. |
|:---:|:---:|:---:|:---:|:---:|
| ↓ | ↓ | ↓ | ↓ | ↓ |
| 1 | + | 0.3m | = | 5.20 |

*Carry out*. We solve the equation.

$$1 + 0.3m = 5.20$$
$$0.3m = 4.20$$
$$m = 14$$

If the taxi travels the first $\frac{1}{2}$ mi plus 14 additional $\frac{1}{4}$-mi units, then it travels a total of $\frac{1}{2} + 14 \cdot \frac{1}{4}$, or $\frac{1}{2} + \frac{7}{2}$, or 4 mi.

*Check*.  We have 4 mi $= \frac{1}{2}$ mi $+ \frac{7}{2}$ mi $= \frac{1}{2}$ mi $+ 14 \cdot \frac{1}{4}$ mi. The fare for traveling this distance is $\$1.00 + \$0.30(14) = \$1.00 + \$4.20 = \$5.20$. The answer checks.

*State*. It is 4 mi from Johnson Street to Elm Street.

**59.**  *Familiarize*. Let $a$ = the amount spent to remodel bathrooms, in billions of dollars. Then $2a$ = the amount spent to remodel kitchens. The sum of these two amounts is \$35 billion.

*Translate*.

| Amount spent on bathrooms | plus | amount spent on kitchens | is | \$35 billion. |
|:---:|:---:|:---:|:---:|:---:|
| ↓ | ↓ | ↓ | ↓ | ↓ |
| a | + | 2a | = | 35 |

*Carry out*. We solve the equation.

$$a + 2a = 35$$
$$3a = 35 \qquad \text{Combining like terms}$$
$$a = \frac{35}{3}, \text{ or } 11\frac{2}{3}$$

If $a = \frac{35}{3}$, then $2a = 2 \cdot \frac{35}{3} = \frac{70}{3} = 23\frac{1}{3}$.

*Check*.  $\frac{70}{3}$ is twice $\frac{35}{3}$, and $\frac{35}{3} + \frac{70}{3} = \frac{105}{3} = 35$. The answer checks.

*State*. $\$11\frac{2}{3}$ billion was spent to remodel bathrooms, and $\$23\frac{1}{3}$ billion was spent to remodel kitchens.

**61.**  *Familiarize*. The total cost is the daily charge plus the mileage charge. The mileage charge is the cost per mile times the number of miles driven. Let $m$ = the number of miles that can be driven for \$80.

*Translate*. We reword the problem.

| Daily rate | plus | Cost per mile | times | Number of miles driven | is | Amount. |
|:---:|:---:|:---:|:---:|:---:|:---:|:---:|
| ↓ | ↓ | ↓ | ↓ | ↓ | ↓ | ↓ |
| 34.95 | + | 0.10 | · | m | = | 80 |

*Carry out*. We solve the equation.

$$34.95 + 0.10m = 80$$
$$100(34.95 + 0.10m) = 100(80) \quad \text{Clearing decimals}$$
$$3495 + 10m = 8000$$
$$10m = 4505$$
$$m = 450.5$$

*Check*.  The mileage cost is found by multiplying 450.5 by \$0.10 obtaining \$45.05. Then we add \$45.05 to \$34.95, the daily rate, and get \$80.

*State*. The businessperson can drive 450.5 mi on the car-rental allotment.

**63.**  *Writing Exercise*

**65.**  First write $f(x) = mx + b$ as $y = mx + b$. Then substitute 1 for $x$ and 2 for $y$ to get one equation and also substitute $-3$ for $x$ and 4 for $y$ to get a second equation:

$$2 = m \cdot 1 + b$$
$$4 = m(-3) + b$$

Solve the resulting system of equations.

$$2 = m + b$$
$$4 = -3m + b$$

Multiply the second equation by $-1$ and add.

$$\begin{array}{l} 2 = \phantom{-}m + b \\ \underline{-4 = 3m - b} \\ -2 = 4m \\ -\dfrac{1}{2} = \phantom{-}m \end{array}$$

Substitute $-\dfrac{1}{2}$ for $m$ in the first equation and solve for $b$.

$$2 = -\frac{1}{2} + b$$
$$\frac{5}{2} = b$$

Thus, $m = -\dfrac{1}{2}$ and $b = \dfrac{5}{2}$.

**67.** Substitute $-4$ for $x$ and $-3$ for $y$ in both equations and solve for $a$ and $b$.

$$-4a - 3b = -26, \quad (1)$$
$$-4b + 3a = 7 \qquad (2)$$

$$\begin{array}{ll} -12a - \phantom{1}9b = -78 & \text{Multiplying (1) by 3} \\ \underline{12a - 16b = \phantom{-}28} & \text{Multiplying (2) by 4} \\ \phantom{12a}-25b = -50 \\ \phantom{12a-25}b = \phantom{-}2 \end{array}$$

Substitute 2 for $b$ in Equation (2).

$$-4 \cdot 2 + 3a = 7$$
$$3a = 15$$
$$a = 5$$

Thus, $a = 5$ and $b = 2$.

**69.** $\dfrac{x+y}{2} - \dfrac{x-y}{5} = 1,$

$\dfrac{x-y}{2} + \dfrac{x+y}{6} = -2$

After clearing fractions we have:

$$3x + 7y = 10, \quad (1)$$
$$4x - 2y = -12 \quad (2)$$

$$\begin{array}{ll} 6x + 14y = \phantom{-}20 & \text{Multiplying (1) by 2} \\ \underline{28x - 14y = -84} & \text{Multiplying (2) by 7} \\ 34x \phantom{- 14y} = -64 \\ \phantom{34}x = -\dfrac{32}{17} \end{array}$$

Substitute $-\dfrac{32}{17}$ for $x$ in Equation (1).

$$3\left(-\frac{32}{17}\right) + 7y = 10$$
$$7y = \frac{266}{17}$$
$$y = \frac{38}{17}$$

The solution is $\left(-\dfrac{32}{17}, \dfrac{38}{17}\right)$.

**71.** $\dfrac{2}{x} + \dfrac{1}{y} = 0, \qquad 2 \cdot \dfrac{1}{x} + \dfrac{1}{y} = 0,$

$$\text{or}$$

$\dfrac{5}{x} + \dfrac{2}{y} = -5 \qquad 5 \cdot \dfrac{1}{x} + 2 \cdot \dfrac{1}{y} = -5$

Substitute $u$ for $\dfrac{1}{x}$ and $v$ for $\dfrac{1}{y}$.

$$2u + \phantom{2}v = 0, \quad (1)$$
$$5u + 2v = -5 \quad (2)$$

$$\begin{array}{ll} -4u - 2v = \phantom{-}0 & \text{Multiplying (1) by } -2 \\ \underline{5u + 2v = -5} & (2) \\ \phantom{-4}u \phantom{-2v} = -5 \end{array}$$

Substitute $-5$ for $u$ in Equation (1).

$$2(-5) + v = 0$$
$$-10 + v = 0$$
$$v = 10$$

If $u = -5$, then $\dfrac{1}{x} = -5$. Thus $x = -\dfrac{1}{5}$.

If $v = 10$, then $\dfrac{1}{y} = 10$. Thus $y = \dfrac{1}{10}$.

The solution is $\left(-\dfrac{1}{5}, \dfrac{1}{10}\right)$.

**73.** *Writing Exercise*

## Exercise Set 3.3

**1.** The Familiarize and Translate steps were done in Exercise 41 of Exercise Set 3.1.

   *Carry out*. We solve the system of equations

$$x + y = 50, \quad (1)$$
$$x = 0.25y, \quad (2)$$

   where $x$ is the first number and $y$ is the second number. We use substitution.

   Substitute $0.25y$ for $x$ in (1) and solve for $y$.

$$0.25y + y = 50$$
$$1.25y = 50$$
$$y = 40$$

   Now substitute 40 for $y$ in (2).

$$x = 0.25(40) = 10$$

   *Check*. The sum of the numbers is $10 + 40$, or 50, and 0.25 times the second number, 40, is the first number, 10. The answer checks.

   *State*. The first number is 10, and the second number is 40.

**3.** The Familiarize and Translate steps were done in Exercise 43 of Exercise Set 3.1.

   *Carry out*. We solve the system of equations

$$m + v = 16, \quad (1)$$
$$m = 2v + 4, \quad (2)$$

where $m$ and $v$ represent the number of ounces of mineral oil and vinegar to be used, respectively. We use substitution.

Substitute $2v + 4$ for $m$ in (1) and solve for $v$.

$$(2v + 4) + v = 16$$
$$3v + 4 = 16$$
$$3v = 12$$
$$v = 4$$

Now substitute 4 for $v$ in (2).

$$m = 2 \cdot 4 + 4 = 8 + 4 = 12$$

**Check**. The mixture contains 12+4, or 16 oz. The amount of mineral oil, 12 oz, is 4 oz more than twice the amount of vinegar, 4 oz: $2 \cdot 4 + 4 = 12$. The answer checks.

**State**. 12 oz of mineral oil and 4 oz of vinegar should be used.

5. The Familiarize and Translate steps were done in Exercise 45 of Exercise Set 3.1

**Carry out**. We solve the system of equations

$$x + y = 180, \quad (1)$$
$$x = 2y - 3 \quad (2)$$

where $x$ = the measure of one angle and $y$ = the measure of the other angle. We use substitution.

Substitute $2y - 3$ for $x$ in (1) and solve for $y$.

$$2y - 3 + y = 180$$
$$3y - 3 = 180$$
$$3y = 183$$
$$y = 61$$

Now substitute 61 for $y$ in (2).

$$x = 2 \cdot 61 - 3 = 122 - 3 = 119$$

**Check**. The sum of the angle measures is $119° + 61°$, or $180°$, so the angles are supplementary. Also $2 \cdot 61° - 3° = 122° - 3° = 119°$. The answer checks.

**State**. The measures of the angles are $119°$ and $61°$.

7. The Familiarize and Translate steps were done in Exercise 47 of Exercise Set 3.1

**Carry out**. We solve the system of equations

$$g + t = 64, \quad (1)$$
$$2g + t = 100 \quad (2)$$

where $g$ = the number of field goals and $t$ = the number of foul shots Chamberlain made. We use elimination.

$$\begin{array}{rl} -g - t = -64 & \text{Multiplying (1) by } -1 \\ \underline{2g + t = 100} & \\ g \phantom{+ t} = 36 & \end{array}$$

Substitute 36 for $g$ in (1) and solve for $t$.

$$36 + t = 64$$
$$t = 28$$

**Check**. The total number of scores was 36+28, or 64. The total number of points was $2 \cdot 36 + 28 = 72 + 28 = 100$. The answer checks.

**State**. Chamberlain made 36 field goals and 28 foul shots.

9. The Familiarize and Translate steps were done in Exercise 49 of Exercise Set 3.1

**Carry out**. We solve the system of equations

$$x + y = 45, \quad (1)$$
$$850x + 975y = 39,875 \quad (2)$$

where $x$ = the number of less expensive brushes sold and $y$ = the number of more expensive brushes sold. We use elimination. Begin by multiplying Equation (1) by $-850$.

$$\begin{array}{rl} -850x - 850y = -38,250 & \text{Multiplying (1)} \\ \underline{850x + 975y = \phantom{-}39,875} & \\ 125y = \phantom{-}1625 & \\ y = \phantom{-}13 & \end{array}$$

Substitute 13 for $y$ in (1) and solve for $x$.

$$x + 13 = 45$$
$$x = 32$$

**Check**. The number of brushes sold is $32 + 13$, or 45. The amount taken in was $\$8.50(32) + \$9.75(13) = \$272 + \$126.75 = \$398.75$. The answer checks.

**State**. 32 of the less expensive brushes were sold, and 13 of the more expensive brushes were sold.

11. The Familiarize and Translate steps were done in Exercise 51 of Exercise Set 3.1

**Carry out**. We solve the system of equations

$$h + n = 50, \quad (1)$$
$$2706h + 3439n = 156,557 \quad (2)$$

where $h$ = the number of vials of Humulin Insulin sold and $n$ = the number of vials of Novolin Velosulin Insulin sold. We use elimination.

$$\begin{array}{rl} -2706h - 2706n = -135,300 & \text{Multiplying (1)} \\ & \text{by } -2706 \\ \underline{2706h + 3439n = \phantom{-}156,557} & \\ 733n = \phantom{-}21,257 & \\ n = \phantom{-}29 & \end{array}$$

Substitute 29 for $n$ in (1) and solve for $h$.

$$h + 29 = 50$$
$$h = 21$$

**Check**. A total of $21 + 29$, or 50 vials, were sold. The amount collected was $\$27.06(21) + \$34.39(29) = \$568.26 + \$997.31 = \$1565.57$. The answer checks.

**State**. 21 vials of Humulin Insulin and 29 vials of Novolin Velosulin Insulin were sold.

13. The Familiarize and Translate steps were done in Exercise 53 of Exercise Set 3.1

**Carry out**. We solve the system of equations

$$2l + 2w = 288, \quad (1)$$
$$l = w + 44 \quad (2)$$

where $l$ = the length, in feet, and $w$ = the width, in feet, of the basketball court. We use substitution.

Substitute $w + 44$ for $l$ in (1) and solve for $w$.

$$2(w + 44) + 2w = 288$$
$$2w + 88 + 2w = 288$$
$$4w + 88 = 288$$
$$4w = 200$$
$$w = 50$$

Now substitute 50 for $w$ in (2).

$$l = 50 + 44 = 94$$

**Check.** The perimeter is $2 \cdot 94$ ft $+ 2 \cdot 50$ ft $= 188$ ft $+ 100$ ft $= 288$ ft. The length, 94 ft, is 44 ft more than the width, 50 ft. The answer checks.

**State.** The length of the basketball court is 94 ft, and the width is 50 ft.

**15. Familiarize.** Let $x =$ the number of sheets of nonrecycled paper used and $y =$ the number of recycled sheets.

**Translate.** We organize the information in a table.

|  | Nonrecycled sheets | Recycled sheets | Total |
|---|---|---|---|
| Number used | $x$ | $y$ | 150 |
| Price | 1.9¢ | 2.4¢ | |
| Total cost | $1.9x$ | $2.4y$ | $3.41, or 341¢ |

We get one equation from the "Number used" row of the table:

$$x + y = 150$$

The "Total cost" row yields a second equation. We express all costs in cents:

$$1.9x + 2.4y = 341$$

After clearing decimals, we have the problem translated to a system of equations.

$$x + y = 150, \quad (1)$$
$$19x + 24y = 3410 \quad (2)$$

**Carry out.** We use the elimination method to solve the system of equations.

$$-19x - 19y = -2850 \quad \text{Multiplying (1) by } -19$$
$$\underline{19x + 24y = \phantom{0}3410}$$
$$5y = \phantom{00}560$$
$$y = \phantom{00}112$$

Substitute 112 for $y$ in (1) and solve for $x$.

$$x + 112 = 150$$
$$x = 38$$

**Check.** A total of $38 + 112$, or 150 sheets of paper were used. The total cost was 1.9¢(38) + 2.4¢(112) = 72.2¢ + 268.8¢ = 341¢, or $3.41. The answer checks.

**State.** 38 sheets of nonrecycled paper and 112 sheets of recycled paper were used.

**17. Familiarize.** Let $g =$ the number of General Electric bulbs purchased and $s =$ the number of SLi bulbs purchased.

**Translate.** We organize the information in a table.

|  | GE bulbs | SLi bulbs | Total |
|---|---|---|---|
| Number purchased | $g$ | $s$ | 200 |
| Price | $7.50 | $5 | |
| Total cost | $7.5g$ | $5s$ | 1150 |

We get our equation from the "Number purchased" row of the table:

$$g + s = 200$$

The "Total cost" row yields a second equation:

$$7.5g + 5s = 1150$$

After clearing decimals, we have the problem translated to a system of equations:

$$g + s = 200, \quad (1)$$
$$75g + 50s = 11,500 \quad (2)$$

**Carry out.** We use the elimination method to solve the system of equations.

$$-50g - 50s = -10,000 \quad \text{Multiplying (1) by } -50$$
$$\underline{75g + 50s = \phantom{0}11,500}$$
$$25g = \phantom{00}1500$$
$$g = \phantom{000}60$$

Substitute 60 for $g$ in (1) and solve for $s$.

$$60 + s = 200$$
$$s = 140$$

**Check.** A total of $60 + 140$, or 200 bulbs, was purchased. The total cost was $7.50(60) + $5(140) = $450 + $700 = $1150. The answer checks.

**State.** 60 General Electric bulbs and 140 SLi bulbs were purchased.

**19. Familiarize.** Let $a =$ the number of Apple cartridges purchased and $h =$ the number of HP cartridges.

**Translate.** We organize the information in a table.

|  | Apple | HP | Total |
|---|---|---|---|
| Number purchased | $a$ | $h$ | 50 |
| Price | $30.86 | $43.58 | |
| Total cost | $30.86a$ | $43.58h$ | 1733.80 |

We get one equation from the "Number purchased" row of the table:

$$a + h = 50$$

The "Total cost" row yields a second equation:

$$30.86a + 43.58h = 1733.80$$

After clearing decimals, we have the problem translated to a system of equations:

$$a + h = 50, \quad (1)$$
$$3086a + 4358h = 173,380 \quad (2)$$

**Carry out.** We use the elimination method to solve the system of equations.

$-3086a-3086h=-154,300$   Multiplying (1) by $-3086$

$\underline{3086a+4358h=\ \ 173,380}$

$\phantom{3086a+}1272h=\ \ \ 19,080$

$\phantom{3086a+1272}h=\ \ \ \ \ \ \ 15$

Substitute 15 for $h$ in (1) and solve for $a$.

$a + 15 = 50$

$a = 35$

**Check**. A total of $35+15$, or 50 cartridges, was sold. The total cost was $\$30.86(35) + \$43.58(15) =$ $\$1080.10 + \$653.70 = \$1733.80$. The answer checks.

**State**. 35 Apple cartridges and 15 HP cartridges were purchased.

**21. Familiarize**. Let $k =$ the number of pounds of Kenyan French Roast coffee and $s =$ the number of pounds of Sumatran coffee to be used in the mixture. The value of the mixture will be $\$8.40(20)$, or $\$168$.

**Translate**. We organize the information in a table.

|  | Kenyan | Sumatran | Mixture |
|---|---|---|---|
| Number of pounds | $k$ | $s$ | 20 |
| Price per pound | $\$9$ | $\$8$ | $\$8.40$ |
| Value of coffee | $9k$ | $8s$ | 168 |

The "Number of pounds" row of the table gives us one equation:

$k + s = 20$

The "Value of coffee" row yields a second equation:

$9k + 8s = 168$

We have translated to a system of equations:

$k + s = 20,\quad (1)$

$9k + 8s = 168\quad (2)$

**Carry out**. We use the elimination method to solve the system of equations.

$-8k - 8s = -160$   Multiplying (1) by $-8$

$\underline{9k + 8s =\ \ \ 168}$

$\phantom{-8}k\phantom{+8s} =\ \ \ \ \ \ 8$

Substitute 8 for $k$ in (1) and solve for $s$.

$8 + s = 20$

$s = 12$

**Check**. The total mixture contains $8 \text{ lb} + 12 \text{ lb}$, or 20 lb. Its value is $\$9 \cdot 8 + \$8 \cdot 12 = \$72 + \$96 = \$168$. The answer checks.

**State**. 8 lb of Kenyan French Roast coffee and 12 lb of Sumatran coffee should be used.

**23.** Observe that the average of 40% and 10% is 25%: $\frac{40\% + 10\%}{2} = \frac{50\%}{2} = 25\%$. Thus, the caterer should use equal parts of the 40% and 10% mixtures. Since a 20-lb mixture is desired, the caterer should use 10 lb each of the 40% and the 10% mixture.

**25. Familiarize**. Let $x =$ the number of pounds of Deep Thought Granola and $y =$ the number of pounds of Oat Dream Granola to be used in the mixture. The amount of nuts and dried fruit in the mixture is 19%(20 lb), or $0.19(20 \text{ lb}) = 3.8 \text{ lb}$.

**Translate**. We organize the information in a table.

|  | Deep Thought | Oat Dream | Mixture |
|---|---|---|---|
| Number of pounds | $x$ | $y$ | 20 |
| Percent of nuts and dried fruit | 25% | 10% | 19% |
| Amount of nuts and dried fruit | $0.25x$ | $0.1y$ | 3.8 lb |

We get one equation from the "Number of pounds" row of the table:

$x + y = 20$

The last row of the table yields a second equation:

$0.25x + 0.1y = 3.8$

After clearing decimals, we have the problem translated to a system of equations:

$x + y = 20,\quad (1)$

$25x + 10y = 380\quad (2)$

**Carry out**. We use the elimination method to solve the system of equations.

$-10x - 10y = -200$   Multiplying (1) by $-10$

$\underline{25x + 10y =\ \ \ 380}$

$15x\phantom{+ 10y} =\ \ \ \ 180$

$\phantom{15}x =\ \ \ \ \ 12$

Substitute 12 for $x$ in (1) and solve for $y$.

$12 + y = 20$

$y = 8$

**Check**. The amount of the mixture is $12 \text{ lb} + 8 \text{ lb}$, or 20 lb. The amount of nuts and dried fruit in the mixture is $0.25(12 \text{ lb}) + 0.1(8 \text{ lb}) = 3 \text{ lb} + 0.8 \text{ lb} = 3.8 \text{ lb}$. The answer checks.

**State**. 12 lb of Deep Thought Granola and 8 lb of Oat Dream Granola should be mixed.

**27. Familiarize**. Let $x =$ the amount of the 6% loan and $y =$ the amount of the 9% loan. Recall that the formula for simple interest is

$\text{Interest} = \text{Principal} \cdot \text{Rate} \cdot \text{Time}.$

*Translate*. We organize the information in a table.

| | 6% loan | 9% loan | Total |
|---|---|---|---|
| Principal | $x$ | $y$ | $12,000 |
| Interest Rate | 6% | 9% | |
| Time | 1 yr | 1 yr | |
| Interest | 0.06x | 0.09y | $855 |

The "Principal" row of the table gives us one equation:

$x + y = 12,000$

The last row of the table yields another equation:

$0.06x + 0.09y = 855$

After clearing decimals, we have the problem translated to a system of equations:

$$x + y = 12,000 \quad (1)$$
$$6x + 9y = 85,500 \quad (2)$$

*Carry out*. We use the elimination method to solve the system of equations.

$$-6x - 6y = -72,000 \quad \text{Multiplying (1) by } -6$$
$$\underline{6x + 9y = 85,500}$$
$$3y = 13,500$$
$$y = 4500$$

Substitute 4500 for $y$ in (1) and solve for $x$.

$$x + 4500 = 12,000$$
$$x = 7500$$

*Check*. The loans total $7500 + $4500, or $12,000. The total interest is $0.06(\$7500) + 0.09(\$4500) = \$450 + \$405 = \$855$. The answer checks.

*State*. The 6% loan was for $7500, and the 9% loan was for $4500.

**29.** *Familiarize*. Let $x =$ the number of liters of Arctic Antifreeze and $y =$ the number of liters of Frost-No-More in the mixture. The amount of alcohol in the mixture is $0.15(20 \text{ L}) = 3 \text{ L}$.

*Translate*. We organize the information in a table.

| | 18% solution | 10% solution | Mixture |
|---|---|---|---|
| Number of liters | $x$ | $y$ | 20 |
| Percent of alcohol | 18% | 10% | 15% |
| Amount of alcohol | 0.18x | 0.1y | 3 |

We get one equation from the "Number of liters" row of the table:

$x + y = 20$

The last row of the table yields a second equation:

$0.18x + 0.1y = 3$

After clearing decimals we have the problem translated to a system of equations:

$$x + y = 20, \quad (1)$$
$$18x + 10y = 300 \quad (2)$$

*Carry out*. We use the elimination method to solve the system of equations.

$$-10x - 10y = -200 \quad \text{Multiplying (1) by } -10$$
$$\underline{18x + 10y = 300}$$
$$8x = 100$$
$$x = 12.5$$

Substitute 12.5 for $x$ in (1) and solve for $y$.

$$12.5 + y = 20$$
$$y = 7.5$$

*Check*. The total amount of the mixture is 12.5 L + 7.5 L or 20 L. The amount of alcohol in the mixture is $0.18(12.5 \text{ L}) + 0.1(7.5 \text{ L}) = 2.25 \text{ L} + 0.75 \text{ L} = 3 \text{ L}$. The answer checks.

*State*. 12.5 L of Arctic Antifreeze and 7.5 L of Frost-No-More should be used.

**31.** *Familiarize*. Let $x =$ the number of gallons of 87-octane gas and $y =$ the number of gallons of 93-octane gas in the mixture. The amount of octane in the mixture can be expressed as $91(12)$, or 1092.

*Translate*. We organize the information in a table.

| | 87-octane | 93-octane | Mixture |
|---|---|---|---|
| Number of gallons | $x$ | $y$ | 12 |
| Octane rating | 87 | 93 | 91 |
| Total octane | 87x | 93y | 1092 |

We get one equation from the "Number of gallons" row of the table:

$x + y = 12$

The last row of the table yields a second equation:

$87x + 93y = 1092$

We have a system of equations:

$$x + y = 12, \quad (1)$$
$$87x + 93y = 1092 \quad (2)$$

*Carry out*. We use the elimination method to solve the system of equations.

$$-87x - 87y = -1044 \quad \text{Multiplying (1) by } -87$$
$$\underline{87x + 93y = 1092}$$
$$6y = 48$$
$$y = 8$$

Substitute 8 for $y$ in (1) and solve for $x$.

$$x + 8 = 12$$
$$x = 4$$

*Check*. The total amount of the mixture is 4 gal + 8 gal, or 12 gal. The amount of octane can be expressed as $87(4) + 93(8) = 348 + 744 = 1092$. The answer checks.

*State*. 4 gal of 87-octane gas and 8 gal of 93-octane gas should be used.

**33.** *Familiarize*. From the bar graph we see that whole milk is 4% milk fat, milk for cream cheese is 8% milk fat, and cream is 30% milk fat. Let $x =$ the number of pounds of whole milk and $y =$ the number of pounds of cream to be used. The mixture contains 8%(200 lb), or 0.08(200 lb) = 16 lb of milk fat.

*Translate*. We organize the information in a table.

|  | Whole milk | Cream | Mixture |
|---|---|---|---|
| Number of pounds | $x$ | $y$ | 200 |
| Percent of milk fat | 4% | 30% | 8% |
| Amount of milk fat | $0.04x$ | $0.3y$ | 16 lb |

We get one equation from the " Number of pounds" row of the table:

$$x + y = 200$$

The last row of the table yields a second equation:

$$0.04x + 0.3y = 16$$

After clearing decimals, we have the problem translated to a system of equations:

$$x + y = 200, \quad (1)$$
$$4x + 30y = 1600 \quad (2)$$

*Carry out*. We use the elimination method to solve the system of equations.

$$
\begin{array}{ll}
-4x - \ 4y = -800 & \text{Multiplying (1) by } -4 \\
\underline{4x + 30y = 1600} & \\
\qquad 26y = 800 &
\end{array}
$$

$$y = \frac{400}{13}, \text{ or } 30\frac{10}{13}$$

Substitute $\dfrac{400}{13}$ for $y$ in (1) and solve for $x$.

$$x + \frac{400}{13} = 200$$

$$x = \frac{2200}{13}, \text{ or } 169\frac{3}{13}$$

*Check*. The total amount of the mixture is $\dfrac{2200}{13}$ lb $+ \dfrac{400}{13}$ lb $= \dfrac{2600}{13}$ lb $= 200$ lb. The amount of milk fat in the mixture is $0.04\left(\dfrac{2200}{13}\text{ lb}\right) +$

$0.3\left(\dfrac{400}{13}\text{ lb}\right) = \dfrac{88}{13}$ lb $+ \dfrac{120}{13}$ lb $= \dfrac{208}{13}$ lb $= 16$ lb.
The answer checks.

*State*. $169\frac{3}{13}$ lb of whole milk and $30\frac{10}{13}$ lb of cream should be mixed.

**35.** *Familiarize*. We first make a drawing.

| Slow train | | | |
|---|---|---|---|
| $d$ kilometers | 75 km/h | $(t + 2)$ hr | |

| Fast train | | | |
|---|---|---|---|
| $d$ kilometers | 125 km/h | $t$ hr | |

From the drawing we see that the distances are the same. Now complete the chart.

$$d = r \cdot t$$

|  | Distance | Rate | Time |  |
|---|---|---|---|---|
| Slow train | $d$ | 75 | $t + 2$ | $\rightarrow d = 75(t+2)$ |
| Fast train | d | 125 | $t$ | $\rightarrow d = 125t$ |

*Translate*. Using $d = rt$ in each row of the table, we get a system of equations:

$$d = 75(t + 2),$$
$$d = 125t$$

*Carry out*. We solve the system of equations.

$$
\begin{array}{ll}
125t = 75(t + 2) & \text{Using substitution} \\
125t = 75t + 150 & \\
50t = 150 & \\
t = 3 &
\end{array}
$$

Then $d = 125t = 125 \cdot 3 = 375$

*Check*. At 125 km/h, in 3 hr the fast train will travel $125 \cdot 3 = 375$ km. At 75 km/h, in 3 + 2, or 5 hr the slow train will travel $75 \cdot 5 = 375$ km. The numbers check.

*State*. The trains will meet 375 km from the station.

**37.** *Familiarize*. We first make a drawing. Let $d =$ the distance and $r =$ the speed of the boat in still water. Then when the boat travels downstream its speed is $r + 6$, and its speed upstream is $r - 6$. From the drawing we see that the distances are the same.

Downstream, 6 mph current

$d$ mi, $r + 6$, 3 hr

Upstream, 6 mph current

$d$ mi, $r - 6$, 5 hr

Organize the information in a table.

|  | Distance | Rate | Time |
|---|---|---|---|
| Down-stream | $d$ | $r + 6$ | 3 |
| Up-stream | $d$ | $r - 6$ | 5 |

*Translate*. Using $d = rt$ in each row of the table, we get a system of equations:

$$d = 3(r + 6), \qquad d = 3r + 18,$$
$$\text{or}$$
$$d = 5(r - 6) \qquad d = 5r - 30$$

*Carry out*. Solve the system of equations.

$3r + 18 = 5r - 30$    Using substitution

$18 = 2r - 30$

$48 = 2r$

$24 = r$

**Check**. When $r = 24$, then $r + 6 = 24 + 6 = 30$, and the distance traveled in 3 hr is $3 \cdot 30 = 90$ mi. Also, $r - 6 = 24 - 6 = 18$, and the distance traveled in 5 hr is $18 \cdot 5 = 90$ mi. The answer checks.

**State**. The speed of the boat in still water is 24 mph.

**39. Familiarize**. We make a drawing. Note that the plane's speed traveling toward London is $360 + 50$, or 410 mph, and the speed traveling toward New York City is $360 - 50$, or 310 mph. Also, when the plane is $d$ mi from New York City, it is $3458 - d$ mi from London.

New York City          London

310 mph    $t$ hours      $t$ hours    410 mph

|——————— 3458 mi ———————|

|——— $d$ ———|——— 3458 mi $-d$ ———|

Organize the information in a table.

|  | Distance | Rate | Time |
|---|---|---|---|
| Toward NYC | $d$ | 310 | $t$ |
| Toward London | $3458 - d$ | 410 | $t$ |

**Translate**. Using $d = rt$ in each row of the table, we get a system of equations:

$d = 310t$,    (1)

$3458 - d = 410t$    (2)

**Carry out**. We solve the system of equations.

$3458 - 310t = 410t$    Using substitution

$3458 = 720t$

$4.8028 \approx t$

Substitute 4.8028 for $t$ in (1).

$d \approx 310(4.8028) \approx 1489$

**Check**. If the plane is 1489 mi from New York City, it can return to New York City, flying at 310 mph, in $1489/310 \approx 4.8$ hr. If the plane is $3458 - 1489$, or 1969 mi from London, it can fly to London, traveling at 410 mph, in $1969/410 \approx 4.8$ hr. Since the times are the same, the answer checks.

**State**. The point of no return is about 1489 mi from New York City.

**41. Familiarize**. Let $l$ = the length, in feet, and $w$ = the width, in feet. Recall that the formula for the perimeter $P$ of a rectangle with length $l$ and width $w$ is $P = 2l + 2w$.

**Translate**.

The perimeter   is   860 ft.

$\downarrow$       $\downarrow$    $\downarrow$

$2l + 2w$    $=$    $860$

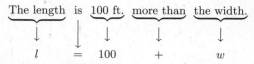

The length   is   100 ft.   more than   the width.

$\downarrow$    $\downarrow$    $\downarrow$      $\downarrow$      $\downarrow$

$l$    $=$    $100$    $+$    $w$

We have translated to a system of equations:

$2l + 2w = 860$,    (1)

$l = 100 + w$    (2)

**Carry out**. We use the substitution method to solve the system of equations.

Substitute $100 + w$ for $l$ in (1) and solve for $w$.

$2(100 + w) + 2w = 860$

$200 + 2w + 2w = 860$

$200 + 4w = 860$

$4w = 660$

$w = 165$

Now substitute 165 for $w$ in (2).

$l = 100 + 165 = 265$

**Check**. The perimeter is $2 \cdot 265$ ft $+ 2 \cdot 165$ ft $= 530$ ft $+ 330$ ft $= 860$ ft. The length, 265 ft, is 100 ft more than the width, 165 ft. The answer checks.

**State**. The length is 265 ft, and the width is 165 ft.

**43. Familiarize**. Let $d$ = the number of properties DeBartolo owned before the merger and $s$ = the number of properties Simon owned.

**Translate**.

The total number of properties is 183.

$d + s = 183$

Simon owned twice as many properties as DeBartolo.

$s = 2d$

We have a system of equations.

$d + s = 183$,    (1)

$s = 2d$      (2)

**Carry out**. We use the substitution method to solve the system of equations.

Substitute $2d$ for $s$ in (1) and solve for $d$.

$d + 2d = 183$

$3d = 183$

$d = 61$

Now substitute 61 for $d$ in (2).

$s = 2 \cdot 61 = 122$

**Check**. The total number of properties is $61 + 122$, or 183. The number of Simon properties, 122, is twice 61, the number of DeBartolo properties. The answer checks.

**State**. Before the merger DeBartolo owned 61 properties and Simon owned 122 properties.

**45. Familiarize**. Let $x$ = the number of 30-sec commercials and $y$ = the number of 60-sec commercials. Then the 30-sec commercials play for $30x$ sec and the 60-sec commercials play for $60y$ sec. Note that 10 min = $10 \times 60$ sec = 600 sec.

*Translate*.

The total number of commercials is 12.

$$x + y = 12$$

The total playing time for the commercials is 600 sec.

$$30x + 60y = 600$$

We have a system of equations.

$$x + y = 12, \quad (1)$$
$$30x + 60y = 600 \quad (2)$$

*Carry out*.   We use elimination to solve the system of equations.

$$\begin{array}{r} -30x - 30y = -360 \quad \text{Multiplying (1) by } -30 \\ \underline{30x + 60y = \phantom{-}600} \\ 30y = \phantom{-}240 \\ y = \phantom{-000}8 \end{array}$$

Now substitute 8 for $y$ in (1) and solve for $x$.

$$x + 8 = 12$$
$$x = 4$$

*Check*.   There is a total of $4 + 8$, or 12 commercials. They play for $30 \cdot 4 + 60 \cdot 8 = 120 + 480 = 600$ sec. The answer checks.

*State*.   Roscoe played 4 30-sec commercials and 8 60-sec commercials.

**47.** *Familiarize*.   The change from the \$9.25 purchase is $20 - \$9.25$, or \$10.75. Let $x =$ the number of quarters and $y =$ the number of fifty-cent pieces.   The total value of the quarters, in dollars, is $0.25x$ and the total value of the fifty-cent pieces, in dollars, is $0.50y$.

*Translate*.

The total number of coins is 30.

$$x + y = 30$$

The total value of the coins is \$10.75.

$$0.25x + 0.50y = 10.75$$

After clearing decimals we have the following system of equations:

$$x + \phantom{0}y = \phantom{00}30, \quad (1)$$
$$25x + 50y = 1075 \quad (2)$$

*Carry out*.   We use the elimination method to solve the system of equations.

$$\begin{array}{r} -25x - 25y = -750 \quad \text{Multiplying (1) by } -25 \\ \underline{25x + 50y = \phantom{-}1075} \\ 25y = \phantom{-0}325 \\ y = \phantom{-000}13 \end{array}$$

Substitute 13 for $y$ in (1) and solve for $x$.

$$x + 13 = 30$$
$$x = 17$$

*Check*.   The total number of coins is $17 + 13$, or 30. The total value of the coins is $\$0.25(17) + \$0.50(13) = \$4.25 + \$6.50 = \$10.75$. The answer checks.

*State*.   There were 17 quarters and 13 fifty-cent pieces.

**49.** *Writing Exercise*

**51.**
$$\begin{aligned} 2x - 3y + 12 &= 2 \cdot 5 - 3 \cdot 2 + 12 \\ &= 10 - 6 + 12 \\ &= 4 + 12 \\ &= 16 \end{aligned}$$

**53.**
$$\begin{aligned} 5a - 7b + 3c &= 5(-2) - 7(3) + 3 \cdot 1 \\ &= -10 - 21 + 3 \\ &= -31 + 3 \\ &= -28 \end{aligned}$$

**55.**
$$\begin{aligned} 4 - 2y + 3z &= 4 - 2 \cdot \frac{1}{3} + 3 \cdot \frac{1}{4} \\ &= 4 - \frac{2}{3} + \frac{3}{4} \\ &= \frac{48}{12} - \frac{8}{12} + \frac{9}{12} \\ &= \frac{40}{12} + \frac{9}{12} \\ &= \frac{49}{12} \end{aligned}$$

**57.** *Writing Exercise*

**59.** *Familiarize*.   Let $x =$ the number of reams of 0% post-consumer fiber paper purchased and $y =$ the number of reams of 30% post-consumer fiber paper.

*Translate*.   We organize the information in a table.

|  | 0% post-consumer | 30% post-consumer | Total |
|---|---|---|---|
| Reams purchased | $x$ | $y$ | 60 |
| Percent of post-consumer fiber | 0% | 30% | 20% |
| Total post-consumer fiber | $0 \cdot x$, or $0$ | $0.3y$ | 0.2(60), or 12 |

We get one equation from the "Reams purchased" row of the table:

$$x + y = 60$$

The last row of the table yields a second equation:

$$0x + 0.3y = 12, \text{ or } 0.3y = 12$$

After clearing the decimal we have the problem translated to a system of equations.

$$x + y = 60, \quad (1)$$
$$3y = 120 \quad (2)$$

*Carry out*.   First we solve (2) for $y$.

$$3y = 120$$
$$y = 40$$

Now substitute 40 for $y$ in (1) and solve for $x$.

$$x + 40 = 60$$
$$x = 20$$

**Check**. The total purchase is $20 + 40$, or 60 reams. The post-consumer fiber can be expressed as $0 \cdot 20 + 0.3(40) = 12$. The answer checks.

**State**. 20 reams of 0% post-consumer fiber paper and 40 reams of 30% post-consumer fiber paper would have to be purchased.

**61. Familiarize**. Let $x =$ the amount of the original solution that remains after some of the original solution is drained and replaced with pure antifreeze. Let $y =$ the amount of the original solution that is drained and replaced with pure antifreeze.

**Translate**. We organize the information in a table. Keep in mind that the table contains information regarding the solution *after* some of the original solution is drained and replaced with pure antifreeze.

|  | Original Solution | Pure Anti-freeze | New Mixture |
|---|---|---|---|
| Amount of solution | $x$ | $y$ | 6.3 L |
| Percent of antifreeze | 30% | 100% | 50% |
| Amount of antifreeze in solution | $0.3x$ | $1 \cdot y$, or $y$ | $0.5(6.3)$, or $3.15$ |

The "Amount of solution" row gives us one equation:

$x + y = 6.3$   (1)

The last row gives us a second equation:

$0.3x + y = 3.15$   (2)

After clearing the decimals we have the following system of equations:

$10x + 10y = 63,$   (3)

$30x + 100y = 315$   (4)

**Carry out**. We use the elimination method.

$-30x - 30y = -189$   Multiplying (3) by $-3$

$\underline{30x + 100y = \phantom{-}315}$

$\phantom{30x + }70y = \phantom{-}126$

$\phantom{30x + 70}y = \phantom{-}1.8$

Although the problem only asks for the amount of pure antifreeze added, we will also find $x$ in order to check.

$x + 1.8 = 6.3$   Substituting 1.8 for $y$ in (1)

$x = 4.5$

**Check**. Total amount of new mixture:

$4.5 + 1.8 = 6.3$ L

Amount of antifreeze in new mixture:

$0.3(4.5) + 1(1.8) = 1.35 + 1.8 = 3.15$ L

The numbers check.

**State**. Michelle should drain 1.8 L of the original solution and replace it with pure antifreeze.

**63. Familiarize**. Let $x =$ the number of members who ordered one book and $y =$ the number of members who ordered two books. Note that the $y$ members ordered a total of $2y$ books.

**Translate**.

The number of books sold was 880.

$x + 2y \qquad = \quad 880$

Total sales were $9840.

$12x + 20y \quad = \quad 9840$

We have a system of equations.

$x + 2y = 880,$   (1)

$12x + 20y = 9840$   (2)

**Carry out**. We use the elimination method.

$-10x - 20y = -8800$   Multiplying (1) by $-10$

$\underline{12x + 20y = \phantom{-}9840}$

$\phantom{-}2x \phantom{+ 20y} = \phantom{-}1040$

$\phantom{-2}x \phantom{+ 20y} = \phantom{-}520$

Substitute 520 for $x$ in (1) and solve for $y$.

$520 + 2y = 880$

$2y = 360$

$y = 180$

**Check**. Total number of books sold: $520 + 2 \cdot 180 = 520 + 360 = 880$

Total sales: $\$12 \cdot 520 + \$20 \cdot 180 = \$6240 + \$3600 = \$9840$

The answer checks.

**State**. 180 members ordered two books

**65. Familiarize**. Let $x =$ the number of gallons of pure brown and $y =$ the number of gallons of neutral stain that should be added to the original 0.5 gal. Note that a total of 1 gal of stain needs to be added to bring the amount of stain up to 1.5 gal. The original 0.5 gal of stain contains $20\%(0.5$ gal), or $0.2(0.5$ gal) $= 0.1$ gal of brown stain. The final solution contains $60\%(1.5$ gal), or $0.6(1.5$ gal) $= 0.9$ gal of brown stain. This is composed of the original 0.1 gal and the $x$ gal that are added.

**Translate**.

The amount of stain added was 1 gal.

$x + y \qquad = \qquad 1$

The amount of brown stain in the final solution is 0.9 gal.

$0.1 + x \qquad = \qquad 0.9$

We have a system of equations.

$x + y = 1,$   (1)

$0.1 + x = 0.9$   (2)

***Carry out***. First we solve (2) for $x$.

$$0.1 + x = 0.9$$
$$x = 0.8$$

Then substitute 0.8 for $x$ in (1) and solve for $y$.

$$0.8 + y = 1$$
$$y = 0.2$$

***Check***. Total amount of stain: $0.5 + 0.8 + 0.2 = 1.5$ gal

Total amount of brown stain: $0.1 + 0.8 = 0.9$ gal

Total amount of neutral stain: $0.8(0.5) + 0.2 = 0.4 + 0.2 = 0.6$ gal $= 0.4(1.5$ gal$)$

The answer checks.

***State***. 0.8 gal of pure brown and 0.2 gal of neutral stain should be added.

**67.** ***Familiarize***. Let $x$ and $y$ represent the number of city miles and highway miles that were driven, respectively. Then in city driving, $\dfrac{x}{18}$ gallons of gasoline are used; in highway driving, $\dfrac{y}{24}$ gallons are used.

***Translate***. We organize the information in a table.

| Type of driving | City | Highway | Total |
|---|---|---|---|
| Number of miles | $x$ | $y$ | 465 |
| Gallons of gasoline used | $\dfrac{x}{18}$ | $\dfrac{y}{24}$ | 23 |

The first row of the table gives us one equation:

$$x + y = 465$$

The second row gives us another equation:

$$\frac{x}{18} + \frac{y}{24} = 23$$

After clearing fractions, we have the following system of equations:

$$x + y = 465, \qquad (1)$$
$$24x + 18y = 9936 \qquad (2)$$

***Solve***. We solve the system of equations using the elimination method.

$$
\begin{array}{ll}
-18x - 18y = -8370 & \text{Multiplying (1) by } -18 \\
\underline{\phantom{-}24x + 18y = \phantom{-}9936} & \\
\phantom{-}6x \phantom{ + 18y} = \phantom{-}1566 & \\
\phantom{-6}x \phantom{ + 18y} = \phantom{-}261 &
\end{array}
$$

Now substitute 261 for $x$ in Equation (1) and solve for $y$.

$$261 + y = 465$$
$$y = 204$$

***Check***. The total mileage is $261 + 204$, or 465. In 216 city miles, $261/18$, or 14.5 gal of gasoline are used; in 204 highway miles, $204/24$, or 8.5 gal are used. Then a total of $14.5 + 8.5$ or 23 gal of gasoline are used. The answer checks.

***State***. 261 miles were driven in the city, and 204 miles were driven on the highway.

**69.** The 1.5 gal mixture contains $0.1 + x$ gal of pure brown stain. (See Exercise 65.). Thus, the function
$$P(x) = \frac{0.1 + x}{1.5}$$
gives the percentage of brown in the mixture as a decimal quantity. Using the Intersect feature, we confirm that when $x = 0.8$, then $P(x) = 0.6$ or 60%.

## Exercise Set 3.4

**1.** The equation is equivalent to one in the form $Ax + By + Cz = D$, so the statement is true.

**3.** False; see Example 5.

**5.** True; see Example 6.

**7.** Substitute $(2, -1, -2)$ into the three equations, using alphabetical order.

$$
\begin{array}{c|c}
\multicolumn{2}{l}{x + y - 2z = 5} \\
\hline
2 + (-1) - 2(-2) & 5 \\
2 - 1 + 4 & \\
& \overset{?}{} \\
\multicolumn{2}{c}{5 = 5 \quad \text{TRUE}}
\end{array}
$$

$$
\begin{array}{c|c}
\multicolumn{2}{l}{2x - y - z = 7} \\
\hline
2 \cdot 2 - (-1) - (-2) & 7 \\
4 + 1 + 2 & \\
& \overset{?}{} \\
\multicolumn{2}{c}{7 = 7 \quad \text{TRUE}}
\end{array}
$$

$$
\begin{array}{c|c}
\multicolumn{2}{l}{-x - 2y + 3z = 6} \\
\hline
-2 - 2(-1) + 3(-2) & 6 \\
-2 + 2 - 6 & \\
& \overset{?}{} \\
\multicolumn{2}{c}{-6 = 6 \quad \text{FALSE}}
\end{array}
$$

The triple $(2, -1, -2)$ does not make the third equation true, so it is not a solution of the system.

**9.** 
$$
\begin{array}{ll}
2x - y + z = 10, & (1) \\
4x + 2y - 3z = 10, & (2) \\
x - 3y + 2z = 8 & (3)
\end{array}
$$

1., 2. The equations are already in standard form with no fractions or decimals.

3. Use Equations (1) and (2) to eliminate $y$:

$$
\begin{array}{ll}
4x - 2y + 2z = 20 & \text{Multiplying (1) by 2} \\
\underline{4x + 2y - 3z = 10} & (2) \\
8x \phantom{ + 2y} - z = 30 & (4)
\end{array}
$$

4. Use a different pair of equations and eliminate $y$:

$$
\begin{array}{ll}
-6x + 3y - 3z = -30 & \text{Multiplying (1) by } -3 \\
\underline{\phantom{-}x - 3y + 2z = \phantom{-}8} & (3) \\
-5x \phantom{ + 3y} - z = -22 & (5)
\end{array}
$$

5. Now solve the system of Equations (4) and (5).

$$8x - z = 30 \quad (4)$$
$$-5x - z = -22 \quad (5)$$

$$8x - z = 30 \quad (4)$$
$$\underline{5x + z = 22} \quad \text{Multiplying (5) by } -1$$
$$13x \phantom{+z} = 52$$
$$x = 4$$

$$8 \cdot 4 - z = 30 \quad \text{Substituting in (4)}$$
$$32 - z = 30$$
$$-z = -2$$
$$z = 2$$

6. Substitute in one of the original equations to find $y$.

$$2 \cdot 4 - y + 2 = 10 \quad \text{Substituting in (1)}$$
$$10 - y = 10$$
$$-y = 0$$
$$y = 0$$

We obtain $(4, 0, 2)$. This checks, so it is the solution.

**11.**
$$x - y + z = 6, \quad (1)$$
$$2x + 3y + 2z = 2, \quad (2)$$
$$3x + 5y + 4z = 4 \quad (3)$$

1., 2. The equations are already in standard form with no fractions or decimals.

3., 4. We eliminate $y$ from two different pairs of equations.

$$3x - 3y + 3z = 18 \quad \text{Multiplying (1) by 3}$$
$$\underline{2x + 3y + 2z = 2} \quad (2)$$
$$5x \phantom{+3y} + 5z = 20 \quad (4)$$

$$5x - 5y + 5z = 30 \quad \text{Multiplying (1) by 5}$$
$$\underline{3x + 5y + 4z = 4} \quad (3)$$
$$8x \phantom{+5y} + 9z = 34 \quad (5)$$

5. Now solve the system of Equations (4) and (5).

$$5x + 5z = 20 \quad (4)$$
$$8x + 9z = 34 \quad (5)$$

$$45x + 45z = 180 \quad \text{Multiplying (4) by 9}$$
$$\underline{-40x - 45z = -170} \quad \text{Multiplying (5) by } -5$$
$$5x \phantom{-45z} = 10$$
$$x = 2$$

$$5 \cdot 2 + 5z = 20 \quad \text{Substituting in (4)}$$
$$10 + 5z = 20$$
$$5z = 10$$
$$z = 2$$

6. Substitute in one of the original equations to find $y$.

$$2 - y + 2 = 6 \quad \text{Substituting in (1)}$$
$$4 - y = 6$$
$$-y = 2$$
$$y = -2$$

We obtain $(2, -2, 2)$. This checks, so it is the solution.

**13.**
$$6x - 4y + 5z = 31, \quad (1)$$
$$5x + 2y + 2z = 13, \quad (2)$$
$$x + y + z = 2 \quad (3)$$

1., 2. The equations are already in standard form with no fractions or decimals.

3., 4. We eliminate $y$ from two different pairs of equations.

$$6x - 4y + 5z = 31 \quad (1)$$
$$\underline{4x + 4y + 4z = 8} \quad \text{Multiplying (3) by 4}$$
$$10x \phantom{+4y} + 9z = 39 \quad (4)$$

$$5x + 2y + 2z = 13 \quad (2)$$
$$\underline{-2x - 2y - 2z = -4} \quad \text{Multiplying (3) by } -2$$
$$3x \phantom{-2y} = 9$$
$$x = 3$$

5. When we used Equations (2) and (3) to eliminate $y$, we also eliminated $z$ and found that $x = 3$. Substitute 3 for $x$ in Equation (4) to find $z$.

$$10 \cdot 3 + 9z = 39 \quad \text{Substituting in (4)}$$
$$30 + 9z = 39$$
$$9z = 9$$
$$z = 1$$

6. Substitute in one of the original equations to find $y$.

$$3 + y + 1 = 2 \quad \text{Substituting in (3)}$$
$$y + 4 = 2$$
$$y = -2$$

We obtain $(3, -2, 1)$. This checks, so it is the solution.

**15.**
$$x + y + z = 0, \quad (1)$$
$$2x + 3y + 2z = -3, \quad (2)$$
$$-x + 2y - z = 1 \quad (3)$$

1., 2. The equations are already in standard form with no fractions or decimals.

3., 4. We eliminate $x$ from two different pairs of equations.

$$-2x - 2y - 2z = 0 \quad \text{Multiplying (1) by } -2$$
$$\underline{2x + 3y + 2z = -3} \quad (2)$$
$$y \phantom{+3y} = -3$$

We eliminated not only $x$ but also $z$ and found that $y = -3$.

5., 6. Substitute $-3$ for $y$ in two of the original equations to produce a system of two equations in two variables. Then solve this system.

$x - 3 + z = 0$   Substituting in (1)

$-x + 2(-3) - z = 1$   Substituting in (3)

Simplifying we have

$$x + z = 3$$
$$\underline{-x - z = 7}$$
$$0 = 10$$

We get a false equation, so there is no solution.

**17.**   $2x + y - 3z = -4,$   (1)

      $4x - 2y + z = 9,$   (2)

      $3x + 5y - 2z = 5$   (3)

1., 2.   The equations are already in standard form with no fractions or decimals.

3., 4.   We eliminate $z$ from two different pairs of equations.

$$2x + y - 3z = -4 \quad (1)$$
$$\underline{12x - 6y + 3z = 27} \quad \text{Multiplying (2) by 3}$$
$$14x - 5y \quad\quad = 23 \quad (4)$$

$$8x - 4y + 2z = 18 \quad \text{Multiplying (2) by 2}$$
$$\underline{3x + 5y - 2z = 5} \quad (3)$$
$$11x + y \quad\quad = 23 \quad (5)$$

5.   Now solve the system of Equations (4) and (5).

$$14x - 5y = 23 \quad (4)$$
$$11x + y = 23 \quad (5)$$

$$14x - 5y = 23 \quad (4)$$
$$\underline{55x + 5y = 115} \quad \text{Multiplying (5) by 5}$$
$$69x \quad\quad = 138$$
$$x = 2$$

$11 \cdot 2 + y = 23$   Substituting in (5)

$$22 + y = 23$$
$$y = 1$$

6.   Substitute in one of the original equations to find $z$.

$4 \cdot 2 - 2 \cdot 1 + z = 9$   Substituting in (2)

$$6 + z = 9$$
$$z = 3$$

We obtain $(2, 1, 3)$. This checks, so it is the solution.

**19.**   $2x + y + 2z = 11,$   (1)

      $3x + 2y + 2z = 8,$   (2)

      $x + 4y + 3z = 0$   (3)

1., 2.   The equations are already in standard form with no fractions or decimals.

3., 4.   We eliminate $x$ from two different pairs of equations.

$$2x + y + 2z = 11 \quad (1)$$
$$\underline{-2x - 8y - 6z = 0} \quad \text{Multiplying (3) by } -2$$
$$-7y - 4z = 11 \quad (4)$$

$$3x + 2y + 2z = 8 \quad (2)$$
$$\underline{-3x - 12y - 9z = 0} \quad \text{Multiplying (3) by } -3$$
$$-10y - 7z = 8 \quad (5)$$

5.   Now solve the system of Equations (4) and (5).

$$-7y - 4z = 11 \quad (4)$$
$$-10y - 7z = 8 \quad (5)$$

$$-49y - 28z = 77 \quad \text{Multiplying (4) by 7}$$
$$\underline{40y + 28z = -32} \quad \text{Multiplying (5) by } -4$$
$$-9y \quad\quad = 45$$
$$y = -5$$

$-7(-5) - 4z = 11$   Substituting in (4)

$$35 - 4z = 11$$
$$-4z = -24$$
$$z = 6$$

6.   Substitute in one of the original equations to find $x$.

$x + 4(-5) + 3 \cdot 6 = 0$   Substituting in (3)

$$x - 2 = 0$$
$$x = 2$$

We obtain $(2, -5, 6)$. This checks, so it is the solution.

**21.**   $-2x + 8y + 2z = 4,$   (1)

      $x + 6y + 3z = 4,$   (2)

      $3x - 2y + z = 0$   (3)

1., 2.   The equations are already in standard form with no fractions or decimals.

3., 4.   We eliminate $z$ from two different pairs of equations.

$$-2x + 8y + 2z = 4 \quad (1)$$
$$\underline{-6x + 4y - 2z = 0} \quad \text{Multiplying (3) by } -2$$
$$-8x + 12y \quad = 4 \quad (4)$$

$$x + 6y + 3z = 4 \quad (2)$$
$$\underline{-9x + 6y - 3z = 0} \quad \text{Multiplying (3) by } -3$$
$$-8x + 12y \quad = 4 \quad (5)$$

5.   Now solve the system of Equations (4) and (5).

$$-8x + 12y = 4 \quad (4)$$
$$-8x + 12y = 4 \quad (5)$$

$$-8x + 12y = 4 \quad (4)$$
$$\underline{8x - 12y = -4} \quad \text{Multiplying (5) by } -1$$
$$0 = 0 \quad (6)$$

Equation (6) indicates that Equations (1), (2), and (3) are dependent. (Note that if Equation (1) is subtracted from Equation (2), the result is Equation (3).) We could also have concluded that the equations are dependent by observing that Equations (4) and (5) are identical.

**23.**
$$4x - y - z = 4, \quad (1)$$
$$2x + y + z = -1, \quad (2)$$
$$6x - 3y - 2z = 3 \quad (3)$$

1., 2. The equations are already in standard form with no fractions or decimals.

3. Add Equations (1) and (2) to eliminate $y$.

$$\begin{array}{ll} 4x - y - z = 4 & (1) \\ 2x + y + z = -1 & (2) \\ \hline 6x \qquad\quad = 3 & (4) \quad \text{Adding} \end{array}$$

4. At this point we can either continue by eliminating $y$ from a second pair of equations or we can solve (4) for $x$ and substitute that value in a different pair of the original equations to obtain a system of two equations in two variables. We take the second option.

$$6x = 3 \quad (4)$$
$$x = \frac{1}{2}$$

Substitute $\frac{1}{2}$ for $x$ in (1):

$$4\left(\frac{1}{2}\right) - y - z = 4$$
$$2 - y - z = 4$$
$$-y - z = 2 \quad (5)$$

Substitute $\frac{1}{2}$ for $x$ in (3):

$$6\left(\frac{1}{2}\right) - 3y - 2z = 3$$
$$3 - 3y - 2z = 3$$
$$-3y - 2z = 0 \quad (6)$$

5. Now solve the system of Equations (5) and (6).

$$\begin{array}{ll} 2y + 2z = -4 & \text{Multiplying (5) by } -2 \\ -3y - 2z = 0 & (6) \\ \hline -y \qquad = -4 & \\ y = 4 & \end{array}$$

6. Substitute to find $z$.
$$-4 - z = 2 \quad \text{Substituting 4 for } y \text{ in (5)}$$
$$-z = 6$$
$$z = -6$$

We obtain $\left(\frac{1}{2}, 4, -6\right)$. This checks, so it is the solution.

**25.**
$$r + \frac{3}{2}s + 6t = 2,$$
$$2r - 3s + 3t = 0.5,$$
$$r + s + t = 1$$

1. All equations are already in standard form.

2. Multiply the first equation by 2 to clear the fraction. Also, multiply the second equation by 10 to clear the decimal.

---

$$2r + 3s + 12t = 4, \quad (1)$$
$$20r - 30s + 30t = 5, \quad (2)$$
$$r + s + t = 1 \quad (3)$$

3., 4. We eliminate $s$ from two different pairs of equations.

$$\begin{array}{ll} 20r + 30s + 120t = 40 & \text{Multiplying (1) by 10} \\ 20r - 30s + 30t = 5 & (2) \\ \hline 40r \qquad + 150t = 45 & (4) \quad \text{Adding} \end{array}$$

$$\begin{array}{ll} 20r - 30s + 30t = 5 & (2) \\ 30r + 30s + 30t = 30 & \text{Multiplying (3) by 30} \\ \hline 50r \qquad + 60t = 35 & (5) \quad \text{Adding} \end{array}$$

5. Solve the system of Equations (4) and (5).

$$40r + 150t = 45 \quad (4)$$
$$50r + 60t = 35 \quad (5)$$

$$\begin{array}{ll} 200r + 750t = 225 & \text{Multiplying (4) by 5} \\ -200r - 240t = -140 & \text{Multiplying (5) by } -4 \\ \hline 510t = 85 & \end{array}$$
$$t = \frac{85}{510}$$
$$t = \frac{1}{6}$$

$$40r + 150\left(\frac{1}{6}\right) = 45 \quad \text{Substituting } \frac{1}{6} \text{ for } t \text{ in (4)}$$
$$40r + 25 = 45$$
$$40r = 20$$
$$r = \frac{1}{2}$$

6. Substitute in one of the original equations to find $s$.

$$\frac{1}{2} + s + \frac{1}{6} = 1 \quad \text{Substituting } \frac{1}{2} \text{ for } r \text{ and } \frac{1}{6} \text{ for } t \text{ in (3)}$$
$$s + \frac{2}{3} = 1$$
$$s = \frac{1}{3}$$

We obtain $\left(\frac{1}{2}, \frac{1}{3}, \frac{1}{6}\right)$. This checks, so it is the solution.

**27.**
$$4a + 9b = 8, \quad (1)$$
$$8a + 6c = -1, \quad (2)$$
$$6b + 6c = -1 \quad (3)$$

1., 2. The equations are already in standard form with no fractions or decimals.

3., 4. Note that there is no $c$ in Equation (1). We will use Equations (2) and (3) to obtain another equation with no $c$-term.

$$\begin{array}{ll} 8a \qquad + 6c = -1 & (2) \\ -6b - 6c = 1 & \text{Multiplying (3) by } -1 \\ \hline 8a - 6b \qquad = 0 & (4) \text{ Adding} \end{array}$$

5. Now solve the system of Equations (1) and (4).

$$-8a - 18b = -16 \quad \text{Multiplying (1) by } -2$$
$$\underline{8a - 6b = 0}$$
$$-24b = -16$$
$$b = \frac{2}{3}$$

$$8a - 6\left(\frac{2}{3}\right) = 0 \quad \text{Substituting } \frac{2}{3} \text{ for } b$$
$$8a - 4 = 0 \qquad \text{in (4)}$$
$$8a = 4$$
$$a = \frac{1}{2}$$

6. Substitute in Equation (2) or (3) to find $c$.

$$8\left(\frac{1}{2}\right) + 6c = -1 \quad \text{Substituting } \frac{1}{2} \text{ for } a$$
$$4 + 6c = -1 \qquad \text{in (2)}$$
$$6c = -5$$
$$c = -\frac{5}{6}$$

We obtain $\left(\frac{1}{2}, \frac{2}{3}, -\frac{5}{6}\right)$. This checks, so it is the solution.

**29.**
$$x + y + z = 57, \quad (1)$$
$$-2x + y \qquad = 3, \quad (2)$$
$$x \qquad - z = 6 \quad (3)$$

1., 2. The equations are already in standard form with no fractions or decimals.

3., 4. Note that there is no $z$ in Equation (2). We will use Equations (1) and (3) to obtain another equation with no $z$-term.

$$x + y + z = 57 \qquad (1)$$
$$\underline{x \qquad - z = 6 \qquad (3)}$$
$$2x + y \qquad = 63 \qquad (4)$$

5. Now solve the system of Equations (2) and (4).

$$-2x + y = 3 \qquad (2)$$
$$\underline{2x + y = 63 \qquad (4)}$$
$$2y = 66$$
$$y = 33$$

$$2x + 33 = 63 \quad \text{Substituting 33 for y in (4)}$$
$$2x = 30$$
$$x = 15$$

6. Substitute in Equation (1) or (3) to find $z$.

$$15 - z = 6 \quad \text{Substituting 15 for } x \text{ in (3)}$$
$$9 = z$$

We obtain $(15, 33, 9)$. This checks, so it is the solution.

**31.**
$$a \qquad - 3c = 6, \quad (1)$$
$$b + 2c = 2, \quad (2)$$
$$7a - 3b - 5c = 14 \quad (3)$$

1., 2. The equations are already in standard form with no fractions or decimals.

3., 4. Note that there is no $b$ in Equation (1). We will use Equations (2) and (3) to obtain another equation with no $b$-term.

$$3b + 6c = 6 \quad \text{Multiplying (2) by 3}$$
$$\underline{7a - 3b - 5c = 14 \quad (3)}$$
$$7a \qquad + c = 20 \quad (4)$$

5. Now solve the system of Equations (1) and (4).

$$a - 3c = 6 \quad (1)$$
$$7a + c = 20 \quad (4)$$

$$a - 3c = 6 \quad (1)$$
$$\underline{21a + 3c = 60 \quad \text{Multiplying (4) by 3}}$$
$$22a \qquad = 66$$
$$a = 3$$

$$3 - 3c = 6 \quad \text{Substituting in (1)}$$
$$-3c = 3$$
$$c = -1$$

6. Substitute in Equation (2) or (3) to find $b$.

$$b + 2(-1) = 2 \quad \text{Substituting in (2)}$$
$$b - 2 = 2$$
$$b = 4$$

We obtain $(3, 4, -1)$. This checks, so it is the solution.

**33.**
$$x + y + z = 83, \quad (1)$$
$$y = 2x + 3, \qquad (2)$$
$$z = 40 + x \qquad (3)$$

Observe, from Equations (2) and (3), that we can substitute $2 + 3x$ for $y$ and $40 + x$ for $z$ in Equation (1) and solve for $x$.

$$x + y + x = 83$$
$$x + (2x + 3) + (40 + x) = 83$$
$$4x + 43 = 83$$
$$4x = 40$$
$$x = 10$$

Now substitute 10 for $x$ in Equation (2).

$$y = 2x + 3 = 2 \cdot 10 + 3 = 20 + 3 = 23$$

Finally, substitute 10 for $x$ in Equation (3).

$$z = 40 + x = 40 + 10 = 50.$$

We obtain $(10, 23, 50)$. This checks, so it is the solution.

**35.**
$$x \qquad + z = 0, \quad (1)$$
$$x + y + 2z = 3, \quad (2)$$
$$y + z = 2 \quad (3)$$

1., 2. The equations are already in standard form with no fractions or decimals.

3., 4.  Note that there is no $y$ in Equation (1). We use Equations (2) and (3) to obtain another equation with no $y$-term.

$$x + y + 2z = 3 \quad (2)$$
$$\underline{\quad - y - \ z = -2 \quad \text{Multiplying (3)}}$$
$$\phantom{xxx}\text{by } -1$$
$$x \quad\ \ + z = 1 \quad (4) \quad \text{Adding}$$

5. Now solve the system of Equations (1) and (4).

$$x + z = 0 \quad (1)$$
$$\underline{-x - z = -1 \quad \text{Multiplying (4) by } -1}$$
$$0 = -1 \quad \text{Adding}$$

We get a false equation, or contradiction. There is no solution.

**37.**
$$x + \ y + z = \ 1, \quad (1)$$
$$-x + 2y + z = \ 2, \quad (2)$$
$$2x - \ y \quad\ \ = -1 \quad (3)$$

1., 2. The equations are already in standard form with no fractions or decimals.

3. Note that there is no $z$ in Equation (3). We will use Equations (1) and (2) to eliminate $z$:

$$x + \ y + z = \ 1 \quad (1)$$
$$\underline{x - 2y - z = -2 \quad \text{Multiplying (2) by } -1}$$
$$2x - \ y \quad\ \ = -1 \quad \text{Adding}$$

Equations (3) and (4) are identical, so Equations (1), (2), and (3) are dependent. (We have seen that if Equation (2) is multiplied by $-1$ and added to Equation (1), the result is Equation (3).)

**39.** *Writing Exercise*

**41.** Let $x$ represent the larger number and $y$ represent the smaller number. Then we have $x = 2y$.

**43.** Let $x$, $x + 1$, and $x + 2$ represent the numbers. Then we have $x + (x + 1) + (x + 2) = 45$.

**45.** Let $x$ and $y$ represent the first two numbers and let $z$ represent the third number. Then we have $x + y = 5z$.

**47.** *Writing Exercise*

**49.**
$$\frac{x + 2}{3} - \frac{y + 4}{2} + \frac{z + 1}{6} = 0,$$
$$\frac{x - 4}{3} + \frac{y + 1}{4} - \frac{z - 2}{2} = -1,$$
$$\frac{x + 1}{2} + \frac{y}{2} + \frac{z - 1}{4} = \frac{3}{4}$$

1., 2. We clear fractions and write each equation in standard form.

To clear fractions, we multiply both sides of each equation by the LCM of its denominators. The LCM's are 6, 12, and 4, respectively.

$$6\left(\frac{x + 2}{3} - \frac{y + 4}{2} + \frac{z + 1}{6}\right) = 6 \cdot 0$$
$$2(x + 2) - 3(y + 4) + (z + 1) = 0$$
$$2x + 4 - 3y - 12 + z + 1 = 0$$
$$2x - 3y + z = 7$$

$$12\left(\frac{x - 4}{3} + \frac{y + 1}{4} - \frac{z - 2}{2}\right) = 12 \cdot (-1)$$
$$4(x - 4) + 3(y + 1) - 6(z - 2) = -12$$
$$4x - 16 + 3y + 3 - 6z + 12 = -12$$
$$4x + 3y - 6z = -11$$

$$4\left(\frac{x + 1}{2} + \frac{y}{2} + \frac{z - 1}{4}\right) = 4 \cdot \frac{3}{4}$$
$$2(x + 1) + 2(y) + (z - 1) = 3$$
$$2x + 2 + 2y + z - 1 = 3$$
$$2x + 2y + z = 2$$

The resulting system is

$$2x - 3y + \ z = \ 7, \quad (1)$$
$$4x + 3y - 6z = -11, \quad (2)$$
$$2x + 2y + \ z = \ \ 2 \quad (3)$$

3., 4. We eliminate $z$ from two different pairs of equations.

$$12x - 18y + 6z = \ 42 \quad \text{Multiplying (1) by 6}$$
$$\underline{4x + \ 3y - 6z = -11 \quad (2)}$$
$$16x - 15y \quad\ \ = \ 31 \quad (4) \quad \text{Adding}$$

$$2x - 3y + z = \ 7 \quad (1)$$
$$\underline{-2x - 2y - z = -2 \quad \text{Multiplying (3) by } -1}$$
$$-5y \quad\ \ = \ 5 \quad (5) \quad \text{Adding}$$

5. Solve (5) for $y$:
$$-5y = 5$$
$$y = -1$$

Substitute $-1$ for $y$ in (4):
$$16x - 15(-1) = 31$$
$$16x + 15 = 31$$
$$16x = 16$$
$$x = 1$$

6. Substitute 1 for $x$ and $-1$ for $y$ in (1):
$$2 \cdot 1 - 3(-1) + z = 7$$
$$5 + z = 7$$
$$z = 2$$

We obtain $(1, -1, 2)$. This checks, so it is the solution.

**51.**
$$w + \ x - \ y + \ z = 0, \quad (1)$$
$$w - 2x - 2y - \ z = -5, \quad (2)$$
$$w - 3x - \ y + \ z = 4, \quad (3)$$
$$2w - \ x - \ y + 3z = 7 \quad (4)$$

The equations are already in standard form with no fractions or decimals.

Start by eliminating $z$ from three different pairs of equations.

$$w + x - y + z = 0 \quad (1)$$
$$\underline{w - 2x - 2y - z = -5 \quad (2)}$$
$$2w - x - 3y \phantom{+ z} = -5 \quad (5) \text{ Adding}$$

$$w - 2x - 2y - z = -5 \quad (2)$$
$$\underline{w - 3x - y + z = 4 \quad (3)}$$
$$2w - 5x - 3y \phantom{+ z} = -1 \quad (6) \text{ Adding}$$

$$3w - 6x - 6y - 3z = -15 \quad \text{Multiplying (2) by 3}$$
$$\underline{2w - x - y + 3z = 7 \quad (4)}$$
$$5w - 7x - 7y \phantom{+ 3z} = -8 \quad (7) \text{ Adding}$$

Now solve the system of equations (5), (6), and (7).

$$2w - x - 3y = -5, \quad (5)$$
$$2w - 5x - 3y = -1, \quad (6)$$
$$5w - 7x - 7y = -8. \quad (7)$$

$$2w - x - 3y = -5 \quad (5)$$
$$\underline{-2w + 5x + 3y = 1} \quad \text{Multiplying (6) by } -1$$
$$4x \phantom{+ 3y} = -4$$
$$x \phantom{+ 3y} = -1$$

Substituting $-1$ for $x$ in (5) and (7) and simplifying, we have

$$2w - 3y = -6, \quad (8)$$
$$5w - 7y = -15. \quad (9)$$

Now solve the system of Equations (8) and (9).

$$10w - 15y = -30 \quad \text{Multiplying (8) by 5}$$
$$\underline{-10w + 14y = 30} \quad \text{Multiplying (9) by } -2$$
$$-y = 0$$
$$y = 0$$

Substitute 0 for $y$ in Equation (8) or (9) and solve for $w$.

$$2w - 3 \cdot 0 = -6 \quad \text{Substituting in (8)}$$
$$2w = -6$$
$$w = -3$$

Substitute in one of the original equations to find $z$.

$$-3 - 1 - 0 + z = 0 \quad \text{Substituting in (1)}$$
$$-4 + z = 0$$
$$z = 4$$

We obtain $(-3, -1, 0, 4)$. This checks, so it is the solution.

**53.**
$$\frac{2}{x} + \frac{2}{y} - \frac{3}{z} = 3,$$
$$\frac{1}{x} - \frac{2}{y} - \frac{3}{z} = 9,$$
$$\frac{7}{x} - \frac{2}{y} + \frac{9}{z} = -39$$

Let $u$ represent $\frac{1}{x}$, $v$ represent $\frac{1}{y}$, and $w$ represent $\frac{1}{z}$. Substituting, we have

$$2u + 2v - 3w = 3, \quad (1)$$
$$u - 2v - 3w = 9, \quad (2)$$
$$7u - 2v + 9w = -39 \quad (3)$$

1., 2.  The equations in $u$, $v$, and $w$ are in standard form with no fractions or decimals.

3., 4.  We eliminate $v$ from two different pairs of equations.

$$2u + 2v - 3w = 3 \quad (1)$$
$$\underline{u - 2v - 3w = 9 \quad (2)}$$
$$3u \phantom{- 2v} - 6w = 12 \quad (4) \text{ Adding}$$

$$2u + 2v - 3w = 3 \quad (1)$$
$$\underline{7u - 2v + 9w = -39 \quad (3)}$$
$$9u \phantom{- 2v} + 6w = -36 \quad (5) \text{ Adding}$$

5.  Now solve the system of Equations (4) and (5).

$$3u - 6w = 12, \quad (4)$$
$$\underline{9u + 6w = -36} \quad (5)$$
$$12u \phantom{+ 6w} = -24$$
$$u = -2$$

$$3(-2) - 6w = 12 \quad \text{Substituting in (4)}$$
$$-6 - 6w = 12$$
$$-6w = 18$$
$$w = -3$$

6.  Substitute in Equation (1), (2), or (3) to find $v$.

$$2(-2) + 2v - 3(-3) = 3 \quad \text{Substituting in (1)}$$
$$2v + 5 = 3$$
$$2v = -2$$
$$v = -1$$

Solve for $x$, $y$, and $z$. We substitute $-2$ for $u$, $-1$ for $v$, and $-3$ for $w$.

$$u = \frac{1}{x} \qquad v = \frac{1}{y} \qquad w = \frac{1}{z}$$
$$-2 = \frac{1}{x} \qquad -1 = \frac{1}{y} \qquad -3 = \frac{1}{z}$$
$$x = \frac{1}{2} \qquad y = -1 \qquad z = -\frac{1}{3}$$

We obtain $\left(-\frac{1}{2}, -1, -\frac{1}{3}\right)$. This checks, so it is the solution.

**55.**
$$5x - 6y + kz = -5, \quad (1)$$
$$x + 3y - 2z = 2, \quad (2)$$
$$2x - y + 4z = -1 \quad (3)$$

Eliminate $y$ from two different pairs of equations.

$$5x - 6y + kz = -5 \quad (1)$$
$$\underline{2x + 6y - 4z = 4} \quad \text{Multiplying (2) by 2}$$
$$7x \phantom{+ 6y} + (k-4)z = -1 \quad (4)$$

$$x + 3y - 2z = 2 \quad (2)$$
$$\underline{6x - 3y + 12z = -3} \quad \text{Multiplying (3) by 3}$$
$$7x \phantom{+ 3y} + 10z = -1 \quad (5)$$

Solve the system of Equations (4) and (5).

$$7x + (k-4)z = -1 \quad (4)$$
$$7x + 10z = -1 \quad (5)$$

$$-7x - \quad (k-4)z = \quad 1 \quad \text{Multiplying (4) by } -1$$
$$\underline{7x + \qquad 10z = -1 \quad (5)}$$
$$(-k+14)z = \quad 0 \quad (6)$$

The system is dependent for the value of $k$ that makes Equation (6) true. This occurs when $-k+14$ is 0. We solve for $k$:

$$-k+14=0$$
$$14=k$$

**57.** $z = b - mx - ny$

Three solutions are $(1, 1, 2)$, $(3, 2, -6)$, and $\left(\frac{3}{2}, 1, 1\right)$. We substitute for $x$, $y$, and $z$ and then solve for $b$, $m$, and $n$.

$$2 = b - m - n,$$
$$-6 = b - 3m - 2n,$$
$$1 = b - \frac{3}{2}m - n$$

**1., 2.** Write the equations in standard form. Also, clear the fraction in the last equation.

$$b - \quad m - \quad n = \quad 2, \quad (1)$$
$$b - 3m - 2n = -6, \quad (2)$$
$$2b - 3m - 2n = \quad 2 \quad (3)$$

**3., 4.** Eliminate $b$ from two different pairs of equations.

$$b - \quad m - \quad n = 2 \quad (1)$$
$$\underline{-b + 3m + 2n = 6 \quad \text{Multiplying (2) by } -1}$$
$$2m + \quad n = 8 \quad (4) \quad \text{Adding}$$

$$-2b + 2m + 2n = -4 \quad \text{Multiplying (1) by } -2$$
$$\underline{2b - 3m - 2n = \quad 2 \quad (3)}$$
$$-m \qquad = -2 \quad (5) \quad \text{Adding}$$

**5.** We solve Equation (5) for $m$:

$$-m = -2$$
$$m = 2$$

Substitute in Equation (4) and solve for $n$.

$$2 \cdot 2 + n = 8$$
$$4 + n = 8$$
$$n = 4$$

**6.** Substitute in one of the original equations to find $b$.

$$b - 2 - 4 = 2 \quad \text{Substituting 2 for } m$$
$$\qquad\qquad\qquad \text{and 4 for } n \text{ in (1)}$$
$$b - 6 = 2$$
$$b = 8$$

The solution is $(8, 2, 4)$, so the equation is $z = 8 - 2x - 4y$.

**Exercise Set 3.5**

**1. *Familiarize*.** Let $x =$ the first number, $y =$ the second number, and $z =$ the third number.

***Translate*.**

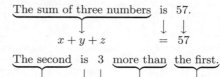
$$x + y + z = 57$$

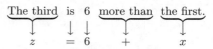
$$y = 3 + x$$

The third is 6 more than the first.
$$z = 6 + x$$

We now have a system of equations.

$$\begin{array}{ll} x + y + z = 57, \quad \text{or} & x + y + z = 57, \\ y = 3 + x & -x + y \quad = 3, \\ z = 6 + x & -x \quad + z = 6 \end{array}$$

***Carry out*.** Solving the system we get $(16, 19, 22)$.

***Check*.** The sum of the three numbers is $16 + 19 + 22$, or 57. The second number, 19, is three more than the first number, 16. The third number, 22, is 6 more than the first number, 16. The numbers check.

***State*.** The numbers are 16, 19, and 22.

**3. *Familiarize*.** Let $x =$ the first number, $y =$ the second number, and $z =$ the third number.

***Translate*.**

The sum of three numbers is 26.
$$x + y + z = 26$$

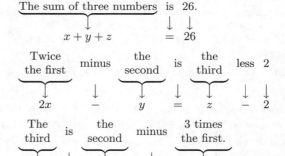
$$2x - y = z - 2$$

The third is the second minus 3 times the first.
$$z = y - 3x$$

We now have a system of equations.

$$\begin{array}{ll} x + y + z = 26, \quad \text{or} & x + y + z = 26, \\ 2x - y = z - 2, & 2x - y - z = -2, \\ z = y - 3x & 3x - y + z = 0 \end{array}$$

***Carry out*.** Solving the system we get $(8, 21, -3)$.

***Check*.** The sum of the numbers is $8 + 21 - 3$, or 26. Twice the first minus the second is $2 \cdot 8 - 21$, or $-5$, which is 2 less than the third. The second minus three times the first is $21 - 3 \cdot 8$, or $-3$, which is the third. The numbers check.

***State*.** The numbers are 8, 21, and $-3$.

**5. Familiarize.** We first make a drawing.

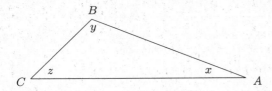

We let $x$, $y$, and $z$ represent the measures of angles $A$, $B$, and $C$, respectively. The measures of the angles of a triangle add up to $180°$.

**Translate.**

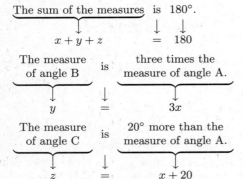

The sum of the measures is $180°$.

$$x + y + z = 180$$

The measure of angle B is three times the measure of angle A.

$$y = 3x$$

The measure of angle C is $20°$ more than the measure of angle A.

$$z = x + 20$$

We now have a system of equations.

$$x + y + z = 180,$$
$$y = 3x,$$
$$z = x + 20$$

**Carry out.** Solving the system we get $(32, 96, 52)$.

**Check.** The sum of the measures is $32° + 96° + 52°$, or $180°$. Three times the measure of angle $A$ is $3 \cdot 32°$, or $96°$, the measure of angle $B$. $20°$ more than the measure of angle $A$ is $32° + 20°$, or $52°$, the measure of angle $C$. The numbers check.

**State.** The measures of angles $A$, $B$, and $C$ are $32°$, $96°$, and $52°$, respectively.

**7. Familiarize.** Let $x$, $y$, and $z$ represent the monthly rates for an individual adult, a spouse, and a child, respectively.

**Translate.** The monthly rate for an individual adult and a spouse (a couple) is $121, so we have

$$x + y = 121.$$

The monthly rate for an individual adult and one child is $107, so we have

$$x + z = 107.$$

The monthly rate for a couple and one child is $164, so we have

$$x + y + z = 164.$$

We now have a system of equations.

$$x + y = 121,$$
$$x + z = 107,$$
$$x + y + z = 164$$

**Carry out.** Solving the system, we get $(64, 57, 43)$.

**Check.** The monthly rate for a couple is $64 + $57, or $121. For an individual adult and one child the monthly rate is $64 + $43, or $107, and for a couple and one child the rate is $64 + $57 + $43, or $164. The answer checks.

**State.** The monthly rate for an individual adult is $64, for a spouse it is $57, and for a child it is $43.

**9. Familiarize.** Let $x$, $y$, and $z$ represent the number of grams of fiber in 1 bran muffin, 1 banana, and a 1-cup serving of Wheaties, respectively.

**Translate.**

Two bran muffins, 1 banana, and a 1-cup serving of Wheaties contain 9 g of fiber, so we have

$$2x + y + z = 9.$$

One bran muffin, 2 bananas, and a 1-cup serving of Wheaties contain 10.5 g of fiber, so we have

$$x + 2y + z = 10.5$$

Two bran muffins and a 1-cup serving of Wheaties contain 6 g of fiber, so we have

$$2x + z = 6.$$

We now have a system of equations.

$$2x + y + z = 9,$$
$$x + 2y + z = 10.5,$$
$$2x + z = 6$$

**Carry out.** Solving the system, we get $(1.5, 3, 3)$.

**Check.** Two bran muffins 1 banana, and a 1-cup serving of Wheaties contain $2(1.5) + 3 + 3$, or 9 g of fiber. One bran muffin, 2 bananas, and a 1-cup serving of Wheaties contain $1.5 + 2 \cdot 3 + 3$, or 10.5 g of fiber. Two bran muffins and a 1-cup serving of Wheaties contain $2(1.5) + 3$, or 6 g of fiber. The answer checks.

**State.** A bran muffin has 1.5 g of fiber, a banana has 3 g, and a 1-cup serving of Wheaties has 3 g.

**11.** Observe that the basic model with a sunroof costs $25,495 and when 4WD is added the price rises to $27,465. This tells us that the price of 4WD is $27,465 − $25,495, or $1970. Now observe that the basic model with 4WD costs $26,665 so the basic model costs $26,665 − $1970, or $24,695. Finally, we know that the sunroof costs $25,495 − $24,695, or $800.

**13.** We know that Elrod, Dot, and Wendy can weld 74 linear feet per hour when working together. We also know that Elrod and Dot together can weld 44 linear feet per hour, which leads to the conclusion that Wendy can weld $74−44$, or 30 linear feet per hour alone. We also know that Elrod and Wendy together can weld 50 linear feet per hour. This, along with the earlier conclusion that Wendy can weld 30 linear feet per hour alone, leads to two conclusions: Elrod can weld $50 − 30$, or 20 linear feet per hour alone and Dot can weld $74 − 50$, or 24 linear feet per hour alone.

**15. Familiarize.** Let $x =$ the number of 12-oz cups, $y =$ the number of 16-oz cups, and $z =$ the number of 20-oz cups that Roz filled. Note that six 144-oz brewers contain $6 \cdot 144$, or 864 oz of coffee. Also, $x$ 12-oz cups contain a total of

12x oz of coffee and bring in $1.40x, y 16-oz cups contain 16y oz and bring in $1.60y, and z 20-oz cups contain 20z oz and bring in $1.70z.

**Translate**.

The total number of coffees served  was  55.

$$x + y + z \qquad = \qquad 55$$

The total amount of coffee served  was  864 oz.

$$12x + 16y + 20z \qquad = \qquad 864$$

The total amount collected  was  $85.90.

$$1.40x + 1.60y + 1.70z \qquad = \qquad 85.90$$

Now we have a system of equations.

$$x + \quad y + \quad z = \quad 55,$$
$$12x + \quad 16y + \quad 20z = \quad 864,$$
$$1.40x + 1.60y + 1.70z = 85.90$$

**Carry out**. Solving the system we get $(17, 25, 13)$.

**Check**. The total number of coffees served was $17+25+13$, or 55. The total amount of coffee served was $12 \cdot 17 + 16 \cdot 25 + 20 \cdot 13 = 204 + 400 + 260 = 864$ oz. The total amount collected was $\$1.40(17) + \$1.60(25) + \$1.70(13) = \$23.80 + \$40 + \$22.10 = \$85.90$. The numbers check.

**State**. Roz filled 17 12-oz cups, 25 16-oz cups, and 13 20-oz cups.

**17. Familiarize**. Let $x$, $y$, and $z$ represent the number of small, medium, and large soft drinks sold, respectively. Then $x$ small drinks brought in $\$1 \cdot x$, or $\$x$; $y$ medium drinks brought in $\$1.15y$; and $z$ large drinks brought in $\$1.30z$.

**Translate**.

The total number of drinks served  was  40.

$$x + y + z \qquad = \qquad 40$$

The total amount collected  was  $45.25.

$$x + 1.15y + 1.30z \qquad = \qquad 45.25$$

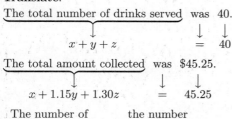

The number of small and large drinks sold  was  the number of medium drinks sold  less 10.

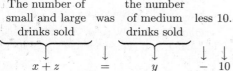

$$x + z \qquad = \qquad y \qquad - 10$$

Now we have a system of equations.

$$x + y + z = 40,$$
$$x + 1.15y + 1.30z = 45.25,$$
$$x + z = y - 10$$

**Carry out**. Solving the system, we get $(10, 25, 5)$.

**Check**. The total number of drinks sold was $10 + 25 + 5$, or 40. The total amount they brought in was $\$1(10)+\$1.15(25)+\$1.30(5)=\$10+\$28.75+\$6.50=\$45.25$. The number of small and large drinks sold, $10 + 5$, or 15,

is 10 less than 25, the number of medium drinks sold. The numbers check.

**State**. 10 small drinks, 25 medium drinks, and 5 large drinks were sold.

**19. Familiarize**. Let $r =$ the number of servings of roast beef, $p =$ the number of baked potatoes, and $b =$ the number of servings of broccoli. Then $r$ servings of roast beef contain $300r$ Calories, $20r$ g of protein, and no vitamin C. In $p$ baked potatoes there are $100p$ Calories, $5p$ g of protein, and $20p$ mg of vitamin C. And $b$ servings of broccoli contain $50b$ Calories, $5b$ g of protein, and $100b$ mg of vitamin C. The patient requires 800 Calories, 55 g of protein, and 220 mg of vitamin C.

**Translate**. Write equations for the total number of calories, the total amount of protein, and the total amount of vitamin C.

$$300r + 100p + 50b = 800 \quad \text{(Calories)}$$
$$20r + \quad 5p + \quad 5b = 55 \quad \text{(protein)}$$
$$20p + 100b = 220 \quad \text{(vitamin C)}$$

We now have a system of equations.

**Carry out**. Solving the system we get $(2, 1, 2)$.

**Check**. Two servings of roast beef provide 600 Calories, 40 g of protein, and no vitamin C. One baked potato provides 100 Calories, 5 g of protein, and 20 mg of vitamin C. And 2 servings of broccoli provide 100 Calories, 10 g of protein, and 200 mg of vitamin C. Together, then, they provide 800 Calories, 55 g of protein, and 220 mg of vitamin C. The values check.

**State**. The dietician should prepare 2 servings of roast beef, 1 baked potato, and 2 servings of broccoli.

**21. Familiarize**. Let $x$, $y$, and $z$ represent the populations of Asia, Africa, and the rest of the world respectively, in billions, in 2050.

**Translate**.

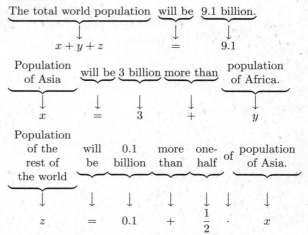

Now we have a system of equations.

$$x + y + z = 9.1,$$
$$x = 3 + y,$$
$$z = 0.1 + \frac{1}{2}x$$

***Carry out***. Solving the system, we get $(4.8, 1.8, 2.5)$.

***Check***. The total population will be $4.8 + 1.8 + 2.5$, or 9.1 billion. The population of Asia, 4.8 billion, is 3 billion more than the population of Africa, 1.8 billion. Also, 0.1 billion more than half the population of Asia is $0.1 + \frac{1}{2}(4.8) = 0.1 + 2.4 = 2.5$ billion, the population of the rest of the world. The numbers check.

***State***. In 2050 the population of Asia will be 4.8 billion, the population of Africa will be 1.8 billion, and the population of the rest of the world will be 2.5 billion.

**23.** ***Familiarize***. Let $x$, $y$, and $z$ represent the number of 2-point field goals, 3-point field goals, and 1-point foul shots made, respectively. The total number of points scored from each of these types of goals is $2x$, $3y$, and $z$.

***Translate***.

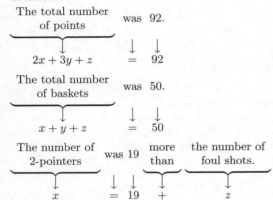

Now we have a system of equations.

$$2x + 3y + z = 92,$$
$$x + y + z = 50,$$
$$x = 19 + z$$

***Carry out***. Solving the system we get $(32, 5, 13)$.

***Check***. The total number of points was $2 \cdot 32 + 3 \cdot 5 + 13 = 64 + 15 + 13 = 92$. The number of baskets was $32 + 5 + 13$, or 50. The number of 2-pointers, 32, was 19 more than the number of foul shots, 13. The numbers check.

***State***. The Knicks made 32 two-point field goals, 5 three-point field goals, and 13 foul shots.

**25.** *Writing Exercise*

**27.** $5(-3) + 7 = -15 + 7 = -8$

**29.** $-6(8) + (-7) = -48 + (-7) = -55$

**31.** $-7(2x - 3y + 5z) = -7 \cdot 2x - 7(-3y) - 7(5z)$
$$= -14x + 21y - 35z$$

**33.** $\qquad -4(2a + 5b) + 3a + 20b$
$$= -8a - 20b + 3a + 20b$$
$$= -8a + 3a - 20b + 20b$$
$$= -5a$$

**35.** *Writing Exercise*

**37.** ***Familiarize***. Let $w$, $x$, $y$, and $z$ represent the monthly rates for an applicant, a spouse, the first child, and the second child respectively.

***Translate***.

The rate for the applicant and his or her spouse is \$160/month, so we have

$$w + x = 160.$$

The rate for the applicant, a spouse, and one child is \$203/month, so we have

$$w + x + y = 203.$$

The rate for the applicant, a spouse, and two children is \$243/month, so we have

$$w + x + y + z = 243.$$

The rate for the applicant and one child is \$145/month, so we have

$$w + y = 145.$$

Now we have a system of equations.

$$w + x = 160, \qquad (1)$$
$$w + x + y = 203, \qquad (2)$$
$$w + x + y + z = 243, \qquad (3)$$
$$w + y = 145 \qquad (4)$$

***Carry out***. We solve the system of equations. First substitute 203 for $w + x + y$ in (3) and solve for $z$.

$$203 + z = 243$$
$$z = 40$$

Next substitute 160 for $w + x$ in (2) and solve for $y$.

$$160 + y = 203$$
$$y = 43$$

Substitute 43 for $y$ in (4) and solve for $w$.

$$w + 43 = 145$$
$$w = 102$$

Finally substitute 102 for $w$ in (1) and solve for $x$.

$$102 + x = 160$$
$$x = 58$$

The solution is $(102, 58, 43, 40)$.

***Check***. The check is left to the student.

***State***. The separate monthly rates for an applicant, a spouse, the first child, and the second child are \$102, \$58, \$43, and \$40, respectively.

**39.** ***Familiarize***. Let $w$, $x$, $y$, and $z$ represent the ages of Tammy, Carmen, Dennis, and Mark respectively.

***Translate***.

Tammy's age is the sum of the ages of Carmen and Dennis, so we have

$$w = x + y.$$

Carmen's age is 2 more than the sum of the ages of Dennis and Mark, so we have

$$x = 2 + y + z.$$

Dennis's age is four times Mark's age, so we have

$y = 4z.$

The sum of all four ages is 42, so we have

$w + x + y + z = 42.$

Now we have a system of equations.

$w = x + y,$       (1)

$x = 2 + y + z,$       (2)

$y = 4z,$       (3)

$w + x + y + z = 42$    (4)

*Carry out*. We solve the system of equations. First we will express $w$, $x$, and $y$ in terms of $z$ and then solve for $z$. From (3) we know that $y = 4z$. Substitute $4z$ for $y$ in (2):

$x = 2 + 4z + z = 2 + 5z.$

Substitute $2 + 5z$ for $x$ and $4z$ for $y$ in (1):

$w = 2 + 5z + 4z = 2 + 9z.$

Now substitute $2 + 9z$ for $w$, $2 + 5z$ for $x$, and $4z$ for $y$ in (4) and solve for $z$.

$$2 + 9z + 2 + 5z + 4z + z = 42$$
$$19z + 4 = 42$$
$$19z = 38$$
$$z = 2$$

Then we have:

$w = 2 + 9z = 2 + 9 \cdot 2 = 20,$

$x = 2 + 5z = 2 + 5 \cdot 2 = 12,$ and

$y = 4z = 4 \cdot 2 = 8$

Although we were asked to find only Tammy's age, we found all of the ages so that we can check the result.

*Check*. The check is left to the student.

*State*. Tammy is 20 years old.

**41.** Let $T$, $G$, and $H$ represent the number of tickets Tom, Gary, and Hal begin with, respectively. After Hal gives tickets to Tom and Gary, each has the following number of tickets:

Tom:    $T + T$, or $2T$,

Gary:    $G + G$, or $2G$,

Hal:    $H - T - G$.

After Tom gives tickets to Gary and Hal, each has the following number of tickets:

Gary:    $2G + 2G$, or $4G$,

Hal:    $(H - T - G) + (H - T - G)$, or

       $2(H - T - G),$

Tom:    $2T - 2G - (H - T - G)$, or

       $3T - H - G$

After Gary gives tickets to Hal and Tom, each has the following number of tickets:

Hal:    $2(H - T - G) + 2(H - T - G)$, or

       $4(H - T - G)$

Tom:    $(3T - H - G) + (3T - H - G)$, or

       $2(3T - H - G),$

Gary:    $4G - 2(H - T - G) - (3T - H - G)$, or

       $7G - H - T.$

Since Hal, Tom, and Gary each finish with 40 tickets, we write the following system of equations:

$$4(H - T - G) = 40,$$
$$2(3T - H - G) = 40,$$
$$7G - H - T = 40$$

Solving the system we find that $T = 35$, so Tom started with 35 tickets.

## Exercise Set 3.6

**1.** horizontal; columns; see page 196 in the text.

**3.** entry; see page 196 in the text.

**5.** multiple; see page 199 in the text.

**7.** $9x - 2y = 5,$

    $3x - 3y = 11$

Write a matrix using only the constants.

$$\begin{bmatrix} 9 & -2 & \vdots & 5 \\ 3 & -3 & \vdots & 11 \end{bmatrix}$$

Multiply row 2 by 3 to make the first number in row 2 a multiple of 9.

$$\begin{bmatrix} 9 & -2 & \vdots & 5 \\ 9 & -9 & \vdots & 33 \end{bmatrix}$$ New Row 2 = 3(Row 2)

Multiply row 1 by $-1$ and add it to row 2.

$$\begin{bmatrix} 9 & -2 & \vdots & 5 \\ 0 & -7 & \vdots & 28 \end{bmatrix}$$ New Row 2 = $-1$(Row 1) + Row 2

Reinserting the variables, we have

$9x - 2y = 5,$    (1)

     $-7y = 28.$    (2)

Solve Equation (2) for $y$.

$$-7y = 28$$
$$y = -4$$

Substitute $-4$ for $y$ in Equation (1) and solve for $x$.

$$9x - 2y = 5$$
$$9x - 2(-4) = 5$$
$$9x + 8 = 5$$
$$9x = -3$$
$$x = -\frac{1}{3}$$

The solution is $\left( -\frac{1}{3}, -4 \right).$

**9.** $x + 4y = 8,$

$\quad 3x + 5y = 3$

We first write a matrix using only the constants.

$$\begin{bmatrix} 1 & 4 & \vdots & 8 \\ 3 & 5 & \vdots & 3 \end{bmatrix}$$

Multiply the first row by $-3$ and add it to the second row.

$$\begin{bmatrix} 1 & 4 & \vdots & 8 \\ 0 & -7 & \vdots & -21 \end{bmatrix} \text{New Row } 2 = -3(\text{Row } 1) + \text{Row } 2$$

Reinserting the variables, we have

$\quad x + 4y = 8, \qquad (1)$

$\qquad -7y = -21. \quad (2)$

Solve Equation (2) for $y$.

$\quad -7y = -21$

$\qquad y = 3$

Substitute 3 for $y$ in Equation (1) and solve for $x$.

$\quad x + 4 \cdot 3 = 8$

$\quad x + 12 = 8$

$\qquad x = -4$

The solution is $(-4, 3)$.

**11.** $6x - 2y = 4,$

$\quad 7x + \quad y = 13$

Write a matrix using only the constants.

$$\begin{bmatrix} 6 & -2 & \vdots & 4 \\ 7 & 1 & \vdots & 13 \end{bmatrix}$$

Multiply the second row by 6 to make the first number in row 2 a multiple of 6.

$$\begin{bmatrix} 6 & -2 & \vdots & 4 \\ 42 & 6 & \vdots & 78 \end{bmatrix} \text{New Row } 2 = 6(\text{Row } 2)$$

Now multiply the first row by $-7$ and add it to the second row.

$$\begin{bmatrix} 6 & -2 & \vdots & 4 \\ 0 & 20 & \vdots & 50 \end{bmatrix} \text{New Row } 2 = -7(\text{Row } 1) + \text{Row } 2$$

Reinserting the variables, we have

$\quad 6x - 2y = 4, \qquad (1)$

$\qquad 20y = 50. \quad (2)$

Solve Equation (2) for $y$.

$\quad 20y = 50$

$\qquad y = \dfrac{5}{2}$

Substitute $\dfrac{5}{2}$ for $y$ in Equation (1) and solve for $x$.

$6x - 2y = 4$

$6x - 2\left(\dfrac{5}{2}\right) = 4$

$6x - 5 = 4$

$6x = 9$

$x = \dfrac{3}{2}$

The solution is $\left(\dfrac{3}{2}, \dfrac{5}{2}\right)$.

**13.** $3x + 2y + 2z = 3,$

$\quad x + 2y - \quad z = 5,$

$\quad 2x - 4y + \quad z = 0$

We first write a matrix using only the constants.

$$\begin{bmatrix} 3 & 2 & 2 & \vdots & 3 \\ 1 & 2 & -1 & \vdots & 5 \\ 2 & -4 & 1 & \vdots & 0 \end{bmatrix}$$

First interchange rows 1 and 2 so that each number below the first number in the first row is a multiple of that number.

$$\begin{bmatrix} 1 & 2 & -1 & \vdots & 5 \\ 3 & 2 & 2 & \vdots & 3 \\ 2 & -4 & 1 & \vdots & 0 \end{bmatrix}$$

Multiply row 1 by $-3$ and add it to row 2.

Multiply row 1 by $-2$ and add it to row 3.

$$\begin{bmatrix} 1 & 2 & -1 & \vdots & 5 \\ 0 & -4 & 5 & \vdots & -12 \\ 0 & -8 & 3 & \vdots & -10 \end{bmatrix}$$

Multiply row 2 by $-2$ and add it to row 3.

$$\begin{bmatrix} 1 & 2 & -1 & \vdots & 5 \\ 0 & -4 & 5 & \vdots & -12 \\ 0 & 0 & -7 & \vdots & 14 \end{bmatrix}$$

Reinserting the variables, we have

$\quad x + 2y - z = 5, \qquad (1)$

$\qquad -4y + 5z = -12, \quad (2)$

$\qquad\qquad -7z = 14. \qquad (3)$

Solve (3) for $z$.

$\quad -7z = 14$

$\qquad z = -2$

Substitute $-2$ for $z$ in (2) and solve for $y$.

$\quad -4y + 5(-2) = -12$

$\quad -4y - 10 = -12$

$\qquad -4y = -2$

$\qquad\quad y = \dfrac{1}{2}$

Substitute $\dfrac{1}{2}$ for $y$ and $-2$ for $z$ in (1) and solve for $x$.

$x + 2 \cdot \dfrac{1}{2} - (-2) = 5$

$x + 1 + 2 = 5$

$x + 3 = 5$

$x = 2$

The solution is $\left(2, \frac{1}{2}, -2\right)$.

**15.** $p - 2q - 3r = 3,$

$2p - q - 2r = 4,$

$4p + 5q + 6r = 4$

We first write a matrix using only the constants.

$$\begin{bmatrix} 1 & -2 & -3 & | & 3 \\ 2 & -1 & -2 & | & 4 \\ 4 & 5 & 6 & | & 4 \end{bmatrix}$$

$$\begin{bmatrix} 1 & -2 & -3 & | & 3 \\ 0 & 3 & 4 & | & -2 \\ 0 & 13 & 18 & | & -8 \end{bmatrix}$$
New Row 2 = $-2$(Row 1) + Row 2
New Row 3 = $-4$(Row 1) + Row 3

$$\begin{bmatrix} 1 & -2 & -3 & | & 3 \\ 0 & 3 & 4 & | & -2 \\ 0 & 39 & 54 & | & -24 \end{bmatrix}$$ New Row 3 = 3(Row 3)

$$\begin{bmatrix} 1 & -2 & -3 & | & 3 \\ 0 & 3 & 4 & | & -2 \\ 0 & 0 & 2 & | & 2 \end{bmatrix}$$ New Row 3 = $-13$(Row 2)+ Row 3

Reinserting the variables, we have

$p - 2q - 3r = 3,$ (1)

$3q + 4r = -2,$ (2)

$2r = 2$ (3)

Solve (3) for $r$.

$2r = 2$

$r = 1$

Substitute 1 for $r$ in (2) and solve for $q$.

$3q + 4 \cdot 1 = -2$

$3q + 4 = -2$

$3q = -6$

$q = -2$

Substitute $-2$ for $q$ and 1 for $r$ in (1) and solve for $p$.

$p - 2(-2) - 3 \cdot 1 = 3$

$p + 4 - 3 = 3$

$p + 1 = 3$

$p = 2$

The solution is $(2, -2, 1)$.

**17.** $3p \qquad + 2r = 11,$

$q - 7r = 4,$

$p - 6q \qquad = 1$

We first write a matrix using only the constants.

$$\begin{bmatrix} 3 & 0 & 2 & | & 11 \\ 0 & 1 & -7 & | & 4 \\ 1 & -6 & 0 & | & 1 \end{bmatrix}$$

$$\begin{bmatrix} 1 & -6 & 0 & | & 1 \\ 0 & 1 & -7 & | & 4 \\ 3 & 0 & 2 & | & 11 \end{bmatrix}$$ Interchange Row 1 and Row 3

$$\begin{bmatrix} 1 & -6 & 0 & | & 1 \\ 0 & 1 & -7 & | & 4 \\ 0 & 18 & 2 & | & 8 \end{bmatrix}$$ New Row 3 = $-3$(Row 1) + Row 3

$$\begin{bmatrix} 1 & -6 & 0 & | & 1 \\ 0 & 1 & -7 & | & 4 \\ 0 & 0 & 128 & | & -64 \end{bmatrix}$$ New Row 3 = $-18$(Row 2) + Row 3

Reinserting the variables, we have

$p - 6q \qquad = 1,$ (1)

$q - 7r = 4,$ (2)

$128r = -64.$ (3)

Solve (3) for $r$.

$128r = -64$

$r = -\frac{1}{2}$

Substitute $-\frac{1}{2}$ for $r$ in (2) and solve for $q$.

$q - 7r = 4$

$q - 7\left(-\frac{1}{2}\right) = 4$

$q + \frac{7}{2} = 4$

$q = \frac{1}{2}$

Substitute $\frac{1}{2}$ for $q$ in (1) and solve for $p$.

$p - 6 \cdot \frac{1}{2} = 1$

$p - 3 = 1$

$p = 4$

The solution is $\left(4, \frac{1}{2}, -\frac{1}{2}\right)$.

**19.** We will rewrite the equations with the variables in alphabetical order:

$-2w + 2x + 2y - 2z = -10,$

$w + x + y + z = -5,$

$3w + x - y + 4z = -2,$

$w + 3x - 2y + 2z = -6$

Write a matrix using only the constants.

$$\begin{bmatrix} -2 & 2 & 2 & -2 & | & -10 \\ 1 & 1 & 1 & 1 & | & -5 \\ 3 & 1 & -1 & 4 & | & -2 \\ 1 & 3 & -2 & 2 & | & -6 \end{bmatrix}$$

$$\begin{bmatrix} -1 & 1 & 1 & -1 & | & -5 \\ 1 & 1 & 1 & 1 & | & -5 \\ 3 & 1 & -1 & 4 & | & -2 \\ 1 & 3 & -2 & 2 & | & -6 \end{bmatrix}$$ New Row 1 = $\frac{1}{2}$(Row 1)

$$\begin{bmatrix} -1 & 1 & 1 & -1 & | & -5 \\ 0 & 2 & 2 & 0 & | & -10 \\ 0 & 4 & 2 & 1 & | & -17 \\ 0 & 4 & -1 & 1 & | & -11 \end{bmatrix}$$ New Row 2 = Row 1 + Row 2
New Row 3 = 3(Row 1) + Row 3
New Row 4 = Row 1 + Row 4

$$\begin{bmatrix} -1 & 1 & 1 & -1 & | & -5 \\ 0 & 2 & 2 & 0 & | & -10 \\ 0 & 0 & -2 & 1 & | & 3 \\ 0 & 0 & -5 & 1 & | & 9 \end{bmatrix}$$ New Row 3 = −2(Row 2) + Row 3
New Row 4 = −2(Row 2) + Row 4

$$\begin{bmatrix} -1 & 1 & 1 & -1 & | & -5 \\ 0 & 2 & 2 & 0 & | & -10 \\ 0 & 0 & -2 & 1 & | & 3 \\ 0 & 0 & -10 & 2 & | & 18 \end{bmatrix}$$ New Row 4 = 2(Row 4)

$$\begin{bmatrix} -1 & 1 & 1 & -1 & | & -5 \\ 0 & 2 & 2 & 0 & | & -10 \\ 0 & 0 & -2 & 1 & | & 3 \\ 0 & 0 & 0 & -3 & | & 3 \end{bmatrix}$$ New Row 4 = −5(Row 3) + Row 4

Reinserting the variables, we have

$$-w + x + y - z = -5, \quad (1)$$
$$2x + 2y \qquad\quad = -10, \quad (2)$$
$$- 2y + z = 3, \quad (3)$$
$$- 3z = 3. \quad (4)$$

Solve (4) for $z$.

$$-3z = 3$$
$$z = -1$$

Substitute −1 for $z$ in (3) and solve for $y$.

$$-2y + (-1) = 3$$
$$-2y = 4$$
$$y = -2$$

Substitute −2 for $y$ in (2) and solve for $x$.

$$2x + 2(-2) = -10$$
$$2x - 4 = -10$$
$$2x = -6$$
$$x = -3$$

Substitute −3 for $x$, −2 for $y$, and −1 for $z$ in (1) and solve for $w$.

$$-w + (-3) + (-2) - (-1) = -5$$
$$-w - 3 - 2 + 1 = -5$$
$$-w - 4 = -5$$
$$-w = -1$$
$$w = 1$$

The solution is $(1, -3, -2, -1)$.

**21.** ***Familiarize***. Let $d$ = the number of dimes and $n$ = the number of nickels. The value of $d$ dimes is $\$0.10d$, and the value of $n$ nickels is $\$0.05n$.

***Translate***.

Total number of coins is 42.

$$d + n = 42$$

Total value of coins is $3.

$$0.10d + 0.05n = 3$$

After clearing decimals, we have this system.

$$d + n = 42,$$
$$10d + 5n = 300$$

***Carry out***. Solve using matrices.

$$\begin{bmatrix} 1 & 1 & | & 42 \\ 10 & 5 & | & 300 \end{bmatrix}$$

$$\begin{bmatrix} 1 & 1 & | & 42 \\ 0 & -5 & | & -120 \end{bmatrix}$$ New Row 2 = −10(Row 1) + Row 2

Reinserting the variables, we have

$$d + n = 42, \quad (1)$$
$$-5n = -120 \quad (2)$$

Solve (2) for $n$.

$$-5n = -120$$
$$n = 24$$

$$d + 24 = 42 \quad \text{Back-substituting}$$
$$d = 18$$

***Check***. The sum of the two numbers is 42. The total value is $\$0.10(18) + \$0.05(24) = \$1.80 + \$1.20 = \$3$. The numbers check.

***State***. There are 18 dimes and 24 nickels.

**23.** ***Familiarize***. We let $x$ represent the number of pounds of the $\$4.05$ kind and $y$ represent the number of pounds of the $\$2.70$ kind of granola. We organize the information in a table.

| Granola | Number of pounds | Price per pound | Value |
|---|---|---|---|
| $4.05 kind | $x$ | $4.05 | $4.05x$ |
| $2.70 kind | $y$ | $2.70 | $2.70y$ |
| Mixture | 15 | $3.15 | $3.15 \times 15$ or $47.25 |

*Translate.*

Total number of pounds is 15.

$$x + y = 15$$

Total value of mixture is $47.25.

$$4.05x + 2.70y = 47.25$$

After clearing decimals, we have this system:

$$x + y = 15,$$
$$405x + 270y = 4725$$

*Carry out.* Solve using matrices.

$$\begin{bmatrix} 1 & 1 & | & 15 \\ 405 & 270 & | & 4725 \end{bmatrix}$$

$$\begin{bmatrix} 1 & 1 & | & 15 \\ 0 & -135 & | & -1350 \end{bmatrix} \text{New Row 2} =$$
$$-405(\text{Row 1}) + \text{Row 2}$$

Reinserting the variables, we have

$$x + y = 15, \quad (1)$$
$$-135y = -1350 \quad (2)$$

Solve (2) for $y$.

$$-135y = -1350$$
$$y = 10$$

Back-substitute 10 for $y$ in (1) and solve for $x$.

$$x + 10 = 15$$
$$x = 5$$

*Check.* The sum of the numbers is 15. The total value is $4.05(5) + $2.70(10)$, or $20.25 + $27.00$, or $47.25$. The numbers check.

*State.* 5 pounds of the $4.05 per lb granola and 10 pounds of the $2.70 per lb granola should be used.

**25.** *Familiarize.* We let $x$, $y$, and $z$ represent the amounts invested at 7%, 8%, and 9%, respectively. Recall the formula for simple interest:

$$\text{Interest} = \text{Principal} \times \text{Rate} \times \text{Time}$$

*Translate.* We organize the information in a table.

| | First Investment | Second Investment | Third Investment | Total |
|---|---|---|---|---|
| P | $x$ | $y$ | $z$ | $2500 |
| R | 7% | 8% | 9% | |
| T | 1 yr | 1 yr | 1 yr | |
| I | $0.07x$ | $0.08y$ | $0.09z$ | $212 |

The first row gives us one equation:

$$x + y + z = 2500$$

The last row gives a second equation:

$$0.07x + 0.08y + 0.09z = 212$$

Amount invested at 9% is $1100 more than amount invested at 8%.

$$z = $1100 + y$$

After clearing decimals, we have this system:

$$x + y + z = 2500,$$
$$7x + 8y + 9z = 21,200,$$
$$-y + z = 1100$$

*Carry out.* Solve using matrices.

$$\begin{bmatrix} 1 & 1 & 1 & | & 2500 \\ 7 & 8 & 9 & | & 21,200 \\ 0 & -1 & 1 & | & 1100 \end{bmatrix}$$

$$\begin{bmatrix} 1 & 1 & 1 & | & 2500 \\ 0 & 1 & 2 & | & 3700 \\ 0 & -1 & 1 & | & 1100 \end{bmatrix} \begin{array}{l} \text{New Row 2} = \\ -7(\text{Row 1}) + \text{Row 2} \end{array}$$

$$\begin{bmatrix} 1 & 1 & 1 & | & 2500 \\ 0 & 1 & 2 & | & 3700 \\ 0 & 0 & 3 & | & 4800 \end{bmatrix} \begin{array}{l} \text{New Row 3} = \\ \text{Row 2} + \text{Row 3} \end{array}$$

Reinserting the variables, we have

$$x + y + z = 2500, \quad (1)$$
$$y + 2z = 3700, \quad (2)$$
$$3z = 4800 \quad (3)$$

Solve (3) for $z$.

$$3z = 4800$$
$$z = 1600$$

Back-substitute 1600 for $z$ in (2) and solve for $y$.

$$y + 2 \cdot 1600 = 3700$$
$$y + 3200 = 3700$$
$$y = 500$$

Back-substitute 500 for $y$ and 1600 for $z$ in (1) and solve for $x$.

$$x + 500 + 1600 = 2500$$
$$x + 2100 = 2500$$
$$x = 400$$

*Check.* The total investment is $400 + $500 + $1600$, or $2500$. The total interest is $0.07($400) + 0.08($500) + 0.09($1600) = $28 + $40 + $144 = $212$. The amount invested at 9%, $1600, is $1100 more than the amount invested at 8%, $500. The numbers check.

*State.* $400 is invested at 7%, $500 is invested at 8%, and $1600 is invested at 9%.

**27.** *Writing Exercise*

**29.** $5(-3) - (-7)4 = -15 - (-28) = -15 + 28 = 13$

**31.**
$$-2(5 \cdot 3 - 4 \cdot 6) - 3(2 \cdot 7 - 15) + 4(3 \cdot 8 - 5 \cdot 4)$$
$$= -2(15 - 24) - 3(14 - 15) + 4(24 - 20)$$
$$= -2(-9) - 3(-1) + 4(4)$$
$$= 18 + 3 + 16$$
$$= 21 + 16$$
$$= 37$$

**33.** *Writing Exercise*

**35.** ***Familiarize***. Let $w$, $x$, $y$, and $z$ represent the thousand's, hundred's, ten's, and one's digits, respectively.

***Translate***.

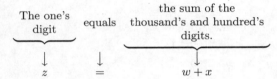

We have a system of equations which can be written as

$$w + x + y + z = 10,$$
$$2w - x + 2y - z = -1,$$
$$-2w \quad\quad + y \quad\quad = 0,$$
$$w + x \quad\quad - z = 0.$$

***Carry out***. We can use matrices to solve the system. We get $(1, 3, 2, 4)$.

***Check***. The sum of the digits is 10. Twice the sum of 1 and 2 is 6. This is one less than the sum of 3 and 4. The ten's digit, 2, is twice the thousand's digit, 1. The one's digit, 4, equals $1 + 3$. The numbers check.

***State***. The number is 1324.

## Exercise Set 3.7

**1.** True; see page 201 in the text.

**3.** False; see page 203 in the text.

**5.** False; it is the value of the denominator, not the numerator, that indicates whether the equations are dependent. See page 202 in the text.

**7.** $\begin{vmatrix} 5 & 1 \\ 2 & 4 \end{vmatrix} = 5 \cdot 4 - 2 \cdot 1 = 20 - 2 = 18$

**9.** $\begin{vmatrix} 6 & -9 \\ 2 & 3 \end{vmatrix} = 6 \cdot 3 - 2(-9) = 18 + 18 = 36$

**11.** $\begin{vmatrix} 1 & 4 & 0 \\ 0 & -1 & 2 \\ 3 & -2 & 1 \end{vmatrix}$

$= 1 \begin{vmatrix} -1 & 2 \\ -2 & 1 \end{vmatrix} - 0 \begin{vmatrix} 4 & 0 \\ -2 & 1 \end{vmatrix} + 3 \begin{vmatrix} 4 & 0 \\ -1 & 2 \end{vmatrix}$

$= 1[-1 \cdot 1 - (-2) \cdot 2] - 0 + 3[4 \cdot 2 - (-1) \cdot 0]$

$= 1 \cdot 3 - 0 + 3 \cdot 8$

$= 3 - 0 + 24$

$= 27$

**13.** $\begin{vmatrix} -1 & -2 & -3 \\ 3 & 4 & 2 \\ 0 & 1 & 2 \end{vmatrix}$

$= -1 \begin{vmatrix} 4 & 2 \\ 1 & 2 \end{vmatrix} - 3 \begin{vmatrix} -2 & -3 \\ 1 & 2 \end{vmatrix} + 0 \begin{vmatrix} -2 & -3 \\ 4 & 2 \end{vmatrix}$

$= -1[4 \cdot 2 - 1 \cdot 2] - 3[-2 \cdot 2 - 1(-3)] + 0$

$= -1 \cdot 6 - 3 \cdot (-1) + 0$

$= -6 + 3 + 0$

$= -3$

**15.** $\begin{vmatrix} -4 & -2 & 3 \\ -3 & 1 & 2 \\ 3 & 4 & -2 \end{vmatrix}$

$= -4 \begin{vmatrix} 1 & 2 \\ 4 & -2 \end{vmatrix} - (-3) \begin{vmatrix} -2 & 3 \\ 4 & -2 \end{vmatrix} + 3 \begin{vmatrix} -2 & 3 \\ 1 & 2 \end{vmatrix}$

$= -4[1(-2) - 4 \cdot 2] + 3[-2(-2) - 4 \cdot 3] + $
$\quad\quad 3(-2 \cdot 2 - 1 \cdot 3)$

$= -4(-10) + 3(-8) + 3(-7)$

$= 40 - 24 - 21 = -5$

**17.** $5x + 8y = 1,$
$3x + 7y = 5$

We compute $D$, $D_x$, and $D_y$.

$D = \begin{vmatrix} 5 & 8 \\ 3 & 7 \end{vmatrix} = 35 - 24 = 11$

$D_x = \begin{vmatrix} 1 & 8 \\ 5 & 7 \end{vmatrix} = 7 - 40 = -33$

$D_y = \begin{vmatrix} 5 & 1 \\ 3 & 5 \end{vmatrix} = 25 - 3 = 22$

Then,

$$x = \frac{D_x}{D} = \frac{-33}{11} = -3$$

and

$$y = \frac{D_y}{D} = \frac{22}{11} = 2.$$

The solution is $(-3, 2)$.

**19.** $5x - 4y = -3,$
$7x + 2y = 6$

We compute $D$, $D_x$, and $D_y$.

$$D = \begin{vmatrix} 5 & -4 \\ 7 & 2 \end{vmatrix} = 10 - (-28) = 38$$

$$D_x = \begin{vmatrix} -3 & -4 \\ 6 & 2 \end{vmatrix} = -6 - (-24) = 18$$

$$D_y = \begin{vmatrix} 5 & -3 \\ 7 & 6 \end{vmatrix} = 30 - (-21) = 51$$

Then,

$$x = \frac{D_x}{D} = \frac{18}{38} = \frac{9}{19}$$

and

$$y = \frac{D_y}{D} = \frac{51}{38}.$$

The solution is $\left( \dfrac{9}{19}, \dfrac{51}{38} \right).$

**21.** $3x - y + 2z = 1,$
$x - y + 2z = 3,$
$-2x + 3y + z = 1$

We compute $D$, $D_x$, and $D_y$.

$$D = \begin{vmatrix} 3 & -1 & 2 \\ 1 & -1 & 2 \\ -2 & 3 & 1 \end{vmatrix}$$

$$= 3\begin{vmatrix} -1 & 2 \\ 3 & 1 \end{vmatrix} - 1\begin{vmatrix} -1 & 2 \\ 3 & 1 \end{vmatrix} - 2\begin{vmatrix} -1 & 2 \\ -1 & 2 \end{vmatrix}$$

$$= 3(-7) - 1(-7) - 2(0)$$
$$= -21 + 7 - 0$$
$$= -14$$

$$D_x = \begin{vmatrix} 1 & -1 & 2 \\ 3 & -1 & 2 \\ 1 & 3 & 1 \end{vmatrix}$$

$$= 1\begin{vmatrix} -1 & 2 \\ 3 & 1 \end{vmatrix} - 3\begin{vmatrix} -1 & 2 \\ 3 & 1 \end{vmatrix} + 1\begin{vmatrix} -1 & 2 \\ -1 & 2 \end{vmatrix}$$

$$= 1(-7) - 3(-7) + 1(0)$$
$$= -7 + 21 + 0$$
$$= 14$$

$$D_y = \begin{vmatrix} 3 & 1 & 2 \\ 1 & 3 & 2 \\ -2 & 1 & 1 \end{vmatrix}$$

$$= 3\begin{vmatrix} 3 & 2 \\ 1 & 1 \end{vmatrix} - 1\begin{vmatrix} 1 & 2 \\ 1 & 1 \end{vmatrix} - 2\begin{vmatrix} 1 & 2 \\ 3 & 2 \end{vmatrix}$$

$$= 3 \cdot 1 - 1(-1) - 2(-4)$$
$$= 3 + 1 + 8$$
$$= 12$$

Then,

$$x = \frac{D_x}{D} = \frac{14}{-14} = -1$$

and

$$y = \frac{D_y}{D} = \frac{12}{-14} = -\frac{6}{7}.$$

Substitute in the third equation to find $z$.

$$-2(-1) + 3\left( -\frac{6}{7} \right) + z = 1$$

$$2 - \frac{18}{7} + z = 1$$

$$-\frac{4}{7} + z = 1$$

$$z = \frac{11}{7}$$

The solution is $\left( -1, -\dfrac{6}{7}, \dfrac{11}{7} \right).$

**23.** $2x - 3y + 5z = 27,$
$x + 2y - z = -4,$
$5x - y + 4z = 27$

We compute $D$, $D_x$, and $D_y$.

$$D = \begin{vmatrix} 2 & -3 & 5 \\ 1 & 2 & -1 \\ 5 & -1 & 4 \end{vmatrix}$$

$$= 2\begin{vmatrix} 2 & -1 \\ -1 & 4 \end{vmatrix} - 1\begin{vmatrix} -3 & 5 \\ -1 & 4 \end{vmatrix} + 5\begin{vmatrix} -3 & 5 \\ 2 & -1 \end{vmatrix}$$

$$= 2(7) - 1(-7) + 5(-7)$$
$$= 14 + 7 - 35 = -14$$

$$D_x = \begin{vmatrix} 27 & -3 & 5 \\ -4 & 2 & -1 \\ 27 & -1 & 4 \end{vmatrix}$$

$$= 27\begin{vmatrix} 2 & -1 \\ -1 & 4 \end{vmatrix} - (-4)\begin{vmatrix} -3 & 5 \\ -1 & 4 \end{vmatrix} + 27\begin{vmatrix} -3 & 5 \\ 2 & -1 \end{vmatrix}$$

$$= 27(7) + 4(-7) + 27(-7)$$
$$= 189 - 28 - 189$$
$$= -28$$

$$D_y = \begin{vmatrix} 2 & 27 & 5 \\ 1 & -4 & -1 \\ 5 & 27 & 4 \end{vmatrix}$$

$$= 2\begin{vmatrix} -4 & -1 \\ 27 & 4 \end{vmatrix} - 1\begin{vmatrix} 27 & 5 \\ 27 & 4 \end{vmatrix} + 5\begin{vmatrix} 27 & 5 \\ -4 & -1 \end{vmatrix}$$

$$= 2(11) - 1(-27) + 5(-7)$$
$$= 22 + 27 - 35 = 14$$
$$= 14$$

Then,

$$x = \frac{D_x}{D} = \frac{-28}{-14} = 2,$$

and

$$y = \frac{D_y}{D} = \frac{14}{-14} = -1.$$

We substitute in the second equation to find $z$.

$$2 + 2(-1) - z = -4$$
$$2 - 2 - z = -4$$
$$-z = -4$$
$$z = 4$$

The solution is $(2, -1, 4)$.

**25.**    $r - 2s + 3t = 6,$
$\phantom{}2r - \phantom{2}s - \phantom{2}t = -3,$
$\phantom{2}r + \phantom{2}s + \phantom{2}t = 6$

We compute $D$, $D_r$, and $D_s$.

$$D = \begin{vmatrix} 1 & -2 & 3 \\ 2 & -1 & -1 \\ 1 & 1 & 1 \end{vmatrix}$$

$$= 1\begin{vmatrix} -1 & -1 \\ 1 & 1 \end{vmatrix} - 2\begin{vmatrix} -2 & 3 \\ 1 & 1 \end{vmatrix} + 1\begin{vmatrix} -2 & 3 \\ -1 & -1 \end{vmatrix}$$

$$= 1(0) - 2(-5) + 1(5)$$
$$= 0 + 10 + 5$$
$$= 15$$

$$D_r = \begin{vmatrix} 6 & -2 & 3 \\ -3 & -1 & -1 \\ 6 & 1 & 1 \end{vmatrix}$$

$$= 6\begin{vmatrix} -1 & -1 \\ 1 & 1 \end{vmatrix} - (-3)\begin{vmatrix} -2 & 3 \\ 1 & 1 \end{vmatrix} + 6\begin{vmatrix} -2 & 3 \\ -1 & -1 \end{vmatrix}$$

$$= 6(0) + 3(-5) + 6(5)$$
$$= 0 - 15 + 30$$
$$= 15$$

$$D_s = \begin{vmatrix} 1 & 6 & 3 \\ 2 & -3 & -1 \\ 1 & 6 & 1 \end{vmatrix}$$

$$= 1\begin{vmatrix} -3 & -1 \\ 6 & 1 \end{vmatrix} - 2\begin{vmatrix} 6 & 3 \\ 6 & 1 \end{vmatrix} + 1\begin{vmatrix} 6 & 3 \\ -3 & -1 \end{vmatrix}$$

$$= 1(3) - 2(-12) + 1(3)$$
$$= 3 + 24 + 3$$
$$= 30$$

Then,

$$r = \frac{D_r}{D} = \frac{15}{15} = 1,$$

and

$$s = \frac{D_s}{D} = \frac{30}{15} = 2.$$

Substitute in the third equation to find $t$.

$$1 + 2 + t = 6$$
$$3 + t = 6$$
$$t = 3$$

The solution is $(1, 2, 3)$.

**27.** *Writing Exercise*

**29.**   $0.5x - 2.34 + 2.4x = 7.8x - 9$
$\phantom{0.5x}2.9x - 2.34 = 7.8x - 9$
$\phantom{0.5x - 2.34}6.66 = 4.9x$
$$\frac{6.66}{4.9} = x$$
$$\frac{666}{490} = x$$
$$\frac{333}{245} = x$$

The solution is $\dfrac{333}{245}$.

**31. Familiarize.** We first make a drawing.

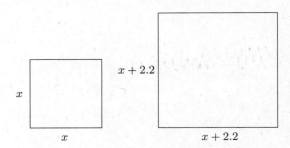

Let $x$ represent the length of a side of the smaller square and $x + 2.2$ the length of a side of the larger square. The perimeter of the smaller square is $4x$. The perimeter of the larger square is $4(x + 2.2)$.

**Translate.**

$\underbrace{\text{The sum of the perimeters}}$ is 32.8 ft.

$$\underbrace{4x + 4(x + 2.2)}\phantom{xx} = \phantom{x} 32.8$$

**Carry out.** We solve the equation.

$$4x + 4x + 8.8 = 32.8$$
$$8x = 24$$
$$x = 3$$

**Check.** If $x = 3$ ft, then $x + 2.2 = 5.2$ ft. The perimeters are $4 \cdot 3$, or 12 ft, and $4(5.2)$, or 20.8 ft. The sum of the two perimeters is $12 + 20.8$, or 32.8 ft. The values check.

**State.** The wire should be cut into two pieces, one measuring 12 ft and the other 20.8 ft.

**33. Familiarize.** Let $x$ represent the number of rolls of insulation required for the Mazzas' attic and let $y$ represent the number of rolls required for the Kranepools' attic.

**Translate.**

Insulation for Mazzas' attic    is    three and a half    times    insulation for Kranepools' attic.

$$\downarrow \qquad \downarrow \quad \downarrow \qquad \downarrow \qquad \downarrow$$
$$x \qquad = \quad 3.5 \qquad \cdot \qquad y$$

Total number of rolls is 36.

$$\downarrow \qquad\qquad \downarrow\ \downarrow$$
$$x + y \qquad = 36$$

We have a system of equations:

$$x = 3.5y, \quad (1)$$
$$x + y = 36 \quad (2)$$

**Carry out.** We use the substitution method to solve the system of equations. First we substitute 3.5y for x in Equation (2).

$$x + y = 36 \quad (2)$$
$$3.5y + y = 36 \quad \text{Substituting}$$
$$4.5y = 36$$
$$y = 8$$

Now substitute 8 for y in Equation (1).

$$x = 3.5(8) = 28$$

**Check.** The number 28 is three and a half times 8. Also, the total number of rolls is 28 + 8, or 36. The answer checks.

**State.** The Mazzas' attic requires 28 rolls of insulation, and the Kranepools' attic requires 8 rolls.

**35.** *Writing Exercise*

**37.** $\begin{vmatrix} y & -2 \\ 4 & 3 \end{vmatrix} = 44$

$\quad y \cdot 3 - 4(-2) = 44 \quad$ Evaluating the determinant

$\quad\quad 3y + 8 = 44$

$\quad\quad\quad 3y = 36$

$\quad\quad\quad y = 12$

**39.** $\begin{vmatrix} m+1 & -2 \\ m-2 & 1 \end{vmatrix} = 27$

$\quad (m+1)(1) - (m-2)(-2) = 27 \quad$ Evaluating the determinant

$\quad\quad m + 1 + 2m - 4 = 27$

$\quad\quad\quad 3m = 30$

$\quad\quad\quad m = 10$

## Exercise Set 3.8

**1.** (b); see page 207 in the text.

**3.** (e); see page 207 in the text.

**5.** (h); see page 207 in the text.

**7.** (g); see page 210 in the text.

**9.** $C(x) = 45x + 300,000 \qquad R(x) = 65x$

a) $P(x) = R(x) - C(x)$

$\quad = 65x - (45x + 300,000)$

$\quad = 65x - 45x - 300,000$

$\quad = 20x - 300,000$

b) To find the break-even point we solve the system

$$R(x) = 65x,$$
$$C(x) = 45x + 300,000.$$

Since $R(x) = C(x)$ at the break-even point, we can rewrite the system:

$$R(x) = 65x, \qquad (1)$$
$$R(x) = 45x + 300,000 \quad (2)$$

We solve using substitution.

$65x = 45x + 300,000 \quad$ Substituting 65x for $R(x)$ in (2)

$20x = 300,000$

$x = 15,000$

Thus, 15,000 units must be produced and sold in order to break even. The revenue will be $R(65) = 65(15,000)$, or 975,000. The break-even point is (15,000 units, \$975,000).

**11.** $C(x) = 10x + 120,000 \qquad R(x) = 60x$

a) $P(x) = R(x) - C(x)$

$\quad = 60x - (10x + 120,000)$

$\quad = 60x - 10x - 120,000$

$\quad = 50x - 120,000$

b) Solve the system

$$R(x) = 60x,$$
$$C(x) = 10x + 120,000.$$

Since both $R(x)$ and $C(x)$ are in dollars and they are equal at the break-even point, we can rewrite the system:

$$d = 60x, \qquad (1)$$
$$d = 10x + 120,000 \quad (2)$$

We solve using substitution.

$60x = 10x + 120,000 \quad$ Substituting 60x for d in (2)

$50x = 120,000$

$x = 2400$

Thus, 2400 units must be produced and sold in order to break even. The revenue will be $R(2400) = 60 \cdot 2400 = 144,000$. The break-even point is (2400 units, \$144,000).

**13.** $C(x) = 40x + 22,500 \qquad R(x) = 85x$

a) $P(x) = R(x) - C(x)$

$\quad = 85x - (40x + 22,500)$

$\quad = 85x - 40x - 22,500$

$\quad = 45x - 22,500$

b) Solve the system

$$R(x) = 85x,$$

$$C(x) = 40x + 22,500.$$

Since both $R(x)$ and $C(x)$ are in dollars and they are equal at the break-even point, we can rewrite the system:

$$d = 85x, \qquad (1)$$

$$d = 40x + 22,500 \quad (2)$$

We solve using substitution.

$85x = 40x + 22,500$  Substituting $85x$ for $d$ in (2)

$$45x = 22,500$$

$$x = 500$$

Thus, 500 units must be produced and sold in order to break even. The revenue will be $R(500) = 85 \cdot 500 = 42,500$. The break-even point is (500 units, $42,500).

**15.** $C(x) = 22x + 16,000 \qquad R(x) = 40x$

a) $P(x) = R(x) - C(x)$

$$= 40x - (22x + 16,000)$$

$$= 40x - 22x - 16,000$$

$$= 18x - 16,000$$

b) Solve the system

$$R(x) = 40x,$$

$$C(x) = 22x + 16,000.$$

Since both $R(x)$ and $C(x)$ are in dollars and they are equal at the break-even point, we can rewrite the system:

$$d = 40x, \qquad (1)$$

$$d = 22x + 16,000 \quad (2)$$

We solve using substitution.

$40x = 22x + 16,000$  Substituting $40x$ for $d$ in (2)

$$18x = 16,000$$

$$x \approx 889 \text{ units}$$

Thus, 889 units must be produced and sold in order to break even. The revenue will be $R(889) = 40 \cdot 889 = 35,560$. The break-even point is (889 units, $35,560).

**17.** $C(x) = 75x + 100,000 \qquad R(x) = 125x$

a) $P(x) = R(x) - C(x)$

$$= 125x - (75x + 100,000)$$

$$= 125x - 75x - 100,000$$

$$= 50x - 100,000$$

b) Solve the system

$$R(x) = 125x,$$

$$C(x) = 75x + 100,000.$$

Since $R(x) = C(x)$ at the break-even point, we can rewrite the system:

$$R(x) = 125x, \qquad (1)$$

$$R(x) = 75x + 100,000 \quad (2)$$

We solve using substitution.

$125x = 75x + 100,000$  Substituting $125x$ for $R(x)$ in (2)

$$50x = 100,000$$

$$x = 2000$$

To break even 2000 units must be produced and sold. The revenue will be $R(2000) = 125 \cdot 2000 = 250,000$. The break-even point is (2000 units, $250,000).

**19.** $D(p) = 1000 - 10p,$

$$S(p) = 230 + p$$

Since both demand and supply are quantities, the system can be rewritten:

$$q = 1000 - 10p, \quad (1)$$

$$q = 230 + p \qquad (2)$$

Substitute $1000 - 10p$ for $q$ in (2) and solve.

$$1000 - 10p = 230 + p$$

$$770 = 11p$$

$$70 = p$$

The equilibrium price is $70 per unit. To find the equilibrium quantity we substitute $70 into either $D(p)$ or $S(p)$.

$$D(70) = 1000 - 10 \cdot 70 = 1000 - 700 = 300$$

The equilibrium quantity is 300 units.

The equilibrium point is ($70, 300).

**21.** $D(p) = 760 - 13p,$

$$S(p) = 430 + 2p$$

Rewrite the system:

$$q = 760 - 13p, \quad (1)$$

$$q = 430 + 2p \qquad (2)$$

Substitute $760 - 13p$ for $q$ in (2) and solve.

$$760 - 13p = 430 + 2p$$

$$330 = 15p$$

$$22 = p$$

The equilibrium price is $22 per unit.

To find the equilibrium quantity we substitute $22 into either $D(p)$ or $S(p)$.

$$S(22) = 430 + 2(22) = 430 + 44 = 474$$

The equilibrium quantity is 474 units.

The equilibrium point is ($22, 474).

**23.** $D(p) = 7500 - 25p,$

$$S(p) = 6000 + 5p$$

Rewrite the system:

$$q = 7500 - 25p, \quad (1)$$

$$q = 6000 + 5p \qquad (2)$$

Substitute $7500 - 25p$ for $q$ in (2) and solve.

$7500 - 25p = 6000 + 5p$

$1500 = 30p$

$50 = p$

The equilibrium price is $50 per unit.

To find the equilibrium quantity we substitute $50 into either $D(p)$ or $S(p)$.

$D(50) = 7500 - 25(50) = 7500 - 1250 = 6250$

The equilibrium quantity is 6250 units.

The equilibrium point is ($50, 6250$).

**25.** $D(p) = 1600 - 53p,$

$S(p) = 320 + 75p$

Rewrite the system:

$q = 1600 - 53p,$   (1)

$q = 320 + 75p$   (2)

Substitute $1600 - 53p$ for $q$ in (2) and solve.

$1600 - 53p = 320 + 75p$

$1280 = 128p$

$10 = p$

The equilibrium price is $10 per unit.

To find the equilibrium quantity we substitute $10 into either $D(p)$ or $S(p)$.

$S(10) = 320 + 75(10) = 320 + 750 = 1070$

The equilibrium quantity is 1070 units.

The equilibrium point is ($10, 1070$).

**27.** a)   $C(x) = $ Fixed costs $+$ Variable costs

$C(x) = 125,300 + 450x,$

where $x$ is the number of computers produced.

b)   Each computer sells for $800. The total revenue is 800 times the number of computers sold. We assume that all computers produced are sold.

$R(x) = 800x$

c)   $P(x) = R(x) - C(x)$

$P(x) = 800x - (125,300 + 450x)$

$= 800x - 125,300 - 450x$

$= 350x - 125,300$

d)   $P(x) = 350x - 125,300$

$P(100) = 350(100) - 125,300$

$= 35,000 - 125,300$

$= -90,300$

The company will realize a $90,300 loss when 100 computers are produced and sold.

$P(400) = 350(400) - 125,300$

$= 140,000 - 125,300$

$= 14,700$

The company will realize a profit of $14,700 from the production and sale of 400 computers.

e)   Solve the system

$R(x) = 800x,$

$C(x) = 125,300 + 450x.$

Since both $R(x)$ and $C(x)$ are in dollars and they are equal at the break-even point, we can rewrite the system:

$d = 800x,$       (1)

$d = 125,300 + 450x$   (2)

We solve using substitution.

$800x = 125,300 + 450x$  Substituting $800x$ for $d$ in (2)

$350x = 125,300$

$x = 358$

The firm will break even if it produces and sells 358 computers and takes in a total of $R(358) = 800 \cdot 358 = \$286,400$ in revenue. Thus, the break-even point is (358 computers, $286,400).

**29.** a)   $C(x) = $ Fixed costs $+$ Variable costs

$C(x) = 16,404 + 6x,$

where $x$ is the number of caps produced, in dozens.

b)   Each dozen caps sell for $18. The total revenue is 18 times the number of caps sold, in dozens. We assume that all caps produced are sold.

$R(x) = 18x$

c)   $P(x) = R(x) - C(x)$

$P(x) = 18x - (16,404 + 6x)$

$= 18x - 16,404 - 6x$

$= 12x - 16,404$

d)   $P(3000) = 12(3000) - 16,404$

$= 36,000 - 16,404$

$= 19,596$

The company will realize a profit of $19,596 when 3000 dozen caps are produced and sold.

$P(1000) = 12(1000) - 16,404$

$= 12,000 - 16,404$

$= -4404$

The company will realize a $4404 loss when 1000 dozen caps are produced and sold.

e)   Solve the system

$R(x) = 18x,$

$C(x) = 16,404 + 6x.$

Since both $R(x)$ and $C(x)$ are in dollars and they are equal at the break-even point, we can rewrite the system:

$d = 18x,$       (1)

$d = 16,404 + 6x$   (2)

We solve using substitution.

$18x = 16,404 + 6x$   Substituting $18x$ for $d$ in (2)

$12x = 16,404$

$x = 1367$

The firm will break even if it produces and sells 1367 dozen caps and takes in a total of $R(1367) = 18 \cdot 1367 = \$24,606$ in revenue. Thus, the break-even point is (1367 dozen caps, $24,606).

**31.** *Writing Exercise*

**33.**   $3x - 9 = 27$

$\quad 3x = 36$   Adding 9 to both sides

$\quad x = 12$   Dividing both sides by 3

The solution is 12.

**35.**   $4x - 5 = 7x - 13$

$\quad -5 = 3x - 13$   Subtracting $4x$ from both sides

$\quad 8 = 3x$   Adding 13 to both sides

$\quad \dfrac{8}{3} = x$   Dividing both sides by 3

The solution is $\dfrac{8}{3}$.

**37.**   $7 - 2(x - 8) = 14$

$\quad 7 - 2x + 16 = 14$   Removing parentheses

$\quad -2x + 23 = 14$   Collecting like terms

$\quad -2x = -9$   Subtracting 23 from both sides

$\quad x = \dfrac{9}{2}$   Dividing both sides by $-2$

The solution is $\dfrac{9}{2}$.

**39.** *Writing Exercise*

**41.** The supply function contains the points ($2, 100) and ($8, 500). We find its equation:

$$m = \frac{500 - 100}{8 - 2} = \frac{400}{6} = \frac{200}{3}$$

$y - y_1 = m(x - x_1)$   Point-slope form

$y - 100 = \dfrac{200}{3}(x - 2)$

$y - 100 = \dfrac{200}{3}x - \dfrac{400}{3}$

$y = \dfrac{200}{3}x - \dfrac{100}{3}$

We can equivalently express supply $S$ as a function of price $p$:

$$S(p) = \frac{200}{3}p - \frac{100}{3}$$

The demand function contains the points ($1, 500) and ($9, 100). We find its equation:

$$m = \frac{100 - 500}{9 - 1} = \frac{-400}{8} = -50$$

$y - y_1 = m(x - x_1)$

$y - 500 = -50(x - 1)$

$y - 500 = -50x + 50$

$y = -50x + 550$

We can equivalently express demand $D$ as a function of price $p$:

$$D(p) = -50p + 550$$

We have a system of equations

$$S(p) = \frac{200}{3}p - \frac{100}{3},$$

$$D(p) = -50p + 550.$$

Rewrite the system:

$q = \dfrac{200}{3}p - \dfrac{100}{3},$   (1)

$q = -50p + 550$   (2)

Substitute $\dfrac{200}{3}p - \dfrac{100}{3}$ for $q$ in (2) and solve.

$\dfrac{200}{3}p - \dfrac{100}{3} = -50p + 550$

$200p - 100 = -150p + 1650$   Multiplying by 3 to clear fractions

$350p - 100 = 1650$

$350p = 1750$

$p = 5$

The equilibrium price is $5 per unit.

To find the equilibrium quantity, we substitute $5 into either $S(p)$ or $D(p)$.

$$D(5) = -50(5) + 550 = -250 + 550 = 300$$

The equilibrium quantity is 300 yo-yo's.

The equilibrium point is ($5, 300 yo-yo's).

**43.** a) Use a graphing calculator to find the first coordinate of the point of intersection of $y_1 = -14.97x + 987.35$ and $y_2 = 98.55x - 5.13$, to the nearest hundredth. It is 8.74, so the price per unit that should be charged is $8.74.

b) Use a graphing calculator to find the first coordinate of the point of intersection of $y_1 = 87,985 + 5.15x$ and $y_2 = 8.74x$. It is about 24,508.4, so 24,509 units must be sold in order to break even.

# Chapter 4

# Inequalities and Problem Solving

## Exercise Set 4.1

1. If we add 7 to both sides of the inequality $x - 7 > -2$, we get the inequality $x > 5$ so these are equivalent inequalities.

3. If we add $3x$ to both sides of the equation $5x + 7 = 6 - 3x$, we get the equation $8x + 7 = 6$, so these are equivalent equations.

5. The solution set of $-4t \leq 12$ is $\{t|t \geq -3\}$ and the solution set of $t \leq -3$ is $\{t|t \leq -3\}$. The solution sets are not the same, so the inequalities are not equivalent.

7. The expressions are equivalent by the distributive law.

9. The solution set of $-\frac{1}{2}x < 7$ is $\{x|x > -14\}$ and the solution set of $x > 14$ is $\{x|x > 14\}$. The solution sets are not the same, so the inequalities are not equivalent.

11. $x - 3 \geq 5$

 $-4$ : We substitute and get $-4 - 3 \geq 5$, or $-7 \geq 5$, a false sentence. Therefore, $-4$ is not a solution.

 $0$ : We substitute and get $0 - 3 \geq 5$, or $-3 \geq 5$, a false false sentence. Therefore, 0 is not a solution.

 $8$ : We substitute and get $8 - 3 \geq 5$, or $5 \geq 5$, a true sentence. Therefore, 8 is a solution.

 $13$ : We substitute and get $13 - 3 \geq 5$, or $10 \geq 5$, a true sentence. Therefore, 13 is a solution.

13. $t - 6 > 2t - 1$

 $0$ : We substitute and get $0 - 6 > 2 \cdot 0 - 1$, or $-6 > -1$, a false sentence. Therefore, 0 is not a solution.

 $-8$ : We substitute and get $-8 - 6 > 2(-8) - 1$, or $-14 > -17$, a true sentence. Therefore, $-8$ is a solution.

 $-9$ : We substitute and get $-9 - 6 > 2(-9) - 1$, or $-15 > -19$, a true sentence. Therefore, $-9$ is a solution.

 $-3$ : We substitute and get $-3 - 6 > 2(-3) - 1$, or $-9 > -7$, a false sentence. Therefore, $-3$ is not a solution.

15. $y < 6$

 Graph: The solutions consist of all real numbers less than 6, so we shade all numbers to the left of 5 and use an open circle at 6 to indicate that it is not a solution.

 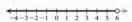

 Set builder notation: $\{y|y < 6\}$

 Interval notation: $(-\infty, 6)$

17. $x \geq -4$

 Graph: We shade all numbers to the right of $-4$ and use a solid endpoint at $-4$ to indicate that it is also a solution.

 Set builder notation: $\{x|x \geq -4\}$

 Interval notation: $[-4, \infty)$

19. $t > -3$

 Graph: We shade all numbers to the right of $-3$ and use an open circle at $-3$ to indicate that it is not a solution.

 Set builder notation: $\{t|t > -3\}$

 Interval notation: $(-3, \infty)$

21. $x \leq -7$

 Graph: We shade all numbers to the left of $-7$ and use a solid endpoint at $-7$ to indicate that it is also a solution.

 Set builder notation: $\{x|x \leq -7\}$

 Interval notation: $(-\infty, -7]$

23.
$$x + 8 > 2$$
$$x + 8 + (-8) > 2 + (-8) \qquad \text{Adding } -8$$
$$x > -6$$

 The solution set is $\{x|x > -6\}$, or $(-6, \infty)$.

25.
$$a + 7 \leq -13$$
$$a + 7 + (-7) \leq -13 + (-7) \qquad \text{Adding } -7$$
$$a \leq -20$$

 The solution set is $\{a|a \leq -20\}$, or $(-\infty, -20]$.

27.
$$x - 8 \leq 9$$
$$x - 8 + 8 \leq 9 + 8 \qquad \text{Adding } 8$$
$$x \leq 17$$

 The solution set is $\{x|x \leq 17\}$, or $(-\infty, 17]$.

29.
$$y - 9 > -18$$
$$y - 9 + 9 > -18 + 9 \qquad \text{Adding } 9$$
$$y > -9$$

The solution set is $\{y|y > -9\}$, or $(-9, \infty)$.

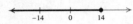

**31.**    $y - 20 \le -6$

$y - 20 + 20 \le -6 + 20$   Adding 20

$y \le 14$

The solution set is $\{y|y \le 14\}$, or $(-\infty, 14]$.

**33.**    $9t < -81$

$\dfrac{1}{9} \cdot 9t < \dfrac{1}{9}(-81)$   Multiplying by $\dfrac{1}{9}$

$t < -9$

The solution set is $\{t|t < -9\}$, or $(-\infty, -9)$.

**35.**    $0.3x < -18$

$\dfrac{1}{0.3}(0.3x) < \dfrac{1}{0.3}(-18)$   Multiplying by $\dfrac{1}{0.3}$

$x < -\dfrac{18}{0.3}$

$x < -60$

The solution set is $\{x|x < -60\}$, or $(-\infty, 60)$.

**37.**    $-9x \ge -8.1$

$-\dfrac{1}{9}(-9x) \le -\dfrac{1}{9}(-8.1)$   Multiplying by $-\dfrac{1}{9}$ and reversing the inequality symbol

$x \le \dfrac{8.1}{9}$

$x \le 0.9$

The solution set is $\{x|x \le 0.9\}$, or $(-\infty, 0.9]$.

**39.**    $-\dfrac{3}{4}x \ge -\dfrac{5}{8}$

$-\dfrac{4}{3}\left(-\dfrac{3}{4}x\right) \le -\dfrac{4}{3}\left(-\dfrac{5}{8}\right)$   Multiplying by $-\dfrac{4}{3}$ and reversing the inequality symbol

$x \le \dfrac{20}{24}$

$x \le \dfrac{5}{6}$

The solution set is $\left\{x\middle|x \le \dfrac{5}{6}\right\}$, or $\left(-\infty, \dfrac{5}{6}\right]$.

**41.**    $\dfrac{2x + 7}{5} < -9$

$5 \cdot \dfrac{2x + 7}{5} < 5(-9)$   Multiplying by 5

$2x + 7 < -45$

$2x < -52$   Adding $-7$

$x < -26$   Dividing by 2

The solution set is $\{x|x < -26\}$, or $(-\infty, -26)$.

**43.**    $\dfrac{3t - 7}{-4} \le 5$

$-4 \cdot \dfrac{3t - 7}{-4} \ge -4 \cdot 5$   Multiplying by $-4$ and reversing the inequality symbol

$3t - 7 \ge -20$

$3t \ge -13$   Adding 7

$t \ge -\dfrac{13}{3}$   Dividing by 3

The solution set is $\left\{t\middle|t \ge -\dfrac{13}{3}\right\}$, or $\left[-\dfrac{13}{3}, \infty\right)$.

**45.** $f(x) = 2x + 1$, $g(x) = x + 7$

$f(x) \ge g(x)$

$2x + 1 \ge x + 7$

$x + 1 \ge 7$      Adding $-x$

$x \ge 6$      Adding $-1$

The solution set is $\{x|x \ge 6\}$, or $[6, \infty)$.

**47.** $f(x) = 7 - 3x$, $g(x) = 2x - 3$

$f(x) \le g(x)$

$7 - 3x \le 2x - 3$

$7 - 5x \le -3$      Adding $-2x$

$-5x \le -10$      Adding $-7$

$x \ge 2$      Multiplying by $-\dfrac{1}{5}$ and reversing the inequality symbol

The solution set is $\{x|x \ge 2\}$, or $[2, \infty)$.

**49.** $f(x) = 2x - 7$, $g(x) = 5x - 9$

$f(x) \le g(x)$

$2x - 7 < 5x - 9$

$-3x - 7 < -9$      Adding $-5x$

$-3x < -2$      Adding 7

$x > \dfrac{2}{3}$      Dividing by $-3$

The solution set is $\left\{ x \middle| x > \dfrac{2}{3} \right\}$, or $\left( \dfrac{2}{3}, \infty \right)$.

**51.** $f(x) = \dfrac{3}{8} + 2x$, $g(x) = 3x - \dfrac{1}{8}$

$g(x) \geq f(x)$

$3x - \dfrac{1}{8} \geq \dfrac{3}{8} + 2x$

$x - \dfrac{1}{8} = \dfrac{3}{8}$     Adding $-2x$

$x \geq \dfrac{1}{2}$     Adding $\dfrac{1}{8}$

The solution set is $\left\{ x \middle| x \geq \dfrac{1}{2} \right\}$, or $\left[ \dfrac{1}{2}, \infty \right)$.

**53.** $4(3y - 2) \geq 9(2y + 5)$

$12y - 8 \geq 18y + 45$

$-6y - 8 \geq 45$

$-6y \geq 53$

$y \leq -\dfrac{53}{6}$

The solution set is $\left\{ y \middle| y \leq -\dfrac{53}{6} \right\}$, or $\left( -\infty, -\dfrac{53}{6} \right]$.

**55.** $5(t - 3) + 4t < 2(7 + 2t)$

$5t - 15 + 4t < 14 + 4t$

$9t - 15 < 14 + 4t$

$5t - 15 < 14$

$5t < 29$

$t < \dfrac{29}{5}$

The solution set is $\left\{ t \middle| t < \dfrac{29}{5} \right\}$, or $\left( -\infty, \dfrac{29}{5} \right)$.

**57.** $5[3m - (m + 4)] > -2(m - 4)$

$5(3m - m - 4) > -2(m - 4)$

$5(2m - 4) > -2(m - 4)$

$10m - 20 > -2m + 8$

$12m - 20 > 8$

$12m > 28$

$m > \dfrac{28}{12}$

$m > \dfrac{7}{3}$

The solution set is $\left\{ m \middle| m > \dfrac{7}{3} \right\}$, or $\left( \dfrac{7}{3}, \infty \right)$.

**59.** $19 - (2x + 3) \leq 2(x + 3) + x$

$19 - 2x - 3 \leq 2x + 6 + x$

$16 - 2x \leq 3x + 6$

$16 - 5x \leq 6$

$-5x \leq -10$

$x \geq 2$

The solution set is $\{x | x \geq 2\}$, or $[2, \infty)$.

**61.** $\dfrac{1}{4}(8y + 4) - 17 < -\dfrac{1}{2}(4y - 8)$

$2y + 1 - 17 < -2y + 4$

$2y - 16 < -2y + 4$

$4y - 16 < 4$

$4y < 20$

$y < 5$

The solution set is $\{y | y < 5\}$, or $(-\infty, 5)$.

**63.** $2[8 - 4(3 - x)] - 2 \geq 8[2(4x - 3) + 7] - 50$

$2[8 - 12 + 4x] - 2 \geq 8[8x - 6 + 7] - 50$

$2[-4 + 4x] - 2 \geq 8[8x + 1] - 50$

$-8 + 8x - 2 \geq 64x + 8 - 50$

$8x - 10 \geq 64x - 42$

$-56x - 10 \geq -42$

$-56x \geq -32$

$x \leq \dfrac{32}{56}$

$x \leq \dfrac{4}{7}$

The solution set is $\left\{ x \middle| x \leq \dfrac{4}{7} \right\}$, or $\left( -\infty, \dfrac{4}{7} \right]$.

**65. *Familiarize*.** Let $m$ = the mileage. Then the mileage charge is $0.20m$ and the total cost of the rental is $45 + 0.20m$.

***Translate***. We write an inequality stating that the cost of the rental is at most $75.

$45 + 0.20m \leq 75$

***Carry out***.

$45 + 0.20m \leq 75$

$0.20m \leq 30$

$m \leq 150$

***Check***. We can do a partial check by substituting a value for $m$ greater than 150. When $m = 151$, the rental cost is $45 + $0.20(151) = $75.20$. This is more than the budget amount, $75. We cannot check all possible values for $m$, so we stop here.

***State***. The budget will not be exceeded for mileages less than or equal to 150 mi.

**67. *Familiarize*.** Let $v$ = the blue book value of the car. Since the car was not replaced, we know that $9200 does not exceed 80% of the blue book value.

*Translate*. We write an inequality stating that $9200 does not exceed 80% of the blue book value.

$$9200 \leq 0.8v$$

**Carry out**.

$$9200 \leq 0.8v$$

$$11,500 \leq v \qquad \text{Multiplying by } \frac{1}{0.8}$$

**Check**. We can do a partial check by substituting a value for $v$ greater than 11,500. When $v = 11,525$, then 80% of $v$ is $0.8(11,525)$, or $9220. This is greater than $9200; that is, $9200 does not exceed this amount. We cannot check all possible values for $v$, so we stop here.

**State**. The blue book value of the car is $11,500 or more.

69. **Familiarize**. Let $m =$ the number of peak local minutes used. Then the charge for the minutes used is $0.022m$ and the total monthly charge is $13.55 + \$0.022m$.

*Translate*. We write an inequality stating that the monthly charge is at least $39.40.

$$13.55 + 0.022m \geq 39.40$$

**Carry out**.

$$13.55 + 0.022m \geq 39.40$$

$$0.022m \geq 25.85$$

$$m \geq 1175$$

**Check**. We can do a partial check by substituting a value for $m$ less than 1175. When $m = 1174$, the monthly charge is $13.55 + \$0.022(1174) \approx \$39.38$. This is less than the maximum charge of $39.40. We cannot check all possible values for $m$, so we stop here.

**State**. A customer must speak on the phone for 1175 local peak minutes or more if the maximum charge is to apply.

71. **Familiarize**. Let $c =$ the number of checks per month. Then the Anywhere plan will cost $0.20c$ per month and the Acu-checking plan will cost $2 + \$0.12c$ per month.

*Translate*. We write an inequality stating that the Acu-checking plan costs less than the Anywhere plan.

$$2 + 0.12c < 0.20c$$

**Carry out**.

$$2 + 0.12c < 0.20c$$

$$2 < 0.08c$$

$$25 < c$$

**Check**. We can do a partial check by substituting a value for $c$ less than 25 and a value for $c$ greater than 25. When $c = 24$, the Acu-checking plan costs $2 + \$0.12(24)$, or $4.88, and the Anywhere plan costs $0.20(24)$, or $4.80, so the Anywhere plan is less expensive. When $c = 26$, the Acu-checking plan costs $2 + \$0.12(26)$, or $5.12, and the Anywhere plan costs $0.20(26)$, or $5.20, so Acu-checking is less expensive. We cannot check all possible values for $c$, so we stop here.

**State**. The Acu-checking plan costs less for more than 25 checks per month.

73. **Familiarize**. We list the given information in a table.

| Plan A: Monthly Income | Plan B: Monthly Income |
|---|---|
| $400 salary | $610 |
| 8% of sales | 5% of sales |
| Total: 400 + 8% of sales | Total: 610 + 5% of sales |

Suppose Toni had gross sales of $5000 one month. Then under plan A she would earn

$$\$400 + 0.08(\$5000), \text{ or } \$800.$$

Under plan B she would earn

$$\$610 + 0.05(\$5000), \text{ or } \$860.$$

This shows that, for gross sales of $5000, plan B is better. If Toni had gross sales of $10,000 one month, then under plan A she would earn

$$\$400 + 0.08(\$10,000), \text{ or } \$1200.$$

Under plan B she would earn

$$\$610 + 0.05(\$10,000), \text{ or } \$1110.$$

This shows that, for gross sales of $10,000, plan A is better. To determine all values for which plan A is better we solve an inequality.

*Translate*.

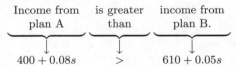

$$\begin{array}{ccc} \text{Income from} & \text{is greater} & \text{income from} \\ \text{plan A} & \text{than} & \text{plan B.} \\ \downarrow & \downarrow & \downarrow \\ 400 + 0.08s & > & 610 + 0.05s \end{array}$$

**Carry out**.

$$400 + 0.08s > 610 + 0.05s$$

$$400 + 0.03s > 610$$

$$0.03s > 210$$

$$s > 7000$$

**Check**. For $s = \$7000$, the income from plan A is

$$\$400 + 0.08(\$7000), \text{ or } \$960$$

and the income from plan B is

$$\$610 + 0.05(\$7000), \text{ or } \$960.$$

This shows that for sales of $7000 Toni's income is the same from each plan. In the Familiarize step we shows that, for a value less than $7000, plan B is better and, for a value greater than $7000, plan A is better. Since we cannot check all possible values, we stop here.

**State**. Toni should select plan A for gross sales greater than $7000.

75. **Familiarize**. Let $p =$ the number of people in the party. Then plan A will cost $30p and plan B will cost $1300 + \$20(p - 25)$.

*Translate*. We write an inequality stating that the cost of plan B is less than the cost of plan A.

$$1300 + 20(p - 25) < 30p$$

**Carry out.**

$$1300 + 20(p - 25) < 30p$$
$$1300 + 20p - 500 < 30p$$
$$800 + 20p < 30p$$
$$800 < 10p$$
$$80 < p$$

**Check.** We can do a partial check by substituting a value for $p$ less than 80 and a value for $p$ greater than 80. When $p = 79$, plan A costs $30 \cdot 79$, or \$2370, and plan B costs \$1300+\$20(79−25), or \$2380, so plan A costs less. When $p = 81$, plan A costs $30 \cdot 81$, or \$2430, and plan B costs \$1300 + \$20(81 − 25), or \$2420, so plan B costs less. We cannot check all possible values of $p$, so we stop here.

**State.** Plan B costs less for parties of more than 80 people.

**77. Familiarize.** Let $n =$ the number of people who attend. Then the total receipts are $\$6 \cdot n$, and the amount of receipts over \$750 is $\$6 \cdot n - \$750$. The band will receive \$750 plus 15% of $\$6 \cdot n - \$750$, or $\$750 + 0.15(\$6 \cdot n - \$750)$.

**Translate.** We write an inequality stating that the amount the band receives is at least \$1200.

$$750 + 0.15(6n - 750) \geq 1200$$

**Carry out.**

$$750 + 0.15(6n - 750) \geq 1200$$
$$750 + 0.9n - 112.5 \geq 1200$$
$$0.9n + 637.5 \geq 1200$$
$$0.9n \geq 562.5$$
$$n \geq 625$$

**Check.** When $n = 625$, the band receives \$75+ $0.15(\$6 \cdot 625 - \$750)$, or \$750+\$0.15(\$3000), or \$750+\$450, or \$1200. When $n = 626$, the band receives \$750 + $0.15(\$6 \cdot 626 - \$750)$, or \$750 + 0.15(\$3006), or \$750 + \$450.90, or \$1200.90. Since the band receives exactly \$1200 when 625 people attend and more than \$1200 when 626 people attend, we have performed a partial check. We cannot check all possible solutions, so we stop here.

**State.** At least 625 people must attend in order for the band to receive at least \$1200.

**79. a) Familiarize.** Find the values of $x$ for which $R(x) < C(x)$.

**Translate.**

$$26x < 90,000 + 15x$$

**Carry out.**

$$11x < 90,000$$
$$x < 8181\frac{9}{11}$$

**Check.** $R\left(8181\frac{9}{11}\right) = \$212,727.27 = C\left(8181\frac{9}{11}\right)$.

Calculate $R(x)$ and $C(x)$ for some $x$ greater than $8181\frac{9}{11}$ and for some $x$ less than $8181\frac{9}{11}$.

Suppose $x = 8200$ :

$$R(x) = 26(8200) = 213,200 \quad \text{and}$$
$$C(x) = 90,000 + 15(8200) = 213,000.$$

In this case $R(x) > C(x)$.

Suppose $x = 8000$ :

$$R(x) = 26(8000) = 208,000 \quad \text{and}$$
$$C(x) = 90,000 + 15(8000) = 210,000.$$

In this case $R(x) < C(x)$.

Then for $x < 8181\frac{9}{11}$, $R(x) < C(x)$.

**State.** We will state the result in terms of integers, since the company cannot sell a fraction of a lamp. For 8181 or fewer lamps the company loses money.

b) Our check in part a) shows that for $x > 8181\frac{9}{11}$, $R(x) > C(x)$ and the company makes a profit. Again, we will state the result in terms of an integer. For more than 8181 lamps the company makes money.

**81.** *Writing Exercise*

**83.** $f(x) = \dfrac{3}{x - 2}$

Since $\dfrac{3}{x - 2}$ cannot be computed when $x - 2$ is 0, we solve an equation:

$$x - 2 = 0$$
$$x = 2$$

The domain is $\{x | x$ is a real number *and* $x \neq 2\}$.

**85.** $f(x) = \dfrac{5x}{7 - 2x}$

Since $\dfrac{5x}{7 - 2x}$ cannot be computed when $7 - 2x$ is 0, we solve an equation:

$$7 - 2x = 0$$
$$7 = 2x$$
$$\frac{7}{2} = x$$

The domain is $\left\{x \middle| x$ is a real number *and* $x \neq \dfrac{7}{2}\right\}$.

**87.** $9x - 2(x - 5) = 9x - 2x + 10 = 7x + 10$

**89.** *Writing Exercise*

**91.**

$$3ax + 2x \geq 5ax - 4$$
$$2x - 2ax \geq -4$$
$$2x(1 - a) \geq -4$$
$$x(1 - a) \geq -2$$
$$x \leq -\frac{2}{1 - a}, \text{ or } \frac{2}{a - 1}$$

We reversed the inequality symbol when we divided because when $a > 1$, then $1 - a < 0$.

The solution set is $\left\{x \middle| x \leq \dfrac{2}{a - 1}\right\}$.

**93.** $a(by - 2) \geq b(2y + 5)$

$aby - 2a \geq 2by + 5b$

$aby - 2by \geq 2a + 5b$

$y(ab - 2b) \geq 2a + 5b$

$y \geq \dfrac{2a + 5b}{ab - 2b}, \text{ or } \dfrac{2a + 5b}{b(a - 2)}$

The inequality symbol remained unchanged when we divided because when $a > 2$ and $b > 0$, then $ab - 2b > 0$.

The solution set is $\left\{ y \middle| y \geq \dfrac{2a + 5b}{b(a - 2)} \right\}$.

**95.** $c(2 - 5x) + dx > m(4 + 2x)$

$2c - 5cx + dx > 4m + 2mx$

$-5cx + dx - 2mx > 4m - 2c$

$x(-5c + d - 2m) > 4m - 2c$

$x[d - (5c + 2m)] > 4m - 2c$

$x > \dfrac{4m - 2c}{d - (5c + 2m)}$

The inequality symbol remained unchanged when we divided because when $5c + 2m < d$, then $d - (5c + 2m) > 0$.

The solution set is $\left\{ x \middle| x > \dfrac{4m - 2c}{d - (5c + 2m)} \right\}$.

**97.** False. If $a = 2$, $b = 3$, $c = 4$, and $d = 5$, then $2 < 3$ and $4 < 5$ but $2 - 4 = 3 - 5$.

**99.** *Writing Exercise*

**101.** $x + 5 \leq 5 + x$

$\quad\quad 5 \leq 5 \quad\quad$ Subtracting $x$

We get an inequality that is true for all real numbers $x$. Thus the solution set is all real numbers.

**103.** $0^2 = 0$, $x^2 > 0$ for $x \neq 0$

The solution is $\{x | x \text{ is a real number and } x \neq 0\}$.

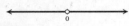

**105.** ▨

## Exercise Set 4.2

**1.** h

**3.** f

**5.** e

**7.** b

**9.** c

**11.** $\{5, 9, 11\} \cap \{9, 11, 18\}$

The numbers 9 and 11 are common to both sets, so the intersection is $\{9, 11\}$.

**13.** $\{0, 5, 10, 15\} \cup \{5, 15, 20\}$

The numbers in either or both sets are 0, 5, 10, 15, and 20, so the union is $\{0, 5, 10, 15, 20\}$.

**15.** $\{a, b, c, d, e, f\} \cap \{b, d, f\}$

The letters b, d, and f are common to both sets, so the intersection is $\{b, d, f\}$.

**17.** $\{r, s, t\} \cup \{r, u, t, s, v\}$

The letters in either or both sets are r, s, t, u, and v, so the union is $\{r, s, t, u, v\}$.

**19.** $\{3, 6, 9, 12\} \cap \{5, 10, 15\}$

There are no numbers common to both sets, so the solution set has no members. It is $\emptyset$.

**21.** $\{3, 5, 7\} \cup \emptyset$

The numbers in either or both sets are 3, 5, and 7, so the union is $\{3, 5, 7\}$.

**23.** $3 < x < 7$

This inequality is an abbreviation for the conjunction $3 < x \text{ and } x < 7$. The graph is the intersection of two separate solution sets: $\{x | 3 < x\} \cap \{x | x < 7\} = \{x | 3 < x < 7\}$.

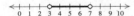

Interval notation: $(3, 7)$

**25.** $-6 \leq y \leq -2$

This inequality is an abbreviation for the conjunction $-6 \leq y \text{ and } y \leq -2$.

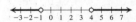

Interval notation: $[-6, -2]$

**27.** $x < -1 \text{ or } x > 4$

The graph of this disjunction is the union of the graphs of the individual solution sets $\{x | x < -1\}$ and $\{x | x > 4\}$.

Interval notation: $(-\infty, -2) \cup (3, \infty)$

**29.** $x \leq -2 \text{ or } x > 1$

Interval notation: $(-\infty, -2] \cup (1, \infty)$

**31.** $-4 \leq -x < 2$

$\quad 4 \geq x > -2 \quad$ Multiplying by $-1$ and reversing the inequality symbols

$-2 < x \leq 4 \quad$ Rewriting

Interval notation: $(-2, 4]$

**33.** $x > -2$ *and* $x < 4$

This conjunction can be abbreviated as $-2 < x < 4$.

Interval notation: $(-2, 4)$

**35.** $5 > a$ *or* $a > 7$

Interval notation: $(-\infty, 5) \cup (7, \infty)$

**37.** $x \geq 5$ *or* $-x \geq 4$

Multiplying the second inequality by $-1$ and reversing the inequality symbols, we get $x \geq 5$ *or* $x \leq -4$.

Interval notation: $(-\infty - 4] \cup [5, \infty)$

**39.** $7 > y$ *and* $y \geq -3$

This conjunction can be abbreviated as $-3 \leq y < 7$.

Interval notation: $[-3, 7)$

**41.** $x < 7$ *and* $x \geq 3$

This conjunction can be abbreviated as $3 \leq x < 7$.

Interval notation: $[3, 7)$

**43.** $t < 2$ *or* $t < 5$

Observe that every number that is less than 2 is also less than 5. Then $t < 2$ *or* $t < 5$ is equivalent to $t < 5$ and the graph of this disjunction is the set $\{t | t < 5\}$.

Interval notation: $(-\infty, 5)$

**45.**
$$-2 < t + 1 < 8$$
$$-2 - 1 < t < 8 - 1$$
$$-3 < t < 7$$

The solution set is $\{t | -3 < t < 7\}$, or $(-3, 7)$.

**47.**
$$2 < x + 3 \quad \text{and} \quad x + 1 \leq 5$$
$$-1 < x \quad \text{and} \quad x \leq 4$$

We can abbreviate the answer as $-1 < x \leq 4$. The solution set is $\{x | -1 < x \leq 4\}$, or $(-1, 4]$.

**49.** $-7 \leq 2a - 3$ *and* $3a + 1 < 7$
$$-4 \leq 2a \quad \text{and} \quad 3a < 6$$
$$-2 \leq a \quad \text{and} \quad a < 2$$

We can abbreviate the answer as $-2 \leq a < 2$. The solution set is $\{a | -2 \leq a < 2\}$, or $[-2, 2)$.

**51.** $x + 7 \leq -2$ *or* $x + 7 \geq -3$

Observe that any real number is either less than or equal to $-2$ or greater than or equal to $-3$. Then the solution set is $\{x | x$ is a real number$\}$, or $(-\infty, \infty)$.

**53.**
$$5 > \frac{x - 3}{4} > 1$$
$$20 > x - 3 > 4 \quad \text{Multiplying by 4}$$
$$23 > x > 7, \text{ or}$$
$$7 < x < 23$$

The solution set is $\{x | 7 < x < 23\}$, or $(7, 23)$.

**55.**
$$-7 \leq 4x + 5 \leq 13$$
$$-12 \leq 4x \leq 8$$
$$-3 \leq x \leq 2$$

The solution set is $\{x | -3 \leq x \leq 2\}$, or $[-3, 2]$.

**57.**
$$2 \leq 3x - 1 \leq 8$$
$$3 \leq 3x \leq 9$$
$$1 \leq x \leq 3$$

The solution set is $\{x | 1 \leq x \leq 3\}$, or $[1, 3]$.

**59.**
$$-21 \leq -2x - 7 < 0$$
$$-14 \leq -2x < 7$$
$$7 \geq x > -\frac{7}{2}, \text{ or}$$
$$-\frac{7}{2} < x \leq 7$$

The solution set is $\left\{x \left| -\frac{7}{2} < x \leq 7 \right.\right\}$, or $\left(-\frac{7}{2}, 7\right]$.

**61.** $3x - 1 \leq 2$ *or* $3x - 1 \geq 8$
$$3x \leq 3 \quad \text{or} \quad 3x \geq 9$$
$$x \leq 1 \quad \text{or} \quad x \geq 3$$

The solution set is $\{x | x \leq 1$ *or* $x \geq 3\}$, or $(-\infty, 1] \cup [3, \infty)$.

**63.** $2x - 7 < -3 \quad or \quad 2x - 7 > 5$

$\qquad 2x < 4 \quad or \qquad 2x > 12$

$\qquad x < 2 \quad or \qquad x > 6$

The solution set is $\{x|x < 2 \ or \ x > 6\}$, or $(-\infty, 2) \cup (6, \infty)$.

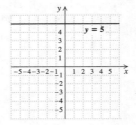

**65.** $6 > 2a - 1 \quad or \quad -4 \leq -3a + 2$

$\qquad 7 > 2a \qquad or \quad -6 \leq -3a$

$\qquad \dfrac{7}{2} > a \qquad or \quad 2 \geq a$

The solution set is $\left\{a \left| \dfrac{7}{2} > a\right.\right\} \cup \{a|2 \geq a\} =$ $\left\{a \left| \dfrac{7}{2} > a\right.\right\}$, or $\left\{a \left| a < \dfrac{7}{2}\right.\right\}$, or $\left(-\infty, \dfrac{7}{2}\right)$.

**67.** $a + 3 < -2 \quad and \quad 3a - 4 < 8$

$\qquad a < -5 \quad and \qquad 3a < 12$

$\qquad a < -5 \quad and \qquad a < 4$

The solution set is $\{a|a < -5\} \cap \{a|a < 4\} =$ $\{a|a < -5\}$, or $(-\infty, -5)$.

**69.** $3x + 2 < 2 \quad or \quad 4 - 2x < 14$

$\qquad 3x < 0 \quad or \qquad -2x < 10$

$\qquad x < 0 \quad or \qquad x > -5$

The solution set is $\{x|x < 0\} \cup \{x|x > -5\} =$ the set of all real numbers, or $(-\infty, \infty)$.

**71.** $2t - 7 \leq 5 \quad or \quad 5 - 2t > 3$

$\qquad 2t \leq 12 \quad or \qquad -2t > -2$

$\qquad t \leq 6 \quad or \qquad t < 1$

The solution set is $\{t|t \leq 6\} \cup \{t|t < 1\} = \{t|t \leq 6\}$, or $(-\infty, 6]$.

**73.** $f(x) = \dfrac{9}{x + 8}$

$f(x)$ cannot be computed when the denominator is 0. Since $x + 8 = 0$ is equivalent to $x = -8$, we have Domain of $f = \{x|x$ is a real number $and$ $x \neq -8\} =$ $(-\infty, -8) \cup (-8, \infty)$.

**75.** $f(x) = \sqrt{x - 6}$

The expression $\sqrt{x - 6}$ is not a real number when $x - 6$ is negative. Thus, the domain of $f$ is the set of all $x$-values for which $x - 6 \geq 0$. Since $x - 6 \geq 0$ is equivalent to $x \geq 6$, we have Domain of $f = [6, \infty)$.

**77.** $f(x) = \dfrac{x + 3}{2x - 8}$

$f(x)$ cannot be computed when the denominator is 0. Since $2x - 8 = 0$ is equivalent to $x = 4$, we have Domain of $f = \{x|x$ is a real number and $x \neq 4\}$, or $(-\infty, 4) \cup (4, \infty)$.

**79.** $f(x) = \sqrt{2x + 7}$

The expression $\sqrt{2x + 7}$ is not a real number when $2x + 7$ is negative. Thus, the domain of $f$ is the set of all $x$-values for which $2x + 7 \geq 0$. Since $2x + 7 \geq 0$ is equivalent to $x \geq -\dfrac{7}{2}$, we have Domain of $f = \left[-\dfrac{7}{2}, \infty\right)$.

**81.** $f(x) = \sqrt{8 - 2x}$

The expression $\sqrt{8 - 2x}$ is not a real number when $8 - 2x$ is negative. Thus, the domain of $f$ is the set of all $x$-values for which $8 - 2x \geq 0$. Since $8 - 2x \geq 0$ is equivalent to $x \leq 4$, we have Domain of $f = (-\infty, 4]$.

**83.** *Writing Exercise*

**85.** Graph: $y = 5$

The graph of any constant function $y = c$ is a horizontal line that crosses the vertical axis at $(0, c)$. Thus, the graph of $y = 5$ is a horizontal line that crosses the vertical axis at $(0, 5)$.

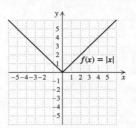

**87.** Graph $f(x) = |x|$

We make a table of values, plot points, and draw the graph.

| $x$ | $f(x)$ |
|-----|--------|
| $-5$ | $5$ |
| $-2$ | $2$ |
| $0$ | $0$ |
| $1$ | $1$ |
| $4$ | $4$ |

**89.** Graph both equations.

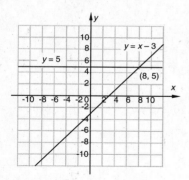

The solution (point of intersection) is apparently $(8, 5)$.

$$
\begin{array}{c|c}
y = x - 3 & y = 5 \\
\hline
5 \mid 8 - 3 & ? \\
\quad ? & 5 = 5 \quad \text{TRUE} \\
5 = 5 \qquad \text{TRUE} &
\end{array}
$$

The solution is $(8, 5)$.

**91.** *Writing Exercise*

**93.** From the graph we observe that the values of $x$ for which $2x - 5 > -7$ *and* $2x - 5 < 7$ are $\{x \mid -1 < x < 6\}$, or $(-1, 6)$.

**95.** Solve $19 < P(t) < 30$, or $19 < 0.44t + 10.2 < 30$.

$$19 < 0.44t + 10.2 < 30$$
$$8.8 < 0.44t < 19.8$$
$$20 < t < 45$$

The percentage of childless 40-44-year-old women will be between 19% and 30% from 20 years after 1980 to 45 years after 1980 or from 2000 to 2025.

**97.** Solve $32 < f(x) < 46$, or $32 < 2(x + 10) < 46$.

$$32 < 2(x + 10) < 46$$
$$32 < 2x + 20 < 46$$
$$12 < 2x < 26$$
$$6 < x < 13$$

For U.S. dress sizes between 6 and 13, dress sizes in Italy will be between 32 and 46.

**99.** $10.8 < -0.0433t + 10.49 < 11.5$

$$0.31 < -0.0433t < 1.01$$
$$-7.16 > t > -23.33$$

$1988.5 - 7.16 = 1981.34$ and $1988.5 - 23.33 = 1965.17$

Thus, records of 11.5 sec and 10.8 sec were set during 1965 and 1981, respectively. Thus, we have $1965 \le y \le 1981$.

**101.** Let $c$ = the number of crossings per year. Then at the $3 per crossing rate, the total cost of $c$ crossings is $3c$. Two six-month passes cost $2 \cdot \$15$, or $30. The additional $0.50 per crossing toll brings the total cost of $c$ crossings to $30 + \$0.50c$. A one-year pass costs $150 regardless of the number of crossings.

We write an inequality that states that the cost of $c$ crossings per year using the six-month passes is less than the cost using the $3 per crossing toll and is less than the cost using the one-year pass. Solve:

$$30 + 0.50c < 3c \text{ and } 30 + 0.50c < 150$$

We get $12 < c$ *and* $c < 240$, or $12 < c < 240$.

For more than 12 crossings but less than 240 crossings per year the six-month passes are the most economical choice.

**103.** $4m - 8 > 6m + 5$ *or* $5m - 8 < -2$

$$-13 > 2m \qquad \text{or} \qquad 5m < 6$$
$$-\frac{13}{2} > m \qquad \text{or} \qquad m < \frac{6}{5}$$

$\left\{ m \mid m < \dfrac{6}{5} \right\}$, or $\left( -\infty, \dfrac{6}{5} \right)$

**105.** $3x < 4 - 5x < 5 + 3x$

$$0 < 4 - 8x < 5$$
$$-4 < -8x < 1$$
$$\frac{1}{2} > x > -\frac{1}{8}$$

The solution set is $\left\{ x \mid -\dfrac{1}{8} < x < \dfrac{1}{2} \right\}$, or $\left( -\dfrac{1}{8}, \dfrac{1}{2} \right)$.

**107.** Let $a = b = c = 2$. Then $a \le c$ and $c \le b$, but $b \not> a$. The given statement is false.

**109.** If $-a < c$, then $-1(-a) > -1 \cdot c$, or $a > -c$. Then if $a > -c$ and $-c > b$, we have $a > -c > b$, so $a > b$ and the given statement is true.

**111.** $f(x) = \dfrac{\sqrt{3 - 4x}}{x + 7}$

$3 - 4x \ge 0$ is equivalent to $x \le \dfrac{3}{4}$ and $x + 7 = 0$ is equivalent to $x = -7$. Then we have Domain of $f = \left\{ x \mid x \le \dfrac{3}{4} \text{ and } x \ne -7 \right\}$, or $(-\infty, -7) \cup \left( -7, \dfrac{3}{4} \right]$.

**113.** Observe that the graph of $y_1$ lies below the graph of $y_2$ for $x$ in the interval $\left( -\dfrac{3}{2}, \infty \right)$. Also, the graph of $y_3$ lies below the graph of $y_4$ for $x$ in the interval $(-\infty, -7)$. Thus, the solution set is $(-\infty, -7) \cup \left( -\dfrac{3}{2}, \infty \right)$.

**115.**

## Exercise Set 4.3

**1.** True; see page 245 in the text.

**3.** $|0| = 0$, so the statement is false.

**5.** True; see page 245 in the text.

**7.** False; see page 250 in the text.

**9.** $|x| = 7$

$x = -7 \ or \ x = 7$   Using the absolute-value
principle

The solution set is $\{-7, 7\}$.

**11.** $|x| = -6$

The absolute value of a number is always nonnegative. Therefore, the solution set is $\emptyset$.

**13.** $|p| = 0$

The only number whose absolute value is 0 is 0. The solution set is $\{0\}$.

**15.** $|t| = 5.5$

$t = -5.5 \ or \ t = 5.5$   Absolute-value principle

The solution set is $\{-5.5, 5.5\}$.

**17.** $|2x - 3| = 4$

$2x - 3 = -4 \ or \ 2x - 3 = 4$   Absolute-value
principle

$2x = -1 \ or \qquad 2x = 7$

$x = -\dfrac{1}{2} \ or \qquad x = \dfrac{7}{2}$

The solution set is $\left\{ -\dfrac{1}{2}, \dfrac{7}{2} \right\}$.

**19.** $|3x - 5| = -8$

Absolute value is always nonnegative, so the equation has no solution. The solution set is $\emptyset$.

**21.** $|x - 2| = 6$

$x - 2 = -6 \ or \ x - 2 = 6$   Absolute-value principle

$x = -4 \ or \qquad x = 8$

The solution set is $\{-4, 8\}$.

**23.** $|x - 5| = 3$

$x - 5 = -3 \ or \ x - 5 = 3$

$x = 2 \ or \qquad x = 8$

The solution set is $\{2, 8\}$.

**25.** $|x - 7| = 9$

$x - 7 = -9 \ or \ x - 7 = 9$

$x = -2 \ or \qquad x = 16$

The solution set is $\{-2, 16\}$.

**27.** $|5x| - 3 = 37$

$|5x| = 40$   Adding 3

$5x = -40 \ or \ 5x = 40$

$x = -8 \ or \quad x = 8$

The solution set is $\{-8, 8\}$.

**29.** $7|q| - 2 = 9$

$7|q| = 11$   Adding 2

$|q| = \dfrac{11}{7}$   Multiplying by $\dfrac{1}{7}$

$q = -\dfrac{11}{7} \ or \ q = \dfrac{11}{7}$

The solution set is $\left\{ -\dfrac{11}{7}, \dfrac{11}{7} \right\}$.

**31.** $\left| \dfrac{2x-1}{3} \right| = 5$

$\dfrac{2x-1}{3} = -5 \ or \ \dfrac{2x-1}{3} = 5$

$2x - 1 = -15 \ or \ 2x - 1 = 15$

$2x = -14 \ or \qquad 2x = 16$

$x = -7 \ or \qquad x = 8$

The solution set is $\{-7, 8\}$.

**33.** $|m + 5| + 9 = 16$

$|m + 5| = 7$   Adding $-9$

$m + 5 = -7 \ or \ m + 5 = 7$

$m = -12 \ or \qquad m = 2$

The solution set is $\{-12, 2\}$.

**35.** $5 - 2|3x - 4| = -5$

$-2|3x - 4| = -10$

$|3x - 4| = 5$

$3x - 4 = -5 \ or \ 3x - 4 = 5$

$3x = -1 \ or \qquad 3x = 9$

$x = -\dfrac{1}{3} \ or \qquad x = 3$

The solution set is $\left\{ -\dfrac{1}{3}, 3 \right\}$.

**37.** $|2x + 6| = 8$

$2x + 6 = -8 \ or \ 2x + 6 = 8$

$2x = -14 \ or \qquad 2x = 2$

$x = -7 \ or \qquad x = 1$

The solution set is $\{-7, 1\}$.

**39.** $|x| - 3 = 5.7$

$|x| = 8.7$

$x = -8.7 \ or \ x = 8.7$

The solution set is $\{-8.7, 8.7\}$.

**41.** $\left|\dfrac{3x-2}{5}\right| = 2$

$\dfrac{3x-2}{5} = -2 \quad or \quad \dfrac{3x-2}{5} = 2$

$3x - 2 = -10 \quad or \quad 3x - 2 = 10$

$3x = -8 \quad or \qquad 3x = 12$

$x = -\dfrac{8}{3} \quad or \qquad x = 4$

The solution set is $\left\{-\dfrac{8}{3}, 4\right\}$.

**43.** $|x+4| = |2x-7|$

$x + 4 = 2x - 7 \quad or \quad x + 4 = -(2x-7)$

$4 = x - 7 \quad or \quad x + 4 = -2x + 7$

$11 = x \qquad or \quad 3x + 4 = 7$

$\qquad\qquad\qquad 3x = 3$

$\qquad\qquad\qquad x = 1$

The solution set is $\{1, 11\}$.

**45.** $|x+4| = |x-3|$

$x + 4 = x - 3 \quad or \quad x + 4 = -(x-3)$

$4 = -3 \quad or \quad x + 4 = -x + 3$

$\text{False} \qquad\qquad 2x = -1$

$\qquad\qquad\qquad x = -\dfrac{1}{2}$

The solution set is $\left\{-\dfrac{1}{2}\right\}$.

**47.** $|3a-1| = |2a+4|$

$3a - 1 = 2a + 4 \quad or \quad 3a - 1 = -(2a+4)$

$a - 1 = 4 \qquad or \quad 3a - 1 = -2a - 4$

$a = 5 \qquad or \quad 5a - 1 = -4$

$\qquad\qquad\qquad 5a = -3$

$\qquad\qquad\qquad a = -\dfrac{3}{5}$

The solution set is $\left\{-\dfrac{3}{5}, 5\right\}$.

**49.** $|n-3| = |3-n|$

$n - 3 = 3 - n \quad or \quad n - 3 = -(3-n)$

$2n - 3 = 3 \qquad or \quad n - 3 = -3 + n$

$2n = 6 \qquad or \qquad -3 = -3$

$n = 3 \qquad \text{True for all real values of } n$

The solution set is the set of all real numbers.

**51.** $|7-a| = |a+5|$

$7 - a = a + 5 \quad or \quad 7 - a = -(a+5)$

$7 = 2a + 5 \quad or \quad 7 - a = -a - 5$

$2 = 2a \qquad or \qquad 7 = -5$

$1 = a \qquad\qquad\qquad \text{False}$

The solution set is $\{1\}$.

**53.** $\left|\dfrac{1}{2}x - 5\right| = \left|\dfrac{1}{4}x + 3\right|$

$\dfrac{1}{2}x - 5 = \dfrac{1}{4}x + 3 \quad or \quad \dfrac{1}{2}x - 5 = -\left(\dfrac{1}{4}x + 3\right)$

$\dfrac{1}{4}x - 5 = 3 \qquad\qquad or \quad \dfrac{1}{2}x - 5 = -\dfrac{1}{4}x - 3$

$\dfrac{1}{4}x = 8 \qquad\qquad or \quad \dfrac{3}{4}x - 5 = -3$

$x = 32 \qquad\qquad or \qquad \dfrac{3}{4}x = 2$

$\qquad\qquad\qquad\qquad\qquad x = \dfrac{8}{3}$

The solution set is $\left\{32, \dfrac{8}{3}\right\}$.

**55.** $|a| \leq 9$

$-9 \leq a \leq 9 \qquad \text{Part (b)}$

The solution set is $\{a | -9 \leq a \leq 9\}$, or $[-9, 9]$.

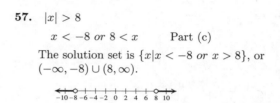

**57.** $|x| > 8$

$x < -8 \; or \; 8 < x \qquad \text{Part (c)}$

The solution set is $\{x | x < -8 \; or \; x > 8\}$, or $(-\infty, -8) \cup (8, \infty)$.

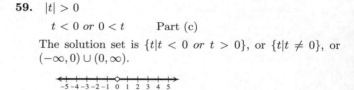

**59.** $|t| > 0$

$t < 0 \; or \; 0 < t \qquad \text{Part (c)}$

The solution set is $\{t | t < 0 \; or \; t > 0\}$, or $\{t | t \neq 0\}$, or $(-\infty, 0) \cup (0, \infty)$.

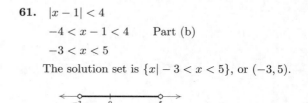

**61.** $|x-1| < 4$

$-4 < x - 1 < 4 \qquad \text{Part (b)}$

$-3 < x < 5$

The solution set is $\{x | -3 < x < 5\}$, or $(-3, 5)$.

**63.** $|x+2| \leq 6$

$-6 \leq x + 2 \leq 6 \quad \text{Part (b)}$

$-8 \leq x \leq 4 \qquad \text{Adding } -2$

The solution set is $\{x | -8 \leq x \leq 4\}$, or $[-8, 4]$.

**65.** $|x-3| + 2 > 7$

$|x-3| > 5 \quad \text{Adding } -2$

$x - 3 < -5 \; or \; 5 < x - 3 \quad \text{Part (c)}$

$x < -2 \; or \; 8 < x$

The solution set is $\{x | x < -2 \; or \; x > 8\}$, or $(-\infty, -2) \cup (8, \infty)$.

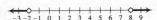

**67.** $|2y - 9| > -5$

Since absolute value is never negative, any value of $2y - 9$, and hence any value of $y$, will satisfy the inequality. The solution set is the set of all real numbers, or $(-\infty, \infty)$.

**69.**  $|3a - 4| + 2 \geq 8$

   $|3a - 4| \geq 6$    Adding $-2$

$3a - 4 \leq -6$  *or*  $6 \leq 3a - 4$   Part (c)

   $3a \leq -2$  *or*  $10 \leq 3a$

   $a \leq -\dfrac{2}{3}$  *or*  $\dfrac{10}{3} \leq a$

The solution set is $\left\{a \middle| a \leq -\dfrac{2}{3} \ or \ a \geq \dfrac{10}{3}\right\}$, or

$\left(-\infty, -\dfrac{2}{3}\right] \cup \left[\dfrac{10}{3}, \infty\right)$.

**71.**  $|y - 3| < 12$

   $-12 < y - 3 < 12$    Part (b)

   $-9 < y < 15$        Adding 3

The solution set is $\{y| -9 < y < 15\}$, or $(-9, 15)$.

**73.**  $9 - |x + 4| \leq 5$

   $-|x + 4| \leq -4$

   $|x + 4| \geq 4$      Multiplying by $-1$

$x + 4 \leq -4$  *or*  $4 \leq x + 4$   Part (c)

   $x \leq -8$  *or*  $0 \leq x$

The solution set is $\{x| x \leq -8 \ or \ x \geq 0\}$, or $(-\infty, -8] \cup [0, \infty)$.

**75.**  $|4 - 3y| > 8$

$4 - 3y < -8$   *or*   $8 < 4 - 3y$   Part (c)

   $-3y < -12$  *or*   $4 < -3y$    Adding $-4$

      $y > 4$    *or*  $-\dfrac{4}{3} > y$  Multiplying by $-\dfrac{1}{3}$

The solution set is $\left\{y \middle| y < -\dfrac{4}{3} \ or \ y > 4\right\}$, or

$\left(-\infty, -\dfrac{4}{3}\right) \cup (4, \infty)$.

**77.**  $|5 - 4x| < -6$

Absolute value is always nonnegative, so the inequality has no solution. The solution set is $\emptyset$.

**79.**  $\left|\dfrac{2 - 5x}{4}\right| \geq \dfrac{2}{3}$

$\dfrac{2 - 5x}{4} \leq -\dfrac{2}{3}$  *or*  $\dfrac{2}{3} \leq \dfrac{2 - 5x}{4}$  Part (c)

$2 - 5x \leq -\dfrac{8}{3}$  *or*  $\dfrac{8}{3} \leq 2 - 5x$  Multiplying by 4

$-5x \leq -\dfrac{14}{3}$  *or*  $\dfrac{2}{3} \leq -5x$    Adding $-2$

$x \geq \dfrac{14}{15}$   *or*  $-\dfrac{2}{15} \geq x$   Multiplying by $-\dfrac{1}{5}$

The solution set is $\left\{x \middle| x \leq -\dfrac{2}{15} \ or \ x \geq \dfrac{14}{15}\right\}$, or

$\left(-\infty, -\dfrac{2}{15}\right] \cup \left[\dfrac{14}{15}, \infty\right)$.

**81.**  $|m + 3| + 8 \leq 14$

   $|m + 3| \leq 6$    Adding $-8$

   $-6 \leq m + 3 \leq 6$

   $-9 \leq m \leq 3$

The solution set is $\{m| -9 \leq m \leq 3\}$, or $[-9, 3]$.

**83.**  $25 - 2|a + 3| > 19$

   $-2|a + 3| > -6$

   $|a + 3| < 3$     Multiplying by $-\dfrac{1}{2}$

   $-3 < a + 3 < 3$   Part (b)

   $-6 < a < 0$

The solution set is $\{a| -6 < a < 0\}$, or $(-6, 0)$.

**85.**  $|2x - 3| \leq 4$

   $-4 \leq 2x - 3 \leq 4$   Part (b)

   $-1 \leq 2x \leq 7$      Adding 3

   $-\dfrac{1}{2} \leq x \leq \dfrac{7}{2}$     Multiplying by $\dfrac{1}{2}$

The solution set is $\left\{x \middle| -\dfrac{1}{2} \leq x \leq \dfrac{7}{2}\right\}$, or $\left[-\dfrac{1}{2}, \dfrac{7}{2}\right]$.

**87.**  $5 + |3x - 4| \geq 16$

   $|3x - 4| \geq 11$

$3x - 4 \leq -11$  *or*  $11 \leq 3x - 4$  Part (c)

   $3x \leq -7$   *or*  $15 \leq 3x$

   $x \leq -\dfrac{7}{3}$   *or*   $5 \leq x$

The solution set is $\left\{x \middle| x \le -\dfrac{7}{3} \text{ or } x \ge 5\right\}$, or $\left(-\infty, -\dfrac{7}{3}\right] \cup [5, \infty)$.

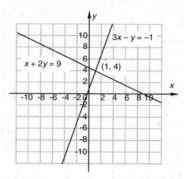

**89.** $7 + |2x - 1| < 16$

$|2x - 1| < 9$

$-9 < 2x - 1 < 9$   Part (b)

$-8 < 2x < 10$

$-4 < x < 5$

The solution set is $\{x | -4 < x < 5\}$, or $(-4, 5)$.

**91.** *Writing Exercise*

**93.** $2x - 3y = 7,$   (1)

$3x + 2y = -10$   (2)

We will use the elimination method. First, multiply equation (1) by 2 and equation (2) by 3 and add to eliminate a variable.

$4x - 6y = \phantom{-}14$

$\underline{9x + 6y = -30}$

$13x \phantom{+6y} = -16$

$x = -\dfrac{16}{13}$

Now substitute $-\dfrac{16}{13}$ for $x$ in either of the original equations and solve for $y$.

$3x + 2y = -10$   (2)

$3\left(-\dfrac{16}{13}\right) + 2y = -10$

$-\dfrac{48}{13} + 2y = -10$

$2y = -\dfrac{130}{13} + \dfrac{48}{13}$

$2y = -\dfrac{82}{13}$

$y = -\dfrac{41}{13}$

The solution is $\left(-\dfrac{16}{13}, -\dfrac{41}{13}\right)$.

**95.** $x = -2 + 3y,$   (1)

$x - 2y = 2$   (2)

We will use the substitution method. We substitute $-2 + 3y$ for $x$ in equation (2).

$x - 2y = 2$   (2)

$(-2 + 3y) - 2y = 2$   Substituting

$-2 + y = 2$

$y = 4$

Now substitute 4 for $y$ in equation (1) and find $x$.

$x = -2 + 3 \cdot 4 = -2 + 12 = 10$

The solution is $(10, 4)$.

**97.** Graph both equations.

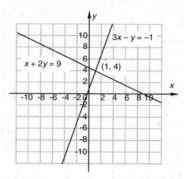

The solution (point of intersection) is apparently $(1, 4)$.

Check:

| $x + 2y = 9$ | | $3x - y = -1$ | |
|---|---|---|---|
| $1 + 2 \cdot 4$ | $9$ | $3 \cdot 1 - 4$ | $-1$ |
| $1 + 8$ | | $3 - 4$ | |
| | $9 \overset{?}{=} 9$ TRUE | | $-1 \overset{?}{=} -1$ TRUE |

The solution is $(1, 4)$.

**99.** *Writing Exercise*

**101.** From the definition of absolute value, $|3t - 5| = 3t - 5$ only when $3t - 5 \ge 0$. Solve $3t - 5 \ge 0$.

$3t - 5 \ge 0$

$3t \ge 5$

$t \ge \dfrac{5}{3}$

The solution set is $\left\{t \middle| t \ge \dfrac{5}{3}\right\}$, or $\left[\dfrac{5}{3}, \infty\right)$.

**103.** $|x + 2| > x$

The inequality is true for all $x < 0$ (because absolute value must be nonnegative). The solution set in this case is $\{x | x < 0\}$. If $x = 0$, we have $|0 + 2| > 0$, which is true. The solution set in this case is $\{0\}$. If $x > 0$, we have the following:

$x + 2 < -x$ *or* $x < x + 2$

$2x < -2$ *or* $0 < 2$

$x < -1$

Although $x > 0$ *and* $x < -1$ yields no solution, $x > 0$ and $2 > 0$ (true for all $x$) yields the solution set $\{x | x > 0\}$ in this case. The solution set for the inequality is $\{x | x < 0\} \cup \{0\} \cup \{x | x > 0\}$, or $\{x | x$ is a real number$\}$, or $(-\infty, \infty)$.

**105.** $|5t - 3| = 2t + 4$

From the definition of absolute value, we know that $2t + 4 \geq 0$, or $t \geq -2$. So we have

$t \geq -2$ *and*

$5t - 3 = -(2t + 4)$  *or*  $5t - 3 = 2t + 4$

$5t - 3 = -2t - 4$  *or*  $3t = 7$

$7t = -1$  *or*  $t = \dfrac{7}{3}$

$t = -\dfrac{1}{7}$  *or*  $t = \dfrac{7}{3}$

Since $-\dfrac{1}{7} \geq -2$ and $\dfrac{7}{3} \geq -2$, the solution set is $\left\{ -\dfrac{1}{7}, \dfrac{7}{3} \right\}$.

**107.** Using part (b), we find that $-3 < x < 3$ is equivalent to $|x| < 3$.

**109.**   $x \leq -6$  *or*  $6 \leq x$

$|x| \geq 6$   Using part (c)

**111.**   $x < -8$  *or*  $2 < x$

$x + 3 < -5$  *or*  $5 < x + 3$   Adding 3

$|x + 3| > 5$   Using part (c)

**113.** The distance from $x$ to 7 is $|x - 7|$ or $|7 - x|$, so we have $|x - 7| < 2$, or $|7 - x| < 2$.

**115.** The length of the segment from $-1$ to 7 is $|-1 - 7| = |-8| = 8$ units. The midpoint of the segment is $\dfrac{-1 + 7}{2} = \dfrac{6}{2} = 3$. Thus, the interval extends $8/2$, or 4, units on each side of 3. An inequality for which the closed interval is the solution set is then $|x - 3| \leq 4$.

**117.** The length of the segment from $-7$ to $-1$ is $|-7 - (-1)| = |-6| = 6$ units. The midpoint of the segment is $\dfrac{-7 + (-1)}{2} = \dfrac{-8}{2} = -4$. Thus, the interval extends $6/2$, or 3, units on each side of $-4$. An inequality for which the open interval is the solution set is $|x - (-4)| < 3$, or $|x + 4| < 3$.

**119.**   $|d - 60 \text{ ft}| \leq 10 \text{ ft}$

$-10 \text{ ft} \leq d - 60 \text{ ft} \leq 10 \text{ ft}$

$50 \text{ ft} \leq d \leq 70 \text{ ft}$

When the bungee jumper is 50 ft above the river, she is $150 - 50$, or 100 ft, from the bridge. When she is 70 ft above the river, she is $150 - 70$, or 80 ft, from the bridge. Thus, at any given time, the bungee jumper is between 80 ft and 100 ft from the bridge.

**121.** Graph $g(x) = 4$ on the same axes as $f(x) = |2x - 6|$.

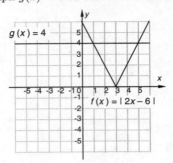

The solution set consists of the $x$-values for which $(x, f(x))$ is on or below the horizontal line $g(x) = 4$. These $x$-values comprise the interval $[1, 5]$.

**123.**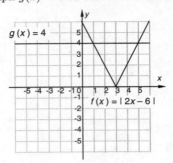

**125.** *Writing Exercise*

## Exercise Set 4.4

**1.** e; see pages 253 and 261 in the text.

**3.** d; see pages 258-261 in the text.

**5.** b; see page 259 in the text.

**7.** We replace $x$ with $-4$ and $y$ with 2.

$$\begin{array}{c|c}
2x + 3y < -1 & \\
\hline
2(-4) + 3 \cdot 2 & -1 \\
-8 + 6 & \\
& \overset{?}{} \\
-2 \overset{?}{=} -1 & \text{TRUE}
\end{array}$$

Since $-2 < -1$ is true, $(-4, 2)$ is a solution.

**9.** We replace $x$ with 8 and $y$ with 14.

$$\begin{array}{c|c}
2y - 3x \geq 9 & \\
\hline
2 \cdot 14 - 3 \cdot 8 & 9 \\
28 - 24 & \\
& \overset{?}{} \\
4 \overset{?}{=} 9 & \text{FALSE}
\end{array}$$

Since $4 > 9$ is false, $(8, 14)$ is not a solution.

**11.** Graph: $y > \dfrac{1}{2}x$

We first graph the line $y = \dfrac{1}{2}x$. We draw the line dashed since the inequality symbol is $>$. To determine which half-plane to shade, test a point not on the line. We try $(0, 1)$:

$$\begin{array}{c|c}
y > \dfrac{1}{2}x & \\
\hline
1 & \dfrac{1}{2} \cdot 0 \\
& \overset{?}{} \\
1 \overset{?}{=} 0 & \text{TRUE}
\end{array}$$

Since $1 > 0$ is true, $(0, 1)$ is a solution as are all of the points in the half-plane containing $(0, 1)$. We shade that half-plane and obtain the graph.

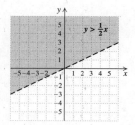

**13.** Graph: $y \geq x - 3$

First graph the line $y = x - 3$. Draw it solid since the inequality symbol is $\geq$. Test the point $(0,0)$ to determine if it is a solution.

$$\begin{array}{c|c} y \geq x - 3 \\ \hline 0 & 0 - 3 \\ & \overset{?}{} \\ 0 & = -3 \qquad \text{TRUE} \end{array}$$

Since $0 \geq -3$ is true, we shade the half-plane that contains $(0,0)$ and obtain the graph.

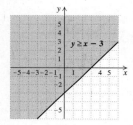

**15.** Graph: $y \leq x + 5$

First graph the line $y = x + 5$. Draw it solid since the inequality symbol is $\leq$. Test the point $(0,0)$ to determine if it is a solution.

$$\begin{array}{c|c} y \leq x + 5 \\ \hline 0 & 0 + 5 \\ & \overset{?}{} \\ 0 & = 5 \qquad \text{TRUE} \end{array}$$

Since $0 \leq 5$ is true, we shade the half-plane that contains $(0,0)$ and obtain the graph.

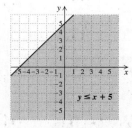

**17.** Graph: $x - y \leq 4$

First graph the line $x - y = 4$. Draw a solid line since the inequality symbol is $\leq$. Test the point $(0,0)$ to determine if it is a solution.

$$\begin{array}{c|c} x - y \leq 4 \\ \hline 0 - 0 & 4 \\ & \overset{?}{} \\ 0 & = 4 \qquad \text{TRUE} \end{array}$$

Since $0 \leq 4$ is true, we shade the half-plane that contains $(0,0)$ and obtain the graph.

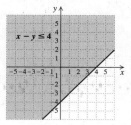

**19.** Graph: $2x + 3y < 6$

First graph $2x + 3y = 6$. Draw the line dashed since the inequality symbol is $<$. Test the point $(0,0)$ to determine if it is a solution.

$$\begin{array}{c|c} 2x + 3y < 6 \\ \hline 2 \cdot 0 + 3 \cdot 0 & 6 \\ & \overset{?}{} \\ 0 & = 6 \qquad \text{TRUE} \end{array}$$

Since $0 < 6$ is true, we shade the half-plane containing $(0,0)$ and obtain the graph.

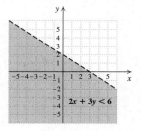

**21.** Graph: $2x - y \leq 4$

We first graph $2x - y = 4$. Draw the line solid since the inequality symbol is $\leq$. Test the point $(0,0)$ to determine if it is a solution.

$$\begin{array}{c|c} 2x - y \leq 4 \\ \hline 2 \cdot 0 - 0 & 4 \\ & \overset{?}{} \\ 0 & = 4 \qquad \text{TRUE} \end{array}$$

Since $0 \leq 4$ is true, we shade the half-plane containing $(0,0)$ and obtain the graph.

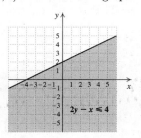

**23.** Graph: $2x - 2y \geq 8 + 2y$

$$2x - 4y \geq 8$$

First graph $2x - 4y = 8$. Draw the line solid since the inequality symbol is $\geq$. Test the point $(0,0)$ to determine if it is a solution.

$$\frac{2x - 4y \geq 8}{2 \cdot 0 - 4 \cdot 0 \;\Big|\; 8}$$

$$\overset{?}{0 = 8} \qquad \text{FALSE}$$

Since $0 \geq 8$ is false, we shade the half-plane that does not contain $(0, 0)$ and obtain the graph.

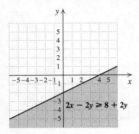

**25.** Graph: $y \geq 3$

We first graph $y = 3$. Draw the line solid since the inequality symbol is $\geq$. Test the point $(0, 0)$ to determine if it is a solution.

$$\frac{y \geq 3}{\overset{?}{0 = 3}} \qquad \text{FALSE}$$

Since $0 \geq 3$ is false, we shade the half-plane that does not contain $(0, 0)$ and obtain the graph.

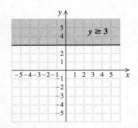

**27.** Graph: $x \leq 6$

We first graph $x = 6$. We draw the line solid since the inequality symbol is $\leq$. Test the point $(0, 0)$ to determine if it is a solution.

$$\frac{x \leq 6}{\overset{?}{0 = 6}} \qquad \text{TRUE}$$

Since $0 \leq 6$ is true, we shade the half-plane containing $(0, 0)$ and obtain the graph.

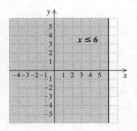

**29.** Graph: $-2 < y < 7$

This is a system of inequalities:

$$-2 < y,$$
$$y < 7$$

The graph of $-2 < y$ is the half-plane above the line $-2 = y$; the graph of $y < 7$ is the half-plane below the line $y = 7$. We shade the intersection of these graphs.

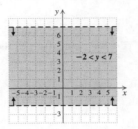

**31.** Graph: $-4 \leq x \leq 2$

This is a system of inequalities:

$$-4 \leq x,$$
$$x \leq 2$$

Graph $-4 \leq x$ and $x \leq 2$. Then shade the intersection of these graphs.

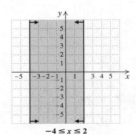

**33.** Graph: $0 \leq y \leq 3$

This is a system of inequalities:

$$0 \leq y,$$
$$y \leq 3$$

Graph $0 \leq y$ and $y \leq 3$.

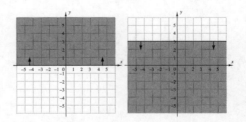

Then we shade the intersection of these graphs.

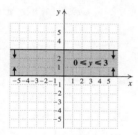

**35.** Graph: $y < -x,$

$\qquad\quad y > x + 2$

We graph the lines $y = -x$ and $y = x + 2$, using dashed lines. We indicate the region for each inequality by the arrows at the ends of the lines. Note where the regions overlap and shade the region of solutions.

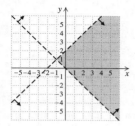

**37.** Graph: $y \geq x,$

$\qquad\quad y \leq 2x - 4$

Graph $y = x$ and $y = 2x - 4$, using solid lines. Indicate the region for each inequality by arrows, and shade the region where they overlap.

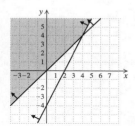

**39.** Graph: $y \leq -3,$

$\qquad\quad x \geq -1$

Graph $y = -3$ and $x = -1$ using solid lines. Indicate the region for each inequality by arrows, and shade the region where they overlap.

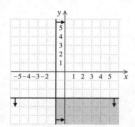

**41.** Graph: $x > -4,$

$\qquad\quad y < -2x + 3$

Graph the lines $x = -4$ and $y = -2x + 3$, using dashed lines. Indicate the region for each inequality by arrows, and shade the region where they overlap.

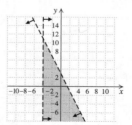

**43.** Graph: $y \leq 5,$

$\qquad\quad y \geq -x + 4$

Graph the lines $y = 5$ and $y = -x + 4$, using solid lines. Indicate the region for each inequality by arrows, and shade the region where they overlap.

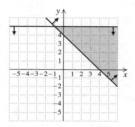

**45.** Graph: $x + y \leq 6,$

$\qquad\quad x - y \leq 4$

Graph the lines $x + y = 6$ and $x - y = 4$, using solid lines. Indicate the region for each inequality by arrows, and shade the region where they overlap.

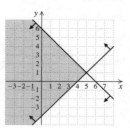

**47.** Graph: $y + 3x > 0,$

$\qquad\quad y + 3x < 2$

Graph the lines $y + 3x = 0$ and $y + 3x = 2$, using dashed lines. Indicate the region for each inequality by arrows, and shade the region where they overlap.

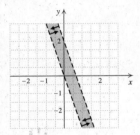

**49.** Graph:   $y \le 2x - 3,$     (1)
  $y \ge -2x + 1,$   (2)
  $x \le 5$        (3)

Graph the lines $y = 2x - 3$, $y = -2x + 1$, and $x = 5$ using solid lines. Indicate the region for each inequality by arrows, and shade the region where they overlap.

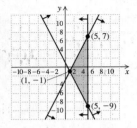

To find the vertex we solve three different systems of related equations.

From (1) and (2) we have   $y = 2x - 3,$
  $y = -2x + 1.$

Solving, we obtain the vertex $(1, -1)$.

From (1) and (3) we have   $y = 2x - 3,$
  $x = 5.$

Solving, we obtain the vertex $(5, 7)$.

From (2) and (3) we have   $y = -2x + 1,$
  $x = 5.$

Solving, we obtain the vertex $(5, -9)$.

**51.** Graph:   $x + 2y \le 12,$   (1)
  $2x + y \le 12$   (2)
  $x \ge 0,$        (3)
  $y \ge 0$         (4)

Graph the lines $x + 2y = 12$, $2x + y = 12$, $x = 0$, and $y = 0$ using solid lines. Indicate the region for each inequality by arrows, and shade the region where they overlap.

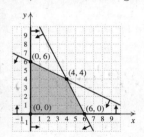

To find the vertices we solve four different systems of equations.

From (1) and (2) we have   $x + 2y = 12,$
  $2x + y = 12.$

Solving, we obtain the vertex $(4, 4)$.

From (1) and (3) we have   $x + 2y = 12,$
  $x = 0.$

Solving, we obtain the vertex $(0, 6)$.

From (2) and (4) we have   $2x + y = 12,$
  $y = 0.$

Solving, we obtain the vertex $(6, 0)$.

From (3) and (4) we have   $x = 0,$
  $y = 0.$

Solving, we obtain the vertex $(0, 0)$.

**53.** Graph:   $8x + 5y \le 40,$   (1)
  $x + 2y \le 8$    (2)
  $x \ge 0,$        (3)
  $y \ge 0$         (4)

Graph the lines $8x + 5y = 40$, $x + 2y = 8$, $x = 0$, and $y = 0$ using solid lines. Indicate the region for each inequality by arrows, and shade the region where they overlap.

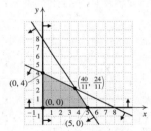

To find the vertices we solve four different systems of equations.

From (1) and (2) we have   $8x + 5y = 40,$
  $x + 2y = 8.$

Solving, we obtain the vertex $\left(\dfrac{40}{11}, \dfrac{24}{11}\right)$.

From (1) and (4) we have   $8x + 5y = 40,$
  $y = 0.$

Solving, we obtain the vertex $(5, 0)$.

From (2) and (3) we have   $x + 2y = 8,$
  $x = 0.$

Solving, we obtain the vertex $(0, 4)$.

From (3) and (4) we have   $x = 0,$
  $y = 0.$

Solving, we obtain the vertex $(0, 0)$.

**55.** Graph: $y - x \geq 2$, (1)

$\qquad y - x \leq 4$, (2)

$\qquad 2 \leq x \leq 5$ (3)

Think of (3) as two inequalities:

$\qquad 2 \leq x$, (4)

$\qquad x \leq 5$ (5)

Graph the lines $y - x = 2$, $y - x = 4$, $x = 2$, and $x = 5$, using solid lines. Indicate the region for each inequality by arrows, and shade the region where they overlap.

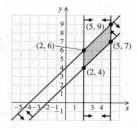

To find the vertices we solve four different systems of equations.

From (1) and (4) we have $\quad y - x = 2$,

$\qquad\qquad\qquad\qquad\qquad x = 2$.

Solving, we obtain the vertex $(2, 4)$.

From (1) and (5) we have $\quad y - x = 2$,

$\qquad\qquad\qquad\qquad\qquad x = 5$.

Solving, we obtain the vertex $(5, 7)$.

From (2) and (4) we have $\quad y - x = 4$,

$\qquad\qquad\qquad\qquad\qquad x = 2$.

Solving, we obtain the vertex $(2, 6)$.

From (2) and (5) we have $\quad y - x = 4$,

$\qquad\qquad\qquad\qquad\qquad x = 5$.

Solving, we obtain the vertex $(5, 9)$.

**57.** *Writing Exercise*

**59.** *Familiarize.* We let $x$ and $y$ represent the number of pounds of peanuts and fancy nuts in the mixture, respectively. We organize the given information in a table.

| Type of nuts | Peanuts | Fancy | Mixture |
|---|---|---|---|
| Amount | $x$ | $y$ | 10 |
| Price per pound | \$2.50 | \$7 | |
| Value | $2.5x$ | $7y$ | 40 |

*Translate.* We get a system of equations from the first and third rows of the table.

$\qquad x + y = 10$,

$\qquad 2.5x + 7y = 40$

Clearing decimals we have

$\qquad x + y = 10$, (1)

$\qquad 25x + 70y = 400$. (2)

*Carry out*. We use the elimination method. Multiply Equation (1) by $-25$ and add.

$\qquad -25x - 25y = -250$

$\qquad \underline{25x + 70y = 400}$

$\qquad\qquad\quad 45y = 150$

$\qquad\qquad\qquad y = \dfrac{10}{3}$, or $3\dfrac{1}{3}$

Substitute $\dfrac{10}{3}$ for $y$ in Equation (1) and solve for $x$.

$\qquad x + y = 10$

$\qquad x + \dfrac{10}{3} = 10$

$\qquad\qquad x = \dfrac{20}{3}$, or $6\dfrac{2}{3}$

*Check*. The sum of $6\dfrac{2}{3}$ and $3\dfrac{1}{3}$ is 10. The value of the mixture is $2.5\left(\dfrac{20}{3}\right) + 7\left(\dfrac{10}{3}\right)$, or $\dfrac{50}{3} + \dfrac{70}{3}$, or \$40. These numbers check.

*State*. $6\dfrac{2}{3}$ lb of peanuts and $3\dfrac{1}{3}$ lb of fancy nuts should be used.

**61.** *Familiarize.* Let $x =$ the number of card holders tickets that were sold and $y =$ the number of noncard holders tickets. We arrange the information in a table.

| | Card holders | Noncard holders | Total |
|---|---|---|---|
| Price | \$2.50 | \$4 | |
| Number sold | $x$ | $y$ | 203 |
| Money taken in | $2.5x$ | $4y$ | \$620 |

*Translate.* The last two rows of the table give us two equations. The total number of tickets sold was 203, so we have

$\qquad x + y = 203$.

The total amount of money collected was \$620, so we have

$\qquad 2.5x + 4y = 620$.

We can multiply the second equation on both sides by 10 to clear the decimal. The resulting system is

$\qquad x + y = 203$, (1)

$\qquad 25x + 40y = 6200$. (2)

*Carry out*. We use the elimination method. We multiply on both sides of Equation (1) by $-25$ and then add.

$\qquad -25x - 25y = -5075 \quad$ Multiplying by $-25$

$\qquad \underline{25x + 40y = \phantom{-}6200}$

$\qquad\qquad\quad 15y = \phantom{-}1125$

$\qquad\qquad\qquad y = \phantom{-}75$

We go back to Equation (1) and substitute 75 for $y$.

$\qquad x + y = 203$

$\qquad x + 75 = 203$

$\qquad\qquad x = 128$

***Check***. The number of tickets sold was $128 + 75$, or 203. The money collected was $\$2.50(128) + \$4(75)$, or $\$320 + \$300$, or $\$620$. These numbers check.

***State***.  128 card holders tickets and 75 noncard holders tickets were sold.

**63.** ***Familiarize***. The formula for the area of a triangle with base $b$ and height $h$ is $A = \frac{1}{2}bh$.

***Translate***. Substitute 200 for $A$ and 16 for $b$ in the formula.

$$A = \frac{1}{2}bh$$

$$200 = \frac{1}{2} \cdot 16 \cdot h$$

***Carry out***. We solve the equation.

$$200 = \frac{1}{2} \cdot 16 \cdot h$$

$$200 = 8h \qquad \text{Multiplying}$$

$$25 = h \qquad \text{Dividing by 8 on both sides}$$

***Check***. The area of a triangle with base 16 ft and height 25 ft is $\frac{1}{2} \cdot 16 \cdot 25$, or 200 ft$^2$. The answer checks.

***State***. The seed can fill a triangle that is 25 ft tall.

**65.** *Writing Exercise*

**67.**  Graph:  $x + y > 8$,
$\qquad\qquad\quad x + y \leq -2$

Graph the line $x + y = 8$ using a dashed line and graph $x + y = -2$, using a solid line. Indicate the region for each inequality by arrows. The regions do not overlap (the solution set is $\emptyset$), so we do not shade any portion of the graph.

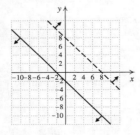

**69.**  Graph:  $x - 2y \leq 0$,
$\qquad\qquad -2x + y \leq 2$,
$\qquad\qquad\qquad x \leq 2$,
$\qquad\qquad\qquad y \leq 2$,
$\qquad\qquad\quad x + y \leq 4$

Graph the five inequalities above, and shade the region where they overlap.

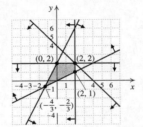

**71.** Both the width and the height must be positive, so we have

$$w > 0,$$
$$h > 0.$$

To be checked as luggage, the sum of the width, height, and length cannot exceed 62 in., so we have

$$w + h + 30 \leq 62, \text{ or}$$
$$w + h \leq 32.$$

The girth is represented by $2w + 2h$ and the length is 30 in. In order to meet postal regulations the sum of the girth and the length cannot exceed 130 in., so we have:

$$2w + 2h + 30 < 130, \text{ or}$$
$$2w + 2h \leq 100, \text{ or}$$
$$w + h \leq 50$$

Thus, have a system of inequalities:

$$w > 0,$$
$$h > 0$$
$$w + h < 32,$$
$$w + h \leq 50$$

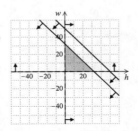

**73.**  Graph:  $35c + 75a > 1000$,
$\qquad\qquad\qquad\quad c \geq 0$,
$\qquad\qquad\qquad\quad a \geq 0$

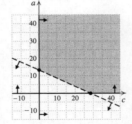

**75.** a)

$3x + 6y > 2$

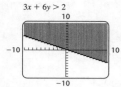

b)

$x - 5y \leq 10$

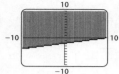

c)

$13x - 25y + 10 \leq 0$

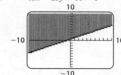

d)

$2x + 5y > 0$

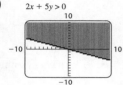

# Exercise Set 4.5

**1.** minimized; see page 266 in the text.

**3.** constraints; see page 266 in the text.

**5.** feasible; see page 267 in the text.

**7.** Find the maximum and minimum values of
$F = 2x + 14y$,

subject to

$$5x + 3y \leq 34, \quad (1)$$
$$3x + 5y \leq 30, \quad (2)$$
$$x \geq 0, \quad (3)$$
$$y \geq 0. \quad (4)$$

Graph the system of inequalities and find the coordinates of the vertices.

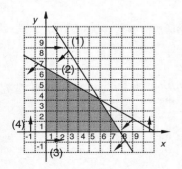

To find one vertex we solve the system

$$x = 0,$$
$$y = 0.$$

This vertex is $(0, 0)$.

To find a second vertex we solve the system

$$5x + 3y = 34,$$
$$y = 0.$$

This vertex is $\left(\dfrac{34}{5}, 0\right)$.

To find a third vertex we solve the system

$$5x + 3y = 34,$$
$$3x + 5y = 30.$$

This vertex is $(5, 3)$.

To find the fourth vertex we solve the system

$$3x + 5y = 30,$$
$$x = 0.$$

This vertex is $(0, 6)$.

Now find the value of $F$ at each of these points.

| Vertex $(x, y)$ | $F = 2x + 14y$ | |
|---|---|---|
| $(0, 0)$ | $2 \cdot 0 + 14 \cdot 0 = 0 + 0 = 0$ | ← Minimum |
| $\left(\dfrac{34}{5}, 0\right)$ | $2 \cdot \dfrac{34}{5} + 14 \cdot 0 = \dfrac{68}{5} + 0 = 13\dfrac{3}{5}$ | |
| $(5, 3)$ | $2 \cdot 5 + 14 \cdot 3 = 10 + 42 = 52$ | |
| $(0, 6)$ | $2 \cdot 0 + 14 \cdot 6 = 0 + 84 = 84$ | ← Maximum |

The maximum value of $F$ is 84 when $x = 0$ and $y = 6$.
The minimum value of $F$ is 0 when $x = 0$ and $y = 0$.

**9.** Find the maximum and minimum values of
$P = 8x - y + 20$,

subject to

$$6x + 8y \leq 48, \quad (1)$$
$$0 \leq y \leq 4, \quad (2)$$
$$0 \leq x \leq 7. \quad (3)$$

Think of (2) as $\quad 0 \leq y, \quad (4)$
$$y \leq 4. \quad (5)$$

Think of (3) as $\quad 0 \leq x, \quad (6)$
$$x \leq 7. \quad (7)$$

Graph the system of inequalities.

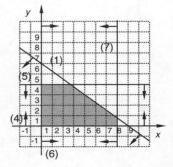

To determine the coordinates of the vertices, we solve the following systems:

$$x = 0, \qquad x = 7, \qquad 6x + 8y = 48,$$
$$y = 0; \qquad y = 0; \qquad x = 7;$$

$$6x + 8y = 48, \qquad x = 0,$$
$$y = 4; \qquad\qquad y = 4$$

The vertices are $(0,0)$, $(7,0)$, $\left(7, \dfrac{3}{4}\right)$, $\left(\dfrac{8}{3}, 4\right)$, and $(0,4)$, respectively. Compute the value of $P$ at each of these points.

| Vertex $(x, y)$ | $P = 8x - y + 20$ |
|---|---|
| $(0,0)$ | $8 \cdot 0 - 0 + 20 =$ <br> $0 - 0 + 20 = 20$ |
| $(7,0)$ | $8 \cdot 7 - 0 + 20 =$ <br> $56 - 0 + 20 =$ <br> $76$ ←——Maximum |
| $\left(7, \dfrac{3}{4}\right)$ | $8 \cdot 7 - \dfrac{3}{4} + 20 =$ <br> $56 - \dfrac{3}{4} + 20 = 75\dfrac{1}{4}$ |
| $\left(\dfrac{8}{3}, 4\right)$ | $8 \cdot \dfrac{8}{3} - 4 + 20 =$ <br> $\dfrac{64}{3} - 4 + 20 = 37\dfrac{1}{3}$ |
| $(0,4)$ | $8 \cdot 0 - 4 + 20 =$ <br> $0 - 4 + 20 =$ <br> $16$ ←——Minimum |

The maximum is 76 when $x = 7$ and $y = 0$. The minimum is 16 when $x = 0$ and $y = 4$.

**11.** Find the maximum and minimum values of
$$F = 2y - 3x,$$
subject to
$$y \le 2x + 1, \qquad (1)$$
$$y \ge -2x + 3, \qquad (2)$$
$$x \le 3 \qquad\qquad (3)$$

Graph the system of inequalities and find the coordinates of the vertices.

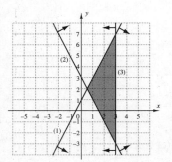

To determine the coordinates of the vertices, we solve the following systems:

$$y = 2x + 1, \qquad y = 2x + 1, \qquad y = -2x + 3,$$
$$y = -2x + 3; \qquad x = 3; \qquad x = 3$$

The solutions of the systems are $\left(\dfrac{1}{2}, 2\right)$, $(3,7)$, and $(3, -3)$, respectively. Now find the value of $F$ at each of these points.

| Vertex $(x, y)$ | $F = 2y - 3x$ |
|---|---|
| $\left(\dfrac{1}{2}, 2\right)$ | $2 \cdot 2 - 3 \cdot \dfrac{1}{2} = \dfrac{5}{2}$ |
| $(3,7)$ | $2 \cdot 7 - 3 \cdot 3 = 5$ ←——Maximum |
| $(3, -3)$ | $2(-3) - 3 \cdot 3 = -15$ ←Minimum |

The maximum value is 5 when $x = 3$ and $y = 7$. The minimum value is $-15$ when $x = 3$ and $y = -3$.

**13. *Familiarize*.** Let $x =$ the number of gallons of gasoline the car uses and $y =$ the number of gallons the moped uses.

***Translate*.** The mileage $m$ is given by
$$M = 20x + 100y.$$

We wish to maximize $M$ subject to these constraints:
$$x + y \le 12,$$
$$0 \le x \le 10,$$
$$0 \le y \le 3.$$

***Carry out*.** We graph the system of inequalities, determine the vertices, and evaluate $M$ at each vertex.

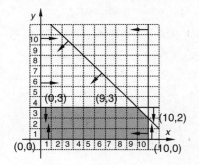

| Vertex | $M = 20x + 100y$ |
|---|---|
| $(0,0)$ | $0$ |
| $(0,3)$ | $300$ |
| $(9,3)$ | $480$ |
| $(10,2)$ | $400$ |
| $(10,0)$ | $200$ |

The largest mileage in the table is 480, obtained when the car uses 9 gallons of gasoline and the moped uses 3 gallons.

***Check*.** Go over the algebra and arithmetic.

***State*.** The maximum number of miles is 480 when the car uses 9 gal of gasoline and the moped uses 3 gal.

**15.** *Familiarize*. Let $x =$ the number of units of lumber and $y =$ the number of units of plywood produced per week.

*Translate*. The profit $P$ is given by

$P = \$20x + \$30y$.

We wish to maximize $P$ subject to these constraints:

$x + y \leq 400$,

$x \geq 100$,

$y \geq 150$.

*Carry out*. We graph the system of inequalities, determine the vertices, and evaluate $P$ at each vertex.

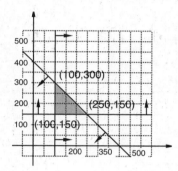

| Vertex | $P = \$20x + \$30y$ |
|--------|---------------------|
| $(100, 150)$ | $\$20 \cdot 100 + \$30 \cdot 150 = \$6500$ |
| $(100, 300)$ | $\$20 \cdot 100 + \$30 \cdot 300 = \$11,000$ |
| $(250, 150)$ | $\$20 \cdot 250 + \$30 \cdot 150 = \$9500$ |

The greatest profit in the table is $\$11,000$, obtained when 100 units of lumber and 300 units of plywood are produced.

*Check*. Go over the algebra and arithmetic.

*State*. The maximum profit is achieved by producing 100 units of lumber and 300 units of plywood.

**17.** In order to earn the most interest Rosa should invest the entire $\$40,000$. She should also invest as much as possible in the type of investment that has the higher interest rate. Thus, she should invest $\$22,000$ in corporate bonds and the remaining $\$18,000$ in municipal bonds. The maximum income is $0.08(\$22,000) + 0.075(\$18,000) = \$3110$.

We can also solve this problem as follows.

Let $x =$ the amount invested in corporate bonds and $y =$ the amount invested in municipal bonds. Find the maximum value of

$I = 0.08x + 0.075y$

subject to

$x + y \leq \$40,000$,

$\$6000 \leq x \leq \$22,000$

$0 \leq y \leq \$30,000$.

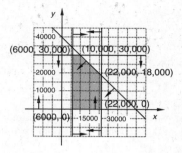

| Vertex | $I = 0.08x + 0.075y$ |
|--------|----------------------|
| $(\$6000, \$0)$ | $\$480$ |
| $(\$6000, \$30,000)$ | $\$2730$ |
| $(\$10,000, \$30,000)$ | $\$3050$ |
| $(\$22,000, \$18,000)$ | $\$3110$ |
| $(\$22,000, \$0)$ | $\$1760$ |

The maximum income of $\$3110$ occurs when $\$22,000$ is invested in corporate bonds and $\$18,000$ is invested in municipal bonds.

**19.** *Familiarize*. We organize the information in a table. Let $x =$ the number of matching questions and $y =$ the number of essay questions you answer.

| Type | Number of points for each | Number answered | Total points |
|------|---------------------------|-----------------|--------------|
| Matching | 10 | $3 \leq x \leq 12$ | $10x$ |
| Essay | 25 | $4 \leq y \leq 15$ | $25y$ |
| Total | | $x + y \leq 20$ | $10x + 25y$ |

Since Phil can answer no more than a total of 20 questions, we have the inequality $x + y \leq 20$ in the "Number answered" column. The expression $10x + 25y$ in the "Total points" column gives the total score on the test.

*Translate*. The score $S$ is given by

$S = 10x + 25y$.

We wish to maximize $S$ subject to these facts (constraints) about $x$ and $y$.

$3 \leq x \leq 12$,

$4 \leq y \leq 15$,

$x + y \leq 20$.

*Carry out*. We graph the system of inequalities, determine the vertices, and evaluate $S$ at each vertex.

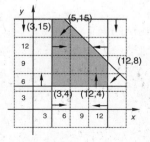

| Vertex | $S = 4x + 7y$ |
|--------|---------------|
| $(3,4)$ | $10 \cdot 3 + 25 \cdot 4 = 130$ |
| $(3,15)$ | $10 \cdot 3 + 25 \cdot 15 = 405$ |
| $(5,15)$ | $10 \cdot 5 + 25 \cdot 15 = 425$ |
| $(12,8)$ | $10 \cdot 12 + 25 \cdot 8 = 320$ |
| $(12,4)$ | $10 \cdot 12 + 25 \cdot 4 = 220$ |

The greatest score in the table is 425, obtained when 5 matching questions and 15 essay questions are answered correctly.

**Check**. Go over the algebra and arithmetic.

**State**. The maximum score is 425 points when 5 matching questions and 15 essay questions are answered correctly.

**21. Familiarize**. Let $x =$ the Merlot acreage and $y =$ the Cabernet acreage.

**Translate**. The profit $P$ is given by

$P = \$400x + \$300y.$

We wish to maximize $P$ subject to these constraints:

$x + y \le 240,$

$2x + y \le 320,$

$x \ge 0,$

$y \ge 0.$

**Carry out**. We graph the system of inequalities, determine the vertices, and evaluate $P$ at each vertex.

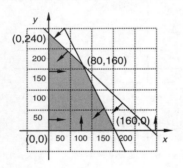

| Vertex | $P = \$400x + \$300y$ |
|--------|------------------------|
| $(0,0)$ | $\$0$ |
| $(0,240)$ | $\$72,000$ |
| $(80,160)$ | $\$80,000$ |
| $(160,0)$ | $\$64,000$ |

**Check**. Go over the algebra and arithmetic.

**State**. The maximum profit occurs by planting 80 acres of Merlot grapes and 160 acres of Cabernet graphs.

**23. Familiarize**. Let $x =$ the number of knit suits and $y =$ the number of worsted suits made per day.

**Translate**. The profit $P$ is given by

$P = \$68x + \$62y.$

We wish to maximize $P$ subject to these constraints.

$2x + 4y \le 20,$

$4x + 2y \le 16,$

$x \ge 0,$

$y \ge 0.$

**Carry out**. Graph the system of inequalities, determine the vertices, the evaluate $P$ at each vertex.

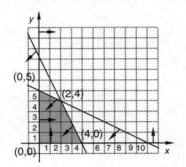

| Vertex | $P = \$68x + \$62y$ |
|--------|----------------------|
| $(0,0)$ | $\$68 \cdot 0 + \$62 \cdot 0 = \$0$ |
| $(0,5)$ | $\$68 \cdot 0 + \$62 \cdot 5 = \$310$ |
| $(2,4)$ | $\$68 \cdot 2 + \$62 \cdot 4 = \$384$ |
| $(4,0)$ | $\$68 \cdot 4 + \$62 \cdot 0 = \$272$ |

**Check**. Go over the algebra and arithmetic.

**State**. The maximum profit per day is $384 when 2 knit suits and 4 worsted suits are made.

**25.** *Writing Exercise*

**27.** $5x^3 - 4x^2 - 7x + 2$

$= 5(-2)^3 - 4(-2)^2 - 7(-2) + 2$

$= 5(-8) - 4(4) - 7(-2) + 2$

$= -40 - 16 + 14 + 2$

$= -40$

**29.** $3(2x - 5) + 4(x + 5) = 6x - 15 + 4x + 20 = 10x + 5$

**31.** $6x - 3(x + 2) = 6x - 3x - 6 = 3x - 6$

**33.** *Writing Exercise*

**35. Familiarize**. Let $x$ represent the number of T3 planes and $y$ represent the number of S5 planes. Organize the information in a table.

| Plane | Number of planes | Passengers | | |
|-------|------------------|------------|---|---|
|       |                  | First | Tourist | Economy |
| T3 | $x$ | $40x$ | $40x$ | $120x$ |
| S5 | $y$ | $80y$ | $30y$ | $40y$ |

| Plane | Cost per mile |
|-------|---------------|
| T3 | $30x$ |
| S5 | $25y$ |

**Translate.** Suppose $C$ is the total cost per mile. Then $C = 30x + 25y$. We wish to minimize $C$ subject to these facts (constraints) about $x$ and $y$.

$$40x + 80y \geq 2000,$$
$$40x + 30y \geq 1500,$$
$$120x + 40y \geq 2400,$$
$$x \geq 0,$$
$$y \geq 0$$

**Carry out.** Graph the system of inequalities, determine the vertices, and evaluate $C$ at each vertex.

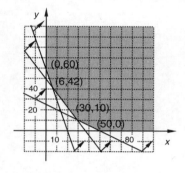

| Vertex | $C = 30x + 25y$ |
|--------|-----------------|
| $(0, 60)$ | $30(0) + 25(60) = 1500$ |
| $(6, 42)$ | $30(6) + 25(42) = 1230$ |
| $(30, 10)$ | $30(30) + 25(10) = 1150$ |
| $(50, 0)$ | $30(50) + 25(0) = 1500$ |

**Check.** Go over the algebra and arithmetic.

**State.** In order to minimize the operating cost, 30 T3 planes and 10 S5 planes should be used.

**37. Familiarize.** Let $x =$ the number of chairs and $y =$ the number of sofas produced.

**Translate.** Find the maximum value of

$$I = \$80x + \$1200y$$

subject to

$$20x + 100y \leq 1900,$$
$$x + 50y \leq 500,$$
$$2x + 20y \leq 240,$$
$$x \geq 0,$$
$$y \geq 0.$$

**Carry out.** Graph the system of inequalities, determine the vertices, and evaluate $I$ at each vertex.

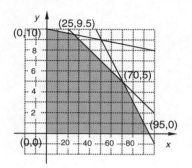

| Vertex | $I = \$80x + \$1200y$ |
|--------|------------------------|
| $(0, 0)$ | $\$0$ |
| $(0, 10)$ | $\$12,000$ |
| $(25, 9.5)$ | $\$13,400$ |
| $(70, 5)$ | $\$11,600$ |
| $(95, 0)$ | $\$7600$ |

**Check.** Go over the algebra and arithmetic.

**State.** The maximum income of $\$13,400$ occurs when 25 chairs and 9.5 sofas are made. A more practical answer is that the maximum income of $\$12,800$ is achieved when 25 chairs and 9 sofas are made.

# Chapter 5

# Polynomials and Polynomial Functions

---

## Exercise Set 5.1

---

**1.** g

**3.** a

**5.** b

**7.** j

**9.** f

**11.** $-6x^5 - 8x^3 + x^2 + 3x - 4$

| Term | $-6x^5$ | $-8x^3$ | $x^2$ | $3x$ | $-4$ |
|---|---|---|---|---|---|
| Degree | 5 | 3 | 2 | 1 | 0 |
| Degree of polynomial | 5 | | | | |

**13.** $y^3 + 2y^7 + x^2y^4 - 8$

| Term | $y^3$ | $2y^7$ | $x^2y^4$ | $-8$ |
|---|---|---|---|---|
| Degree | 3 | 7 | 6 | 0 |
| Degree of polynomial | 7 | | | |

**15.** $a^5 + 4a^2b^4 + 6ab + 4a - 3$

| Term | $a^5$ | $4a^2b^4$ | $6ab$ | $4a$ | $-3$ |
|---|---|---|---|---|---|
| Degree | 5 | 6 | 2 | 1 | 0 |
| Degree of polynomial | 6 | | | | |

**17.** $-18y^4 + 11y^3 + 6y^2 - 5y + 3$; $-18y^4$; $-18$

**19.** $-a^7 + 8a^5 + 5a^3 - 19a^2 + a$; $-a^7$; $-1$

**21.** $-9 + 6x - 5x^2 + 3x^4$

**23.** $-9xy + 5x^2y^2 + 8x^3y^2 - 5x^4$

**25.** $-7x + 5 = -7 \cdot 4 + 5 = -28 + 5 = -23$

**27.** $x^3 - 5x^2 + x = 4^3 - 5 \cdot 4^2 + 4 = 64 - 5 \cdot 16 + 4 =$
$64 - 80 + 4 = -12$

**29.** $f(x) = -5x^3 + 3x^2 - 4x - 3$
$f(-1) = -5(-1)^3 + 3(-1)^2 - 4(-1) - 3$
$\quad = 5 + 3 + 4 - 3$
$\quad = 9$

**31.** $F(x) = 2x^2 - 6x - 9$
$F(2) = 2 \cdot 2^2 - 6 \cdot 2 - 9 = 8 - 12 - 9 = -13$
$F(5) = 2 \cdot 5^2 - 6 \cdot 5 - 9 = 50 - 30 - 9 = 11$

**33.** $Q(y) = -8y^3 + 7y^2 - 4y - 9$
$Q(-3) = -8(-3)^3 + 7(-3)^2 - 4(-3) - 9 =$
$\qquad 216 + 63 + 12 - 9 = 282$
$Q(0) = -8 \cdot 0^3 + 7 \cdot 0^2 - 4 \cdot 0 - 9 =$
$\qquad 0 + 0 + 0 - 9 = -9$

**35.** We evaluate the function for $t = 2.34$.
$\qquad v(t) = 10.9t$
$\qquad v(2.34) = 10.9(2.34) \approx 25.5$
Dumont was going about 25.5 mph.

**37.** We evaluate the function for $t = 1.36$.
$\qquad s(t) = 16t^2$
$\qquad s(1.36) = 16(1.36)^2 \approx 29.6$
Dyani dropped about 29.6 ft.

**39.** $N(p) = p^3 - 3p^2 + 2p$
$N(20) = 20^3 - 3 \cdot 20^2 + 2 \cdot 20$
$\qquad = 8000 - 1200 + 40$
$\qquad = 6840$

A president, vice president, and treasurer can be elected in 6840 ways.

**41.** 2009 is 20 years since 1989. Locate 20 on the horizontal axis. From there move vertically to the graph and then horizontally to the vertical axis. This locates a value of about 26.4. Thus, the attendance at NASCAR races in 2009 will be about 26.4 million.

**43.** Locate 8 on the horizontal axis. From there move vertically to the graph and then horizontally to the vertical axis. This locates a value of about 6.4. Thus, $A(8) \approx 6.4$ million.

**45.** Using this function, we find $N(3)$.
$\qquad N(x) = \frac{1}{3}x^3 + \frac{1}{2}x^2 + \frac{1}{6}x$
$\qquad N(3) = \frac{1}{3} \cdot 3^3 + \frac{1}{2} \cdot 3^2 + \frac{1}{6} \cdot 3$
$\qquad\quad = \frac{1}{3} \cdot 27 + \frac{1}{2} \cdot 9 + \frac{1}{6} \cdot 3$
$\qquad\quad = 9 + \frac{9}{2} + \frac{1}{2} = 14$

From the figure we see that the bottom layer has 9 spheres, the second layer has 4, and the third layer has 1. Thus there are $9 + 4 + 1$, or 14 spheres.

Using either the function or the figure, we find that $N(3) = 14$.

To calculate the number of oranges in a pyramid with 5 layers, we evaluate the function for $x = 5$.

$$N(5) = \frac{1}{3} \cdot 5^3 + \frac{1}{2} \cdot 5^2 + \frac{1}{6} \cdot 5$$

$$= \frac{1}{3} \cdot 125 + \frac{1}{2} \cdot 25 + \frac{1}{6} \cdot 5$$

$$= \frac{125}{3} + \frac{25}{2} + \frac{5}{6}$$

$$= \frac{250}{6} + \frac{75}{6} + \frac{5}{6} = \frac{330}{6}$$

$$= 55 \text{ oranges}$$

**47.** Locate 2 on the horizontal axis. From there move vertically to the graph and then horizontally to the $C(t)$-axis. This locates a value of about 2.3. Thus, about 2.3 mcg/mL of Gentamicin is in the bloodstream 2 hr after injection.

**49.** Locate 2 on the horizontal axis. From there move vertically to the graph and then horizontally to the $C(t)$-axis. This locates a value of about 2.3. Thus, $C(2) \approx 2.3$.

**51.** We evaluate the polynomial for $h = 6.3$ and $r = 1.2$:

$2\pi rh + 2\pi r^2 = 2\pi(1.2)(6.3) + 2\pi(1.2)^2 \approx 56.5$

The surface area is about 56.5 in$^2$.

**53.** Evaluate the polynomial function for $x = 75$:

$$R(x) = 280x - 0.4x^2$$
$$R(75) = 280 \cdot 75 - 0.4(75)^2$$
$$= 21,000 - 0.4(5625)$$
$$= 21,000 - 2250 = 18,750$$

The total revenue is $18,750.

**55.** Evaluate the polynomial function for $x = 75$:

$$C(x) = 5000 + 0.6x^2$$
$$C(75) = 5000 + 0.6(75)^2$$
$$= 5000 + 0.6(5625)$$
$$= 5000 + 3375$$
$$= 8375$$

The total cost is $8375.

**57.**    $5a + 6 - 4 + 2a^3 - 6a + 2$

$= 2a^3 + (5 - 6)a + (6 - 4 + 2)$

$= 2a^3 - a + 4$

**59.**    $3a^2b + 4b^2 - 9a^2b - 7b^2$

$= (3 - 9)a^2b + (4 - 7)b^2$

$= -6a^2b - 3b^2$

**61.**    $9x^2 - 3xy + 12y^2 + x^2 - y^2 + 5xy + 4y^2$

$= (9 + 1)x^2 + (-3 + 5)xy + (12 - 1 + 4)y^2$

$= 10x^2 + 2xy + 15y^2$

**63.**    $(7x - 5y + 3z) + (9x + 12y - 8z)$

$= (7 + 9)x + (-5 + 12)y + (3 - 8)z$

$= 16x + 7y - 5z$

**65.**    $(x^2 + 2x - 3xy - 7) + (-3x^2 - x + 2xy + 6)$

$= (1 - 3)x^2 + (2 - 1)x + (-3 + 2)xy + (-7 + 6)$

$= -2x^2 + x - xy - 1$

**67.**    $(8x^2y - 3xy^2 + 4xy) + (-2x^2y - xy^2 + xy)$

$= (8 - 2)x^2y + (-3 - 1)xy^2 + (4 + 1)xy$

$= 6x^2y - 4xy^2 + 5xy$

**69.**    $(2r^2 + 12r - 11) + (6r^2 - 2r + 4) + (r^2 - r - 2)$

$= (2 + 6 + 1)r^2 + (12 - 2 - 1)r + (-11 + 4 - 2)$

$= 9r^2 + 9r - 9$

**71.**    $\left(\frac{1}{8}xy - \frac{3}{5}x^3y^2 + 4.3y^3\right) +$

$\qquad \left(-\frac{1}{3}xy - \frac{3}{4}x^3y^2 - 2.9y^3\right)$

$= \left(\frac{1}{8} - \frac{1}{3}\right)xy + \left(-\frac{3}{5} - \frac{3}{4}\right)x^3y^2 + (4.3 - 2.9)y^3$

$= \left(\frac{3}{24} - \frac{8}{24}\right)xy + \left(-\frac{12}{20} - \frac{15}{20}\right)x^3y^2 + 1.4y^3$

$= -\frac{5}{24}xy - \frac{27}{20}x^3y^2 + 1.4y^3$

**73.** $5x^3 - 7x^2 + 3x - 9$

a)  $-(5x^3 - 7x^2 + 3x - 9)$    Writing the opposite of $P$ as $-P$

b)  $-5x^3 + 7x^2 - 3x + 9$    Changing the sign of every term

**75.** $-12y^5 + 4ay^4 - 7by^2$

a)  $-(-12y^5 + 4ay^4 - 7by^2)$

b)  $12y^5 - 4ay^4 + 7by^2$

**77.**    $(7x - 5) - (-3x + 4)$

$= (7x - 5) + (3x - 4)$

$= 10x - 9$

**79.**    $(-3x^2 + 2x + 9) - (x^2 + 5x - 4)$

$= (-3x^2 + 2x + 9) + (-x^2 - 5x + 4)$

$= -4x^2 - 3x + 13$

**81.**    $(8a - 3b + c) - (2a + 3b - 4c)$

$= (8a - 3b + c) + (-2a - 3b + 4c)$

$= 6a - 6b + 5c$

**83.**    $(6a^2 + 5ab - 4b^2) - (8a^2 - 7ab + 3b^2)$

$= (6a^2 + 5ab - 4b^2) + (-8a^2 + 7ab - 3b^2)$

$= -2a^2 + 12ab - 7b^2$

**85.**    $(6ab - 4a^2b + 6ab^2) - (3ab^2 - 10ab - 12a^2b)$

$= (6ab - 4a^2b + 6ab^2) + (-3ab^2 + 10ab + 12a^2b)$

$= 8a^2b + 16ab + 3ab^2$

**87.**    $\left(\frac{5}{8}x^4 - \frac{1}{4}x^2 - \frac{1}{2}\right) - \left(-\frac{3}{8}x^4 + \frac{3}{4}x^2 + \frac{1}{2}\right)$

$= \left(\frac{5}{8}x^4 - \frac{1}{4}x^2 - \frac{1}{2}\right) + \left(\frac{3}{8}x^4 - \frac{3}{4}x^2 - \frac{1}{2}\right)$

$= x^4 - x^2 - 1$

**89.** $\quad (5x^2 + 7) - (2x^2 + 1) + (x^2 + 2x)$
$= (5x^2 + 7) + (-2x^2 - 1) + (x^2 + 2x)$
$= (5 - 2 + 1)x^2 + 2x + (7 - 1)$
$= 4x^2 + 2x + 6$

**91.** $\quad (8r^2 - 6r) - (2r - 6) + (5t^2 - 7)$
$= (8r^2 - 6r) + (-2r + 6) + (5r^2 - 7)$
$= (8 + 5)r^2 + (-6 - 2)r + (6 - 7)$
$= 13r^2 - 8r - 1$

**93.** $(x^2 - 4x + 7) + (3x^2 - 9) - (x^2 - 4x + 7)$

Note that $x^2 - 4x + 7$ and $-(x^2 - 4x + 7)$ are opposites so their sum is 0. Then the result is $3x^2 - 9$.

**95.** $\quad P(x) = R(x) - C(x)$
$P(x) = (280x - 0.4x^2) - (5000 + 0.6x^2)$
$P(x) = (280x - 0.4x^2) + (-5000 - 0.6x^2)$
$P(x) = 280x - x^2 - 5000$
$P(70) = 280 \cdot 70 - 70^2 - 5000$
$\quad\quad = 19,600 - 4900 - 5000 = 9700$

The profit is \$9700.

**97.** *Writing Exercise*

**99.** $(x^2)^5 = x^{2 \cdot 5} = x^{10}$

**101.** $(5x^3)^2 = 5^2(x^3)^2 = 25x^{3 \cdot 2} = 25x^6$

**103.** $x^5 \cdot x^4 = x^{5+4} = x^9$

**105.** *Writing Exercise*

**107.** $\quad 2[P(x)]$
$= 2(13x^5 - 22x^4 - 36x^3 + 40x^2 - 16x + 75)$
$= 26x^5 - 44x^4 - 72x^3 + 80x^2 - 32x + 150$

Use columns to add:

$\quad 26x^5 - 44x^4 - 72x^3 + 80x^2 - 32x + 150$
$\quad \underline{42x^5 - 37x^4 + 50x^3 - 28x^2 + 34x + 100}$
$\quad 68x^5 - 81x^4 - 22x^3 + 52x^2 + \phantom{0}2x + 250$

**109.** $\quad 2[Q(x)]$
$= 2(42x^5 - 37x^4 + 50x^3 - 28x^2 + 34x + 100)$
$= 84x^5 - 74x^4 + 100x^3 - 56x^2 + 68x + 200$

$\quad 3[P(x)]$
$= 3(13x^5 - 22x^4 - 36x^3 + 40x^2 - 16x + 75)$
$= 39x^5 - 66x^4 - 108x^3 + 120x^2 - 48x + 225$

Use columns to subtract, adding the opposite of $3[P(x)]$:

$\quad\phantom{-} 84x^5 - 74x^4 + 100x^3 - \phantom{0}56x^2 + \phantom{0}68x + 200$
$\quad \underline{- 39x^5 + 66x^4 + 108x^3 - 120x^2 + \phantom{0}48x - 225}$
$\quad\phantom{-} 45x^5 - \phantom{0}8x^4 + 208x^3 - 176x^2 + 116x - \phantom{0}25$

**111.** First we find the number of truffles in the display.

$N(x) = \dfrac{1}{6}x^3 + \dfrac{1}{2}x^2 + \dfrac{1}{3}x$

$N(5) = \dfrac{1}{6} \cdot 5^3 + \dfrac{1}{2} \cdot 5^2 + \dfrac{1}{3} \cdot 5$

$\quad\quad = \dfrac{1}{6} \cdot 125 + \dfrac{1}{2} \cdot 25 + \dfrac{5}{3}$

$\quad\quad = \dfrac{125}{6} + \dfrac{25}{2} + \dfrac{5}{3}$

$\quad\quad = \dfrac{125}{6} + \dfrac{75}{6} + \dfrac{10}{6}$

$\quad\quad = \dfrac{210}{6} = 35$

There are 35 truffles in the display. Now find the volume of one truffle. Each truffle's diameter is 3 cm, so the radius is $\dfrac{3}{2}$, or 1.5 cm.

$V(r) = \dfrac{4}{3}\pi r^3$

$V(1.5) \approx \dfrac{4}{3}(3.14)(1.5)^3 \approx 14.13 \text{ cm}^3$

Finally, multiply the number of truffles and the volume of a truffle to find the total volume of chocolate.

$35(14.13 \text{ cm}^3) = 494.55 \text{ cm}^3$

The display contains about 494.55 cm$^3$ of chocolate.

**113.** The area of the base is $x \cdot x$, or $x^2$.
The area of each side is $x \cdot (x - 2)$.
The total area of all four sides is $4x(x - 2)$.

The surface area of this box can be expressed as a polynomial function.

$S(x) = x^2 + 4x(x - 2)$
$\quad\quad = x^2 + 4x^2 - 8x$
$\quad\quad = 5x^2 - 8x$

**115.** $\quad (2x^{2a} + 4x^a + 3) + (6x^{2a} + 3x^a + 4)$
$= (2 + 6)x^{2a} + (4 + 3)x^a + (3 + 4)$
$= 8x^{2a} + 7x^a + 7$

**117.** $\quad (2x^{5b} + 4x^{4b} + 3x^{3b} + 8) -$
$\quad\quad (x^{5b} + 2x^{3b} + 6x^{2b} + 9x^b + 8)$
$= (2 - 1)x^{5b} + 4x^{4b} + (3 - 2)x^{3b} - 6x^{2b} - 9x^b + (8 - 8)$
$= x^{5b} + 4x^{4b} + x^{3b} - 6x^{2b} - 9x^b$

**119.** ▨

---

## Exercise Set 5.2

**1.** True; the product of two monomials is a product of constants and variables, so it is a monomial.

**3.** True; see Example 2.

**5.** False; FOIL is intended to be used to multiply two binomials.

**7.** True; see page 295 in the text.

**9.** $8a^2 \cdot 5a = (8 \cdot 5)(a^2 \cdot a) = 40a^3$

**11.** $5x(-4x^2y) = 5(-4)(x \cdot x^2)y = -20x^3y$

**13.** $(4x^3y^2)(-5x^2y^4) = 4(-5)(x^3 \cdot x^2)(y^2 \cdot y^4) = -20x^5y^6$

**15.** $7x(3 - x) = 7x \cdot 3 - 7x \cdot x$
$$= 21x - 7x^2$$

**17.** $\quad 5cd(4c^2d - 5cd^2)$
$$= 5cd \cdot 4c^2d - 5cd \cdot 5cd^2$$
$$= 20c^3d^2 - 25c^2d^3$$

**19.** $\quad (x + 3)(x + 5)$
$$= x^2 + 5x + 3x + 15 \qquad \text{FOIL}$$
$$= x^2 + 8x + 15$$

**21.** $\quad (t - 2)(t + 7)$
$$= t^2 + 7t - 2t - 14 \qquad \text{FOIL}$$
$$= t^2 + 5t - 14$$

**23.** $\quad (2a + 3)(4a + 1)$
$$= 8a^2 + 2a + 12a + 3 \qquad \text{FOIL}$$
$$= 8a^2 + 14a + 3$$

**25.** $\quad (5x - 4y)(2x + y)$
$$= 10x^2 + 5xy - 8xy - 4y^2 \qquad \text{FOIL}$$
$$= 10x^2 - 3xy - 4y^2$$

**27.** $\quad (x + 2)(x^2 - 3x + 1)$
$$= x(x^2 - 3x + 1) + 2(x^2 - 3x + 1)$$
$$= x^3 - 3x^2 + x + 2x^2 - 6x + 2$$
$$= x^3 - x^2 - 5x + 2$$

**29.** $\quad (t - 5)(t^2 + 2t - 3)$
$$= t(t^2 + 2t - 3) - 5(t^2 + 2t - 3)$$
$$= t^3 + 2t^2 - 3t - 5t^2 - 10t + 15$$
$$= t^3 - 3t^2 - 13t + 15$$

**31.**
$$
\begin{array}{l}
\quad\quad\quad\quad\quad a^2 + \ \ a - 1 \\
\quad\quad\quad\quad\quad a^2 + 4a - 5 \\
\hline
\quad\quad\quad\quad -5a^2 - 5a + 5 \quad \text{Multiplying by } -5 \\
\quad\quad 4a^3 + \ \ 4a^2 - 4a \quad\quad \text{Multiplying by } 4a \\
a^4 + \ a^3 - \quad a^2 \quad\quad\quad\quad \text{Multiplying by } a^2 \\
\hline
a^4 + 5a^3 - \ \ 2a^2 - 9a + 5 \quad \text{Adding}
\end{array}
$$

**33.** $\quad (x + 3)(x^2 - 3x + 9)$
$$= (x + 3)(x^2) + (x + 3)(-3x) + (x + 3)(9)$$
$$= x^3 + 3x^2 - 3x^2 - 9x + 9x + 27$$
$$= x^3 + 27$$

**35.** $\quad (a - b)(a^2 + ab + b^2)$
$$= (a - b)(a^2) + (a - b)(ab) + (a - b)(b^2)$$
$$= a^3 - a^2b + a^2b - ab^2 + ab^2 - b^3$$
$$= a^3 - b^3$$

**37.** $\quad \left(t - \dfrac{1}{3}\right)\left(t - \dfrac{1}{4}\right)$
$$= t^2 - \frac{1}{4}t - \frac{1}{3}t + \frac{1}{12} \qquad \text{FOIL}$$
$$= t^2 - \frac{3}{12}t - \frac{4}{12}t + \frac{1}{12}$$
$$= t^2 - \frac{7}{12}t + \frac{1}{12}$$

**39.** $\quad (r + 3)(r + 2)(r - 1)$
$$= (r^2 + 2r + 3r + 6)(r - 1) \qquad \text{FOIL}$$
$$= (r^2 + 5r + 6)(r - 1)$$
$$= (r^2 + 5r + 6) \cdot r + (r^2 + 5r + 6)(-1)$$
$$= r^3 + 5r^2 + 6r - r^2 - 5r - 6$$
$$= r^3 + 4r^2 + r - 6$$

**41.** $\quad (x + 5)^2$
$$= x^2 + 2 \cdot x \cdot 5 + 5^2 \quad (A + B)^2 = A^2 + 2AB + B^2$$
$$= x^2 + 10x + 25$$

**43.** $\quad (1.2t + 3s)(2.5t - 5s)$
$$= 3t^2 - 6st + 7.5st - 15s^2 \qquad \text{FOIL}$$
$$= 3t^2 + 1.5t - 15s^2$$

**45.** $\quad (2y - 7)^2$
$$= (2y)^2 - 2 \cdot 2y \cdot 7 + 7^2 \quad (A - B)^2 = A^2 - 2AB + B^2$$
$$= 4y^2 - 28y + 49$$

**47.** $\quad (5a - 3b)^2$
$$= (5a)^2 - 2 \cdot 5a \cdot 3b + (3b)^2$$
$$\qquad\qquad (A - B)^2 = A^2 - 2AB + B^2$$
$$= 25a^2 - 30ab + 9b^2$$

**49.** $\quad (2a^3 - 3b^2)^2$
$$= (2a^3)^2 - 2(2a^3)(3b^2) + (3b^2)^2$$
$$\qquad\qquad (A - B)^2 = A^2 - 2AB + B^2$$
$$= 4a^6 - 12a^3b^2 + 9b^4$$

**51.** $\quad (x^3y^4 + 5)^2$
$$= (x^3y^4)^2 + 2 \cdot x^3y^4 \cdot 5 + 5^2$$
$$\qquad\qquad (A - B)^2 = A^2 - 2AB + B^2$$
$$= x^6y^8 + 10x^3y^4 + 25$$

**53.** $\quad P(x) \cdot Q(x) = (3x^2 - 5)(4x^2 - 7x + 1)$
$$= (3x^2 - 5)(4x^2) + (3x^2 - 5)(-7x) + (3x^2 - 5)(1)$$
$$= 12x^4 - 20x^2 - 21x^3 + 35x + 3x^2 - 5$$
$$= 12x^4 - 21x^3 - 17x^2 + 35x - 5$$

**55.** $\quad P(x) \cdot P(x)$
$$= (5x - 2)(5x - 2)$$
$$= (5x)^2 - 2 \cdot 5x \cdot 2 + 2^2$$
$$\qquad\qquad (A - B)^2 = A^2 - 2AB + B^2$$
$$= 25x^2 - 20x + 4$$

**57.** $[F(x)]^2$

$= \left(2x - \dfrac{1}{3}\right)^2$

$= (2x)^2 - 2 \cdot 2x \cdot \dfrac{1}{3} + \left(\dfrac{1}{3}\right)^2$

$\qquad\qquad (A - B)^2 = A^2 - 2AB + B^2$

$= 4x^2 - \dfrac{4}{3}x + \dfrac{1}{9}$

**59.** $(c + 7)(c - 7)$

$= c^2 - 7^2 \qquad (A + B)(A - B) = A^2 - B^2$

$= c^2 - 49$

**61.** $(4x + 1)(4x - 1)$

$= (4x)^2 - 1^2 \quad (A + B)(A - B) = A^2 - B^2$

$= 16x^2 - 1$

**63.** $(3m - 2n)(3m + 2n)$

$= (3m)^2 - (2n)^2 \quad (A + B)(A - B) = A^2 - B^2$

$= 9m^2 - 4n^2$

**65.** $(x^3 + yz)(x^3 - yz)$

$= (x^3)^2 - (yz)^2 \quad (A + B)(A - B) = A^2 - B^2$

$= x^6 - y^2z^2$

**67.** $(-mn + m^2)(mn + m^2)$

$= (m^2 - mn)(m^2 + mn)$

$= (m^2)^2 - (mn)^2 \qquad (A + B)(A - B) = A^2 - B^2$

$= m^4 - m^2n^2, \text{ or } -m^2n^2 + m^4$

**69.** $(x + 7)^2 - (x + 3)(x - 3)$

$= x^2 + 2 \cdot x \cdot 7 + 7^2 - (x^2 - 3^2)$

$= x^2 + 14x + 49 - (x^2 - 9)$

$= x^2 + 14x + 49 - x^2 + 9$

$= 14x + 58$

**71.** $(2m - n)(2m + n) - (m - 2n)^2$

$= [(2m)^2 - n^2] - [m^2 - 2 \cdot m \cdot 2n + (2n)^2]$

$= 4m^2 - n^2 - (m^2 - 4mn + 4n^2)$

$= 4m^2 - n^2 - m^2 + 4mn - 4n^2$

$= 3m^2 + 4mn - 5n^2$

**73.** $(a + b + 1)(a + b - 1)$

$= [(a + b) + 1][(a + b) - 1]$

$= (a + b)^2 - 1^2$

$= a^2 + 2ab + b^2 - 1$

**75.** $(2x + 3y + 4)(2x + 3y - 4)$

$= [(2x + 3y) + 4][(2x + 3y) - 4]$

$= (2x + 3y)^2 - 4^2$

$= 4x^2 + 12xy + 9y^2 - 16$

**77.** $A = P(1 + i)^2$

$A = P(1 + 2i + i^2)$

$A = P + 2Pi + Pi^2$

**79.** a) Replace $x$ with $t - 1$.

$\quad f(t - 1) = (t - 1)^2 + 5$

$\qquad\qquad = t^2 - 2t + 1 + 5$

$\qquad\qquad = t^2 - 2t + 6$

b) $\quad f(a + h) - f(a)$

$= [(a + h)^2 + 5] - (a^2 + 5)$

$= a^2 + 2ah + h^2 + 5 - a^2 - 5$

$= 2ah + h^2$

c) $\quad f(a) - f(a - h)$

$= (a^2 + 5) - [(a - h)^2 + 5]$

$= a^2 + 5 - (a^2 - 2ah + h^2 + 5)$

$= a^2 + 5 - a^2 + 2ah - h^2 - 5$

$= 2ah - h^2$

**81.** *Writing Exercise*

**83.** $ab + ac = d$

$a(b + c) = d$

$a = \dfrac{d}{b + c}$

**85.** $mn + m = p$

$m(n + 1) = p$

$m = \dfrac{p}{n + 1}$

**87.** *Familiarize*. Let $d$, $n$, and $q$ represent the number of rolls of dimes, nickels, and quarters, respectively, that Kacie has. The value of a roll of dimes is 50($0.10), or $5.00; the value of a roll of nickels is 40($0.05), or $2.00; and the value of a roll of quarters is 40($0.25), or $10. Then the value of the rolls of dimes, nickels, and quarters is $5d$, $2n$, and $10q$, respectively.

*Translate*.

$\underbrace{\text{Total number of rolls of coins}}$ is 13.

$\qquad\downarrow \qquad\qquad\quad \downarrow \quad \downarrow$

$\quad d + n + q \qquad = \quad 13$

$\underbrace{\text{Total value of the coins}}$ is $89.

$\qquad\downarrow \qquad\qquad \downarrow \quad \downarrow$

$\quad 5d + 2n + 10q \qquad = \quad 89$

$\underbrace{\text{Number of rolls of dimes}}$ is 3 $\underbrace{\text{more than}}$ $\underbrace{\text{number of rolls of quarters.}}$

$\quad\downarrow \qquad\qquad \downarrow\downarrow \quad \downarrow \qquad\qquad \downarrow$

$\quad d \qquad\qquad = 3 \quad + \qquad\qquad n$

We have a system of equations.

$\quad d + n + q = 13,$

$\quad 5d + 2n + 10q = 89,$

$\quad d = 3 + n$

*Carry out*. Solving the system of equations, we get $(5, 2, 6)$.

*Check*. The total number of rolls of coins is $5 + 2 + 6$, or 13. The total value of the coins is $\$5 \cdot 5 + \$2 \cdot 2 + \$10 \cdot 6$, or $25 + $4 + $60, or $89. The number of rolls of dimes, 5, is three more than 2, the number of rolls of nickels. The answer checks.

*State*. Kacie has 5 rolls of dimes, 2 rolls of nickels, and 6 rolls of quarters.

**89.** *Writing Exercise*

**91.** $(x^2 + y^n)(x^2 - y^n) = (x^2)^2 - (y^n)^2 = x^4 - y^{2n}$

**93.** $x^2y^3(5x^n + 4y^n) = x^2y^3 \cdot 5x^n + x^2y^3 \cdot 4y^n =$
$5x^{n+2}y^3 + 4x^2y^{n+3}$

**95.**    $(t^n + 3)(t^{2n} - 2t^n + 1)$
$= t^n(t^{2n} - 2t^n + 1) + 3(t^{2n} - 2t^n + 1)$
$= t^{n+2n} - 2t^{n+n} + t^n + 3t^{2n} - 6t^n + 3$
$= t^{3n} - 2t^{2n} + t^n + 3t^{2n} - 6t^n + 3$
$= t^{3n} + t^{2n} - 5t^n + 3$

**97.**    $(a - b + c - d)(a + b + c + d)$
$= [(a + c) - (b + d)][(a + c) + (b + d)]$
$= (a + c)^2 - (b + d)^2$
$= (a^2 + 2ac + c^2) - (b^2 + 2bd + d^2)$
$= a^2 + 2ac + c^2 - b^2 - 2bd - d^2$

**99.**    $\left(\dfrac{2}{3}x + \dfrac{1}{3}y + 1\right)\left(\dfrac{2}{3}x - \dfrac{1}{3}y - 1\right)$
$= \left[\dfrac{2}{3}x + \left(\dfrac{1}{3}y + 1\right)\right]\left[\dfrac{2}{3}x - \left(\dfrac{1}{3}y + 1\right)\right]$
$= \left(\dfrac{2}{3}x\right)^2 - \left(\dfrac{1}{3}y + 1\right)^2$
$= \dfrac{4}{9}x^2 - \left(\dfrac{1}{9}y^2 + \dfrac{2}{3}y + 1\right)$
$= \dfrac{4}{9}x^2 - \dfrac{1}{9}y^2 - \dfrac{2}{3}y - 1$

**101.**    $(x^a + y^b)(x^a - y^b)(x^{2a} + y^{2b})$
$= (x^{2a} - y^{2b})(x^{2a} + y^{2b})$
$= x^{4a} - y^{4b}$

**103.** $(x^{a-b})^{a+b} = x^{(a-b)(a+b)} = x^{a^2 - b^2}$

**105.**    $(x - a)(x - b)(x - c) \cdots (x - z)$
$= (x - a)(x - b) \cdots (x - x)(x - y)(x - z)$
$= (x - a)(x - b) \cdots 0 \cdot (x - y)(x - z)$
$= 0$

**107.**    $\dfrac{g(a + h) - g(a)}{h}$
$= \dfrac{(a + h)^2 - 9 - (a^2 - 9)}{h}$
$= \dfrac{a^2 + 2ah + h^2 - 9 - a^2 + 9}{h}$
$= \dfrac{2ah + h^2}{h} = \dfrac{\cancel{h}(2a + h)}{\cancel{h}}$
$= 2a + h$

**109.** One method is as follows. For each equation, let $y_1$ represent the left-hand side and $y_2$ represent the right-hand side, and let $y_3 = y_2 - y_1$. Then use a graphing calculator to view the graph of $y_3$ and/or a table of values for $y_3$. If $y_3 = 0$, the equation is an identity. If $y_3 \neq 0$, the equation is not an identity.

a) Not an identity

b) Identity

c) Identity

d) Not an identity

e) Not an identity

## Exercise Set 5.3

**1.** False; the largest common factor is $5x^2$.

**3.** True; see page 302 in the text.

**5.** True; see page 303 in the text.

**7.** True; $-(a - b) = -a + b = b - a$; $-1(a - b) = -a + b = b - a$.

**9.**    $3y^2 + 6y$
$= 3y \cdot y + 3y \cdot 2$
$= 3y(y + 2)$

**11.**    $x^2 + 9x$
$= x \cdot x + 9 \cdot x$
$= x(x + 9)$

**13.**    $x^3 + 8x^2$
$= x^2 \cdot x + x^2 \cdot 8$
$= x^2(x + 8)$

**15.**    $8y^2 + 4y^4$
$= 4y^2 \cdot 2 + 4y^2 \cdot y^2$
$= 4y^2(2 + y^2)$

**17.**    $5x^2y^3 + 15x^3y^2$
$= 5x^2y^2 \cdot y + 5x^2y^2 \cdot 3x$
$= 5x^2y^2(y + 3x)$

**19.**    $5x^2 - 5x + 15$
$= 5 \cdot x^2 - 5 \cdot x + 5 \cdot 3$
$= 5(x^2 - x + 3)$

**21.**    $8xy + 10xz - 14xw$
$= 2x \cdot 4y + 2x \cdot 5z - 2x \cdot 7w$
$= 2x(4y + 5z - 7w)$

**23.**    $9x^3y^6z^2 - 12x^4y^4z^4 + 15x^2y^5z^3$
$= 3x^2y^4z^2 \cdot 3xy^2 - 3x^2y^4z^2 \cdot 4x^2z^2 + 3x^2y^4z^2 \cdot 5yz$
$= 3x^2y^4z^2(3xy^2 - 4x^2z^2 + 5yz)$

**25.** $-5x - 40 = -5(x + 8)$

**27.** $-8t + 72 = -8(t - 9)$

**29.** $-2x^2 + 12x + 40 = -2(x^2 - 6x - 20)$

**31.** $7x - 56y = -7(-x + 8y)$, or $-7(8y - x)$

**33.** $5r - 10s = -5(-r + 2s)$, or $-5(2s - r)$

**35.** $-p^3 - 4p^2 + 11 = -(p^3 + 4p^2 - 11)$

**37.** $-m^3 - m^2 + m - 2 = -(m^3 + m^2 - m + 2)$

**39.**      $a(b - 5) + c(b - 5)$
$= (b - 5)(a + c)$

**41.**      $(x + 7)(x - 1) + (x + 7)(x - 2)$
$= (x + 7)(x - 1 + x - 2)$
$= (x + 7)(2x - 3)$

**43.**      $a^2(x - y) + 5(y - x)$
$= a^2(x - y) + 5(-1)(x - y)$    Factoring out $-1$
                    to reverse the second subtraction
$= a^2(x - y) - 5(x - y)$          Simplifying
$= (x - y)(a^2 - 5)$

**45.**      $xy + xz + wy + wz$
$= x(y + z) + w(y + z)$
$= (y + z)(x + w)$

**47.**      $y^3 - y^2 + 3y - 3$
$= y^2(y - 1) + 3(y - 1)$
$= (y - 1)(y^2 + 3)$

**49.**      $t^3 + 6t^2 - 2t - 12$
$= t^2(t + 6) - 2(t + 6)$
$= (t + 6)(t^2 - 2)$

**51.**      $12a^4 - 21a^3 - 9a^2$
$= 3a^2 \cdot 4a^2 - 3a^2 \cdot 7a - 3a^2 \cdot 3$
$= 3a^2(4a^2 - 7a - 3)$

**53.**      $y^4 - y^3 - y + y^2$
$= y(y^3 - y^2 - 1 + y)$
$= y[y^2(y - 1) + y - 1]$      $(-1 + y = y - 1)$
$= y(y - 1)(y^2 + 1)$

**55.**      $2xy - x^2y - 6 + 3x$
$= xy(2 - x) - 3(2 - x)$
$= (2 - x)(xy - 3)$

**57.** a)   $h(t) = -16t^2 + 72t$
$h(t) = -8t(2t - 9)$

b) Using $h(t) = -16t^2 + 72t$:
$h(1) = -16 \cdot 1^2 + 72 \cdot 1 = -16 \cdot 1 + 72$
$= -16 + 72 = 56$ ft

Using $h(t) = -8t(2t - 9)$:
$h(1) = -8(1)(2 \cdot 1 - 9) = -8(1)(-7) = 56$ ft

The expressions have the same value for $t = 1$, so the factorization is probably correct.

**59.**   $R(n) = n^2 - n$
$R(n) = n(n - 1)$

**61.**   $P(t) = t^2 - 5t$
$P(t) = t(t - 5)$

**63.**   $R(x) = 280x - 0.4x^2$
$R(x) = 0.4x(700 - x)$

**65.**   $P(n) = \dfrac{1}{2}n^2 - \dfrac{3}{2}n$
$P(n) = \dfrac{1}{2}(n^2 - 3n)$

**67.**   $H(n) = \dfrac{1}{2}n^2 - \dfrac{1}{2}n$
$H(n) = \dfrac{1}{2}n(n - 1)$

**69.** *Writing Exercise*

**71.** $2(-3) + 4(-5) = -6 - 20 = -26$

**73.** $4(-6) - 3(2) = -24 - 6 = -30$

**75. *Familiarize*.** Let $n$ = the first even number. Then $n + 2$ and $n + 4$ are the next two even numbers. Recall that the perimeter of a triangle is the sum of the lengths of the sides.

**Translate**.

$\underbrace{\text{The perimeter}}$     is 174.
$\quad\quad\downarrow\quad\quad\quad\quad\downarrow\;\downarrow$
$n + (n + 2) + (n + 4) = 174$

***Carry out***. We solve the equation.
$n + (n + 2) + (n + 4) = 174$
$3n + 6 = 174$
$3n = 168$
$n = 56$

When $n = 56$, then $n + 2 = 56 + 2$, or 58, and $n + 4 = 56 + 4$, or 60.

***Check***. The numbers 56, 58, and 60 are consecutive even integers. Also $56 + 58 + 60 = 174$. The answer checks.

***State***. The lengths of the sides of the triangle are 56, 58, and 60.

**77.** *Writing Exercise*

**79.** $x^5y^4 + \underline{\quad} = x^3y(\underline{\quad} + xy^5)$

The term that goes in the first blank is the product of $x^3y$ and $xy^5$, or $x^4y^6$.

The term that goes in the second blank is the expression that is multiplied with $x^3y$ to obtain $x^5y^4$, or $x^2y^3$. Thus, we have
$x^5y^4 + x^4y^6 = x^3y(x^2y^3 + xy^5)$.

**81.**      $rx^2 - rx + 5r + sx^2 - sx + 5s$
$= r(x^2 - x + 5) + s(x^2 - x + 5)$
$= (x^2 - x + 5)(r + s)$

**83.**
$$a^4x^4 + a^4x^2 + 5a^4 + a^2x^4 + a^2x^2 + 5a^2 +$$
$$5x^4 + 5x^2 + 25$$
$$= a^4(x^4+x^2+5)+a^2(x^4+x^2+5)+5(x^4+x^2+5)$$
$$= (x^4 + x^2 + 5)(a^4 + a^2 + 5)$$

**85.**
$$x^{-6} + x^{-9} + x^{-3}$$
$$= x^{-9} \cdot x^3 + x^{-9} \cdot 1 + x^{-9} \cdot x^6$$
$$= x^{-9}(x^3 + 1 + x^6)$$

**87.**
$$x^{1/3} - 5x^{1/2} + 3x^{3/4}$$
$$= x^{4/12} - 5x^{6/12} + 3x^{9/12}$$
$$= x^{4/12}(1 - 5x^{2/12} + 3x^{5/12})$$
$$= x^{1/3}(1 - 5x^{1/6} + 3x^{5/12})$$

**89.**
$$x^{-5/2} + x^{-3/2}$$
$$= x^{-5/2} \cdot 1 + x^{-5/2} \cdot x$$
$$= x^{-5/2}(1 + x)$$

**91.**
$$x^{-4/5} - x^{-7/5} + x^{-1/3}$$
$$= x^{-7/5} \cdot x^{3/5} - x^{-7/5} \cdot 1 + x^{-7/5} \cdot x^{16/15}$$
$$= x^{-7/5}(x^{3/5} - 1 + x^{16/15})$$

**93.**
$$3a^{n+1} + 6a^n - 15a^{n+2}$$
$$= 3a^n \cdot a + 3a^n \cdot 2 - 3a^n(5a^2)$$
$$= 3a^n(a + 2 - 5a^2)$$

**95.**
$$7y^{2a+b} - 5y^{a+b} + 3y^{a+2b}$$
$$= y^{a+b} \cdot 7y^a - y^{a+b}(5) + y^{a+b} \cdot 3y^b$$
$$= y^{a+b}(7y^a - 5 + 3y^b)$$

**97.** One method is to let $y_1 = (x^2 - 3x + 2)^4$ and let $y_2 = x^8 + 81x^4 + 16$. Then use a table to show that $y_1 \neq y_2$ for all values of $x$.

## Exercise Set 5.4

**1.** True; see Example 3.

**3.** False; whenever the product of a pair of factors is negative, the factors have different signs.

**5.** True; see Example 8.

**7.** False; if $b$ and $c$ are positive, then $p$ and $q$ are both positive.

**9.** $x^2 + 6x + 5$

We look for two numbers whose product is 5 and whose sum is 6. Since 5 and 6 are both positive, we need only consider positive factors. The only positive pair is 1 and 5. They are the numbers we need. The factorization is $(x + 1)(x + 5)$.

**11.** $y^2 + 12y + 27$

Since the constant term is positive and the coefficient of the middle term is also positive, we look for a factorization of 27 in which both factors are positive. Their sum must be 12.

| Pair of Factors | Sum of Factors |
|---|---|
| 1,   27 | 28 |
| 3,   9 | 12 |

The numbers we need are 3 and 9. The factorization is $(y + 3)(y + 9)$.

**13.** $t^2 - 15 - 2t = t^2 - 2t - 15$

Since the constant term is negative, we look for a factorization of $-15$ in which one factor is positive and one factor is negative. Their sum must be $-2$, so the negative factor must have the larger absolute value. Thus we consider only pairs of factors in which the negative factor has the larger absolute value.

| Pair of Factors | Sum of Factors |
|---|---|
| −15,  1 | −14 |
| −5,  3 | −2 |

The numbers we need are $-5$ and 3. The factorization is $(t - 5)(t + 3)$.

**15.**
$$2a^2 - 16a + 32$$
$$= 2(a^2 - 8a + 16) \quad \text{Removing the common factor}$$

We now factor $a^2 - 8a + 16$. We look for two numbers whose product is 16 and whose sum is $-8$. Since the constant term is positive and the coefficient of the middle term is negative, we look for factorization of 16 in which both factors are negative.

| Pair of Factors | Sum of Factors |
|---|---|
| −1, −16 | −17 |
| −2,  −8 | −10 |
| −4,  −4 | −8 |

The numbers we need are $-4$ and $-4$.
$$a^2 - 8a + 16 = (a - 4)(a - 4)$$

We must not forget to include the common factor 2.
$$2a^2 - 16a + 32 = 2(a - 4)(a - 4), \text{ or } 2(a - 4)^2$$

**17.**
$$x^3 + 3x^2 - 54x$$
$$= x(x^2 + 3x - 54) \quad \text{Removing the common factor}$$

We now factor $x^2 + 3x - 54$. Since the constant term is negative, we look for a factorization of $-54$ in which one factor is positive and one factor is negative. We consider only pairs of factors in which the positive factor has the larger absolute value, since the sum of the factors, 3, is positive.

| Pair of Factors | Sum of Factors |
|---|---|
| −1, 54 | 53 |
| −2, 27 | 25 |
| −3, 18 | 15 |
| −6,  9 | 3 |

The numbers we need are $-6$ and 9.
$$x^2 + 3x - 54 = (x - 6)(x + 9)$$

We must not forget to include the common factor $x$.
$$x^3 + 3x^2 - 54x = x(x - 6)(x + 9)$$

**19.** $12y + y^2 + 32 = y^2 + 12y + 32$

Since the constant term and the middle term are both positive, we look for a factorization of 32 in which both factors are positive. Their sum must be 12.

| Pair of Factors | Sum of Factors |
|-----------------|----------------|
| 32, 1 | 33 |
| 16, 2 | 18 |
| 8, 4 | 12 |

The numbers we need are 8 and 4. The factorization is $(y + 8)(y + 4)$.

**21.** $p^2 - 3p - 40$

Since the constant term is negative, we look for a factorization of $-40$ in which one factor is positive and one factor is negative. We consider only pairs of factors in which the negative factor has the larger absolute value, since the sum of the factors, $-3$, is negative.

| Pair of Factors | Sum of Factors |
|-----------------|----------------|
| $-40$, 1 | $-39$ |
| $-20$, 2 | $-18$ |
| $-10$, 4 | $-6$ |
| $-8$, 5 | $-3$ |

The numbers we need are $-8$ and 5. The factorization is $(p - 8)(p + 5)$.

**23.** $a^2 - 11a + 28$

Since the constant term is positive and the coefficient of the middle term is negative, we look for a factorization of 28 in which both factors are negative. Their sum must be $-11$.

| Pair of Factors | Sum of Factors |
|-----------------|----------------|
| $-1$, $-28$ | $-29$ |
| $-2$, $-14$ | $-16$ |
| $-4$, $-7$ | $-11$ |

The numbers we need are $-4$ and $-7$. The factorization is $(a - 4)(a - 7)$.

**25.** $x + x^2 - 6 = x^2 + x - 6$

Since the constant term is negative, we look for a factorization of $-6$ in which one factor is positive and one factor is negative. We consider only pairs of factors in which the positive factor has the larger absolute value, since the sum of the factors, 1, is positive.

| Pair of Factors | Sum of Factors |
|-----------------|----------------|
| 6, $-1$ | 5 |
| 3, $-2$ | 1 |

The numbers we need are 3 and $-2$. The factorization is $(x + 3)(x - 2)$.

**27.** $5y^2 + 40y + 35$
$= 5(y^2 + 8y + 7)$    Removing the common factor

We now factor $y^2 + 8y + 7$. We look for two numbers whose product is 7 and whose sum is 8. Since 7 and 8 are both positive, we need consider only positive factors. The only possible pair is 1 and 7. They are the numbers we need.

$y^2 + 8y + 7 = (y + 1)(y + 7)$

We must not forget to include the common factor 5.

$5y^2 + 40y + 35 = 5(y + 1)(y + 7)$

**29.** $32 + 4y - y^2 = -y^2 + 4y + 32 = -(y^2 - 4y - 32)$

We now factor $y^2 - 4y - 32$. Since the constant term is negative, we look for a factorization of $-32$ in which one factor is positive and one factor is negative. We consider only pairs of factors in which the negative factor has the larger absolute value, since the sum of the factors, $-4$, is negative.

| Pair of Factors | Sum of Factors |
|-----------------|----------------|
| $-32$, 1 | $-31$ |
| $-16$, 2 | $-14$ |
| $-8$, 4 | $-4$ |

The numbers we need are $-8$ and 4. Thus, $y^2 - 4y - 32 = (y - 8)(y + 4)$. We must not forget to include the factor that was factored out earlier:

$32 + 4y - y^2 = -(y - 8)(y + 4)$, or $(-y + 8)(y + 4)$, or $(8 - y)(4 + y)$

**31.** $56x + x^2 - x^3$

There is a common factor, $x$. We also factor out $-1$ in order to make the leading coefficient positive.

$56x + x^2 - x^3 = -x(-56 - x + x^2)$
$= -x(x^2 - x - 56)$

Now we factor $x^2 - x - 56$. Since the constant term is negative, we look for a factorization of $-56$ in which one factor is positive and one factor is negative. We consider only pairs of factors in which the negative factor has the larger absolute value, since the sum of the factors, $-1$, is negative.

| Pair of Factors | Sum of Factors |
|-----------------|----------------|
| $-56$, 1 | $-55$ |
| $-28$, 2 | $-26$ |
| $-14$, 4 | $-10$ |
| $-8$, 7 | $-1$ |

The numbers we need are $-8$ and 7. Thus, $x^2 - x - 56 = (x - 8)(x + 7)$. We must not forget to include the common factor:

$56x + x^2 - x^3 = -x(x - 8)(x + 7)$, or $x(-x + 8)(x + 7)$, or $x(8 - x)(7 + x)$

**33.** $\quad y^4 + 5y^3 - 84y^2$
$= y^2(y^2 + 5y - 84)$    Removing the common factor

We now factor $y^2 + 5y - 84$. We look for pairs of factors of $-84$, one positive and one negative, such that the positive factor has the larger absolute value and the sum of the factors is 5.

| Pair of Factors | Sum of Factors |
|-----------------|----------------|
| 84, $-1$ | 83 |
| 42, $-2$ | 40 |
| 28, $-3$ | 25 |
| 21, $-4$ | 17 |
| 14, $-6$ | 8 |
| 12, $-7$ | 5 |

The numbers we need are 12 and $-7$. Then $y^2 + 5y - 84 = (y + 12)(y - 7)$. We must not forget to include the common factor:

$$y^4 + 5y^3 - 84y^2 = y^2(y + 12)(y - 7)$$

**35.** $x^2 - 3x + 7$

There are no factors of 7 whose sum is $-3$. This trinomial is not factorable into binomials with integer coefficients. The polynomial is prime.

**37.** $x^2 + 12xy + 27y^2$

We look for numbers $p$ and $q$ such that $x^2 + 12xy + 27y^2 = (x + py)(x + qy)$. Our thinking is much the same as if we were factoring $x^2 + 12x + 27$. Since the constant term is positive and the coefficient of the middle term is positive, we look for a factorization of 27 in which both factors are positive. Their sum must be 12.

| Pair of Factors | Sum of Factors |
|-----------------|----------------|
| 1, 27           | 28             |
| 3, 9            | 12             |

The numbers we need are 3 and 9. The factorization is $(x + 3y)(x + 9y)$.

**39.** $x^2 - 14xy + 49y^2$

We look for numbers $p$ and $q$ such that $x^2 - 14xy + 49y^2 = (x + py)(x + qy)$. Our thinking is much the same as if we were factoring $x^2 - 14x + 49$. We look for factors of 49 whose sum is $-14$. Since the constant term is positive and the coefficient of the middle term is negative, both factors must be negative.

| Pair of Factors | Sum of Factors |
|-----------------|----------------|
| $-49$, $-1$     | $-50$          |
| $-7$, $-7$      | $-14$          |

The numbers we need are $-7$ and $-7$. The factorization is $(x - 7y)(x - 7y)$, or $(x - 7y)^2$.

**41.**   $x^4 - 50x^3 + 49x^2$

$= x^2(x^2 - 50x + 49)$   Removing the common factor

Now we factor $x^2 - 50x + 49$. We look for a pair of factors of 49 whose sum is $-50$. The numbers we need are $-1$ and $-49$. Then $x^2 - 50x + 49 = (x - 1)(x - 49)$. We must not forget to include the common factor:

$$x^4 - 50x^3 + 49x^2 = x^2(x - 1)(x - 49)$$

**43.**   $x^6 + 2x^5 - 63x^4$

$= x^4(x^2 + 2x - 63)$   Removing the common factor

We now factor $x^2 + 2x - 63$. We look for a pair of factors of $-63$ whose sum is 2. The numbers we need are 9 and $-7$. Then $x^2 + 2x - 63 = (x + 9)(x - 7)$. We must not forget to include the common factor:

$$x^6 + 2x^5 - 63x^4 = x^4(x + 9)(x - 7)$$

**45.** $3x^2 - 16x - 12$

We will use the FOIL method.

1. There is no common factor (other than 1 or $-1$.)

2. Factor the first term, $3x^2$. The factors are $3x$, $x$. We have this possibility:

$$(3x + \quad)(x + \quad)$$

3. Factor the last term, $-12$. The possibilities are $12(-1)$, $-12 \cdot 1$, $6(-2)$, $-6 \cdot 2$, $4(-3)$, and $-4 \cdot 3$.

4. We need factors for which the sum of the products (the "outer" and "inner" parts of FOIL) is the middle term, $-16x$. Try some possibilities and check by multiplying.

$$(3x + 1)(x - 12) = 3x^2 - 35x - 12$$

We try again.

$$(3x + 2)(x - 6) = 3x^2 - 16x - 12$$

The factorization is $(3x + 2)(x - 6)$.

**47.** $6x^3 - 15x - x^2 = 6x^3 - x^2 - 15x$

We will use the grouping method.

1. Look for a common factor. We factor out $x$:

$$x(6x^2 - x - 15)$$

2. Factor the trinomial $6x^2 - x - 15$. Multiply the leading coefficient, 6, and the constant, $-15$.

$$6(-15) = -90$$

3. Try to factor $-90$ so the sum of the factors is $-1$. We need only consider pairs of factors in which the negative factor has the larger absolute value, since their sum is negative.

| Pair of Factors | Sum of Factors |
|-----------------|----------------|
| $-90$, 1        | $-89$          |
| $-45$, 2        | $-43$          |
| $-30$, 3        | $-27$          |
| $-18$, 5        | $-13$          |
| $-15$, 6        | $-9$           |
| $-10$, 9        | $-1$           |

4. We split the middle term, $-x$, using the results of step (3).

$$-x = -10x + 9x$$

5. Factor by grouping:

$$6x^2 - x - 15 = 6x^2 - 10x + 9x - 15$$
$$= 2x(3x - 5) + 3(3x - 5)$$
$$= (3x - 5)(2x + 3)$$

We must include the common factor to get a factorization of the original trinomial:

$$6x^3 - 15x - x^2 = x(3x - 5)(2x + 3)$$

**49.** $3a^2 - 10a + 8$

We will use the FOIL method.

1. There is no common factor (other than 1 or $-1$).

2. Factor the first term, $3a^2$. The factors are $3a$, $a$. We have this possibility: $(3a + \quad)(a + \quad)$.

3. Factor the last term, 8. The possibilities are $8 \cdot 1$, $-8(-1)$, $4 \cdot 2$, and $-4(-2)$.

4. Look for factors such that the sum of the products is the middle term, $-10a$. Trial and error leads us to the correct factorization:

$$3a^2 - 10a + 8 = (3a - 4)(a - 2)$$

**51.** $9a^2 + 18a + 8$

We will use the grouping method.

1. There is no common factor (other than 1 or $-1$).

2. Multiply the leading coefficient, 9, and the constant, 8: $9(8) = 72$

3. Try to factor 72 so the sum of the factors is 18. We need only consider pairs of positive factors since 72 and 18 are both positive.

| Pair of Factors | Sum of Factors |
|---|---|
| 72, 1 | 73 |
| 36, 2 | 38 |
| 24, 3 | 27 |
| 18, 4 | 22 |
| 12, 6 | 18 |
| 9, 8 | 17 |

4. Split $18a$ using the results of step (3):

$$18a = 12a + 6a$$

5. Factor by grouping:

$$9a^2 + 18a + 8 = 9a^2 + 12a + 6a + 8$$
$$= 3a(3a + 4) + 2(3a + 4)$$
$$= (3a + 4)(3a + 2)$$

**53.** $8x + 30x^2 - 6 = 30x^2 + 8x - 6$

We will use the FOIL method.

1. Factor out the common factor, 2:

$$2(15x^2 + 4x - 3)$$

2. Now we factor the trinomial $15x^2 + 4x - 3$. Factor the first term, $15x^2$. The factors are $15x$, $x$ and $5x$, $3x$. We have these possibilities: $(15x+ \quad)(x+ \quad)$ and $(5x+ \quad)(3x+ \quad)$

3. Factor the last term, $-3$. The possibilities are $(1)(-3)$ and $(-1)3$ as well as $(-3)(1)$ and $3(-1)$.

4. Look for factors such that the sum of the products is the middle term, $4x$. Trial and error leads us to the correct factorization:

$$15x^2 + 4x - 3 = (5x + 3)(3x - 1)$$

We must include the common factor to get a factorization of the original trinomial:

$$8x + 30x^2 - 6 = 2(5x + 3)(3x - 1)$$

**55.** $18x^2 - 24 - 6x = 18x^2 - 6x - 24$

We will use the grouping method.

1. Factor out the common factor, 6:

$$6(3x^2 - x - 4)$$

2. Now we factor the trinomial $3x^2 - x - 4$. Multiply the leading coefficient, 3, and the constant, $-4$: $3(-4) = -12$

3. Factor $-12$ so the sum of the factors is $-1$. We need only consider pairs of factors in which the negative factor has the larger absolute value, since their sum is negative.

| Pair of Factors | Sum of Factors |
|---|---|
| $-12$, 1 | $-11$ |
| $-6$, 2 | $-4$ |
| $-4$, 3 | $-1$ |

4. Split $-x$ using the results of step (3):

$$-x = -4x + 3x$$

5. Factor by grouping:

$$3x^2 - x - 4 = 3x^2 - 4x + 3x - 4$$
$$= x(3x - 4) + (3x - 4)$$
$$= (3x - 4)(x + 1)$$

We must include the common factor to get a factorization of the original trinomial:

$$18x^2 - 24 - 6x = 6(3x - 4)(x + 1)$$

**57.** $\quad t^8 + 5t^7 - 14t^6$

$= t^6(t^2 + 5t - 14)$   Removing the common factor

We now factor $t^2 + 5t - 14$. We look for a pair of factors of $-14$ whose sum is 5. The numbers we need are 7 and $-2$. Then we have $t^2 + 5t - 14 = (t + 7)(t - 2)$. We must not forget to include the common factor:

$$t^8 + 5t^7 - 14t^6 = t^6(t + 7)(t - 2)$$

**59.** $70x^4 - 68x^3 + 16x^2$

We will use the grouping method.

1. Factor out the common factor, $2x^2$:

$$2x^2(35x^2 - 34x + 8)$$

2. Now we factor the trinomial $35x^2 - 34x + 8$. Multiply the leading coefficient, 35, and the constant, 8: $35 \cdot 8 = 280$

3. Factor 280 so the sum of the factors is $-34$. We need only consider pairs of negative factors since the sum is negative.

| Pair of Factors | Sum of Factors |
|---|---|
| $-280$, $-1$ | $-281$ |
| $-140$, $-2$ | $-142$ |
| $-70$, $-4$ | $-74$ |
| $-56$, $-5$ | $-61$ |
| $-40$, $-7$ | $-47$ |
| $-35$, $-8$ | $-43$ |
| $-28$, $-10$ | $-38$ |
| $-20$, $-14$ | $-34$ |

4. Split $-34x$ using the results of step (3):

$$-34x = -20x - 14x$$

5. Factor by grouping:

$$35x^2 - 34x + 8 = 35x^2 - 20x - 14x + 8$$
$$= 5x(7x - 4) - 2(7x - 4)$$
$$= (7x - 4)(5x - 2)$$

We must include the common factor to get a factorization of the original trinomial:

$$70x^4 - 68x^3 + 16x^2 = 2x^2(7x - 4)(5x - 2)$$

**61.** $12a^2 - 14a - 20$

We will use the FOIL method.

1. Factor out the common factor, 2:

$$2(6a^2 - 7a - 10)$$

2. We now factor the trinomial $6a^2 - 7a - 10$. Factor the first term, $6a^2$. The possibilities is $(6a+ \quad)(a+ \quad)$ and $(3a+ \quad)(2a+ \quad)$.

3. Factor the last term, $-10$. The possibilities are $-10 \cdot 1$, $10(-1)$, $-5 \cdot 2$, and $5(-2)$.

4. We need factors for which the sum of the products is the middle term, $-7a$. Trial and error leads us to the correct factorization:

$$6a^2 - 7a - 10 = (6a + 5)(a - 2)$$

We must include the common factor to get a factorization of the original trinomial:

$$12a^2 - 14a - 20 = 2(6a + 5)(a - 2)$$

**63.** $9x^2 + 15x + 4$

We will use the grouping method.

1. There is no common factor (other than 1 or $-1$).

2. Multiply the leading coefficient and constant: $9(4) = 36$

3. Factor 36 so the sum of the factors is 15. We need only consider pairs of positive factors since 36 and 15 are both positive.

| Pair of Factors | Sum of Factors |
|---|---|
| 36, 1 | 37 |
| 18, 2 | 20 |
| 12, 3 | 15 |
| 9, 4 | 13 |
| 6, 6 | 12 |

4. Split $15x$ using the results of step (3):

$$15x = 12x + 3x$$

5. Factor by grouping:

$$9x^2 + 15x + 4 = 9x^2 + 12x + 3x + 4$$
$$= 3x(3x + 4) + 3x + 4$$
$$= (3x + 4)(3x + 1)$$

**65.** $4x^2 + 15x + 9$

We will use the FOIL method.

1. There is no common factor (other than 1 or $-1$).

2. Factor the first term, $4x^2$. The possibilities are $(4x+ \quad)(x+ \quad)$ and $(2x+ \quad)(2x+ \quad)$.

3. Factor the last term, 9. We consider only positive factors since both the middle term and the last term are positive. The possibilities are $9 \cdot 1$ and $3 \cdot 3$.

4. We need factors for which the sum of products is the middle term, $15x$. Trial and error leads us to the correct factorization:

$$(4x + 3)(x + 3)$$

**67.** $-8t^2 - 8t + 30$

We will use the grouping method.

1. Factor out $-2$: $-2(4t^2 + 4t - 15)$

2. Now we factor the trinomial $4t^2 + 4t - 15$. Multiply the leading coefficient and the constant: $4(-15) = -60$

3. Factor $-60$ so the sum of the factors is 4. The desired factorization is $10(-6)$.

4. Split $4t$ using the results of step (3):

$$4t = 10t - 6t$$

5. Factor by grouping:

$$4t^2 + 4t - 15 = 4t^2 + 10t - 6t - 15$$
$$= 2t(2t + 5) - 3(2t + 5)$$
$$= (2t + 5)(2t - 3)$$

We must include the common factor to get a factorization of the original trinomial:

$$-8t^2 - 8t + 30 = -2(2t + 5)(2t - 3)$$

**69.** $18xy^3 + 3xy^2 - 10xy$

We will use the FOIL method.

1. Factor out the common factor, $xy$.

$$xy(18y^2 + 3y - 10)$$

2. We now factor the trinomial $18y^2 + 3y - 10$. Factor the first term, $18y^2$. The possibilities are $(18y+ \quad)(y+ \quad)$, $(9y+ \quad)(2y+ \quad)$, and $(6y+ \quad)(3y+ \quad)$.

3. Factor the last term, $-10$. The possibilities are $-10 \cdot 1$, $-5 \cdot 2$, $10(-1)$ and $5(-2)$.

4. We need factors for which the sum of the products is the middle term, $3y$. Trial and error leads us to the correct factorization.

$$18y^2 + 3y - 10 = (6y + 5)(3y - 2)$$

We must include the common factor to get a factorization of the original trinomial:

$$18xy^3 + 3xy^2 - 10xy = xy(6y + 5)(3y - 2)$$

**71.** $24x^2 - 2 - 47x = 24x^2 - 47x - 2$

We will use the grouping method.

1. There is no common factor (other than 1 or $-1$).

2. Multiply the leading coefficient and the constant: $24(-2) = -48$

3. Factor $-48$ so the sum of the factors is $-47$. The desired factorization is $-48 \cdot 1$.

4. Split $-47x$ using the results of step (3):
$$-47x = -48x + x$$

5. Factor by grouping:
$$24x^2 - 47x - 2 = 24x^2 - 48x + x - 2$$
$$= 24x(x - 2) + (x - 2)$$
$$= (x - 2)(24x + 1)$$

**73.** $63x^3 + 111x^2 + 36x$

We will use the FOIL method.

1. Factor out the common factor, $3x$.
$$3x(21x^2 + 37x + 12)$$

2. Now we will factor the trinomial $21x^2 + 37x + 12$. Factor the first term, $21x^2$. The factors are $21x$, $x$ and $7x$, $3x$. We have these possibilities: $(21x+ \quad)(x+ \quad)$ and $(7x+ \quad)(3x+ \quad)$.

3. Factor the last term, 12. The possibilities are $12 \cdot 1$, $(-12)(-1)$, $6 \cdot 2$, $(-6)(-2)$, $4 \cdot 3$, and $(-4)(-3)$ as well as $1 \cdot 12$, $(-1)(-12)$, $2 \cdot 6$, $(-2)(-6)$, $3 \cdot 4$, and $(-3)(-4)$.

4. Look for factors such that the sum of the products is the middle term, $37x$. Trial and error leads us to the correct factorization:
$$(7x + 3)(3x + 4)$$
We must include the common factor to get a factorization of the original trinomial:
$$63x^3 + 111x^2 + 36x = 3x(7x + 3)(3x + 4)$$

**75.** $48x^4 + 4x^3 - 30x^2$

We will use the grouping method.

1. We factor out the common factor, $2x^2$:
$$2x^2(24x^2 + 2x - 15)$$

2. We now factor $24x^2 + 2x - 15$. Multiply the leading coefficient and the constant:
$$24(-15) = -360$$

3. Factor $-360$ so the sum of the factors is 2. The desired factorization is $-18 \cdot 20$.

4. Split $2x$ using the results of step (3):
$$2x = -18x + 20x$$

5. Factor by grouping:
$$24x^2 + 2x - 15 = 24x^2 - 18x + 20x - 15$$
$$= 6x(4x - 3) + 5(4x - 3)$$
$$= (4x - 3)(6x + 5)$$
We must not forget to include the common factor:
$$48x^4 + 4x^3 - 30x^2 = 2x^2(4x - 3)(6x + 5)$$

**77.** $12a^2 - 17ab + 6b^2$

We will use the FOIL method. (Our thinking is much the same as if we were factoring $12a^2 - 17a + 6$.)

1. There is no common factor (other than 1 or $-1$).

2. Factor the first term, $12a^2$. The factors are $12a$, $a$ and $6a$, $2a$ and $4a$, $3a$. We have these possibilities: $(12a+ \quad)(a+ \quad)$ and $(6a+ \quad)(2a+ \quad)$ and $(4a+ \quad)(3a+ \quad)$.

3. Factor the last term, $6b^2$. The possibilities are $6b \cdot b$, $(-6b)(-b)$, $3b \cdot 2b$, and $(-3b)(-2b)$ as well as $b \cdot 6b$, $(-b)(-6b)$, $2b \cdot 3b$, and $(-2b)(-3b)$.

4. Look for factors such that the sum of the products is the middle term, $-17ab$. Trial and error leads us to the correct factorization:
$$(4a - 3b)(3a - 2b)$$

**79.** $2x^2 + xy - 6y^2$

We will use the grouping method.

1. There is no common factor (other than 1 or $-1$).

2. Multiply the coefficients of the first and last terms: $2(-6) = -12$

3. Factor $-12$ so the sum of the factors is 1. The desired factorization is $4(-3)$.

4. Split $xy$ using the results of step (3):
$$xy = 4xy - 3xy$$

5. Factor by grouping:
$$2x^2 + xy - 6y^2 = 2x^2 + 4xy - 3xy - 6y^2$$
$$= 2x(x + 2y) - 3y(x + 2y)$$
$$= (x + 2y)(2x - 3y)$$

**81.** $6x^2 - 29xy + 28y^2$

We will use the FOIL method.

1. There is no common factor (other than 1 or $-1$).

2. Factor the first term, $6x^2$. The factors are $6x$, $x$ and $3x$, $2x$. We have these possibilities: $(6x+ \quad)(x+ \quad)$ and $(3x+ \quad)(2x+ \quad)$.

3. Factor the last term, $28y^2$. The possibilities are $28y \cdot y$, $(-28y)(-y)$, $14y \cdot 2y$, $(-14y)(-2y)$, $7y \cdot 4y$, and $(-7y)(-4y)$ as well as $y \cdot 28y$, $(-y)(-28y)$, $2y \cdot 14y$, $(-2y)(-14y)$, $4y \cdot 7y$, and $(-4y)(-7y)$.

4. Look for factors such that the sum of the products is the middle term, $-29xy$. Trial and error leads us to the correct factorization:
$$(3x - 4y)(2x - 7y)$$

**83.** $9x^2 - 30xy + 25y^2$

We will use the grouping method.

1. There is no common factor (other than 1 or $-1$).

2. Multiply the coefficients of the first and last terms: $9(25) = 225$

3. Factor 225 so the sum of the factors is $-30$. The desired factorization is $-15(-15)$.

4. Split $-30xy$ using the results of step (3):
$$-30xy = -15xy - 15xy$$

5. Factor by grouping:

$$9x^2 - 30xy + 25y^2 = 9x^2 - 15xy - 15xy + 25y^2$$
$$= 3x(3x - 5y) - 5y(3x - 5y)$$
$$= (3x - 5y)(3x - 5y), \text{ or}$$
$$(3x - 5y)^2$$

**85.** $9x^2y^2 + 5xy - 4$

Let $u = xy$ and $u^2 = x^2y^2$. Factor $9u^2 + 5u - 4$. We will use the FOIL method.

1. There is no common factor (other than 1 or $-1$).

2. Factor the first term, $9u^2$. The factors are $9u, u$ and $3u, 3u$. We have these possibilities: $(9u+\ \ )(u+\ \ )$ and $(3u+\ \ )(3u+\ \ )$.

3. Factor the last term, $-4$. The possibilities are $-4 \cdot 1$, $-2 \cdot 2$, and $-1 \cdot 4$.

4. We need factors for which the sum of the products is the middle term, $5u$. Trial and error leads us to the factorization: $(9u - 4)(u + 1)$. Replace $u$ by $xy$. We have $9x^2y^2 + 5xy - 4 = (9xy - 4)(xy + 1)$.

**87.** *Writing Exercise*

**89.** $(3t)^3 = 3^3 \cdot t^3 = 27t^3$

**91.** $(4x^3)^3 = 4^3 \cdot (x^3)^3 = 64x^{3 \cdot 3} = 64x^9$

**93.**
$$g(x) = -5x^2 - 7x$$
$$g(-3) = -5(-3)^2 - 7(-3) = -5 \cdot 9 + 21 =$$
$$-45 + 21 = -24$$

**95.** *Writing Exercise*

**97.**    $60x^8y^6 + 35x^4y^3 + 5$
$$= 5(12x^8y^6 + 7x^4y^3 + 1) \quad \text{Removing the common factor}$$

To factor the trinomial $12x^8y^6 + 7x^4y^3 + 1$, first note that $(x^4y^3)^2 = x^8y^6$, so the trinomial is of the form $12u^2 + 7u + 1$. Trial and error leads us to the factorization:

$12x^8y^6 + 7x^4y^3 + 1 = (4x^4y^3 + 1)(3x^4y^3 + 1)$

Then the factorization of the original trinomial is

$5(4x^4y^3 + 1)(3x^4y^3 + 1).$

**99.** $y^2 - \dfrac{8}{49} + \dfrac{2}{7}y = y^2 + \dfrac{2}{7}y - \dfrac{8}{49}$

We look for factors of $-\dfrac{8}{49}$ whose sum is $\dfrac{2}{7}$. The factors are $\dfrac{4}{7}$ and $-\dfrac{2}{7}$. The factorization is $\left(y + \dfrac{4}{7}\right)\left(y - \dfrac{2}{7}\right)$.

**101.** $4x^{2a} - 4x^a - 3$

Substitute $u$ for $x^a$ (and $u^2$ for $x^{2a}$). We factor $4u^2 - 4u - 3$. Trial and error leads us to the factorization: $4u^2 - 4u - 3 = (2u + 1)(2u - 3)$. Replace $u$ with $x^a$: $4x^{2a} - 4x^a - 3 = (2x^a + 1)(2x^a - 3)$

**103.** $x^{2a} + 5x^a - 24$

Substitute $u$ for $x^a$ (and $u^2$ for $x^{2a}$). We factor $u^2 + 5u - 24$. We look for factors of $-24$ whose sum is 5. The factors are 8 and $-3$. We have $u^2 + 5u - 24 = (u + 8)(u - 3)$. Replace $u$ with $x^a$: $x^{2a} + 5x^a - 24 = (x^a + 8)(x^a - 3)$.

**105.**    $2ar^2 + 4asr + as^2 - asr$
$$= 2ar^2 + 3asr + as^2$$
$$= a(2r^2 + 3sr + s^2)$$
$$= a(2r + s)(r + s)$$

**107.** $(x + 3)^2 - 2(x + 3) - 35$

Substitute $u$ for $x + 3$ (and $u^2$ for $(x + 3)^2$). We factor $u^2 - 2u - 35$. Look for factors of $-35$ whose sum is $-2$. The factors are $-7$ and 5. We have $u^2 - 2u - 35 = (u - 7)(u + 5)$. Replace $u$ with $x + 3$:

$(x + 3)^2 - 2(x + 3) - 35 = [(x + 3) - 7][(x + 3) + 5]$, or $(x - 4)(x + 8)$

**109.** $x^2 + mx + 75$

All such $m$ are the sums of the factors of 75.

| Pair of Factors | Sum of Factors |
|:---:|:---:|
| 75,   1 | 76 |
| $-75$,  $-1$ | $-76$ |
| 25,   3 | 28 |
| $-25$,  $-3$ | $-28$ |
| 15,   5 | 20 |
| $-15$,  $-5$ | $-20$ |

$m$ can be 76, $-76$, 28, $-28$, 20, or $-20$.

**111.** See the answer section in the text.

**113.**

**115.** *Writing Exercise*

## Exercise Set 5.5

**1.** $9t^2 - 49 = (3t)^2 - 7^2$

This is a difference of squares.

**3.** $36x^2 - 12x + 1 = (6x)^2 - 2 \cdot 6x \cdot 1 + 1^2$

This is a perfect-square trinomial.

**5.** $4r^2$ and 9 are squares but $8r \neq 2 \cdot 2r \cdot 3$ and $8r \neq -2 \cdot 2r \cdot 3$, so this trinomial is classified as none of these.

**7.** $4x^2 + 8x + 10 = 2(2x^2 + 4x + 5)$ and $2x^2 + 4x + 5$ cannot be factored, so this is a polynomial having a common factor.

**9.** $4t^2 + 9s^2 + 12st = 4t^2 + 12st + 9s^2 =$

$(2t)^2 + 2 \cdot 2t \cdot 3s + (3s)^2$

This is a perfect-square trinomial.

**11.**    $(t^2 + 6t + 9) = (t + 3)^2$

Find the square terms and write the square roots with a minus sign between them.

**13.** $a^2 - 14a + 49 = (a - 7)^2$
               Find the square terms and write the square roots with a plus sign between them.

**15.**    $4a^2 - 16a + 16$
$= 4(a^2 - 4a + 4)$   Factoring out the common factor
$= 4(a - 2)^2$   Factoring the perfect-square trinomial

**17.**    $y^2 + 36 + 12y$
$= y^2 + 12y + 36$   Changing order
$= (y + 6)^2$   Factoring the perfect-square trinomial

**19.**    $-18y^2 + y^3 + 81y$
$= y^3 - 18y^2 + 81y$     Changing order
$= y(y^2 - 18y + 81)$    Factoring out the common factor
$= y(y - 9)^2$

**21.**    $2x^2 - 40x + 200$
$= 2(x^2 - 20x + 100)$   Factoring out the common factor
$= 2(x - 10)^2$   Factoring the perfect-square trinomial

**23.**    $1 - 8d + 16d^2$
$= (1 - 4d)^2$   Factoring the perfect-square trinomial

**25.**    $y^3 + 8y^2 + 16y$
$= y(y^2 + 8y + 16)$
$= y(y + 4)^2$

**27.**  $0.25x^2 + 0.30x + 0.09 = (0.5x + 0.3)^2$
               Find the square terms and write the square roots with a plus sign between them.

**29.** $p^2 - 2pq + q^2 = (p - q)^2$

**31.** $25a^2 + 30ab + 9b^2 = (5a + 3b)^2$

**33.**    $5a^2 - 10ab + 5b^2$
$= 5(a^2 - 2ab + b^2)$
$= 5(a - b)^2$

**35.** $y^2 - 100 = (y + 10)(y - 10)$

**37.** $m^2 - 64 = (m + 8)(m - 8)$

**39.** $p^2q^2 - 25 = (pq + 5)(pq - 5)$

**41.**    $8x^2 - 8y^2$
$= 8(x^2 - y^2)$   Factoring out the common factor
$= 8(x + y)(x - y)$   Factoring the difference of squares

**43.**    $7xy^4 - 7xz^4$
$= 7x(y^4 - z^4)$
$= 7x[(y^2)^2 - (z^2)^2]$
$= 7x(y^2 + z^2)(y^2 - z^2)$
$= 7x(y^2 + z^2)(y + z)(y - z)$

**45.** $4a^3 - 49a = a(4a^2 - 49)$
$= a[(2a)^2 - 7^2]$
$= a(2a + 7)(2a - 7)$

**47.**    $3x^8 - 3y^8$
$= 3(x^8 - y^8)$
$= 3[(x^4)^2 - (y^4)^2]$
$= 3(x^4 + y^4)(x^4 - y^4)$
$= 3(x^4 + y^4)[(x^2)^2 - (y^2)^2]$
$= 3(x^4 + y^4)(x^2 + y^2)(x^2 - y^2)$
$= 3(x^4 + y^4)(x^2 + y^2)(x + y)(x - y)$

**49.** $9a^4 - 25a^2b^4 = a^2(9a^2 - 25b^4)$
$= a^2[(3a)^2 - (5b^2)^2]$
$= a^2(3a + 5b^2)(3a - 5b^2)$

**51.** $\dfrac{1}{49} - x^2 = \left(\dfrac{1}{7}\right)^2 - x^2$
$= \left(\dfrac{1}{7} + x\right)\left(\dfrac{1}{7} - x\right)$

**53.** $(a + b)^2 - 9 = (a + b)^2 - 3^2$
$= [(a + b) + 3][(a + b) - 3]$
$= (a + b + 3)(a + b - 3)$

**55.**    $x^2 - 6x + 9 - y^2$
$= (x^2 - 6x + 9) - y^2$   Grouping as a difference of squares
$= (x - 3)^2 - y^2$
$= (x - 3 + y)(x - 3 - y)$

**57.**    $t^3 + 8t^2 - t - 8$
$= t^2(t + 8) - (t + 8)$   Factoring by grouping
$= (t + 8)t^2 - 1)$
$= (t + 8)(t + 1)(t - 1)$   Factoring the difference of squares

**59.**    $r^3 - 3r^2 - 9r + 27$
$= r^2(r - 3) - 9(r - 3)$   Factoring by grouping
$= (r - 3)(r^2 - 9)$
$= (r - 3)(r + 3)(r - 3)$,   Factoring the difference of squares
   or $(r - 3)^2(r + 3)$

**61.**    $m^2 - 2mn + n^2 - 25$
$= (m^2 - 2mn + n^2) - 25$   Grouping as a difference of squares
$= (m - n)^2 - 5^2$
$= (m - n + 5)(m - n - 5)$

**63.** $36 - (x+y)^2 = 6^2 - (x+y)^2$
$$= [6 + (x+y)][6 - (x+y)]$$
$$= (6 + x + y)(6 - x - y)$$

**65.** $r^2 - 2r + 1 - 4s^2$
$$= (r^2 - 2r + 1) - 4s^2 \quad \text{Grouping as a}$$
$$\text{difference of squares}$$
$$= (r-1)^2 - (2s)^2$$
$$= (r - 1 + 2s)(r - 1 - 2s)$$

**67.** $16 - a^2 - 2ab - b^2$
$$= 16 - (a^2 + 2ab + b^2) \quad \text{Grouping as a}$$
$$\text{difference of squares}$$
$$= 4^2 - (a+b)^2$$
$$= [4 + (a+b)][4 - (a+b)]$$
$$= (4 + a + b)(4 - a - b)$$

**69.** $x^3 + 5x^2 - 4x - 20$
$$= x^2(x+5) - 4(x+5)$$
$$= (x+5)(x^2 - 4)$$
$$= (x+5)(x+2)(x-2)$$

**71.** $a^3 - ab^2 - 2a^2 + 2b^2$
$$= a(a^2 - b^2) - 2(a^2 - b^2) \quad \text{Factoring by}$$
$$= (a^2 - b^2)(a - 2) \qquad \text{grouping}$$
$$= (a+b)(a-b)(a-2) \quad \text{Factoring the}$$
$$\text{difference of squares}$$

**73.** *Writing Exercise*

**75.** $(2a^4b^5)^3 = 2^3(a^4)^3(b^5)^3 = 8a^{4\cdot3}b^{5\cdot3} = 8a^{12}b^{15}$

**77.** $(x+y)^3$
$$= (x+y)(x+y)^2$$
$$= (x+y)(x^2 + 2xy + y^2)$$
$$= x(x^2 + 2xy + y^2) + y(x^2 + 2xy + y^2)$$
$$= x^3 + 2x^2y + xy^2 + x^2y + 2xy^2 + y^3$$
$$= x^3 + 3x^2y + 3xy^2 + y^3$$

**79.** $x - y + z = 6, \quad (1)$
$2x + y - z = 0, \quad (2)$
$x + 2y + z = 3 \quad (3)$
Add (1) and (2).
$x - y + z = 6 \quad (1)$
$\underline{2x + y - z = 0 \quad (2)}$
$3x \qquad\quad = 6 \quad \text{Adding}$
$\qquad x = 2$
Add (2) and (3).
$2x + y - z = 0 \quad (2)$
$\underline{x + 2y + z = 3 \quad (3)}$
$3x + 3y \quad\;\; = 3 \quad (4)$

Substitute 2 for $x$ in (4).
$$3(2) + 3y = 3$$
$$6 + 3y = 3$$
$$3y = -3$$
$$y = -1$$
Substitute 2 for $x$ and $-1$ for $y$ in (1).
$$2 - (-1) + z = 6$$
$$3 + z = 6$$
$$z = 3$$
The solution is $(2, -1, 3)$.

**81.** $|5 - 7x| \le 9$
$$-9 \le 5 - 7x \le 9$$
$$-14 \le -7x \le 4$$
$$2 \ge x \ge -\frac{4}{7}$$
The solution set is $\left\{ x \middle| -\frac{4}{7} \le x \le 2 \right\}$, or $\left[ -\frac{4}{7}, 2 \right]$.

**83.** *Writing Exercise*

**85.** $-\frac{8}{27}r^2 - \frac{10}{9}rs - \frac{1}{6}s^2 + \frac{2}{3}rs$
$$= -\frac{8}{27}r^2 - \frac{4}{9}rs - \frac{1}{6}s^2$$
$$= -\frac{1}{54}(16r^2 + 24rs + 9s^2)$$
$$= -\frac{1}{54}(4r + 3s)^2$$

**87.** $0.09x^8 + 0.48x^4 + 0.64 = (0.3x^4 + 0.8)^2$, or
$$\frac{1}{100}(3x^4 + 8)^2$$

**89.** $r^2 - 8r - 25 - s^2 - 10s + 16$
$$= (r^2 - 8r + 16) - (s^2 + 10s + 25)$$
$$= (r-4)^2 - (s+5)^2$$
$$= [(r-4) + (s+5)][(r-4) - (s+5)]$$
$$= (r - 4 + s + 5)(r - 4 - s - 5)$$
$$= (r + s + 1)(r - s - 9)$$

**91.** $x^{4a} - y^{2b} = (x^{2a})^2 - (y^b)^2 = (x^{2a} + y^b)(x^{2a} - y^b)$

**93.** $25y^{2a} - (x^{2b} - 2x^b + 1)$
$$= (5y^a)^2 - (x^b - 1)^2$$
$$= [5y^a + (x^b - 1)][5y^a - (x^b - 1)]$$
$$= (5y^a + x^b - 1)(5y^a - x^b + 1)$$

**95.** $(a-3)^2 - 8(a-3) + 16$ is a perfect-square trinomial of the form $A^2 - 2AB + B^2$ with $A = a - 3$ and $B = 4$. We factor accordingly.
$$(a-3)^2 - 8(a-3) + 16 = (a - 3 - 4)^2, \text{ or } (a-7)^2$$

**97.** $m^2 + 4mn + 4n^2 + 5m + 10n$
$$= (m+2n)(m+2n) + 5(m+2n)$$
$$= (m+2n)(m + 2n + 5)$$

**99.**
$$5c^{100} - 80d^{100}$$
$$= 5(c^{100} - 16d^{100})$$
$$= 5(c^{50} + 4d^{50})(c^{50} - 4d^{50})$$
$$= 5(c^{50} + 4d^{50})(c^{25} + 2d^{25})(c^{25} - 2d^{25})$$

**101.** $c^{2w+1} + 2c^{w+1} + c = c(c^{2w} + 2c^w + 1)$
$$= c(c^w + 1)^2$$

**103.** If $P(x) = x^4$, then
$$P(a + h) - P(a)$$
$$= (a + h)^4 - a^4$$
$$= [(a + h)^2 + a^2][(a + h)^2 - a^2]$$
$$= [(a + h)^2 + a^2][(a + h) + a][(a + h) - a]$$
$$= (a^2 + 2ah + h^2 + a^2)(2a + h)(h)$$
$$= h(2a + h)(2a^2 + 2ah + h^2)$$

**105.**

## Exercise Set 5.6

**1.** $8x^3 - 27 = (2x)^3 - 3^3$

This is a difference of two cubes.

**3.** $9x^4 - 25 = (3x^2)^2 - 5^2$

This is a difference of two squares.

**5.** $1000t^3 + 1 = (10t)^3 + 1^3$

This is a sum of two cubes.

**7.** $25x^2 + 8x$ has a common factor of $x$ so it is not prime but it does not fall into any of the other categories. It is classified as none of these.

**9.** $s^{12} - t^6 = (s^4)^3 - (t^2)^3 = (s^6)^2 - (t^3)^2$

This is both a difference of two cubes and a difference of two squares.

**11.** $x^3 + 64 = x^3 + 4^3$
$$= (x + 4)(x^2 - 4x + 16)$$
$$A^3 + B^3 = (A + B)(A^2 - AB + B^2)$$

**13.** $z^3 - 1 = z^3 - 1^3$
$$= (z - 1)(z^2 + z + 1)$$
$$A^3 - B^3 = (A - B)(A^2 + AB + B^2)$$

**15.** $x^3 - 27 = x^3 - 3^3$
$$= (x - 3)(x^2 + 3x + 9)$$
$$A^3 - B^3 = (A - B)(A^2 + AB + B^2)$$

**17.** $27x^3 + 1 = (3x)^3 + 1^3$
$$= (3x + 1)(9x^2 - 3x + 1)$$
$$A^3 + B^3 = (A + B)(A^2 - AB + B^2)$$

**19.** $64 - 125x^3 = 4^3 - (5x)^3 = (4 - 5x)(16 + 20x + 25x^2)$

**21.** $27y^3 + 64 = (3y)^3 + 4^3 = (3y + 4)(9y^2 - 12y + 16)$

**23.** $x^3 - y^3 = (x - y)(x^2 + xy + y^2)$

**25.** $a^3 + \dfrac{1}{8} = a^3 + \left(\dfrac{1}{2}\right)^3 = \left(a + \dfrac{1}{2}\right)\left(a^2 - \dfrac{1}{2}a + \dfrac{1}{4}\right)$

**27.** $8t^3 - 8 = 8(t^3 - 1) = 8(t^3 - 1^3) = 8(t - 1)(t^2 + t + 1)$

**29.** $54x^3 + 2 = 2(27x^3 + 1) = 2[(3x)^3 + 1^3] =$
$2(3x + 1)(9x^2 - 3x + 1)$

**31.** $ab^3 + 125a = a(b^3 + 125) = a(b^3 + 5^3) =$
$a(b + 5)(b^2 - 5b + 25)$

**33.** $5x^3 - 40z^3 = 5(x^3 - 8z^3)$
$$= 5[x^3 - (2z)^3]$$
$$= 5(x - 2z)(x^2 + 2xz + 4z^2)$$

**35.** $x^3 + 0.001 = x^3 + (0.1)^3 = (x + 0.1)(x^2 - 0.1x + 0.01)$

**37.** $64x^6 - 8t^6 = 8(8x^6 - t^6)$
$$= 8[(2x^2)^3 - (t^2)^3]$$
$$= 8(2x^2 - t^2)(4x^4 + 2x^2t^2 + t^4)$$

**39.** $2y^4 - 128y = 2y(y^3 - 64)$
$$= 2y(y^3 - 4^3)$$
$$= 2y(y - 4)(y^2 + 4y + 16)$$

**41.**
$$z^6 - 1$$
$$= (z^3)^2 - 1^2 \qquad \text{Writing as a difference of squares}$$
$$= (z^3 + 1)(z^3 - 1) \qquad \text{Factoring a difference of squares}$$
$$= (z + 1)(z^2 - z + 1)(z - 1)(z^2 + z + 1)$$
$$\qquad\qquad \text{Factoring a sum and a difference of cubes}$$

**43.** $t^6 + 64y^6 = (t^2)^3 + (4y^2)^3 = (t^2 + 4y^2)(t^4 - 4t^2y^2 + 16y^4)$

**45.** $x^{12} - y^3z^{12} = (x^4)^3 - (yz^4)^3 = (x^4 - yz^4)(x^8 + x^4yz^4 + y^2z^8)$

**47.** *Writing Exercise*

**49. Familiarize.** Let $w =$ the width and $l =$ the length of the rectangle, in feet. Then $w + 2$ represents the width increased by 2 ft. Recall that the perimeter $P$ of a rectangle with length $l$ and width $w$ is given by the formula $P = 2l + 2w$ and the area $A$ is given by $A = lw$.

**Translate.** The width is 7 ft less than the length, so we have one equation:
$$w = l - 7$$
If the width is increased by 2 ft, the perimeter is 66 ft so we have a second equation:
$$2l + 2(w + 2) = 66$$
We have a system of equations:
$$w = l - 7, \qquad\qquad (1)$$
$$2l + 2(w + 2) = 66 \quad (2)$$

**Carry out.** First we substitute $l - 7$ for $w$ in Equation (2) and solve for $l$.

$$2l + 2(l - 7 + 2) = 66$$
$$2l + 2(l - 5) = 66$$
$$2l + 2l - 10 = 66$$
$$4l - 10 = 66$$
$$4l = 76$$
$$l = 19$$

Now substitute 19 for $l$ in Equation (1) and find $w$.

$$w = l - 7 = 19 - 7 = 12$$

For $l = 19$ ft and $w = 12$ ft, $A = 19$ ft $\cdot$ 12 ft $= 228$ ft$^2$.

**Check**. A width of 12 ft is 7 ft less than a length of 19 ft. The width increased by 2 ft is $12 + 2$, or 14 ft, and the perimeter of a rectangle with length 19 ft and width 14 ft is $2 \cdot 19 + 2 \cdot 14$, or $38 + 28$, or 66 ft. The answer checks.

**State**. The area of the original rectangle is 228 ft$^2$.

**51.** Write the equation in slope-intercept form, $y = mx + b$ where $m$ is the slope and $b$ is the $y$-intercept.

$$4x - 3y = 8$$
$$-3y = -4x + 8$$
$$y = \frac{4}{3}x - \frac{8}{3}$$

The slope is $\dfrac{4}{3}$, and the $y$-intercept is $\left(0, -\dfrac{8}{3}\right)$.

**53.** $3x - 5 = 0$

$\quad 3x = 5 \quad$ Adding 5 to both sides

$\quad x = \dfrac{5}{3} \quad$ Dividing both sides by 3

The solution is $\dfrac{5}{3}$.

**55.** *Writing Exercise*

**57.** $x^{6a} - y^{3b} = (x^{2a})^3 - (y^b)^3$
$$= (x^{2a} - y^b)(x^{4a} + x^{2a}y^b + y^{2b})$$

**59.** $(x + 5)^3 + (x - 5)^3 \quad$ Sum of cubes
$$= [(x+5) + (x-5)][(x+5)^2 - (x+5)(x-5) + (x-5)^2]$$
$$= 2x[(x^2 + 10x + 25) - (x^2 - 25) + (x^2 - 10x + 25)]$$
$$= 2x(x^2 + 10x + 25 - x^2 + 25 + x^2 - 10x + 25)$$
$$= 2x(x^2 + 75)$$

**61.** $5x^3y^6 - \dfrac{5}{8}$
$$= 5\left(x^3y^6 - \frac{1}{8}\right)$$
$$= 5\left(xy^2 - \frac{1}{2}\right)\left(x^2y^4 + \frac{1}{2}xy^2 + \frac{1}{4}\right)$$

**63.** $x^{6a} - (x^{2a} + 1)^3$
$$= [x^{2a} - (x^{2a}+1)][x^{4a} + x^{2a}(x^{2a}+1) + (x^{2a}+1)^2]$$
$$= (x^{2a} - x^{2a} - 1)(x^{4a} + x^{4a} + x^{2a} + x^{4a} + 2x^{2a} + 1)$$
$$= -(3x^{4a} + 3x^{2a} + 1)$$

**65.** $t^4 - 8t^3 - t + 8$
$$= t^3(t - 8) - (t - 8)$$
$$= (t - 8)(t^3 - 1)$$
$$= (t - 8)(t - 1)(t^2 + t + 1)$$

**67.** If $Q(x) = x^6$, then
$$Q(a + h) - Q(a)$$
$$= (a + h)^6 - a^6$$
$$= [(a+h)^3 + a^3][(a+h)^3 - a^3]$$
$$= [(a+h) + a] \cdot [(a+h)^2 - (a+h)a + a^2] \cdot$$
$$\qquad [(a+h) - a] \cdot [(a+h)^2 + (a+h)a + a^2]$$
$$= (2a + h) \cdot (a^2 + 2ah + h^2 - a^2 - ah + a^2) \cdot (h) \cdot$$
$$\qquad (a^2 + 2ah + h^2 + a^2 + ah + a^2)$$
$$= h(2a + h)(a^2 + ah + h^2)(3a^2 + 3ah + h^2)$$

**69.** ▩

---

## Exercise Set 5.7

**1.** greatest common factor; see page 330 in the text.

**3.** sum, cubes; see page 330 in the text.

**5.** FOIL, grouping; see page 330 in the text.

**7.** $4m^4 - 100$
$$= 4(m^4 - 25)$$
$$= 4[(m^2)^2 - 5^2] \qquad \text{Difference of squares}$$
$$= 4(m^2 + 5)(m^2 - 5)$$

**9.** $a^2 - 81$
$$= a^2 - 9^2 \qquad \text{Difference of squares}$$
$$= (a + 9)(a - 9)$$

**11.** $8x^2 - 18x - 5$
$$= (4x + 1)(2x - 5) \qquad \text{FOIL or grouping method}$$

**13.** $a^2 + 25 + 10a$
$$= a^2 + 10a + 25 \qquad \text{Perfect-square trinomial}$$
$$= (a + 5)^2$$

**15.** $3x^2 + 15x - 252$
$$= 3(x^2 + 5x - 84)$$
$$= 3(x + 12)(x - 7) \qquad \text{FOIL or grouping method}$$

**17.** $25x^2 - 9y^2$
$$= (5x)^2 - (3y)^2 \qquad \text{Difference of squares}$$
$$= (5x + 3y)(5x - 3y)$$

**19.** $t^6 + 1$
$$= (t^2)^3 + 1^3 \qquad \text{Sum of squares}$$
$$= (t^2 + 1)(t^4 - t^2 + 1)$$

**21.**    $x^2 + 6x - y^2 + 9$

$= x^2 + 6x + 9 - y^2$

$= (x + 3)^2 - y^2$          Difference of squares

$= [(x + 3) + y][(x + 3) - y]$

$= (x + y + 3)(x - y + 3)$

**23.**    $343x^3 + 27y^3$

$= (7x)^3 + (3y)^3$          Sum of cubes

$= (7x + 3y)(49x^2 - 21xy + 9y^2)$

**25.**    $2t^3 + 20t^2 - 48t$

$= 2t(t^2 + 10t - 24)$

$= 2t(t + 12)(t - 2)$    Trial and error

**27.**    $-24x^6 + 6x^4$

$= 6x^4(-4x^2 + 1)$

$= 6x^4(1 - 4x^2)$          Difference of squares

$= 6x^4(1 + 2x)(1 - 2x)$

**29.**    $8m^3 + m^6 - 20$

$= (m^3)^2 + 8m^3 - 20$

$= (m^3 - 2)(m^3 + 10)$    Trial and error

**31.**    $ac + cd - ab - bd$

$= c(a + d) - b(a + d)$    Factoring by grouping

$= (a + d)(c - b)$

**33.**    $4c^2 - 4cd + d^2$    Perfect-square trinomial

$= (2c - d)^2$

**35.**    $24 + 9t^2 + 8t + 3t^3$

$= 3(8 + 3t^2) + t(8 + 3t^2)$      Factoring by
                                                     grouping

$= (8 + 3t^2)(3 + t)$

**37.**    $2x^3 + 6x^2 - 8x - 24$

$= 2(x^3 + 3x^2 - 4x - 12)$

$= 2[x^2(x + 3) - 4(x + 3)]$    Factoring by
                                                      grouping

$= 2(x + 3)(x^2 - 4)$      Difference of squares

$= 2(x + 3)(x + 2)(x - 2)$

**39.**    $54a^3 - 16b^3$

$= 2(27a^3 - 8b^3)$

$= 2[(3a)^3 - (2b)^3]$          Difference of cubes

$= 2(3a - 2b)(9a^2 + 6ab + 4b^2)$

**41.**    $36y^2 - 35 + 12y$

$= 36y^2 + 12y - 35$

$= (6y - 5)(6y + 7)$    FOIL or grouping method

**43.**    $4m^4 - 64n^4$

$= 4(m^4 - 16n^4)$    Difference of squares

$= 4(m^2 + 4n^2)(m^2 - 4n^2)$    Difference of squares

$= 4(m^2 + 4n^2)(m + 2n)(m - 2n)$

**45.**    $a^3b - 16ab^3$

$= ab(a^2 - 16b^2)$    Difference of squares

$= ab(a + 4b)(a - 4b)$

**47.** $34t^3 - 6t = 2t(17t^2 - 3)$

**49.**    $(a - 3)(a + 7) + (a - 3)(a - 1)$

$= (a - 3)(a + 7 + a - 1)$

$= (a - 3)(2a + 6)$

$= (a - 3)(2)(a + 3)$

$= 2(a - 3)(a + 3)$

**51.**    $7a^4 - 14a^3 + 21a^2 - 7a$

$= 7a(a^3 - 2a^2 + 3a - 1)$      Removing a
                                                     common factor

**53.**    $42ab + 27a^2b^2 + 8$

$= 27a^2b^2 + 42ab + 8$

$= (9ab + 2)(3ab + 4)$ FOIL or grouping method

**55.** $-10t^3 + 15t = -5t(2t^2 - 3)$

**57.** $-6x^4 + 8x^3 - 12x = -2x(3x^3 - 4x^2 + 6)$

**59.**    $p - 64p^4$

$= p(1 - 64p^3)$          Sum of cubes

$= p(1 - 4p)(1 + 4p + 16p^2)$

**61.**    $a^2 - b^2 - 6b - 9$

$= a^2 - (b^2 + 6b + 9)$    Factoring out $-1$

$= a^2 - (b + 3)^2$          Difference of squares

$= [a + (b + 3)][a - (b + 3)]$

$= (a + b + 3)(a - b - 3)$

**63.** *Writing Exercise*

**65.**  $5x - 9 = 0$

$5x = 9$    Adding 9 to both sides

$x = \dfrac{9}{5}$    Dividing both sides by 5

The solution is $\dfrac{9}{5}$.

**67.** Graph: $g(x) = 3x - 7$.

We make a table of values, plot the corresponding points, and draw the graph.

| $x$ | $g(x)$ |
|---|---|
| 0 | $-7$ |
| 1 | $-4$ |
| 3 | 2 |
| 4 | 5 |

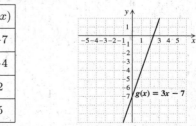

**69.** *Familiarize*. It is helpful to organize the information in a table. We let $x$ represent the number of correct answers and $y$ the number of wrong answers.

|  | Number of questions | Point value |
|---|---|---|
| Correct answers | $x$ | $2x$ |
| Wrong answers | $y$ | $-\dfrac{1}{2}y$ |
| Total | 75 | 100 |

*Translate*. The total number of questions is 75. This gives us one statement.

$$x + y = 75$$

The total point value (score) of 100 gives us a second statement.

$$2x - \frac{1}{2}y = 100$$

We now have a system of equations.

$$x + y = 75, \quad or \quad x + y = 75,$$
$$2x - \frac{1}{2}y = 100 \quad or \quad 4x - y = 200$$

*Carry out*. Solving the system we get $(55, 20)$.

*Check*. The total number of questions is $55 + 20$, or 75. The total score is $2(55) - \frac{1}{2}(20)$, or $110 - 10$, or 100. The numbers check.

*State*. There were 55 correct answers and 20 wrong answers.

**71.** *Writing Exercise*

**73.**
$$28a^3 - 25a^2bc + 3ab^2c^2$$
$$= a(28a^2 - 25abc + 3b^2c^2)$$
$$= a(7a - bc)(4a - 3bc)$$

**75.**
$$(x - p)^2 - p^2$$
$$= (x - p + p)(x - p - p)$$
$$= x(x - 2p)$$

**77.**
$$(y - 1)^4 - (y - 1)^2$$
$$= (y - 1)^2[(y - 1)^2 - 1]$$
$$= (y - 1)^2[(y - 1) + 1][(y - 1) - 1]$$
$$= (y - 1)^2(y)(y - 2), \text{ or } y(y - 1)^2(y - 2)$$

**79.**
$$x^6 - 2x^5 + x^4 - x^2 + 2x - 1$$
$$= x^4(x^2 - 2x + 1) - (x^2 - 2x + 1)$$
$$= (x^2 - 2x + 1)(x^4 - 1)$$
$$= (x - 1)^2(x^2 + 1)(x + 1)(x - 1)$$
$$= (x - 1)^3(x^2 + 1)(x + 1)$$

**81.**
$$(1 - x)^3 - (x - 1)^6$$
$$= [-(x - 1)]^3 - (x - 1)^6$$
$$= -(x - 1)^3[1 + (x - 1)^3]$$
$$= -(x - 1)^3[1 + (x - 1)][1 - (x - 1) + (x - 1)^2]$$
$$= -(x - 1)^3(x)(1 - x + 1 + x^2 - 2x + 1)$$
$$= -(x - 1)^3(x)(x^2 - 3x + 3), \text{ or }$$
$$x(1 - x)^3(x^2 - 3x + 3)$$

**83.**
$$a^{2w+1} + 2a^{w+1} + a$$
$$= a(a^{2w} + 2a^w + 1)$$
$$= a(a^w + 1)^2$$

**85.**
$$a - by^8 + b - ay^8$$
$$= a + b - by^8 - ay^8$$
$$= a + b - y^8(b + a)$$
$$= (b + a)(1 - y^8)$$
$$= (b + a)(1 + y^4)(1 - y^4)$$
$$= (b + a)(1 + y^4)(1 + y^2)(1 - y^2)$$
$$= (b + a)(1 + y^4)(1 + y^2)(1 + y)(1 - y)$$

**87.**
$$3a^2 + 3b^2 - 3c^2 - 3d^2 + 6ab - 6cd$$
$$= 3(a^2 + b^2 - c^2 - d^2 + 2ab - 2cd)$$
$$= 3[(a^2 + 2ab + b^2) - (c^2 + 2cd + d^2)]$$
$$= 3[(a + b)^2 - (c + d)^2]$$
$$= 3(a + b + c + d)(a + b - c - d)$$

**89.**
$$(m - 1)^3 - (m + 1)^3$$
$$= [m - 1 - (m + 1)][(m - 1)^2 + $$
$$(m - 1)(m + 1) + (m + 1)^2]$$
$$= (m - 1 - m - 1)(m^2 - 2m + 1 + m^2 - 1 + $$
$$m^2 + 2m + 1)$$
$$= -2(3m^2 + 1)$$

## Exercise Set 5.8

**1.** True; see page 335 in the text.

**3.** False; see Example 2(b).

**5.** True; see page 335 in the text.

**7.** $(t - 3)(t + 4) = 0$
$$t - 3 = 0 \quad or \quad t + 4 = 0 \quad \text{Principle of zero products}$$
$$t = 3 \quad or \quad t = -4$$

The solutions are 3 and $-4$. The solution set is $\{-4, 3\}$.

**9.** $t^2 + t - 6 = 0$
$$(t + 3)(t - 2) = 0 \quad \text{Factoring}$$
$$t + 3 = 0 \quad or \quad t - 2 = 0 \quad \text{Principle of zero products}$$
$$t = -3 \quad or \quad t = 2$$

The solutions are $-3$ and 2. The solution set is $\{-3, 2\}$.

**11.** $5x(2x - 3) = 0$
$$5x = 0 \quad or \quad 2x - 3 = 0 \quad \text{Principle of zero products}$$
$$x = 0 \quad or \quad 2x = 3$$
$$x = 0 \quad or \quad x = \frac{3}{2}$$

The solutions are 0 and $\frac{3}{2}$. The solution set is $\left\{0, \frac{3}{2}\right\}$.

**13.** $15x^2 - 21x = 0$

$3x(5x - 7) = 0$    Factoring

$3x = 0 \;\; or \;\; 5x - 7 = 0$    Principle of zero products
$x = 0 \;\; or \;\;\;\;\;\;\;\; 5x = 7$
$x = 0 \;\; or \;\;\;\;\;\;\;\;\; x = \dfrac{7}{5}$

The solutions are 0 and $\dfrac{7}{5}$. The solution set is $\left\{0, \dfrac{7}{5}\right\}$.

**15.** $(2t + 5)(t - 7) = 0$

$2t + 5 = 0 \;\;\; or \;\; t - 7 = 0$    Principle of zero products
$2t = -5 \;\;\; or \;\;\;\;\;\; t = 7$
$t = -\dfrac{5}{2} \;\;\; or \;\;\;\;\;\; t = 7$

The solutions are $-\dfrac{5}{2}$ and 7. The solution set is $\left\{-\dfrac{5}{2}, 7\right\}$.

**17.** $x^2 - 7x + 10 = 0$

$(x - 2)(x - 5) = 0$    Factoring

$x - 2 = 0 \;\; or \;\; x - 5 = 0$    Principle of zero products
$x = 2 \;\; or \;\;\;\;\;\; x = 5$

The solutions are 2 and 5. The solution set is $\{2, 5\}$.

**19.** $t^2 - 10t = 0$

$t(t - 10) = 0$    Factoring

$t = 0 \;\; or \;\; t - 10 = 0$    Principle of zero products
$t = 0 \;\; or \;\;\;\;\;\;\; t = 10$

The solutions are 0 and 10. The solution set is $\{0, 10\}$.

**21.** $(3x - 1)(4x - 5) = 0$

$3x - 1 = 0 \;\; or \;\; 4x - 5 = 0$  Principle of zero products
$3x = 1 \;\; or \;\;\;\;\;\; 4x = 5$
$x = \dfrac{1}{3} \;\; or \;\;\;\;\;\;\; x = \dfrac{5}{4}$

The solutions are $\dfrac{1}{3}$ and $\dfrac{5}{4}$. The solution set is $\left\{\dfrac{1}{3}, \dfrac{5}{4}\right\}$.

**23.**     $4a^2 = 10a$

$4a^2 - 10a = 0$    Getting 0 on one side

$2a(2a - 5) = 0$    Factoring

$2a = 0 \;\; or \;\; 2a - 5 = 0$    Principle of zero products
$a = 0 \;\; or \;\;\;\;\;\; 2a = 5$
$a = 0 \;\; or \;\;\;\;\;\;\; a = \dfrac{5}{2}$

The solutions are 0 and $\dfrac{5}{2}$. The solution set is $\left\{0, \dfrac{5}{2}\right\}$.

**25.**     $t^2 - 6t - 16 = 0$

$(t - 8)(t + 2) = 0$    Factoring

$t - 8 = 0 \;\; or \;\; t + 2 = 0$     Principle of zero products
$t = 8 \;\; or \;\;\;\;\;\; t = -2$

The solutions are 8 and $-2$. The solution set is $\{-2, 8\}$.

**27.**       $t^2 - 3t = 28$

$t^2 - 3t - 28 = 0$    Getting 0 on one side

$(t - 7)(t + 4) = 0$    Factoring

$t - 7 = 0 \;\; or \;\; t + 4 = 0$    Principle of zero products
$t = 7 \;\; or \;\;\;\;\;\; t = -4$

The solutions are 7 and $-4$. The solution set is $\{-4, 7\}$.

**29.**       $r^2 + 16 = 8r$

$r^2 - 8r + 16 = 0$    Getting 0 on one side

$(r - 4)(r - 4) = 0$    Factoring

$r - 4 = 0 \;\; or \;\; r - 4 = 0$  Principle of zero products
$r = 4 \;\; or \;\;\;\;\;\; r = 4$

There is only one solution, 4. The solution set is $\{4\}$.

**31.** $y^2 + 16y + 64 = 0$

$(y + 8)(y + 8) = 0$  Factoring

$y + 8 = 0 \;\;\; or \;\; y + 8 = 0$  Principle of zero products
$y = -8 \;\; or \;\;\;\;\;\; y = -8$

The solution is $-8$. The solution set is $\{-8\}$.

**33.**   $8y + y^2 + 15 = 0$

$y^2 + 8y + 15 = 0$  Changing order

$(y + 5)(y + 3) = 0$  Factoring

$y + 5 = 0 \;\;\; or \;\; y + 3 = 0$   Principle of zero products
$y = -5 \;\; or \;\;\;\;\;\; y = -3$

The solutions are $-5$ and $-3$. The solution set is $\{-5, -3\}$.

**35.** $r^2 - 9$

Observe that we can write this equation as $r^2 = 9$. Then the solutions are the numbers which, when squared, are 9. These are the square roots of 9, or $-3$ and 3. The solution set is $\{-3, 3\}$.

We could also use the principle of zero products to solve this equation, as shown below.

$r^2 - 9 = 0$

$(r + 3)(r - 3) = 0$

$r + 3 = 0 \;\;\; or \;\; r - 3 = 0$

$r = -3 \;\; or \;\;\;\;\;\; r = 3$

The solutions are $-3$ and 3. The solution set is $\{-3, 3\}$.

**37.**       $x^3 - 2x^2 = 63x$

$x^3 - 2x^2 - 63x = 0$    Getting 0 on one side

$x(x^2 - 2x - 63) = 0$

$x(x - 9)(x + 7) = 0$

$x = 0 \;\; or \;\; x - 9 = 0 \;\; or \;\; x + 7 = 0$    Principle of
$\;\;\;\;\;\;\;\;\;\;\;\;\;\;\;\;\;\;\;\;\;\;\;\;\;\;\;\;\;\;\;\;\;\;\;\;\;\;\;\;\;\;\;\;\;\;$zero products
$x = 0 \;\; or \;\;\;\;\;\; x = 9 \;\; or \;\;\;\;\;\; x = -7$

The solutions are 0, 9, and $-7$. The solution set is $\{-7, 0, 9\}$.

**39.** $r^2 = 49$

Using the reasoning in Exercise 35, we see that the solution set is composed of the square roots of 49, $\{-7, 7\}$.

We could also do this exercise as follows:

$r^2 = 49$

$r^2 - 49 = 0$

$(r + 7)(r - 7) = 0$

$$r + 7 = 0 \quad or \quad r - 7 = 0$$
$$r = -7 \quad or \quad r = 7$$

The solutions are $-7$ and $7$. The solution set is $\{-7, 7\}$.

**41.** $(a - 4)(a + 4) = 20$
$$a^2 - 16 = 20$$
$$a^2 - 36 = 0$$
$$(a + 6)(a - 6) = 0$$
$$a + 6 = 0 \quad or \quad a - 6 = 0$$
$$a = -6 \quad or \quad a = 6$$

The solutions are $-6$ and $6$. The solution set is $\{-6, 6\}$.

**43.** $-9x^2 + 15x - 4 = 0$
$$9x^2 - 15x + 4 = 0 \quad \text{Multiplying by } -1$$
$$(3x - 4)(3x - 1) = 0$$
$$3x - 4 = 0 \quad or \quad 3x - 1 = 0$$
$$3x = 4 \quad or \quad 3x = 1$$
$$x = \frac{4}{3} \quad or \quad x = \frac{1}{3}$$

The solutions are $\frac{4}{3}$ and $\frac{1}{3}$. The solution set is $\left\{\frac{4}{3}, \frac{1}{3}\right\}$.

**45.** $-8y^3 - 10y^2 - 3y = 0$
$$-y(8y^2 + 10y + 3) = 0$$
$$-y(2y + 1)(4y + 3) = 0$$
$$-y = 0 \quad or \quad 2y + 1 = 0 \quad or \quad 4y + 3 = 0$$
$$y = 0 \quad or \quad 2y = -1 \quad or \quad 4y = -3$$
$$y = 0 \quad or \quad y = -\frac{1}{2} \quad or \quad y = -\frac{3}{4}$$

The solutions are $0$, $-\frac{1}{2}$, and $-\frac{3}{4}$. The solution set is $\left\{0, -\frac{1}{2}, -\frac{3}{4}\right\}$.

**47.** $(z + 4)(z - 2) = -5$
$$z^2 + 2z - 8 = -5 \quad \text{Multiplying}$$
$$z^2 + 2z - 3 = 0$$
$$(z + 3)(z - 1) = 0$$
$$z + 3 = 0 \quad or \quad z - 1 = 0$$
$$z = -3 \quad or \quad z = 1$$

The solutions are $-3$ and $1$. The solution set is $\{-3, 1\}$.

**49.** $x(5 + 12x) = 28$
$$5x + 12x^2 = 28 \quad \text{Multiplying}$$
$$5x + 12x^2 - 28 = 0$$
$$12x^2 + 5x - 28 = 0 \quad \text{Rearranging}$$
$$(4x + 7)(3x - 4) = 0$$
$$4x + 7 = 0 \quad or \quad 3x - 4 = 0$$
$$4x = -7 \quad or \quad 3x = 4$$
$$x = -\frac{7}{4} \quad or \quad x = \frac{4}{3}$$

The solutions are $-\frac{7}{4}$ and $\frac{4}{3}$. The solution set is $\left\{-\frac{7}{4}, \frac{4}{3}\right\}$.

**51.** $a^2 - \frac{1}{64} = 0$
$$\left(a + \frac{1}{8}\right)\left(a - \frac{1}{8}\right) = 0$$
$$a + \frac{1}{8} = 0 \quad or \quad a - \frac{1}{8} = 0$$
$$a = -\frac{1}{8} \quad or \quad a = \frac{1}{8}$$

The solutions are $-\frac{1}{8}$ and $\frac{1}{8}$. The solution set is $\left\{-\frac{1}{8}, \frac{1}{8}\right\}$.

**53.** $t^4 - 26t^2 + 25 = 0$
$$(t^2 - 1)(t^2 - 25) = 0$$
$$(t + 1)(t - 1)(t + 5)(t - 5) = 0$$
$$t + 1 = 0 \quad or \quad t - 1 = 0 \quad or \quad t + 5 = 0 \quad or \quad t - 5 = 0$$
$$t = -1 \quad or \quad t = 1 \quad or \quad t = -5 \quad or \quad t = 5$$

The solutions are $-1$, $1$, $-5$, and $5$. The solution set is $\{-1, 1, -5, 5\}$.

**55.** We set $f(a)$ equal to 8.
$$a^2 + 12a + 40 = 8$$
$$a^2 + 12a + 32 = 0$$
$$(a + 8)(a + 4) = 0$$
$$a + 8 = 0 \quad or \quad a + 4 = 0$$
$$a = -8 \quad or \quad a = -4$$

The values of $a$ for which $f(a) = 8$ are $-8$ and $-4$.

**57.** We set $g(a)$ equal to 12.
$$2a^2 + 5a = 12$$
$$2a^2 + 5a - 12 = 0$$
$$(2a - 3)(a + 4) = 0$$
$$2a - 3 = 0 \quad or \quad a + 4 = 0$$
$$2a = 3 \quad or \quad a = -4$$
$$a = \frac{3}{2} \quad or \quad a = -4$$

The values of $a$ for which $g(a) = 12$ are $\frac{3}{2}$ and $-4$.

**59.** We set $h(a)$ equal to $-27$.
$$12a + a^2 = -27$$
$$12a + a^2 + 27 = 0$$
$$a^2 + 12a + 27 = 0 \quad \text{Rearranging}$$
$$(a + 3)(a + 9) = 0$$
$$a + 3 = 0 \quad or \quad a + 9 = 0$$
$$a = -3 \quad or \quad a = -9$$

The values of $a$ for which $h(a) = -27$ are $-3$ and $-9$.

**61.**
$$f(x) = g(x)$$
$$12x^2 - 15x = 8x - 5$$
$$12x^2 - 23x + 5 = 0$$
$$(4x - 1)(3x - 5) = 0$$
$$4x - 1 = 0 \quad or \quad 3x - 5 = 0$$
$$4x = 1 \quad or \quad 3x = 5$$
$$x = \frac{1}{4} \quad or \quad x = \frac{5}{3}$$

The values of $x$ for which $f(x) = g(x)$ are $\frac{1}{4}$ and $\frac{5}{3}$.

**63.**
$$f(x) = g(x)$$
$$2x^3 - 5x = 10x - 7x^2$$
$$2x^3 + 7x^2 - 15x = 0$$
$$x(2x^2 + 7x - 15) = 0$$
$$x(2x - 3)(x + 5) = 0$$
$$x = 0 \quad or \quad 2x - 3 = 0 \quad or \quad x + 5 = 0$$
$$x = 0 \quad or \quad 2x = 3 \quad or \quad x = -5$$
$$x = 0 \quad or \quad x = \frac{3}{2} \quad or \quad x = -5$$

The values of $x$ for which $f(x) = g(x)$ are $0$, $\frac{3}{2}$, and $-5$.

**65.** $f(x) = \dfrac{3}{x^2 - 4x - 5}$

$f(x)$ cannot be calculated for any $x$-value for which the denominator, $x^2 - 4x - 5$, is 0. To find the excluded values, we solve:
$$x^2 - 4x - 5 = 0$$
$$(x - 5)(x + 1) = 0$$
$$x - 5 = 0 \quad or \quad x + 1 = 0$$
$$x = 5 \quad or \quad x = -1$$

The domain of $f$ is $\{x | x$ is a real number *and* $x \neq 5$ *and* $x \neq -1\}$.

**67.** $f(x) = \dfrac{x}{6x^2 - 54}$

$f(x)$ cannot be calculated for any $x$-value for which the denominator, $6x^2 - 54$, is 0. To find the excluded values, we solve:
$$6x^2 - 54 = 0$$
$$6(x^2 - 9) = 0$$
$$6(x + 3)(x - 3) = 0$$
$$x + 3 = 0 \quad or \quad x - 3 = 0$$
$$x = -3 \quad or \quad x = 3$$

The domain of $f$ is $\{x | x$ is a real number *and* $x \neq -3$ *and* $x \neq 3\}$.

**69.** $f(x) = \dfrac{x - 5}{9x - 18x^2}$

$f(x)$ cannot be calculated for any $x$-value for which the denominator, $9x - 18x^2$, is 0. To find the excluded values, we solve:
$$9x - 18x^2 = 0$$
$$9x(1 - 2x) = 0$$

$$9x = 0 \quad or \quad 1 - 2x = 0$$
$$x = 0 \quad or \quad -2x = -1$$
$$x = 0 \quad or \quad x = \frac{1}{2}$$

The domain of $f$ is $\left\{ x | x$ is a real number *and* $x \neq 0$ *and* $x \neq \dfrac{1}{2} \right\}$.

**71.** $f(x) = \dfrac{7}{5x^3 - 35x^2 + 50x}$

$f(x)$ cannot be calculated for any $x$-value for which the denominator, $5x^3 - 35x^2 + 50x$, is 0. To find the excluded values, we solve:
$$5x^3 - 35x^2 + 50x = 0$$
$$5x(x^2 - 7x + 10) = 0$$
$$5x(x - 2)(x - 5) = 0$$
$$5x = 0 \quad or \quad x - 2 = 0 \quad or \quad x - 5 = 0$$
$$x = 0 \quad or \quad x = 2 \quad or \quad x = 5$$

The domain of $f$ is $\{x | x$ is a real number *and* $x \neq 0$ *and* $x \neq 2$ *and* $x \neq 5\}$.

**73. *Familiarize*.** We let $w$ represent the width and $w + 5$ represent the length. We make a drawing and label it.

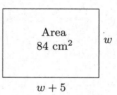

Area
84 cm² $\quad w$

$w + 5$

Recall that the formula for the area of a rectangle is $A = $ length $\times$ width.

***Translate*.**

$$\underbrace{\text{Area}}_{w(w + 5)} \quad \underset{=}{\text{is}} \quad \underbrace{84 \text{ cm}^2}_{84}.$$

***Carry out*.** We solve the equation:
$$w(w + 5) = 84$$
$$w^2 + 5w = 84$$
$$w^2 + 5w - 84 = 0$$
$$(w + 12)(w - 7) = 0$$
$$w + 12 = 0 \quad or \quad w - 7 = 0$$
$$w = -12 \quad or \quad w = 7$$

***Check*.** The number $-12$ is not a solution, because width cannot be negative. If the width is 7 cm and the length is 5 cm more, or 12 cm, then the area is $12 \cdot 7$, or 84 cm². This is a solution.

***State*.** The length is 12 cm, and the width is 7 cm.

**75. *Familiarize*.** We make a drawing and label it. We let $x$ represent the length of a side of the original square, in meters.

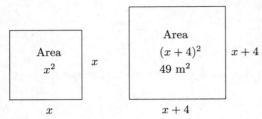

**Translate**.

$$\underbrace{\text{Area of new square}}_{(x+4)^2} \text{ is } \underbrace{49 \text{ m}^2}_{49}.$$

**Carry out**. We solve the equation:

$$(x+4)^2 = 49$$
$$x^2 + 8x + 16 = 49$$
$$x^2 + 8x - 33 = 0$$
$$(x-3)(x+11) = 0$$
$$x - 3 = 0 \ \ or \ \ x + 11 = 0$$
$$x = 3 \ \ or \ \ \ \ \ \ \ x = -11$$

**Check**. We check only 3 since the length of a side cannot be negative. If we increase the length by 4, the new length is $3 + 4$, or 7 m. Then the new area is $7 \cdot 7$, or 49 m$^2$. We have a solution.

**State**. The length of a side of the original square is 3 m.

77. **Familiarize**. We make a drawing and label it with both known and unknown information. We let $x$ represent the width of the frame.

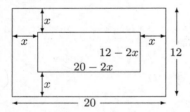

The length and width of the picture that shows are represented by $20 - 2x$ and $12 - 2x$. The area of the picture that shows is 84 cm$^2$.

**Translate**. Using the formula for the area of a rectangle, $A = l \cdot w$, we have

$$84 = (20 - 2x)(12 - 2x).$$

**Carry out**. We solve the equation:

$$84 = 240 - 64x + 4x^2$$
$$84 = 4(60 - 16x + x^2)$$
$$21 = 60 - 16x + x^2 \qquad \text{Dividing by 4}$$
$$0 = x^2 - 16x + 39$$
$$0 = (x-3)(x-13)$$
$$x - 3 = 0 \ \ or \ \ x - 13 = 0$$
$$x = 3 \ \ or \ \ \ \ \ \ \ x = 13$$

**Check**. We see that 13 is not a solution because when $x = 13$, $20 - 2x = -6$ and $12 - 2x = -14$, and the length and width of the frame cannot be negative. We check 3.

When $x = 3$, $20 - 2x = 14$ and $12 - 2x = 6$ and $14 \cdot 6 = 84$. The area is 84. The value checks.

**State**. The width of the frame is 3 cm.

79. **Familiarize**. We let $x$ represent the width of the walkway. We make a drawing and label it with both the known and unknown information.

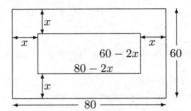

The area of the new lawn is $(80 - 2x)(60 - 2x)$.

**Translate**.

$$\underbrace{\text{Area of new lawn}}_{(80-2x)(60-2x)} \text{ is } \underbrace{2400 \text{ ft}^2}_{2400}.$$

**Carry out**. We solve the equation:

$$(80 - 2x)(60 - 2x) = 2400$$
$$4800 - 280x + 4x^2 = 2400$$
$$4x^2 - 280x + 2400 = 0$$
$$x^2 - 70x + 600 = 0 \qquad \text{Dividing by 4}$$
$$(x-10)(x-60) = 0$$
$$x - 10 = 0 \ \ or \ \ x - 60 = 0$$
$$x = 10 \ \ or \ \ \ \ \ \ \ x = 60$$

**Check**. If the sidewalk is 10 ft wide, the length of the new lawn will be $80 - 2 \cdot 10$, or 60 ft, and its width will be $60 - 2 \cdot 10$, or 40 ft. Then the area of the new lawn will be $60 \cdot 40$, or 2400 ft$^2$. This answer checks.

If the sidewalk is 60 ft wide, the length of the new lawn will be $80 - 2 \cdot 60$, or $-40$ ft. Since the length cannot be negative, 60 is not a solution.

**State**. The sidewalk is 10 ft wide.

81. **Familiarize**. Let $x$ represent the first integer, $x + 2$ the second, and $x + 4$ the third.

**Translate**.

$$\underbrace{\text{Square of}}_{(x+4)^2} \text{ is } 76 \underbrace{\text{more}}_{+} \underbrace{\text{square of}}_{(x+2)^2}$$
the third        than    the second.

**Carry out**. We solve the equation:

$$(x+4)^2 = 76 + (x+2)^2$$
$$x^2 + 8x + 16 = 76 + x^2 + 4x + 4$$
$$x^2 + 8x + 16 = x^2 + 4x + 80$$
$$4x = 64$$
$$x = 16$$

**Check**. We check the integers 16, 18, and 20. The square of 20, or 400, is 76 more than 324, the square of 18. The answer checks.

**State**. The integers are 16, 18, and 20.

**83.** *Familiarize.* Using the labels on the drawing in the text, we let $x$ represent the base of the triangle and $x + 2$ represent the height. Recall that the formula for the area of the triangle with base $b$ and height $h$ is $\frac{1}{2}bh$.

*Translate.*

$$\underbrace{\text{The area}}_{} \quad \text{is} \quad \underbrace{12 \text{ ft}^2}_{}.$$
$$\frac{1}{2}x(x+2) \quad = \quad 12$$

*Carry out.* We solve the equation:

$$\frac{1}{2}x(x+2) = 12$$
$$x(x+2) = 24 \quad \text{Multiplying by 2}$$
$$x^2 + 2x = 24$$
$$x^2 + 2x - 24 = 0$$
$$(x+6)(x-4) = 0$$
$$x + 6 = 0 \quad or \quad x - 4 = 0$$
$$x = -6 \quad or \quad x = 4$$

*Check.* We check only 4 since the length of the base cannot be negative. If the base is 4 ft, then the height is $4+2$, or 6 ft, and the area is $\frac{1}{2} \cdot 4 \cdot 6$, or 12 ft². The answer checks.

*State.* The height is 6 ft, and the base is 4 ft.

**85.** *Familiarize.* Let $h$ represent the height of the sail. Then $h + 9$ represents the base. Recall that the formula for the area of a triangle is $A = \frac{1}{2} \times$ base $\times$ height.

*Translate.*

The area is 56 m².
$$\frac{1}{2}(h+9)h = 56$$

*Carry out.* We solve the equation:

$$\frac{1}{2}(h+9)h = 56$$
$$(h+9)h = 112 \quad \text{Multiplying by 2}$$
$$h^2 + 9h = 112$$
$$h^2 + 9h - 112 = 0$$
$$(h+16)(h-7) = 0$$
$$h + 16 = 0 \quad or \quad h - 7 = 0$$
$$h = -16 \quad or \quad h = 7$$

*Check.* We check only 7, since height cannot be negative. If the height is 7 m, the base is $7 + 9$, or 16 m, and the area is $\frac{1}{2} \cdot 16 \cdot 7$, or 56 m². We have a solution.

*State.* The height is 7 m, and the base is 16 m.

**87.** *Familiarize.* We make a drawing. Let $h =$ the height the ladder reaches on the wall. Then the length of the ladder is $h + 2$.

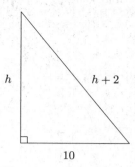

*Translate.* We use the Pythagorean theorem.
$$10^2 + h^2 = (h+2)^2$$

*Carry out.* We solve the equation:
$$100 + h^2 = h^2 + 4h + 4$$
$$96 = 4h$$
$$24 = h$$

*Check.* If $h = 24$, then $h+2 = 26$; $10^2+24^2 = 100+576 = 676 = 26^2$, so the answer checks.

*State.* The ladder is 24 ft long.

**89.** *Familiarize.* Let $w$ represent the width and $w + 3$ represent the length, in meters. Make a drawing.

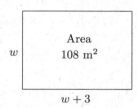

Recall that the formula for the area of a rectangle is $A =$ length $\times$ width.

*Translate.*

$$\underbrace{\text{Area}}_{} \quad \text{is} \quad \underbrace{108 \text{ m}^2}_{}.$$
$$w(w+3) \quad = \quad 108$$

*Carry out.* We solve the equation:

$$w(w+3) = 108$$
$$w^2 + 3w = 108$$
$$w^2 + 3w - 108 = 0$$
$$(w-9)(w+12) = 0$$
$$w - 9 = 0 \quad or \quad w + 12 = 0$$
$$w = 9 \quad or \quad w = -12$$

*Check.* The number $-12$ is not a solution because width cannot be negative. If the width is 9 m and the length is 3 m more, or 12 m, then the area will be $12 \cdot 9$, or 108 m². This is a solution.

*State.* The dimensions will be 12 m by 9 m.

**91. Familiarize**. The firm breaks even when the cost and the revenue are the same. We use the functions given in the text.

**Translate**.

$$\underbrace{\text{Cost}}\quad \text{equals}\quad \underbrace{\text{revenue.}}$$

$$\frac{1}{9}x^2 + 2x + 1 \;=\; \frac{5}{36}x^2 + 2x$$

**Carry out**. We solve the equation:

$$\frac{1}{9}x^2 + 2x + 1 = \frac{5}{36}x^2 + 2x$$

$$0 = \frac{1}{36}x^2 - 1$$

$$0 = \left(\frac{1}{6}x + 1\right)\left(\frac{1}{6}x - 1\right)$$

$$\frac{1}{6}x + 1 = 0 \quad or \quad \frac{1}{6}x - 1 = 0$$

$$\frac{1}{6}x = -1 \quad or \quad \frac{1}{6}x = 1$$

$$x = -6 \quad or \quad x = 6$$

**Check**. We check only 6 since the number of video cameras cannot be negative. If 6 cameras are produced, the cost is $C(6) = \frac{1}{9} \cdot 6^2 + 2 \cdot 6 + 1 = 4 + 12 + 1 = \$17$ thousand. If 6 cameras are sold, the revenue is $R(6) = \frac{5}{36} \cdot 6^2 + 2 \cdot 6 = 5 + 12 = \$17$ thousand. The answer checks.

**State**. The firm breaks even when 6 video cameras are produced and sold.

**93. Familiarize**. We will use the formula in Example 6, $h(t) = -15t^2 + 75t + 10$.

**Translate**. We need to find the value of $t$ for which $h(t) = 10$:

$$-15t^2 + 75t + 10 = 10$$

**Carry out**. We solve the equation.

$$-15t^2 + 75t + 10 = 10$$

$$-15t^2 + 75t = 0$$

$$-15(t - 5) = 0$$

$$-15t = 0 \quad or \quad t - 5 = 0$$

$$t = 0 \quad or \quad t = 5$$

**Check**. We have:

$$h(0) = -15 \cdot 0^2 + 75 \cdot 0 + 10 = 10;$$

$$h(5) = -15 \cdot 5^2 + 75 \cdot 5 + 10 = -375 + 375 + 10 = 10.$$

Both numbers check. However, we reject 0 since this is the moment of launch.

**State**. The tee shirt was airborne for 5 sec before it was caught.

**95. Familiarize**. We will use the formula $N(t) = -2t^2 + 15t + 51$.

**Translate**. We find the value of $t$ for which $N(t) = 69$. We have:

$$-2t^2 + 5t + 51 = 69$$

**Carry out**. We solve the equation.

$$-2t^2 + 15t + 51 = 69$$

$$-2t^2 + 15t - 18 = 0$$

$$2t^2 - 15t + 18 = 0 \quad \text{Multiplying by } -1$$

$$(2t - 3)(t - 6) = 0$$

$$2t - 3 = 0 \quad or \quad t - 6 = 0$$

$$2t = 3 \quad or \quad t = 6$$

$$t = \frac{3}{2} \quad or \quad t = 6$$

**Check**. $N\left(\frac{3}{2}\right) = -2\left(\frac{3}{2}\right)^2 + 15 \cdot \frac{3}{2} + 51 =$

$-\frac{9}{2} + \frac{45}{2} + 51 = 69$; $N(6) = -2 \cdot 6^2 + 15 \cdot 6 + 51 = -72 + 90 + 51 = 69$

Both numbers check

**State**. There were about 69 home-healthcare patients per 10,000 people $\frac{3}{2}$ yr, or $1\frac{1}{2}$ yr, and 6 yr after 1992.

**97.** *Writing Exercise*

**99.** $\dfrac{5 - 10 \cdot 3}{-4 + 11 \cdot 4} = \dfrac{5 - 30}{-4 + 44} = \dfrac{-25}{40} = -\dfrac{5}{8}$

**101. Familiarize**. Let $r$ represent the speed of the faster car and $d$ represent its distance. Then $r - 15$ and $651 - d$ represent the speed and distance of the slower car, respectively. We organize the information in a table.

|  | Speed | Time | Distance |
|---|---|---|---|
| Faster car | $r$ | 7 | $d$ |
| Slower car | $r - 15$ | 7 | $651 - d$ |

**Translate**. We use the formula $rt = d$. Each row of the table gives us an equation.

$$7r = d$$

$$7(r - 15) = 651 - d$$

**Carry out**. We use the substitution method substituting $7r$ for $d$ in the second equation and solving for $r$.

$$7(r - 15) = 651 - 7r \quad \text{Substituting}$$

$$7r - 105 = 651 - 7r$$

$$14r = 756$$

$$r = 54$$

**Check**. If $r = 54$, then the speed of the faster car is 54 mph and the speed of the slower car is $54 - 15$, or 39 mph. The distance the faster car travels is $54 \cdot 7$, or 378 miles. The distance the slower car travels is $39 \cdot 7$, or 273 miles. The total of the two distances is $378 + 273$, or 651 miles. The result checks.

**State**. The speed of the faster car is 54 mph. The speed of the slower car is 39 mph.

**103.** $2x - 14 + 9x > -8x + 16 + 10x$

$\quad\quad 11x - 14 > 2x + 16 \quad$ Collecting like terms

$\quad\quad\quad 9x - 14 > 16 \quad\quad$ Adding $-2x$

$\quad\quad\quad\quad\quad 9x > 30 \quad\quad$ Adding 14

$\quad\quad\quad\quad x > \dfrac{10}{3} \quad\quad$ Multiplying by $\dfrac{1}{9}$

The solution set is $\left\{ x \middle| x > \dfrac{10}{3} \right\}$, or $\left( \dfrac{10}{3}, \infty \right)$.

**105.** *Writing Exercise*

**107.** $(8x + 11)(12x^2 - 5x - 2) = 0$

$\quad (8x + 11)(3x - 2)(4x + 1) = 0$

$\quad 8x + 11 = 0 \quad$ or $\quad 3x - 2 = 0 \quad$ or $\quad 4x + 1 = 0$

$\quad\quad 8x = -11 \quad$ or $\quad\quad 3x = 2 \quad$ or $\quad\quad 4x = -1$

$\quad\quad x = -\dfrac{11}{8} \quad$ or $\quad\quad x = \dfrac{2}{3} \quad$ or $\quad\quad x = -\dfrac{1}{4}$

The solution set is $\left\{ -\dfrac{11}{8}, \dfrac{2}{3}, -\dfrac{1}{4} \right\}$.

**109.**

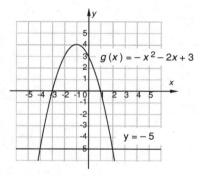

The solutions of $-x^2 - 2x + 3 = 0$ are the first coordinates of the $x$-intercepts. From the graph we see that these are $-3$ and $1$. The solution set is $\{-3, 1\}$.

To solve $-x^2 - 2x + 3 \geq -5$ we find the $x$-values for which $g(x) \geq -5$. From the graph we see that these are the values in the interval $[-4, 2]$. The solution set can also be expressed as $\{x| -4 \leq x \leq 2\}$.

**111.** Answers may vary. A polynomial function of lowest degree that meets the given criteria is of the form $g(x) = ax^3 + bx^2 + cx + d$. Substituting, we have

$a(-3)^3 + b(-3)^2 + c(-3) + d = 0,$

$a \cdot 1^3 + b \cdot 1^2 + c \cdot 1 + d = 0,$

$a \cdot 5^3 + b \cdot 5^2 + c \cdot 5 + d = 0$

$a \cdot 0^3 + b \cdot 0^2 + c \cdot 0 + d = 45,$ or

$-27a + 9b - 3c + d = 0,$

$\quad a + \quad b + \quad c + d = 0,$

$125a + 25b + 5c + d = 0,$

$\quad\quad\quad\quad\quad\quad d = 45.$

Solving the system of equations, we get $(3, -9, -39, 45)$, so the corresponding function is $g(x) = 3x^3 - 9x^2 - 39x + 45$.

**113.** *Familiarize*. Using the labels on the drawing in the text, we let $x$ represent the width of the piece of tin and $2x$ represent the length. Then the width and length of the base of the box are represented by $x - 4$ and $2x - 4$, respectively. Recall that the formula for the volume of a rectangular solid with length $l$, width $w$, and height $h$ is $l \cdot w \cdot h$.

*Translate*.

$\quad$ The volume $\quad$ is $\quad$ 480 cm$^3$.

$\quad\quad\quad \downarrow \quad\quad\quad\quad \downarrow \quad\quad \downarrow$

$(2x - 4)(x - 4)(2) \quad = \quad\quad 480$

*Carry out*. We solve the equation:

$\quad (2x - 4)(x - 4)(2) = 480$

$\quad\quad (2x - 4)(x - 4) = 240 \quad$ Dividing by 2

$\quad\quad\quad 2x^2 - 12x + 16 = 240$

$\quad\quad\quad 2x^2 - 12x - 224 = 0$

$\quad\quad\quad\quad x^2 - 6x - 112 = 0 \quad$ Dividing by 2

$\quad\quad\quad (x + 8)(x - 14) = 0$

$x + 8 = 0 \quad$ or $\quad x - 14 = 0$

$\quad x = -8 \quad$ or $\quad\quad\quad x = 14$

*Check*. We check only 14 since the width cannot be negative. If the width of the piece of tin is 14 cm, then its length is $2 \cdot 14$, or 28 cm, and the dimensions of the base of the box are $14 - 4$, or 10 cm by $28 - 4$, or 24 cm. The volume of the box is $24 \cdot 10 \cdot 2$, or 480 cm$^3$. The answer checks.

*State*. The dimensions of the piece of tin are 14 cm by 28 cm.

**115.** Graph $y_1 = 11.12(x+1)^2$ and $y_2 = 15.4x^2$ in a window that shows the point of intersection of the graphs. The window $[0, 10, 0, 500]$, Xscl $= 1$, Yscl $= 100$ is one good choice. Then find the first coordinate of the point of intersection. It is approximately 5.7, so it will take the camera about 5.7 sec to catch up to the skydiver.

**117.** ▮

**119.** $\{6.90\}$

**121.** $\{3.48\}$

# Chapter 6

# Rational Expressions, Equations, and Functions

## Exercise Set 6.1

**1.** Since $x - 5 = 0$ when $x = 5$, choice (e) is correct.

**3.** Since $x - 2 = 0$ when $x = 2$, choice (g) is correct.

**5.** Since $(x - 2)(x - 5) = 0$ when $x = 2$ or $x = 5$, choice (i) is correct.

**7.** Since $(x - 2)(x + 5) = 0$ when $x = 2$ or $x = -5$, choice (a) is correct.

**9.** Since $x + 3 = 0$ when $x = -3$, choice (d) is correct.

**11.** $H(t) = \dfrac{t^2 + 3t}{2t + 3}$

$H(5) = \dfrac{5^2 + 3 \cdot 5}{2 \cdot 5 + 3} = \dfrac{25 + 15}{10 + 3} = \dfrac{40}{13}$ hr, or $3\dfrac{1}{13}$ hr

**13.** $v(t) = \dfrac{4t^2 - 5t + 2}{t + 3}$

$v(0) = \dfrac{4 \cdot 0^2 - 5 \cdot 0 + 2}{0 + 3} = \dfrac{0 - 0 + 2}{0 + 3} = \dfrac{2}{3}$

$v(-2) = \dfrac{4(-2)^2 - 5(-2) + 2}{-2 + 3} = \dfrac{16 + 10 + 2}{-2 + 3} = 28$

$v(7) = \dfrac{4 \cdot 7^2 - 5 \cdot 7 + 2}{7 + 3} = \dfrac{196 - 35 + 2}{7 + 3} = \dfrac{163}{10}$

**15.** $g(x) = \dfrac{2x^3 - 9}{x^2 - 4x + 4}$

$g(0) = \dfrac{2 \cdot 0^3 - 9}{0^2 - 4 \cdot 0 + 4} = \dfrac{0 - 9}{0 - 0 + 4} = -\dfrac{9}{4}$

$g(2) = \dfrac{2 \cdot 2^3 - 9}{2^2 - 4 \cdot 2 + 4} = \dfrac{16 - 9}{4 - 8 + 4} = \dfrac{7}{0}$

Since division by zero is not defined, $g(2)$ does not exist.

$g(-1) = \dfrac{2(-1)^3 - 9}{(-1)^2 - 4(-1) + 4} = \dfrac{-2 - 9}{1 + 4 + 4} = -\dfrac{11}{9}$

**17.** $\dfrac{4x}{4x} \cdot \dfrac{x - 3}{x + 2} = \dfrac{4x(x - 3)}{4x(x + 2)}$

**19.** $\dfrac{t - 2}{t + 3} \cdot \dfrac{-1}{-1} = \dfrac{(t - 2)(-1)}{(t + 3)(-1)}$

**21.** $\dfrac{15x}{5x^2}$

$= \dfrac{5x \cdot 3}{5x \cdot x}$ Factoring; the greatest common factor is $5x$.

$= \dfrac{5x}{5x} \cdot \dfrac{3}{x}$ Factoring the rational expression

$= 1 \cdot \dfrac{3}{x} \qquad \dfrac{5x}{5x} = 1$

$= \dfrac{3}{x}$ Removing a factor equal to 1

**23.** $\dfrac{18t^3 s^2}{27t^7 s^9}$

$= \dfrac{9t^3 s^2 \cdot 2}{9t^3 s^2 \cdot 3t^4 s^7}$ Factoring the numerator and the denominator

$= \dfrac{9t^3 s^2}{9t^3 s^2} \cdot \dfrac{2}{3t^4 s^7}$ Factoring the rational expression

$= \dfrac{2}{3t^4 s^7}$ Removing a factor equal to 1

**25.** $\dfrac{2a - 10}{2} = \dfrac{2(a - 5)}{2 \cdot 1} = \dfrac{2}{2} \cdot \dfrac{a - 5}{1} = a - 5$

**27.** $\dfrac{15}{25a - 30} = \dfrac{5 \cdot 3}{5(5a - 6)} = \dfrac{5}{5} \cdot \dfrac{3}{5a - 6} = \dfrac{3}{5a - 6}$

**29.** $\dfrac{3x - 12}{3x + 15} = \dfrac{3(x - 4)}{3(x + 5)} = \dfrac{3}{3} \cdot \dfrac{x - 4}{x + 5} = \dfrac{x - 4}{x + 5}$

**31.** $f(x) = \dfrac{3x + 21}{x^2 + 7x}$

$= \dfrac{3(x + 7)}{x(x + 7)}$ Note that $x \neq 0$ and $x \neq -7$.

$= \dfrac{3}{x} \cdot \dfrac{x + 7}{x + 7}$

$= \dfrac{3}{x}$

Thus, $f(x) = \dfrac{3}{x}$, $x \neq 0, -7$.

**33.** $g(x) = \dfrac{x^2 - 9}{5x + 15}$

$= \dfrac{(x + 3)(x - 3)}{5(x + 3)}$ Note that $x \neq -3$.

$= \dfrac{x + 3}{x + 3} \cdot \dfrac{x - 3}{5}$

$= \dfrac{x - 3}{5}$

Thus, $g(x) = \dfrac{x - 3}{5}$, $x \neq -3$.

**35.** $h(x) = \dfrac{4 - x}{5x - 20}$

$= \dfrac{-1(x - 4)}{5(x - 4)}$ Note that $x \neq 4$.

$= \dfrac{-1}{5} \cdot \dfrac{x - 4}{x - 4}$

$= -\dfrac{1}{5}$

Thus, $h(x) = -\dfrac{1}{5}$, $x \neq 4$.

**37.** $f(t) = \dfrac{t^2 - 16}{t^2 - 8t + 16}$

$= \dfrac{(t + 4)(t - 4)}{(t - 4)(t - 4)}$    Note that $t \neq 4$.

$= \dfrac{t + 4}{t - 4} \cdot \dfrac{t - 4}{t - 4}$

$= \dfrac{t + 4}{t - 4}$

Thus, $f(t) = \dfrac{t + 4}{t - 4}$, $t \neq 4$.

**39.** $g(t) = \dfrac{21 - 7t}{3t - 9}$

$= \dfrac{-7(t - 3)}{3(t - 3)}$    Note that $t \neq 3$.

$= \dfrac{-7}{3} \cdot \dfrac{t - 3}{t - 3}$

$= -\dfrac{7}{3}$

Thus, $g(t) = -\dfrac{7}{3}$, $t \neq 3$.

**41.** $h(t) = \dfrac{t^2 + 5t + 4}{t^2 - 8t - 9}$

$= \dfrac{(t + 1)(t + 4)}{(t + 1)(t - 9)}$    Note that $t \neq -1$ and $t \neq 9$.

$= \dfrac{t + 1}{t + 1} \cdot \dfrac{t + 4}{t - 9}$

$= \dfrac{t + 4}{t - 9}$

Thus, $h(t) = \dfrac{t + 4}{t - 9}$, $t \neq -1, 9$.

**43.** $f(x) = \dfrac{9x^2 - 4}{3x - 2}$

$= \dfrac{(3x + 2)(3x - 2)}{3x - 2}$    Note that $x \neq \dfrac{2}{3}$.

$= \dfrac{3x + 2}{1} \cdot \dfrac{3x - 2}{3x - 2}$

$= 3x + 2$

Thus, $f(x) = 3x + 2$, $x \neq \dfrac{2}{3}$.

**45.** $g(t) = \dfrac{16 - t^2}{t^2 - 8t + 16}$

$= \dfrac{(4 + t)(4 - t)}{(t - 4)(t - 4)}$    Note that $t \neq 4$.

$= \dfrac{(4 + t)(4 - t)}{-1(4 - t)(t - 4)}$

$= \dfrac{4 - t}{4 - t} \cdot \dfrac{4 + t}{-1(t - 4)}$

$= \dfrac{4 + t}{4 - t}$

Thus, $g(t) = \dfrac{4 + t}{4 - t}$, $t \neq 4$.  (We could also write this as $g(t) = \dfrac{-t - 4}{t - 4}$, $t \neq 4$.)

**47.** $\dfrac{5a^3}{3b} \cdot \dfrac{7b^3}{10a^7}$

$= \dfrac{5a^3 \cdot 7b^3}{3b \cdot 10a^7}$    Multiplying the numerators and also the denominators

$= \dfrac{5 \cdot a^3 \cdot 7 \cdot b \cdot b^2}{3 \cdot b \cdot 2 \cdot 5 \cdot a^3 \cdot a^4}$    Factoring the numerator and the denominator

$= \dfrac{\cancel{5} \cdot \cancel{a^3} \cdot 7 \cdot \cancel{b} \cdot b^2}{3 \cdot \cancel{b} \cdot 2 \cdot \cancel{5} \cdot \cancel{a^3} \cdot a^4}$    Removing a factor equal to 1

$= \dfrac{7b^2}{6a^4}$

**49.** $\dfrac{8x - 16}{5x} \cdot \dfrac{x^3}{5x - 10} = \dfrac{(8x - 16)(x^3)}{5x(5x - 10)}$

$= \dfrac{8(x - 2)(x)(x^2)}{5 \cdot x \cdot 5(x - 2)}$

$= \dfrac{8(\cancel{x - 2})(\cancel{x})(x^2)}{5 \cdot \cancel{x} \cdot 5(\cancel{x - 2})}$

$= \dfrac{8x^2}{25}$

**51.** $\dfrac{x^2 - 16}{x^2} \cdot \dfrac{x^2 - 4x}{x^2 - x - 12} = \dfrac{(x^2 - 16)(x^2 - 4x)}{x^2(x^2 - x - 12)}$

$= \dfrac{(x + 4)(x - 4)(x)(x - 4)}{x \cdot x(x - 4)(x + 3)}$

$= \dfrac{(x + 4)(\cancel{x - 4})(\cancel{x})(x - 4)}{\cancel{x} \cdot x(\cancel{x - 4})(x + 3)}$

$= \dfrac{(x + 4)(x - 4)}{x(x + 3)}$

**53.** $\dfrac{7a - 14}{4 - a^2} \cdot \dfrac{5a^2 + 6a + 1}{35a + 7}$

$= \dfrac{(7a - 14)(5a^2 + 6a + 1)}{(4 - a^2)(35a + 7)}$

$= \dfrac{7(a - 2)(5a + 1)(a + 1)}{(2 + a)(2 - a)(7)(5a + 1)}$

$= \dfrac{7(-1)(2 - a)(5a + 1)(a + 1)}{(2 + a)(2 - a)(7)(5a + 1)}$

$= \dfrac{\cancel{7}(-1)(\cancel{2 - a})(\cancel{5a + 1})(a + 1)}{(2 + a)(\cancel{2 - a})(\cancel{7})\,(\cancel{5a + 1})}$

$= \dfrac{-1(a + 1)}{2 + a}$

$= \dfrac{-a - 1}{2 + a}$,  or  $-\dfrac{a + 1}{2 + a}$

**55.** $\dfrac{t^3 - 4t}{t - t^4} \cdot \dfrac{t^4 - t}{4t - t^3}$

$= \dfrac{t^3 - 4t}{t - t^4} \cdot \dfrac{-1(t - t^4)}{-1(t^3 - 4t)}$

$= \dfrac{(t^3 - 4t)(-1)(t - t^4)}{(t - t^4)(-1)(t^3 - 4t)}$

$= 1$

**57.** $\dfrac{c^3+8}{c^5-4c^3}\cdot\dfrac{c^6-4c^5+4c^4}{c^2-2c+4}$

$=\dfrac{(c^3+8)(c^6-4c^5+4c^4)}{(c^5-4c^3)(c^2-2c+4)}$

$=\dfrac{(c+2)(c^2-2c+4)(c^4)(c-2)(c-2)}{c^3(c+2)(c-2)(c^2-2c+4)}$

$=\dfrac{c^3(c+2)(c^2-2c+4)(c-2)}{c^3(c+2)(c^2-2c+4)(c-2)}\cdot\dfrac{c(c-2)}{1}$

$=c(c-2)$

**59.** $\dfrac{a^3-b^3}{3a^2+9ab+6b^2}\cdot\dfrac{a^2+2ab+b^2}{a^2-b^2}$

$=\dfrac{(a^3-b^3)(a^2+2ab+b^2)}{(3a^2+9ab+6b^2)(a^2-b^2)}$

$=\dfrac{(a-b)(a^2+ab+b^2)(a+b)(a+b)}{3(a+b)(a+2b)(a+b)(a-b)}$

$=\dfrac{(a-b)(a^2+ab+b^2)(a+b)(a+b)}{3(a+b)(a+2b)(a+b)(a-b)}$

$=\dfrac{a^2+ab+b^2}{3(a+2b)}$

**61.** $\dfrac{9x^5}{8y^2}\div\dfrac{3x}{16y^9}$

$=\dfrac{9x^5}{8y^2}\cdot\dfrac{16y^9}{3x}$     Multiplying by the reciprocal of the divisor

$=\dfrac{9x^5(16y^9)}{8y^2(3x)}$

$=\dfrac{3\cdot3\cdot x\cdot x^4\cdot2\cdot8\cdot y^2\cdot y^7}{8\cdot y^2\cdot3\cdot x}$

$=\dfrac{3\cdot3\cdot x\cdot x^4\cdot2\cdot8\cdot y^2\cdot y^7}{8\cdot y^2\cdot3\cdot x\cdot1}$

$=6x^4y^7$

**63.** $\dfrac{5x+10}{x^8}\div\dfrac{x+2}{x^3}=\dfrac{5x+10}{x^8}\cdot\dfrac{x^3}{x+2}$

$=\dfrac{(5x+10)(x^3)}{x^8(x+2)}$

$=\dfrac{5(x+2)(x^3)}{x^3\cdot x^5(x+2)}$

$=\dfrac{5(x+2)(x^3)}{x^3\cdot x^5(x+2)}$

$=\dfrac{5}{x^5}$

**65.** $\dfrac{25x^2-4}{x^2-9}\div\dfrac{2-5x}{x+3}=\dfrac{25x^2-4}{x^2-9}\cdot\dfrac{x+3}{2-5x}$

$=\dfrac{(25x^2-4)(x+3)}{(x^2-9)(2-5x)}$

$=\dfrac{(5x+2)(5x-2)(x+3)}{(x+3)(x-3)(-1)(5x-2)}$

$=\dfrac{(5x+2)(5x-2)(x+3)}{(x+3)(x-3)(-1)(5x-2)}$

$=\dfrac{5x+2}{-x+3}$, or $-\dfrac{5x+2}{x-3}$

**67.** $\dfrac{5y-5x}{15y^3}\div\dfrac{x^2-y^2}{3x+3y}=\dfrac{5y-5x}{15y^3}\cdot\dfrac{3x+3y}{x^2-y^2}$

$=\dfrac{(5y-5x)(3x+3y)}{15y^3(x^2-y^2)}$

$=\dfrac{5(y-x)(3)(x+y)}{5\cdot3\cdot y^3(x+y)(x-y)}$

$=\dfrac{5(-1)(x-y)(3)(x+y)}{5\cdot3\cdot y^3(x+y)(x-y)}$

$=\dfrac{5(-1)(x-y)(3)(x+y)}{5\cdot3\cdot y^3(x+y)(x-y)}$

$=\dfrac{-1}{y^3}$, or $-\dfrac{1}{y^3}$

**69.** $\dfrac{x^2-16}{x^2-10x+25}\div\dfrac{3x-12}{x^2-3x-10}$

$=\dfrac{x^2-16}{x^2-10x+25}\cdot\dfrac{x^2-3x-10}{3x-12}$

$=\dfrac{(x^2-16)(x^2-3x-10)}{(x^2-10x+25)(3x-12)}$

$=\dfrac{(x+4)(x-4)(x-5)(x+2)}{(x-5)(x-5)(3)(x-4)}$

$=\dfrac{(x+4)(x-4)(x-5)(x+2)}{(x-5)(x-5)(3)(x-4)}$

$=\dfrac{(x+4)(x+2)}{3(x-5)}$

**71.** $\dfrac{x^3-64}{x^3+64}\div\dfrac{x^2-16}{x^2-4x+16}$

$=\dfrac{x^3-64}{x^3+64}\cdot\dfrac{x^2-4x+16}{x^2-16}$

$=\dfrac{(x^3-64)(x^2-4x+16)}{(x^3+64)(x^2-16)}$

$=\dfrac{(x-4)(x^2+4x+16)(x^2-4x+16)}{(x+4)(x^2-4x+16)(x+4)(x-4)}$

$=\dfrac{(x-4)(x^2-4x+16)}{(x-4)(x^2-4x+16)}\cdot\dfrac{x^2+4x+16}{(x+4)(x+4)}$

$=\dfrac{x^2+4x+16}{(x+4)(x+4)}$, or $\dfrac{x^2+4x+16}{(x+4)^2}$

**73.** $f(t)=\dfrac{t^2-16}{4t+12}\cdot\dfrac{t+3}{t-4}$

The denominators of the rational expressions are $4t+12$ and $t-4$, so the domain of $f$ cannot contain $-3$ or $4$.

$f(t)=\dfrac{t^2-16}{4t+12}\cdot\dfrac{t+3}{t-4}$

$=\dfrac{(t^2-16)(t+3)}{(4t+12)(t-4)}$

$=\dfrac{(t+4)(t-4)(t+3)}{4(t+3)(t-4)}$

$=\dfrac{(t+4)(t-4)(t+3)}{4(t+3)(t-4)}$

$=\dfrac{t+4}{4}$, $t\neq-3,4$

**75.** $g(x) = \dfrac{x^2 - 2x - 35}{2x^3 - 3x^2} \cdot \dfrac{4x^3 - 9x}{7x - 49}$

First we find the values of $x$ for which the denominator $2x^3 - 3x^2 = 0$.

$$2x^3 - 3x^2 = 0$$
$$x^2(2x - 3) = 0$$
$$x \cdot x \cdot (2x - 3) = 0$$
$$x = 0 \;\; or \;\; x = 0 \;\; or \;\; 2x - 3 = 0$$
$$x = 0 \;\; or \;\; x = 0 \;\; or \;\;\;\;\;\; x = \frac{3}{2}$$

We see that the domain cannot contain 0 or $\dfrac{3}{2}$. Since $7x - 49 = 0$ when $x = 7$, the number 7 must also be excluded from the domain.

$$g(x) = \frac{x^2 - 2x - 35}{2x^3 - 3x^2} \cdot \frac{4x^3 - 9x}{7x - 49}$$
$$= \frac{(x^2 - 2x - 35)(4x^3 - 9x)}{(2x^3 - 3x^2)(7x - 49)}$$
$$= \frac{(x - 7)(x + 5)(x)(2x + 3)(2x - 3)}{x \cdot x \cdot (2x - 3)(7)(x - 7)}$$
$$= \frac{\cancel{(x - 7)}(x + 5)\cancel{(x)}(2x + 3)\cancel{(2x - 3)}}{\cancel{x} \cdot x \cdot \cancel{(2x - 3)}(7)\cancel{(x - 7)}}$$
$$= \frac{(x + 5)(2x + 3)}{7x}, \;\; x \neq 0, \frac{3}{2}, 7$$

**77.** $f(x) = \dfrac{x^2 - 4}{x^3} \div \dfrac{x^5 - 2x^4}{x + 4}$

$x^3 = 0$ when $x = 0$ and $x + 4 = 0$ when $x = -4$, so the domain cannot contain 0 or $-4$. Also, division by $(x^5 - 2x^4)/(x + 4)$ is defined only when this expression is nonzero. We find the values of $x$ for which $x^5 - 2x^4$ is zero.

$$x^5 - 2x^4 = 0$$
$$x^4(x - 2) = 0$$

The solutions of this equation are 0 and 2 so, in addition to 0 and $-4$, we must also exclude 2 from the domain.

$$f(x) = \frac{x^2 - 4}{x^3} \div \frac{x^5 - 2x^4}{x + 4}$$
$$= \frac{x^2 - 4}{x^3} \cdot \frac{x + 4}{x^5 - 2x^4}$$
$$= \frac{(x^2 - 4)(x + 4)}{x^3(x^5 - 2x^4)}$$
$$= \frac{(x + 2)(x - 2)(x + 4)}{x^3 \cdot x^4(x - 2)}$$
$$= \frac{(x + 2)\cancel{(x - 2)}(x + 4)}{x^7\cancel{(x - 2)}}$$
$$= \frac{(x + 2)(x + 4)}{x^7}, \;\; x \neq -4, 0, 2$$

**79.** $h(n) = \dfrac{n^3 + 3n}{n^2 - 9} \div \dfrac{n^2 + 5n - 14}{n^2 + 4n - 21}$

First we find the values of $n$ for which the denominators of the rational expressions are zero.

$$n^2 - 9 = 0$$
$$(n + 3)(n - 3) = 0$$
$$n + 3 = 0 \;\; or \;\; n - 3 = 0$$
$$x = -3 \;\; or \;\;\;\;\;\; n = 3$$

$$n^2 + 4n - 21 = 0$$
$$(n + 7)(n - 3) = 0$$
$$n + 7 = 0 \;\; or \;\; n - 3 = 0$$
$$n = -7 \;\; or \;\;\;\;\;\; n = 3$$

We see that the domain cannot contain $-7$, $-3$, or 3. Now we find the values of $n$ for which $(n^2 + 5n - 14)/(n^2 + 4n - 21)$ is zero.

$$n^2 + 5n - 14 = 0$$
$$(n + 7)(n - 2) = 0$$
$$n + 7 = 0 \;\; or \;\; n - 2 = 0$$
$$x = -7 \;\; or \;\;\;\;\;\; n = 2$$

We must also exclude 2 from the domain.

$$h(n) = \frac{n^3 + 3n}{n^2 - 9} \div \frac{n^2 + 5n - 14}{n^2 + 4n - 21}$$
$$= \frac{n^3 + 3n}{n^2 - 9} \cdot \frac{n^2 + 4n - 21}{n^2 + 5n - 14}$$
$$= \frac{(n^3 + 3n)(n^2 + 4n - 21)}{(n^2 - 9)(n^2 + 5n - 14)}$$
$$= \frac{n(n^2 + 3)(n + 7)(n - 3)}{(n + 3)(n - 3)(n + 7)(n - 2)}$$
$$= \frac{n(n^2 + 3)\cancel{(n + 7)}\cancel{(n - 3)}}{(n + 3)\cancel{(n - 3)}\cancel{(n + 7)}(n - 2)}$$
$$= \frac{n(n^2 + 3)}{(n + 3)(n - 2)}, \;\; n \neq -7, -3, 2, 3$$

**81.**
$$\frac{4x^2 - 9y^2}{8x^3 - 27y^3} \div \frac{4x + 6y}{3x - 9y} \cdot \frac{4x^2 + 6xy + 9y^2}{4x^2 - 8xy + 3y^2}$$
$$= \frac{4x^2 - 9y^2}{8x^3 - 27y^3} \cdot \frac{3x - 9y}{4x + 6y} \cdot \frac{4x^2 + 6xy + 9y^2}{4x^2 - 8xy + 3y^2}$$
$$= \frac{(4x^2 - 9y^2)(3x - 9y)(4x^2 + 6xy + 9y^2)}{(8x^3 - 27y^3)(4x + 6y)(4x^2 - 8xy + 3y^2)}$$
$$= \frac{(2x + 3y)(2x - 3y)(3)(x - 3y)(4x^2 + 6xy + 9y^2)}{(2x - 3y)(4x^2 + 6xy + 9y^2)(2)(2x + 3y)(2x - y)(2x - 3y)}$$
$$= \frac{(2x + 3y)(2x - 3y)(4x^2 + 6xy + 9y^2)}{(2x + 3y)(2x - 3y)(4x^2 + 6xy + 9y^2)} \cdot \frac{3(x - 3y)}{2(2x - y)(2x - 3y)}$$
$$= \frac{3(x - 3y)}{2(2x - y)(2x - 3y)}$$

**83.**
$$\frac{a^3 - ab^2}{2a^2 + 3ab + b^2} \cdot \frac{4a^2 - b^2}{a^2 - 2ab + b^2} \div \frac{a^2 + a}{a - 1}$$
$$= \frac{a^3 - ab^2}{2a^2 + 3ab + b^2} \cdot \frac{4a^2 - b^2}{a^2 - 2ab + b^2} \cdot \frac{a - 1}{a^2 + a}$$
$$= \frac{(a^3 - ab^2)(4a^2 - b^2)(a - 1)}{(2a^2 + 3ab + b^2)(a^2 - 2ab + b^2)(a^2 + a)}$$
$$= \frac{a(a + b)(a - b)(2a + b)(2a - b)(a - 1)}{(2a + b)(a + b)(a - b)(a - b)(a)(a + 1)}$$
$$= \frac{a(a + b)(a - b)(2a + b)}{a(a + b)(a - b)(2a + b)} \cdot \frac{(2a - b)(a - 1)}{(a - b)(a + 1)}$$
$$= \frac{(2a - b)(a - 1)}{(a - b)(a + 1)}$$

**85.** *Writing Exercise*

**87.** $\dfrac{3}{10} - \dfrac{8}{15} = \dfrac{3}{10} \cdot \dfrac{3}{3} - \dfrac{8}{15} \cdot \dfrac{2}{2}$

$= \dfrac{9}{30} - \dfrac{16}{30}$

$= \dfrac{9-16}{30} = \dfrac{-7}{30}$

$= -\dfrac{7}{30}$

**89.** $\dfrac{2}{3} \cdot \dfrac{5}{7} - \dfrac{5}{7} \cdot \dfrac{1}{6} = \dfrac{10}{21} - \dfrac{5}{42}$

$= \dfrac{10}{21} \cdot \dfrac{2}{2} - \dfrac{5}{42}$

$= \dfrac{20}{42} - \dfrac{5}{42}$

$= \dfrac{15}{42}$

$= \dfrac{3 \cdot 5}{3 \cdot 14} = \dfrac{\cancel{3} \cdot 5}{\cancel{3} \cdot 14}$

$= \dfrac{5}{14}$

**91.** $(8x^3 - 5x^2 + 6x + 2) - (4x^3 + 2x^2 - 3x + 7)$

$= 8x^3 - 5x^2 + 6x + 2 - 4x^3 - 2x^2 + 3x - 7$

$= 4x^3 - 7x^2 + 9x - 5$

**93.** *Writing Exercise*

**95.** $m = \dfrac{f(a+h) - f(a)}{(a+h) - a}$

$= \dfrac{(a+h)^2 + 5 - (a^2 + 5)}{a + h - a}$

$= \dfrac{a^2 + 2ah + h^2 + 5 - a^2 - 5}{h}$

$= \dfrac{2ah + h^2}{h} = \dfrac{h(2a + h)}{h}$

$= 2a + h$

**97.** $g(x) = \dfrac{2x+3}{4x-1}$

a) $g(x+h) = \dfrac{2(x+h) + 3}{4(x+h) - 1} = \dfrac{2x + 2h + 3}{4x + 4h - 1}$

b) $g(2x-2) \cdot g(x) = \dfrac{2(2x-2) + 3}{4(2x-2) - 1} \cdot \dfrac{2x+3}{4x-1}$

$= \dfrac{4x - 1}{8x - 9} \cdot \dfrac{2x+3}{4x-1}$

$= \dfrac{2x+3}{8x-9}$

c) $g\left(\dfrac{1}{2}x + 1\right) \cdot g(x) = \dfrac{2\left(\dfrac{1}{2}x + 1\right) + 3}{4\left(\dfrac{1}{2}x + 1\right) - 1} \cdot \dfrac{2x+3}{4x-1}$

$= \dfrac{x+5}{2x+3} \cdot \dfrac{2x+3}{4x-1}$

$= \dfrac{x+5}{4x-1}$

**99.** $\dfrac{r^2 - 4s^2}{r + 2s} \div (r+2s)^2 \left(\dfrac{2s}{r-2s}\right)^2$

$= \dfrac{r^2 - 4s^2}{r + 2s} \cdot \dfrac{1}{(r+2s)^2} \cdot \dfrac{(2s)^2}{(r-2s)^2}$

$= \dfrac{(r^2 - 4s^2)(4s^2)}{(r+2s)(r+2s)^2(r-2s)^2}$

$= \dfrac{(r+2s)(r-2s)(4s^2)}{(r+2s)(r+2s)^2(r-2s)(r-2s)}$

$= \dfrac{\cancel{(r+2s)}\cancel{(r-2s)}(4s^2)}{\cancel{(r+2s)}(r+2s)^2\cancel{(r-2s)}(r-2s)}$

$= \dfrac{4s^2}{(r+2s)^2(r-2s)}$

**101.** $\dfrac{6t^2 - 26t + 30}{8t^2 - 15t - 21} \cdot \dfrac{5t^2 - 9t - 15}{6t^2 - 14t - 20} \div \dfrac{5t^2 - 9t - 15}{6t^2 - 14t - 20}$

Note that the last two factors represent an expression divided by itself, so their quotient is 1. Then the result is the first factor, $\dfrac{6t^2 - 26t + 30}{8t^2 - 15t - 21}$. (Since $8t^2 - 15t - 21$ cannot be factored, no further simplification is possible.)

**103.** $\dfrac{a^3 - 2a^2 + 2a - 4}{a^3 - 2a^2 - 3a + 6}$

$= \dfrac{a^2(a-2) + 2(a-2)}{a^2(a-2) - 3(a-2)}$

$= \dfrac{(a-2)(a^2 + 2)}{(a-2)(a^2 - 3)}$

$= \dfrac{\cancel{(a-2)}(a^2 + 2)}{\cancel{(a-2)}(a^2 - 3)}$

$= \dfrac{a^2 + 2}{a^2 - 3}$

**105.** $\dfrac{u^6 + v^6 + 2u^3v^3}{u^3 - v^3 + u^2v - uv^2}$

$= \dfrac{u^6 + 2u^3v^3 + v^6}{(u^3 - v^3) + (u^2v - uv^2)}$

$= \dfrac{(u^3 + v^3)^2}{(u-v)(u^2 + uv + v^2) + uv(u-v)}$

$= \dfrac{[(u+v)(u^2 - uv + v^2)]^2}{(u-v)(u^2 + uv + v^2 + uv)}$

$= \dfrac{(u+v)^2(u^2 - uv + v^2)^2}{(u-v)(u^2 + 2uv + v^2)}$

$= \dfrac{(u+v)^2(u^2 - uv + v^2)^2}{(u-v)(u+v)^2}$

$= \dfrac{\cancel{(u+v)^2}(u^2 - uv + v^2)^2}{(u-v)\cancel{(u+v)^2}}$

$= \dfrac{(u^2 - uv + v^2)^2}{u - v}$

**107. a)** $(f \cdot g)(x) = \dfrac{4}{x^2 - 1} \cdot \dfrac{4x^2 + 8x + 4}{x^3 - 1}$

$\qquad = \dfrac{4(4x^2 + 8x + 4)}{(x^2 - 1)(x^3 - 1)}$

$\qquad = \dfrac{4 \cdot 4(x + 1)(x + 1)}{(x + 1)(x - 1)(x - 1)(x^2 + x + 1)}$

$\qquad = \dfrac{4 \cdot 4(\cancel{x + 1})(x + 1)}{(\cancel{x + 1})(x - 1)(x - 1)(x^2 + x + 1)}$

$\qquad = \dfrac{16(x + 1)}{(x - 1)^2(x^2 + x + 1)}$

(Note that $x \neq -1$ is an additional restriction, since $-1$ is not in the domain of $f$.)

**b)** $(f/g)(x) = \dfrac{4}{x^2 - 1} \div \dfrac{4x^2 + 8x + 4}{x^3 - 1}$

$\qquad = \dfrac{4}{x^2 - 1} \cdot \dfrac{x^3 - 1}{4x^2 + 8x + 4}$

$\qquad = \dfrac{4(x^3 - 1)}{(x^2 - 1)(4x^2 + 8x + 4)}$

$\qquad = \dfrac{4(x - 1)(x^2 + x + 1)}{(x + 1)(x - 1)(4)(x + 1)(x + 1)}$

$\qquad = \dfrac{4(\cancel{x - 1})(x^2 + x + 1)}{(x + 1)(\cancel{x - 1})(4)(x + 1)(x + 1)}$

$\qquad = \dfrac{x^2 + x + 1}{(x + 1)^3}$

(Note that $x \neq 1$ is an additional restriction, since 1 is not in the domain of either $f$ or $g$.)

**c)** $(g/f)(x) = \dfrac{1}{(f/g)(x)}$

$\qquad = \dfrac{(x + 1)^3}{x^2 + x + 1}$   (See part (b) above.)

(Note that $x \neq -1$ and $x \neq 1$ are restrictions, since $-1$ is not in the domain of $f$ and 1 is not in the domain of either $f$ or $g$.)

**109.**

**111.** *Writing Exercise*

---

## Exercise Set 6.2

**1.** True; see page 365 in the text.

**3.** False; see page 365 in the text.

**5.** True; see page 365 in the text.

**7.** True; see Examples 5 and 6.

**9.** $\dfrac{3}{2y} + \dfrac{5}{2y} = \dfrac{8}{2y}$   Adding the numerators. The denominator is unchanged.

$\qquad = \dfrac{2 \cdot 4}{2 \cdot y}$

$\qquad = \dfrac{\cancel{2} \cdot 4}{\cancel{2} \cdot y}$

$\qquad = \dfrac{4}{y}$

**11.** $\dfrac{5}{3m^2n^2} - \dfrac{4}{3m^2n^2} = \dfrac{1}{3m^2n^2}$

**13.** $\dfrac{x - 3y}{x + y} + \dfrac{x + 5y}{x + y} = \dfrac{2x + 2y}{x + y}$

$\qquad = \dfrac{2(x + y)}{x + y}$

$\qquad = \dfrac{2(\cancel{x + y})}{1(\cancel{x + y})}$

$\qquad = 2$

**15.** $\dfrac{3t + 2}{t - 4} - \dfrac{t - 2}{t - 4} = \dfrac{3t + 2 - (t - 2)}{t - 4}$

$\qquad = \dfrac{3t + 2 - t + 2}{t - 4}$

$\qquad = \dfrac{2t + 4}{t - 4}$

**17.** $f(x) = \dfrac{9x}{x + 3} + \dfrac{4}{5x}$   Note that $x \neq -3, 0$. The LCD is $5x(x + 3)$.

$\qquad = \dfrac{9x}{x + 3} \cdot \dfrac{5x}{5x} + \dfrac{4}{5x} \cdot \dfrac{x + 3}{x + 3}$

$\qquad = \dfrac{9x \cdot 5x}{5x(x + 3)} + \dfrac{4(x + 3)}{5x(x + 3)}$

$\qquad = \dfrac{45x^2 + 4x + 12}{5x(x + 3)}, \; x \neq -3, 0$

**19.** $f(x) = \dfrac{5}{x^2 - 9} - \dfrac{2}{x^2 - 9}$   Note that $x \neq -3, 3$.

$\qquad = \dfrac{3}{x^2 - 9}, \; x \neq -3, 3$

**21.** $f(x) = \dfrac{7}{3x^2} + \dfrac{2}{x^2 - 1}$

$\qquad = \dfrac{7}{3x^2} + \dfrac{2}{(x + 1)(x - 1)}$   Note that $x \neq 0, -1, 1$.

$\qquad\qquad$ The LCD is $3x^2(x + 1)(x - 1)$.

$\qquad = \dfrac{7}{3x^2} \cdot \dfrac{(x + 1)(x - 1)}{(x + 1)(x - 1)} + \dfrac{2}{(x + 1)(x - 1)} \cdot \dfrac{3x^2}{3x^2}$

$\qquad = \dfrac{7(x + 1)(x - 1) + 2 \cdot 3x^2}{3x^2(x + 1)(x - 1)}$

$\qquad = \dfrac{7x^2 - 7 + 6x^2}{3x^2(x + 1)(x - 1)}$

$\qquad = \dfrac{13x^2 - 7}{3x^2(x + 1)(x - 1)}, \; x \neq -1, 0, 1$

**23.** $f(x) = \dfrac{2}{x - 1} + \dfrac{3}{1 - x}$   Note that $x \neq 1$.

$\qquad = \dfrac{2}{x - 1} + \dfrac{-1}{-1} \cdot \dfrac{3}{1 - x}$

$\qquad = \dfrac{2}{x - 1} + \dfrac{-3}{x - 1}$

$\qquad = \dfrac{-1}{x - 1}, \text{ or } \dfrac{1}{1 - x}, \; x \neq 1$

**25.** $\dfrac{5-4x}{x^2-6x-7} + \dfrac{5x-4}{x^2-6x-7} = \dfrac{x+1}{x^2-6x-7}$

$\qquad\qquad = \dfrac{x+1}{(x-7)(x+1)}$

$\qquad\qquad = \dfrac{\cancel{x+1}}{(x-7)\cancel{(x+1)}}$

$\qquad\qquad = \dfrac{1}{x-7}$

**27.** $\dfrac{2a-5}{a^2-9} - \dfrac{3a-8}{a^2-9} = \dfrac{2a-5-(3a-8)}{a^2-9}$

$\qquad\qquad = \dfrac{2a-5-3a+8}{a^2-9}$

$\qquad\qquad = \dfrac{-a+3}{a^2-9}$

$\qquad\qquad = \dfrac{-1(a-3)}{(a+3)(a-3)}$

$\qquad\qquad = \dfrac{-1\cancel{(a-3)}}{(a+3)\cancel{(a-3)}}$

$\qquad\qquad = \dfrac{-1}{a+3},\ \text{or} \ -\dfrac{1}{a+3}$

**29.** $\dfrac{s^2}{r-s} + \dfrac{r^2}{s-r} = \dfrac{s^2}{r-s} + \dfrac{-r^2}{r-s}$

$\qquad = \dfrac{s^2-r^2}{r-s} = \dfrac{(s+r)(s-r)}{r-s}$

$\qquad = \dfrac{(s+r)(-1)\cancel{(r-s)}}{\cancel{r-s}} = -(s+r)$

**31.** $\dfrac{2}{a} - \dfrac{5}{-a} = \dfrac{2}{a} - \dfrac{-1}{-1}\cdot\dfrac{5}{-a} = \dfrac{2}{a} + \dfrac{5}{a} = \dfrac{7}{a}$

**33.** $\dfrac{y-4}{y^2-25} - \dfrac{9-2y}{25-y^2} = \dfrac{y-4}{y^2-25} - \dfrac{-1}{-1}\cdot\dfrac{9-2y}{25-y^2}$

$\qquad\qquad = \dfrac{y-4}{y^2-25} - \dfrac{2y-9}{y^2-25}$

$\qquad\qquad = \dfrac{y-4-(2y-9)}{y^2-25}$

$\qquad\qquad = \dfrac{y-4-2y+9}{y^2-25}$

$\qquad\qquad = \dfrac{-y+5}{y^2-25}$

$\qquad\qquad = \dfrac{-1(y-5)}{(y+5)(y-5)}$

$\qquad\qquad = \dfrac{-1\cancel{(y-5)}}{(y+5)\cancel{(y-5)}}$

$\qquad\qquad = \dfrac{-1}{y+5},\ \text{or}\ -\dfrac{1}{y+5}$

**35.** $\dfrac{y^2-5}{y^4-81} + \dfrac{4}{81-y^4}$

$= \dfrac{y^2-5}{y^4-81} + \dfrac{-1}{-1}\cdot\dfrac{4}{81-y^4}$

$= \dfrac{y^2-5}{y^4-81} + \dfrac{-4}{y^4-81}$

$= \dfrac{y^2-5+(-4)}{y^4-81}$

$= \dfrac{y^2-9}{y^4-81} = \dfrac{y^2-9}{(y^2+9)(y^2-9)}$

$= \dfrac{y^2-9}{y^2-9}\cdot\dfrac{1}{y^2+9}$

$= \dfrac{1}{y^2+9}$

**37.** $\dfrac{r-6s}{r^3-s^3} - \dfrac{5s}{s^3-r^3} = \dfrac{r-6s}{r^3-s^2} - \dfrac{-1}{-1}\cdot\dfrac{5s}{s^3-r^3}$

$\qquad\qquad = \dfrac{r-6s}{r^3-s^3} - \dfrac{-5s}{r^3-s^3}$

$\qquad\qquad = \dfrac{r-6s-(-5s)}{r^3-s^3}$

$\qquad\qquad = \dfrac{r-s}{(r-s)(r^2+rs+s^2)}$

$\qquad\qquad = \dfrac{\cancel{r-s}}{\cancel{(r-s)}(r^2+rs+s^2)}$

$\qquad\qquad = \dfrac{1}{r^2+rs+s^2}$

**39.** $\dfrac{a+2}{a-4} + \dfrac{a-2}{a+3}$

[LCD is $(a-4)(a+3)$.]

$= \dfrac{a+2}{a-4}\cdot\dfrac{a+3}{a+3} + \dfrac{a-2}{a+3}\cdot\dfrac{a-4}{a-4}$

$= \dfrac{(a^2+5a+6)+(a^2-6a+8)}{(a-4)(a+3)}$

$= \dfrac{2a^2-a+14}{(a-4)(a+3)}$

**41.** $4 + \dfrac{x-3}{x+1} = \dfrac{2}{1} + \dfrac{x-3}{x+1}$

[LCD is $x+1$.]

$= \dfrac{4}{1}\cdot\dfrac{x+1}{x+1} + \dfrac{x-3}{x+1}$

$= \dfrac{(4x+4)+(x-3)}{x+1}$

$= \dfrac{5x+1}{x+1}$

**43.**
$$\frac{4xy}{x^2 - y^2} + \frac{x - y}{x + y}$$

$$= \frac{4xy}{(x + y)(x - y)} + \frac{x - y}{x + y}$$

LCD is $(x + y)(x - y)$.]

$$= \frac{4xy}{(x + y)(x - y)} + \frac{x - y}{x + y} \cdot \frac{x - y}{x - y}$$

$$= \frac{4xy + x^2 - 2xy + y^2}{(x + y)(x - y)}$$

$$= \frac{x^2 + 2xy + y^2}{(x + y)(x - y)} = \frac{(x + y)(x + y)}{(x + y)(x - y)}$$

$$= \frac{(\cancel{x + y})(x + y)}{(\cancel{x + y})(x - y)} = \frac{x + y}{x - y}$$

**45.**
$$\frac{8}{2x^2 - 7x + 5} + \frac{3x + 2}{2x^2 - x - 10}$$

$$= \frac{8}{(2x - 5)(x - 1)} + \frac{3x + 2}{(2x - 5)(x + 2)}$$

[LCD is $(2x - 5)(x - 1)(x + 2)$.]

$$= \frac{8}{(2x-5)(x-1)} \cdot \frac{x+2}{x+2} + \frac{3x+2}{(2x-5)(x+2)} \cdot \frac{x-1}{x-1}$$

$$= \frac{8x + 16 + 3x^2 - x - 2}{(2x - 5)(x - 1)(x + 2)}$$

$$= \frac{3x^2 + 7x + 14}{(2x - 5)(x - 1)(x + 2)}$$

**47.**
$$\frac{4}{x + 1} + \frac{x + 2}{x^2 - 1} + \frac{3}{x - 1}$$

$$= \frac{4}{x + 1} + \frac{x + 2}{(x + 1)(x - 1)} + \frac{3}{x - 1}$$

[LCD is $(x + 1)(x - 1)$.]

$$= \frac{4}{x+1} \cdot \frac{x-1}{x-1} + \frac{x+2}{(x+1)(x-1)} + \frac{3}{x-1} \cdot \frac{x+1}{x+1}$$

$$= \frac{4x - 4 + x + 2 + 3x + 3}{(x + 1)(x - 1)}$$

$$= \frac{8x + 1}{(x + 1)(x - 1)}$$

**49.**
$$\frac{x + 6}{5x + 10} - \frac{x - 2}{4x + 8}$$

$$= \frac{x + 6}{5(x + 2)} - \frac{x - 2}{4(x + 2)}$$

[LCD is $5 \cdot 4(x + 2)$.]

$$= \frac{x + 6}{5(x + 2)} \cdot \frac{4}{4} - \frac{x - 2}{4(x + 2)} \cdot \frac{5}{5}$$

$$= \frac{4(x + 6) - 5(x - 2)}{5 \cdot 4(x + 2)}$$

$$= \frac{4x + 24 - 5x + 10}{5 \cdot 4(x + 2)}$$

$$= \frac{-x + 34}{5 \cdot 4(x + 2)}, \text{ or } \frac{-x + 34}{20(x + 2)}$$

**51.**
$$\frac{5ab}{a^2 - b^2} - \frac{a - b}{a + b}$$

$$= \frac{5ab}{(a + b)(a - b)} - \frac{a - b}{a + b}$$

[LCD is $(a + b)(a - b)$.]

$$= \frac{5ab}{(a + b)(a - b)} - \frac{a - b}{a + b} \cdot \frac{a - b}{a - b}$$

$$= \frac{5ab - (a^2 - 2ab + b^2)}{(a + b)(a - b)}$$

$$= \frac{5ab - a^2 + 2ab - b^2}{(a + b)(a - b)}$$

$$= \frac{-a^2 + 7ab - b^2}{(a + b)(a - b)}$$

**53.**
$$\frac{x}{x^2 + 9x + 20} - \frac{4}{x^2 + 7x + 12}$$

$$= \frac{x}{(x + 5)(x + 4)} - \frac{4}{(x + 3)(x + 4)}$$

[LCD is $(x + 5)(x + 4)(x + 3)$.]

$$= \frac{x}{(x+5)(x+4)} \cdot \frac{x+3}{x+3} - \frac{4}{(x+3)(x+4)} \cdot \frac{x+5}{x+5}$$

$$= \frac{x^2 + 3x - (4x + 20)}{(x + 5)(x + 4)(x + 3)}$$

$$= \frac{x^2 + 3x - 4x - 20}{(x + 5)(x + 4)(x + 3)}$$

$$= \frac{x^2 - x - 20}{(x + 5)(x + 4)(x + 3)}$$

$$= \frac{(x - 5)(x + 4)}{(x + 5)(x + 4)(x + 3)}$$

$$= \frac{(x - 5)(\cancel{x + 4})}{(x + 5)(\cancel{x + 4})(x + 3)}$$

$$= \frac{x - 5}{(x + 5)(x + 3)}$$

**55.**
$$\frac{3y}{y^2 - 7y + 10} - \frac{2y}{y^2 - 8y + 15}$$

$$= \frac{3y}{(y - 5)(y - 2)} - \frac{2y}{(y - 5)(y - 3)}$$

[LCD is $(y - 5)(y - 2)(y - 3)$.]

$$= \frac{3y}{(y-5)(y-2)} \cdot \frac{y-3}{y-3} - \frac{2y}{(y-5)(y-3)} \cdot \frac{y-2}{y-2}$$

$$= \frac{3y^2 - 9y - (2y^2 - 4y)}{(y - 5)(y - 2)(y - 3)}$$

$$= \frac{3y^2 - 9y - 2y^2 + 4y}{(y - 5)(y - 2)(y - 3)}$$

$$= \frac{y^2 - 5y}{(y - 5)(y - 2)(y - 3)} = \frac{y(y - 5)}{(y - 5)(y - 2)(y - 3)}$$

$$= \frac{y(\cancel{y - 5})}{(\cancel{y - 5})(y - 2)(y - 3)} = \frac{y}{(y - 2)(y - 3)}$$

**57.**
$$\frac{2x+1}{x-y}+\frac{5x^2-5xy}{x^2-2xy+y^2}$$
$$=\frac{2x+1}{x-y}+\frac{5x(x-y)}{(x-y)(x-y)}$$
$$=\frac{2x+1}{x-y}+\frac{5x\cancel{(x-y)}}{\cancel{(x-y)}(x-y)}$$
$$=\frac{2x+1}{x-y}+\frac{5x}{x-y}$$
$$=\frac{7x+1}{x-y}$$

**59.**
$$\frac{3y+2}{y^2+5y-24}+\frac{7}{y^2+4y-32}$$
$$=\frac{3y+2}{(y+8)(y-3)}+\frac{7}{(y+8)(y-4)}$$
$$[\text{LCD is }(y+8)(y-3)(y-4).]$$
$$=\frac{3y+2}{(y+8)(y-3)}\cdot\frac{y-4}{y-4}+\frac{7}{(y+8)(y-4)}\cdot\frac{y-3}{y-3}$$
$$=\frac{3y^2-10y-8+7y-21}{(y+8)(y-3)(y-4)}$$
$$=\frac{3y^2-3y-29}{(y+8)(y-3)(y-4)}$$

**61.**
$$\frac{2y-6}{y^2-9}-\frac{y}{y-1}+\frac{y^2+2}{y^2+2y-3}$$
$$=\frac{2(y-3)}{(y+3)(y-3)}-\frac{y}{y-1}+\frac{y^2+2}{(y+3)(y-1)}$$
$$=\frac{2}{y+3}-\frac{y}{y-1}+\frac{y^2+2}{(y+3)(y-1)}$$
$$[\text{LCD is }(y+3)(y-1).]$$
$$=\frac{2}{y+3}\cdot\frac{y-1}{y-1}-\frac{y}{y-1}\cdot\frac{y+3}{y+3}+\frac{y^2+2}{(y+3)(y-1)}$$
$$=\frac{2(y-1)-y(y+3)+y^2+2}{(y+3)(y-1)}$$
$$=\frac{2y-2-y^2-3y+y^2+2}{(y+3)(y-1)}$$
$$=\frac{-y}{(y+3)(y-1)},\text{ or }-\frac{y}{(y+3)(y-1)}$$

**63.**
$$\frac{5y}{1-4y^2}-\frac{2y}{2y+1}+\frac{5y}{4y^2-1}$$

Observe that $\dfrac{5y}{1-4y^2}$ and $\dfrac{5y}{4y^2-1}$ are opposites, so their sum is 0. Then the result is the remaining expression, $-\dfrac{2y}{2y+1}$.

**65.**
$$f(x)=2+\frac{x}{x-3}-\frac{18}{x^2-9}$$
$$=\frac{2}{1}+\frac{x}{x-3}-\frac{18}{(x+3)(x-3)}$$
$$\text{Note that }x\neq-3,3.$$
$$\text{LCD is }(x+3)(x-3).$$
$$=\frac{2}{1}\cdot\frac{(x+3)(x-3)}{(x+3)(x-3)}+\frac{x}{x-3}\cdot\frac{x+3}{x+3}-\frac{18}{(x+3)(x-3)}$$
$$=\frac{2(x+3)(x-3)+x(x+3)-18}{(x+3)(x-3)}$$
$$=\frac{2x^2-18+x^2+3x-18}{(x+3)(x-3)}$$
$$=\frac{3x^2+3x-36}{(x+3)(x-3)}$$
$$=\frac{3(x+4)\cancel{(x-3)}}{(x+3)\cancel{(x-3)}}$$
$$=\frac{3(x+4)}{x+3},\ x\neq-3,3$$

**67.**
$$f(x)=\frac{3x-1}{x^2+2x-3}-\frac{x+4}{x^2-16}$$
$$=\frac{3x-1}{(x+3)(x-1)}-\frac{x+4}{(x+4)(x-4)}$$
$$\text{Note that }x\neq-3,1,-4,4.$$
$$=\frac{3x-1}{(x+3)(x-1)}-\frac{1}{x-4}\quad\begin{array}{l}\text{Removing a factor}\\\text{equal to 1; LCD is}\\(x+3)(x-1)(x-4).\end{array}$$
$$=\frac{3x-1}{(x+3)(x-1)}\cdot\frac{x-4}{x-4}-\frac{1}{x-4}\cdot\frac{(x+3)(x-1)}{(x+3)(x-1)}$$
$$=\frac{(3x-1)(x-4)-(x+3)(x-1)}{(x+3)(x-1)(x-4)}$$
$$=\frac{3x^2-13x+4-(x^2+2x-3)}{(x+3)(x-1)(x-4)}$$
$$=\frac{3x^2-13x+4-x^2-2x+3}{(x+3)(x-1)(x-4)}$$
$$=\frac{2x^2-15x+7}{(x+3)(x-1)(x-4)}$$
$$=\frac{(2x-1)(x-7)}{(x+3)(x-1)(x-4)},\ x\neq-4,-3,1,4$$

**69.**
$$f(x)=\frac{1}{x^2+5x+6}-\frac{2}{x^2+3x+2}-\frac{1}{x^2+5x+6}$$
$$=\frac{1}{(x+3)(x+2)}-\frac{2}{(x+2)(x+1)}-\frac{1}{(x+3)(x+2)}$$
$$\text{Note that }x\neq-3,-2,-1.$$

Observe that the sum of the first and third terms is 0, so the result is the remaining term.
$$f(x)=-\frac{2}{(x+2)(x+1)},\text{ or }\frac{-2}{(x+2)(x+1)},$$
$$x\neq-3,-2,-1$$

**71.** *Writing Exercise*

**73.** $\dfrac{15x^{-7}y^{12}z^4}{35x^{-2}y^6z^{-3}} = \dfrac{15}{35}x^{-7-(-2)}y^{12-6}z^{4-(-3)}$

$\qquad = \dfrac{3}{7}x^{-5}y^6z^7 = \dfrac{3}{7}\cdot\dfrac{1}{x^5}\cdot y^6z^7$

$\qquad = \dfrac{3y^6z^7}{7x^5}$

**75.** $\dfrac{34s^9t^{-40}r^{30}}{10s^{-3}t^{20}r^{-10}} = \dfrac{34}{10}s^{9-(-3)}t^{-40-20}r^{30-(-10)}$

$\qquad = \dfrac{17}{5}s^{12}t^{-60}r^{40}$

$\qquad = \dfrac{17}{5}s^{12}\cdot\dfrac{1}{t^{60}}\cdot r^{40}$

$\qquad = \dfrac{17s^{12}r^{40}}{5t^{60}}$

**77.** *Familiarize.* We let $x$, $y$, and $z$ represent the number of rolls of dimes, nickels, and quarters, respectively.

| Coins | Number of rolls | Value per roll | Total Value |
|-------|-----------------|----------------|-------------|
| Dimes | $x$ | $50 \times 0.10$, or 5.00 | $5x$ |
| Nickels | $y$ | $40 \times 0.05$, or 2.00 | $2y$ |
| Quarters | $z$ | $40 \times 0.25$, or 10.00 | $10z$ |
| Total | 12 | | \$70.00 |

*Translate.* The number of rolls of nickels is three more than the number of rolls of dimes. This gives us one equation.

$\qquad y = x + 3$

From the table we get two more equations.

$\qquad x + y + z = 12$

$\qquad 5x + 2y + 10z = 70$

*Carry out.* Solving the system we get the ordered triple $(2, 5, 5)$.

*Check.*

2 rolls of dimes $= 2 \times \$5$, or \$10

5 rolls of nickels $= 5 \times \$2$, or \$10

5 rolls of quarters $= 5 \times \$10$, or \$50

The total value is $\$10 + \$10 + \$50 = \$70$.

The total number of rolls of coins is $2 + 5 + 5$, or 12, and the number of rolls of nickels, 5, is three more than the number of rolls of dimes, 2. The numbers check.

*State.* Robert has 2 rolls of dimes, 5 rolls of nickels, and 5 rolls of quarters.

**79.** *Writing Exercise*

**81.** We find the least common multiple of 14(2 weeks = 14 days), 20, and 30.

$\qquad 14 = 2\cdot 7$

$\qquad 20 = 2\cdot 2\cdot 5$

$\qquad 30 = 2\cdot 3\cdot 5$

$\qquad \text{LCM} = 2\cdot 2\cdot 3\cdot 5\cdot 7 = 420$

It will be 420 days until Corinna can refill all three prescriptions on the same day.

**83.** The smallest number of parts possible is the least common multiple of 6 and 4.

$\qquad 6 = 2\cdot 3$

$\qquad 4 = 2\cdot 2$

$\qquad \text{LCM} = 2\cdot 3\cdot 2, \text{ or } 12$

A measure should be divided into 12 parts.

**85.** $x^8 - x^4 = x^4(x^2 + 1)(x + 1)(x - 1)$

$\qquad x^5 - x^2 = x^2(x - 1)(x^2 + x + 1)$

$\qquad x^5 - x^3 = x^3(x + 1)(x - 1)$

$\qquad x^5 + x^2 = x^2(x + 1)(x^2 - x + 1)$

The LCM is

$x^4(x^2 + 1)(x + 1)(x - 1)(x^2 + x + 1)(x^2 - x + 1)$.

**87.** The LCM is $8a^4b^7$.

One expression is $2a^3b^7$.

Then the other expression must contain 8, $a^4$, and one of the following:

no factor of $b$, $b$, $b^2$, $b^3$, $b^4$, $b^5$, $b^6$, or $b^7$.

Thus, all the possibilities for the other expression are $8a^4$, $8a^4b$, $8a^4b^2$, $8a^4b^3$, $8a^4b^4$, $8a^4b^5$, $8a^4b^6$, $8a^4b^7$.

**89.** $(f + g)(x) = \dfrac{x^3}{x^2 - 4} + \dfrac{x^2}{x^2 + 3x - 10}$

$\qquad = \dfrac{x^3}{(x + 2)(x - 2)} + \dfrac{x^2}{(x + 5)(x - 2)}$

$\qquad = \dfrac{x^3(x + 5) + x^2(x + 2)}{(x + 2)(x - 2)(x + 5)}$

$\qquad = \dfrac{x^4 + 5x^3 + x^3 + 2x^2}{(x + 2)(x - 2)(x + 5)}$

$\qquad = \dfrac{x^4 + 6x^3 + 2x^2}{(x + 2)(x - 2)(x + 5)}$

**91.** $(f \cdot g)(x) = \dfrac{x^3}{x^2 - 4} \cdot \dfrac{x^2}{x^2 + 3x - 10}$

$\qquad = \dfrac{x^5}{(x^2 - 4)(x^2 + 3x - 10)}$

**93.** $\quad 5(x - 3)^{-1} + 4(x + 3)^{-1} - 2(x + 3)^{-2}$

$\qquad = \dfrac{5}{x - 3} + \dfrac{4}{x + 3} - \dfrac{2}{(x + 3)^2}$

$\qquad \text{[LCD is } (x - 3)(x + 3)^2.]$

$\qquad = \dfrac{5(x + 3)^2 + 4(x - 3)(x + 3) - 2(x - 3)}{(x - 3)(x + 3)^2}$

$\qquad = \dfrac{5x^2 + 30x + 45 + 4x^2 - 36 - 2x + 6}{(x - 3)(x + 3)^2}$

$\qquad = \dfrac{9x^2 + 28x + 15}{(x - 3)(x + 3)^2}$

**95.**

$$\frac{x+4}{6x^2-20x}\left(\frac{x}{x^2-x-20}+\frac{2}{x+4}\right)$$

$$=\frac{x+4}{2x(3x-10)}\left(\frac{x}{(x-5)(x+4)}+\frac{2}{x+4}\right)$$

$$=\frac{x+4}{2x(3x-10)}\left(\frac{x+2(x-5)}{(x-5)(x+4)}\right)$$

$$=\frac{x+4}{2x(3x-10)}\left(\frac{x+2x-10}{(x-5)(x+4)}\right)$$

$$=\frac{(x+4)(3x-10)}{2x(3x-10)(x-5)(x+4)}$$

$$=\frac{(x\!\!\!\!/+4)(3x\!\!\!\!/-10)(1)}{2x(3x\!\!\!\!/-10)(x-5)(x\!\!\!\!/+4)}$$

$$=\frac{1}{2x(x-5)}$$

**97.**

$$\frac{8t^5}{2t^2-10t+12}\div\left(\frac{2t}{t^2-8t+15}-\frac{3t}{t^2-7t+10}\right)$$

$$=\frac{8t^5}{2t^2-10t+12}\div\left(\frac{2t}{(t-5)(t-3)}-\frac{3t}{(t-5)(t-2)}\right)$$

$$=\frac{8t^5}{2t^2-10t+12}\div\left(\frac{2t(t-2)-3t(t-3)}{(t-5)(t-3)(t-2)}\right)$$

$$=\frac{8t^5}{2t^2-10t+12}\div\left(\frac{2t^2-4t-3t^2+9t}{(t-5)(t-3)(t-2)}\right)$$

$$=\frac{8t^5}{2t^2-10t+12}\div\frac{-t^2+5t}{(t-5)(t-3)(t-2)}$$

$$=\frac{8t^5}{2(t-3)(t-2)}\cdot\frac{(t-5)(t-3)(t-2)}{-t(t-5)}$$

$$=\frac{2\cdot4\cdot t\cdot t^4(t\!\!\!\!/-5)(t\!\!\!\!/-3)(t\!\!\!\!/-2)}{2(t\!\!\!\!/-3)(t\!\!\!\!/-2)(-1)(t)(t\!\!\!\!/-5)}$$

$$=-4t^4$$

**99.**

## Exercise Set 6.3

**1.** If we multiply by forms of 1 to get a common denominator in the numerator and in the denominator, we get expression (e).

**3.** When we multiply by 1, writing 1 as the LCD of all the rational expressions divided by itself, we get expression (f).

**5.** If we multiply by forms of 1 to get a common denominator in the numerator and in the denominator, we get expression (b).

**7.**

$$\frac{\dfrac{x+2}{x-1}}{\dfrac{x+4}{x-3}}=\frac{x+2}{x-1}\div\frac{x+4}{x-3}$$

$$=\frac{x+2}{x-1}\cdot\frac{x-3}{x+4}$$

$$=\frac{(x+2)(x-3)}{(x-1)(x+4)}$$

**9.**

$$\frac{\dfrac{5}{a}-\dfrac{4}{b}}{\dfrac{2}{a}+\dfrac{3}{b}}=\frac{\dfrac{5}{a}-\dfrac{4}{b}}{\dfrac{2}{a}+\dfrac{3}{b}}\cdot\frac{ab}{ab}$$ Multiplying by 1, using the LCD

$$=\frac{\left(\dfrac{5}{a}-\dfrac{4}{b}\right)ab}{\left(\dfrac{2}{a}+\dfrac{3}{b}\right)ab}$$ Multiplying the numerators and the denominators

$$=\frac{\dfrac{5}{a}\cdot ab-\dfrac{4}{b}\cdot ab}{\dfrac{2}{a}\cdot ab+\dfrac{3}{b}\cdot ab}$$

$$=\frac{5b-4a}{2b+3a}$$

**11.**

$$\frac{\dfrac{3}{z}+\dfrac{2}{y}}{\dfrac{4}{z}-\dfrac{1}{y}}=\frac{\dfrac{3}{z}+\dfrac{2}{y}}{\dfrac{4}{z}-\dfrac{1}{y}}\cdot\frac{zy}{zy}$$ Multiplying by 1, using the LCD

$$=\frac{\dfrac{3}{z}\cdot zy+\dfrac{2}{y}\cdot zy}{\dfrac{4}{z}\cdot zy-\dfrac{1}{y}\cdot zy}$$

$$=\frac{3y+2z}{4y-z}$$

**13.**

$$\frac{\dfrac{a^2-b^2}{ab}}{\dfrac{a-b}{b}}=\frac{a^2-b^2}{ab}\div\frac{a-b}{b}$$

$$=\frac{a^2-b^2}{ab}\cdot\frac{b}{a-b}$$

$$=\frac{(a+b)(a-b)(b)}{a\cdot b\cdot(a-b)}$$

$$=\frac{(a+b)(a\!\!\!\!/-b)(b\!\!\!/)}{a\cdot b\!\!\!/\cdot(a\!\!\!\!/-b)}$$

$$=\frac{a+b}{a}$$

**15.**

$$\frac{1-\dfrac{2}{3x}}{x-\dfrac{4}{9x}}=\frac{1-\dfrac{2}{3x}}{x-\dfrac{4}{9x}}\cdot\frac{9x}{9x}$$ Multiplying by 1, using the LCD

$$=\frac{1\cdot9x-\dfrac{2}{3x}\cdot9x}{x\cdot9x-\dfrac{4}{9x}\cdot9x}$$

$$=\frac{9x-6}{9x^2-4}$$

$$=\frac{3(3x-2)}{(3x+2)(3x-2)}$$

$$=\frac{3(3x\!\!\!\!/-2)}{(3x+2)(3x\!\!\!\!/-2)}$$

$$=\frac{3}{3x+2}$$

**17.** $\dfrac{x^{-1}+y^{-1}}{\dfrac{x^2-y^2}{xy}} = \dfrac{\dfrac{1}{x}+\dfrac{1}{y}}{\dfrac{x^2-y^2}{xy}}$    Rewriting with positive exponents

$\qquad = \dfrac{\dfrac{1}{x}+\dfrac{1}{y}}{\dfrac{x^2-y^2}{xy}} \cdot \dfrac{xy}{xy}$    Multiplying by 1, using the LCD

$\qquad = \dfrac{\dfrac{1}{x}\cdot xy + \dfrac{1}{y}\cdot xy}{\dfrac{x^2-y^2}{xy}\cdot xy}$

$\qquad = \dfrac{y+x}{x^2-y^2} = \dfrac{y+x}{(x+y)(x-y)}$

$\qquad = \dfrac{\cancel{y+x}}{\cancel{(x+y)}(x-y)} = \dfrac{1}{x-y}$

**19.** $\dfrac{\dfrac{1}{a-h}-\dfrac{1}{a}}{h} = \dfrac{\dfrac{1}{a-h}\cdot\dfrac{a}{a} - \dfrac{1}{a}\cdot\dfrac{a-h}{a-h}}{h}$

          Adding in the numerator

$\qquad = \dfrac{\dfrac{a-(a-h)}{a(a-h)}}{h} = \dfrac{\dfrac{a-a+h}{a(a-h)}}{h}$

$\qquad = \dfrac{\dfrac{h}{a(a-h)}}{h} = \dfrac{h}{a(a-h)}\cdot\dfrac{1}{h}$

          Multiplying by the reciprocal of the divisor

$\qquad = \dfrac{\cancel{h}}{a(a-h)}\cdot\dfrac{1}{\cancel{h}}$

$\qquad = \dfrac{1}{a(a-h)}$

**21.** $\dfrac{\dfrac{a^2-4}{a^2+3a+2}}{\dfrac{a^2-5a-6}{a^2-6a-7}}$

$\qquad = \dfrac{a^2-4}{a^2+3a+2}\cdot\dfrac{a^2-6a-7}{a^2-5a-6}$    Multiplying by the reciprocal of the divisor

$\qquad = \dfrac{(a+2)(a-2)}{(a+2)(a+1)}\cdot\dfrac{(a+1)(a-7)}{(a+1)(a-6)}$

$\qquad = \dfrac{(a+2)(a-2)(a+1)(a-7)}{(a+2)(a+1)(a+1)(a-6)}$

$\qquad = \dfrac{\cancel{(a+2)}(a-2)\cancel{(a+1)}(a-7)}{\cancel{(a+2)}\cancel{(a+1)}(a+1)(a-6)}$

$\qquad = \dfrac{(a-2)(a-7)}{(a+1)(a-6)}$

**23.** $\dfrac{\dfrac{x}{x^2+3x-4}-\dfrac{1}{x^2+3x-4}}{\dfrac{x}{x^2+6x+8}+\dfrac{3}{x^2+6x+8}}$

$\qquad = \dfrac{\dfrac{x-1}{x^2+3x-4}}{\dfrac{x+3}{x^2+6x+8}}$    Adding in the numerator and the denominator

$\qquad = \dfrac{x-1}{x^2+3x-4}\cdot\dfrac{x^2+6x+8}{x+3}$

$\qquad = \dfrac{(x-1)(x+4)(x+2)}{(x+4)(x-1)(x+3)}$

$\qquad = \dfrac{\cancel{(x-1)}\cancel{(x+4)}(x+2)}{\cancel{(x+4)}\cancel{(x-1)}(x+3)} = \dfrac{x+2}{x+3}$

**25.** $\dfrac{\dfrac{1}{y}+2}{\dfrac{1}{y}-3} = \dfrac{\dfrac{1}{y}+2}{\dfrac{1}{y}-3}\cdot\dfrac{y}{y}$    Multiplying by 1, using the LCD

$\qquad = \dfrac{\left(\dfrac{1}{y}+2\right)y}{\left(\dfrac{1}{y}-3\right)y}$

$\qquad = \dfrac{\dfrac{1}{y}\cdot y + 2\cdot y}{\dfrac{1}{y}\cdot y - 3\cdot y}$

$\qquad = \dfrac{1+2y}{1-3y}$

**27.** $\dfrac{y+y^{-1}}{y-y^{-1}} = \dfrac{y+\dfrac{1}{y}}{y-\dfrac{1}{y}}$    Rewriting with positive exponents

$\qquad = \dfrac{y+\dfrac{1}{y}}{y-\dfrac{1}{y}}\cdot\dfrac{y}{y}$    Multiplying by 1, using the LCD

$\qquad = \dfrac{y\cdot y + \dfrac{1}{y}\cdot y}{y\cdot y - \dfrac{1}{y}\cdot y}$

$\qquad = \dfrac{y^2+1}{y^2-1}$

(Although the denominator can be factored, doing so does not lead to further simplification.)

**29.**
$$\frac{\dfrac{1}{x-2}+\dfrac{3}{x-1}}{\dfrac{2}{x-1}+\dfrac{5}{x-2}}$$

$$=\frac{\dfrac{1}{x-2}+\dfrac{3}{x-1}}{\dfrac{2}{x-1}+\dfrac{5}{x-2}}\cdot\frac{(x-2)(x-1)}{(x-2)(x-1)}$$

Multiplying by 1, using the LCD

$$=\frac{\dfrac{1}{x-2}\cdot(x-2)(x-1)+\dfrac{3}{x-1}\cdot(x-2)(x-1)}{\dfrac{2}{x-1}\cdot(x-2)(x-1)+\dfrac{5}{x-2}\cdot(x-2)(x-1)}$$

$$=\frac{x-1+3(x-2)}{2(x-2)+5(x-1)}$$

$$=\frac{x-1+3x-6}{2x-4+5x-5}$$

$$=\frac{4x-7}{7x-9}$$

**31.**
$$\frac{a(a+3)^{-1}-2(a-1)^{-1}}{a(a+3)^{-1}-(a-1)^{-1}}$$

$$=\frac{\dfrac{a}{a+3}-\dfrac{2}{a-1}}{\dfrac{a}{a+3}-\dfrac{1}{a-1}}$$

$$=\frac{\dfrac{a}{a+3}-\dfrac{2}{a-1}}{\dfrac{a}{a+3}-\dfrac{1}{a-1}}\cdot\frac{(a+3)(a-1)}{(a+3)(a-1)}$$

Multiplying by 1, using the LCD

$$=\frac{\dfrac{a}{a+3}\cdot(a+3)(a-1)-\dfrac{2}{a-1}\cdot(a+3)(a-1)}{\dfrac{a}{a+3}\cdot(a+3)(a-1)-\dfrac{1}{a-1}\cdot(a+3)(a-1)}$$

$$=\frac{a(a-1)-2(a+3)}{a(a-1)-(a+3)}$$

$$=\frac{a^2-a-2a-6}{a^2-a-a-3}=\frac{a^2-3a-6}{a^2-2a-3}$$

(Although the denominator can be factored, doing so does not lead to further simplification.)

**33.**
$$\frac{\dfrac{2}{a^2-1}+\dfrac{1}{a+1}}{\dfrac{3}{a^2-1}+\dfrac{2}{a-1}}$$

$$=\frac{\dfrac{2}{(a+1)(a-1)}+\dfrac{1}{a+1}}{\dfrac{3}{(a+1)(a-1)}+\dfrac{2}{a-1}}$$

$$=\frac{\dfrac{2}{(a+1)(a-1)}+\dfrac{1}{a+1}}{\dfrac{3}{(a+1)(a-1)}+\dfrac{2}{a-1}}\cdot\frac{(a+1)(a-1)}{(a+1)(a-1)}$$

Multiplying by 1, using the LCD

$$=\frac{\dfrac{2}{(a+1)(a-1)}\cdot(a+1)(a-1)+\dfrac{1}{a+1}\cdot(a+1)(a-1)}{\dfrac{3}{(a+1)(a-1)}\cdot(a+1)(a-1)+\dfrac{2}{a-1}\cdot(a+1)(a-1)}$$

$$=\frac{2+a-1}{3+2(a+1)}=\frac{a+1}{3+2a+2}=\frac{a+1}{2a+5}$$

**35.**
$$\frac{\dfrac{5}{x^2-4}-\dfrac{3}{x-2}}{\dfrac{4}{x^2-4}-\dfrac{2}{x+2}}$$

$$=\frac{\dfrac{5}{(x+2)(x-2)}-\dfrac{3}{x-2}}{\dfrac{4}{(x+2)(x-2)}-\dfrac{2}{x+2}}$$

$$=\frac{\dfrac{5}{(x+2)(x-2)}-\dfrac{3}{x-2}}{\dfrac{4}{(x+2)(x-2)}-\dfrac{2}{x+2}}\cdot\frac{(x+2)(x-2)}{(x+2)(x-2)}$$

Multiplying by 1, using the LCD

$$=\frac{\dfrac{5}{(x+2)(x-2)}\cdot(x+2)(x-2)-\dfrac{3}{x-2}\cdot(x+2)(x-2)}{\dfrac{4}{(x+2)(x-2)}\cdot(x+2)(x-2)-\dfrac{2}{x+2}\cdot(x+2)(x-2)}$$

$$=\frac{5-3(x+2)}{4-2(x-2)}=\frac{5-3x-6}{4-2x+4}=\frac{-1-3x}{8-2x},\text{ or}$$

$$\frac{3x+1}{2x-8}$$

**37.** 
$$\frac{\dfrac{y}{y^2-4}+\dfrac{5}{4-y^2}}{\dfrac{y^2}{y^2-4}+\dfrac{25}{4-y^2}}$$

$$=\frac{\dfrac{y}{y^2-4}+\dfrac{-1}{-1}\cdot\dfrac{5}{4-y^2}}{\dfrac{y^2}{y^2-4}+\dfrac{-1}{-1}\cdot\dfrac{25}{4-y^2}}$$

$$=\frac{\dfrac{y}{y^2-4}-\dfrac{5}{y^2-4}}{\dfrac{y^2}{y^2-4}-\dfrac{25}{y^2-4}}$$

$$=\frac{\dfrac{y-5}{y^2-4}}{\dfrac{y^2-25}{y^2-4}}\quad\text{Adding in the numerator and the denominator}$$

$$=\frac{y-5}{y^2-4}\cdot\frac{y^2-4}{y^2-25}\quad\text{Multiplying by the reciprocal of the divisor}$$

$$=\frac{(y-5)(y^2-4)}{(y^2-4)(y+5)(y-5)}$$

$$=\frac{\cancel{(y-5)}\cancel{(y^2-4)}(1)}{\cancel{(y^2-4)}(y+5)\cancel{(y-5)}}$$

$$=\frac{1}{y+5}$$

**39.** 
$$\frac{\dfrac{y^2}{y^2-25}-\dfrac{y}{y-5}}{\dfrac{y}{y^2-25}-\dfrac{1}{y+5}}$$

$$=\frac{\dfrac{y^2}{(y+5)(y-5)}-\dfrac{y}{y-5}}{\dfrac{y}{(y+5)(y-5)}-\dfrac{1}{y+5}}\cdot\frac{(y+5)(y-5)}{(y+5)(y-5)}$$

Multiplying by 1, using the LCD

$$=\frac{\dfrac{y^2}{(y+5)(y-5)}\cdot(y+5)(y-5)-\dfrac{y}{y-5}\cdot(y+5)(y-5)}{\dfrac{y}{(y+5)(y-5)}\cdot(y+5)(y-5)-\dfrac{1}{y+5}\cdot(y+5)(y-5)}$$

$$=\frac{y^2-y(y+5)}{y-(y-5)}$$

$$=\frac{y^2-y^2-5y}{y-y+5}=\frac{-5y}{5}=-y$$

**41.** 
$$\frac{\dfrac{a}{a+2}+\dfrac{5}{a}}{\dfrac{a}{2a+4}+\dfrac{1}{3a}}$$

$$=\frac{\dfrac{a}{a+2}+\dfrac{5}{a}}{\dfrac{a}{2(a+2)}+\dfrac{1}{3a}}$$

$$=\frac{\dfrac{a}{a+2}+\dfrac{5}{a}}{\dfrac{a}{2(a+2)}+\dfrac{1}{3a}}\cdot\frac{6a(a+2)}{6a(a+2)}$$

Multiplying by 1, using the LCD

$$=\frac{\dfrac{a}{a+2}\cdot6a(a+2)+\dfrac{5}{a}\cdot6a(a+2)}{\dfrac{a}{2(a+2)}\cdot6a(a+2)+\dfrac{1}{3a}\cdot6a(a+2)}$$

$$=\frac{6a^2+30(a+2)}{3a^2+2(a+2)}$$

$$=\frac{6a^2+30a+60}{3a^2+2a+4}$$

(Although the numerator can be factored, doing so does not lead to further simplification.)

**43.** 
$$\frac{\dfrac{1}{x^2-3x+2}+\dfrac{1}{x^2-4}}{\dfrac{1}{x^2+4x+4}+\dfrac{1}{x^2-4}}$$

$$=\frac{\dfrac{1}{(x-1)(x-2)}+\dfrac{1}{(x+2)(x-2)}}{\dfrac{1}{(x+2)(x+2)}+\dfrac{1}{(x+2)(x-2)}}$$

$$=\frac{\dfrac{1}{(x-1)(x-2)}+\dfrac{1}{(x+2)(x-2)}}{\dfrac{1}{(x+2)(x+2)}+\dfrac{1}{(x+2)(x-2)}}\cdot$$

$$\frac{(x-1)(x-2)(x+2)(x+2)}{(x-1)(x-2)(x+2)(x+2)}$$

Multiplying by 1, using the LCD

$$=\frac{(x+2)(x+2)+(x-1)(x+2)}{(x-1)(x-2)+(x-1)(x+2)}$$

$$=\frac{x^2+4x+4+x^2+x-2}{x^2-3x+2+x^2+x-2}$$

$$=\frac{2x^2+5x+2}{2x^2-2x}$$

$$=\frac{(2x+1)(x+2)}{2x(x-1)}$$

**45.** $\dfrac{\dfrac{3}{a^2 - 4a + 3} + \dfrac{3}{a^2 - 5a + 6}}{\dfrac{3}{a^2 - 3a + 2} + \dfrac{3}{a^2 + 3a - 10}}$

$= \dfrac{\dfrac{3}{(a-1)(a-3)} + \dfrac{3}{(a-2)(a-3)}}{\dfrac{3}{(a-1)(a-2)} + \dfrac{3}{(a+5)(a-2)}}$

$= \dfrac{\dfrac{3}{(a-1)(a-3)} + \dfrac{3}{(a-2)(a-3)}}{\dfrac{3}{(a-1)(a-2)} + \dfrac{3}{(a+5)(a-2)}} \cdot$

$\dfrac{(a-1)(a-3)(a-2)(a+5)}{(a-1)(a-3)(a-2)(a+5)}$

Multiplying by 1, using the LCD

$= \dfrac{3(a-2)(a+5) + 3(a-1)(a+5)}{3(a-3)(a+5) + 3(a-1)(a-3)}$

$= \dfrac{3[(a-2)(a+5) + (a-1)(a+5)]}{3[(a-3)(a+5) + (a-1)(a-3)]}$

$= \dfrac{\cancel{3}[(a-2)(a+5) + (a-1)(a+5)]}{\cancel{3}[(a-3)(a+5) + (a-1)(a-3)]}$

$= \dfrac{a^2 + 3a - 10 + a^2 + 4a - 5}{a^2 + 2a - 15 + a^2 - 4a + 3}$

$= \dfrac{2a^2 + 7a - 15}{2a^2 - 2a - 12}$

$= \dfrac{(2a - 3)(a + 5)}{2(a^2 - a - 6)}$

$= \dfrac{(2a - 3)(a + 5)}{2(a - 3)(a + 2)}$

**47.** $\dfrac{\dfrac{y}{y^2 - 4} - \dfrac{2y}{y^2 + y - 6}}{\dfrac{2y}{y^2 + y - 6} - \dfrac{y}{y^2 - 4}}$

Observe that $\dfrac{y}{y^2 - 4} - \dfrac{2y}{y^2 + y - 6} =$

$-\left( \dfrac{2y}{y^2 + y - 6} - \dfrac{y}{y^2 - 4} \right)$. Then, the numerator and denominator are opposites and thus their quotient is $-1$.

**49.** $\dfrac{\dfrac{3}{x^2 + 2x - 3} - \dfrac{1}{x^2 - 3x - 10}}{\dfrac{3}{x^2 - 6x + 5} - \dfrac{1}{x^2 + 5x + 6}}$

$= \dfrac{\dfrac{3}{(x+3)(x-1)} - \dfrac{1}{(x-5)(x+2)}}{\dfrac{3}{(x-5)(x-1)} - \dfrac{1}{(x+3)(x+2)}}$

$= \dfrac{\dfrac{3}{(x+3)(x-1)} - \dfrac{1}{(x-5)(x+2)}}{\dfrac{3}{(x-5)(x-1)} - \dfrac{1}{(x+3)(x+2)}} \cdot$

$\dfrac{(x+3)(x-1)(x-5)(x+2)}{(x+3)(x-1)(x-5)(x+2)}$

Multiplying by 1, using the LCD

$= \dfrac{3(x-5)(x+2) - (x+3)(x-1)}{3(x+3)(x+2) - (x-1)(x-5)}$

$= \dfrac{3(x^2 - 3x - 10) - (x^2 + 2x - 3)}{3(x^2 + 5x + 6) - (x^2 - 6x + 5)}$

$= \dfrac{3x^2 - 9x - 30 - x^2 - 2x + 3}{3x^2 + 15x + 18 - x^2 + 6x - 5}$

$= \dfrac{2x^2 - 11x - 27}{2x^2 + 21x + 13}$

**51.** *Writing Exercise*

**53.** $2(3x - 1) + 5(4x - 3) = 3(2x + 1)$

$6x - 2 + 20x - 15 = 6x + 3$

$26x - 17 = 6x + 3$

$20x - 17 = 3$

$20x = 20$

$x = 1$

The solution is 1.

**55.** $\dfrac{t}{s + y} = r$

$(s + y) \cdot \dfrac{t}{s + y} = r(s + y)$

$t = rs + ry$

$t - rs = ry$

$\dfrac{t - rs}{r} = y$

**57. *Familiarize*.** We let $l$ and $w$ represent the length and width of the second frame. Then $l - 3$ and $w - 4$ represent the length and width of the first frame. The perimeter of the second frame is $2l + 2w$; the perimeter of the first frame is $2(l - 3) + 2(w - 4)$, or $2l + 2w - 14$.

***Translate*.**

| Perimeter of second | = | 2 | · | perimeter of first | − | 1 |
|---|---|---|---|---|---|---|
| ↓ | ↓ | ↓ | ↓ | ↓ | ↓ | ↓ |
| $2l + 2w$ | = | 2 | · | $(2l + 2w - 14)$ | − | 1 |

*Carry out*. We first solve for $2l + 2w$.

$$2l + 2w = 2(2l + 2w - 14) - 1$$
$$2l + 2w = 4l + 4w - 28 - 1$$
$$2l + 2w = 4l + 4w - 29$$
$$29 = 2l + 2w$$

If $2l + 2w = 29$, then $2l + 2w - 14 = 29 - 14$, or 15.

*Check*. The perimeter of the second frame is 1 less than twice the perimeter of the first frame:

$$29 = 2 \cdot 15 - 1$$

The values check.

*State*. The perimeter of the first frame is 15; the perimeter of the second frame is 29.

**59.** *Writing Exercise*

**61.**

$$\frac{5x^{-2} + 10x^{-1}y^{-1} + 5y^{-2}}{3x^{-2} - 3y^{-2}}$$

$$= \frac{\dfrac{5}{x^2} + \dfrac{10}{xy} + \dfrac{5}{y^2}}{\dfrac{3}{x^2} - \dfrac{3}{y^2}}$$

$$= \frac{\dfrac{5}{x^2} + \dfrac{10}{xy} + \dfrac{5}{y^2}}{\dfrac{3}{x^2} - \dfrac{3}{y^2}} \cdot \frac{x^2 y^2}{x^2 y^2}$$

$$= \frac{5y^2 + 10xy + 5x^2}{3y^2 - 3x^2}$$

$$= \frac{5(y^2 + 2xy + x^2)}{3(y^2 - x^2)}$$

$$= \frac{5(y + x)(y + x)}{3(y + x)(y - x)}$$

$$= \frac{5\cancel{(y + x)}(y + x)}{3\cancel{(y + x)}(y - x)}$$

$$= \frac{5(y + x)}{3(y - x)}$$

**63.** Substitute $\dfrac{c}{4}$ for both $v_1$ and $v_2$.

$$\frac{\dfrac{c}{4} + \dfrac{c}{4}}{1 + \dfrac{\dfrac{c}{4} \cdot \dfrac{c}{4}}{c^2}}$$

$$= \frac{\dfrac{2c}{4}}{1 + \dfrac{\dfrac{c^2}{16}}{c^2}}$$

$$= \frac{\dfrac{c}{2}}{1 + \dfrac{c^2}{16} \cdot \dfrac{1}{c^2}}$$

$$= \frac{\dfrac{c}{2}}{1 + \dfrac{1}{16}}$$

$$= \frac{\dfrac{c}{2}}{\dfrac{17}{16}}$$

$$= \frac{c}{2} \cdot \frac{16}{17}$$

$$= \frac{8c}{17}$$

The observed speed is $\dfrac{8c}{17}$, or $\dfrac{8}{17}$ the speed of light.

**65.** $f(x) = \dfrac{3}{x}$, $f(x + h) = \dfrac{3}{x + h}$

$$\frac{f(x + h) - f(x)}{h} = \frac{\dfrac{3}{x + h} - \dfrac{3}{x}}{h}$$

$$= \frac{\dfrac{3x - 3(x + h)}{x(x + h)}}{h}$$

$$= \frac{3x - 3(x + h)}{x(x + h)} \cdot \frac{1}{h}$$

$$= \frac{3x - 3x - 3h}{xh(x + h)}$$

$$= \frac{-3h}{xh(x + h)}$$

$$= \frac{-3\cancel{h}}{x\cancel{h}(x + h)}$$

$$= \frac{-3}{x(x + h)}$$

**67.** $f(x) = \dfrac{2x}{1+x}$, $f(x+h) = \dfrac{2(x+h)}{1+x+h}$

$\dfrac{f(x+h) - f(x)}{h}$

$= \dfrac{\dfrac{2(x+h)}{1+x+h} - \dfrac{2x}{1+x}}{h}$

$= \dfrac{\dfrac{2(x+h)(1+x) - 2x(1+x+h)}{(1+x+h)(1+x)}}{h}$

$= \dfrac{2(x+h)(1+x) - 2x(1+x+h)}{(1+x+h)(1+x)} \cdot \dfrac{1}{h}$

$= \dfrac{2x + 2x^2 + 2h + 2hx - 2x - 2x^2 - 2xh}{(1+x+h)(1+x)h}$

$= \dfrac{2 \cdot \cancel{h}}{(1+x+h)(1+x)\cancel{h}}$

$= \dfrac{2}{(1+x+h)(1+x)}$

**69.** Division by zero occurs in $\dfrac{1}{x^2 - 1}$ when $x = 1$ or $x = -1$.

Division by zero occurs in $\dfrac{1}{x^2 - 16}$ when $x = 4$ or $x = -4$.

To avoid division in the complex fraction we solve:

$\dfrac{1}{9} - \dfrac{1}{x^2 - 16} = 0$

$x^2 - 16 - 9 = 0$   Multiplying by $9(x^2 - 16)$

$x^2 - 25 = 0$

$(x+5)(x-5) = 0$

$x + 5 = 0$   $or$   $x - 5 = 0$

$x = -5$   $or$   $x = 5.$

The domain of $G = \{x | x$ is a real number $and$ $x \neq 1$ $and$ $x \neq -1$ $and$ $x \neq 4$ $and$ $x \neq -4$ $and$ $x \neq 5$ $and$ $x \neq -5\}$.

**71.**     $f(x) = \dfrac{2}{2+x}$

$f(a) = \dfrac{2}{2+a}$

$f(f(a)) = \dfrac{2}{2 + \dfrac{2}{2+a}}$   Note that $a \neq -2, -3$.

$= \dfrac{2}{2 + \dfrac{2}{2+a}} \cdot \dfrac{2+a}{2+a}$

$= \dfrac{2(2+a)}{2(2+a) + \dfrac{2}{2+a} \cdot 2 + a}$

$= \dfrac{4 + 2a}{4 + 2a + 2}$

$= \dfrac{4 + 2a}{6 + 2a} = \dfrac{2(2+a)}{2(3+a)}$

$= \dfrac{\cancel{2}(2+a)}{\cancel{2}(3+a)} = \dfrac{2+a}{3+a}, a \neq -2, -3$

**73.**  $\left[ \dfrac{\dfrac{x+3}{x-3} + 1}{\dfrac{x+3}{x-3} - 1} \right]^4$

$= \left[ \dfrac{\dfrac{x+3}{x-3} + 1}{\dfrac{x+3}{x-3} - 1} \cdot \dfrac{x-3}{x-3} \right]^4$

$= \left[ \dfrac{x+3+x-3}{x+3-x+3} \right]^4$

$= \left( \dfrac{2x}{6} \right)^4 = \left( \dfrac{x}{3} \right)^4 = \dfrac{x^4}{81}$

Division by zero occurs in both the numerator and the denominator of the original fraction when $x = 3$. To avoid division by zero in the complex fraction we solve:

$\dfrac{x+3}{x-3} - 1 = 0$

$\dfrac{x+3}{x-3} = 1$

$x + 3 = x - 3$

$3 = -3$

The equation has no solution, so the denominator of the complex fraction cannot be zero. Thus, the domain of $f = \{x | x$ is a real number $and$ $x \neq 3\}$.

**75.**

**77.**   $\dfrac{30,000 \cdot \dfrac{0.075}{12}}{\left( 1 + \dfrac{0.075}{12} \right)^{120} - 1} = \dfrac{30,000(0.00625)}{(1 + 0.00625)^{120} - 1}$

$= \dfrac{187.5}{(1.00625)^{120} - 1}$

$\approx \dfrac{187.5}{2.112064637 - 1}$

$\approx \dfrac{187.5}{1.112064637}$

$\approx 168.61$

Alexis' monthly investment is \$168.61.

## Exercise Set 6.4

**1.** Equation; see page 2 in the text.

**3.** Expression; see page 2 in the text.

**5.** Equation; see page 2 in the text.

**7.** Equation; see page 2 in the text.

**9.** Expression; see page 2 in the text.

**11.** Note that there is no value of $t$ that makes a denominator 0.

$$\frac{t}{2} + \frac{t}{3} = 7, \text{ LCD is } 6$$

$$6\left(\frac{t}{2} + \frac{t}{3}\right) = 6 \cdot 7$$

$$6 \cdot \frac{t}{2} + 6 \cdot \frac{t}{3} = 6 \cdot 7$$

$$3t + 2t = 42$$

$$5t = 42$$

$$t = \frac{42}{5}$$

Check: 

$$\frac{t}{2} + \frac{t}{3} = 7$$

$$\begin{array}{c|c} \dfrac{42/5}{2} + \dfrac{42/5}{3} & 7 \\ \dfrac{42}{5} \cdot \dfrac{1}{2} + \dfrac{42}{5} \cdot \dfrac{1}{3} & \\ \dfrac{21}{5} + \dfrac{14}{5} & \\ \dfrac{35}{5} & \\ 7 \overset{?}{=} 7 & \text{TRUE} \end{array}$$

The solution is $\frac{42}{5}$.

**13.** Because $\frac{1}{x}$ is undefined when $x$ is 0, at the outset we state the restriction that $x \neq 0$.

$$\frac{5}{8} - \frac{1}{x} = \frac{3}{4}, \text{ LCD is } 8x$$

$$8x\left(\frac{5}{8} - \frac{1}{x}\right) = 8x \cdot \frac{3}{4}$$

$$8x \cdot \frac{5}{8} - 8x \cdot \frac{1}{x} = 8x \cdot \frac{3}{4}$$

$$5x - 8 = 6x$$

$$-8 = x$$

Check: 

$$\frac{5}{8} - \frac{1}{x} = \frac{3}{4}$$

$$\begin{array}{c|c} \dfrac{5}{8} - \dfrac{1}{-8} & \dfrac{3}{4} \\ \dfrac{5}{8} + \dfrac{1}{8} & \\ \dfrac{3}{4} \overset{?}{=} \dfrac{3}{4} & \text{TRUE} \end{array}$$

The solution is $-8$.

**15.** Note that there is no value of $x$ that makes a denominator 0.

$$\frac{2x+1}{3} + \frac{x+2}{5} = \frac{4x}{15}, \text{ LCD is } 15$$

$$15\left(\frac{2x+1}{3} + \frac{x+2}{5}\right) = 15 \cdot \frac{4x}{15}$$

$$15 \cdot \frac{2x+1}{3} + 15 \cdot \frac{x+2}{5} = 15 \cdot \frac{4x}{15}$$

$$5(2x+1) + 3(x+2) = 4x$$

$$10x + 5 + 3x + 6 = 4x$$

$$13x + 11 = 4x$$

$$11 = -9x$$

$$-\frac{11}{9} = x$$

This number checks. The solution is $-\frac{11}{9}$.

**17.** Because $\frac{1}{t}$ and $\frac{7}{3t}$ are undefined when $t$ is 0, at the outset we state the restriction that $t \neq 0$.

$$\frac{2}{3} - \frac{1}{t} = \frac{7}{3t}, \text{ LCD is } 3t$$

$$3t\left(\frac{2}{3} - \frac{1}{t}\right) = 3t \cdot \frac{7}{3t}$$

$$3t \cdot \frac{2}{3} - 3t \cdot \frac{1}{t} = 3t \cdot \frac{7}{3t}$$

$$2t - 3 = 7$$

$$2t = 10$$

$$t = 5$$

Check: 

$$\frac{2}{3} - \frac{1}{t} = \frac{7}{3t}$$

$$\begin{array}{c|c} \dfrac{2}{3} - \dfrac{1}{5} & \dfrac{7}{3 \cdot 5} \\ \dfrac{7}{15} \overset{?}{=} \dfrac{7}{15} & \text{TRUE} \end{array}$$

The solution is 5.

**19.** To ensure that no denominator is 0, at the outset we state the restriction that $x \neq 1$.

$$\frac{4}{x-1} + \frac{5}{6} = \frac{2}{3x-3}$$

$$\frac{4}{x-1} + \frac{5}{6} = \frac{2}{3(x-1)}, \text{ LCD is } 6(x-1)$$

$$6(x-1)\left(\frac{4}{x-1} + \frac{5}{6}\right) = 6(x-1) \cdot \frac{2}{3(x-1)}$$

$$6(x-1) \cdot \frac{4}{x-1} + 6(x-1) \cdot \frac{5}{6} = 6(x-1) \cdot \frac{2}{3(x-1)}$$

$$6 \cdot 4 + 5(x-1) = 2 \cdot 2$$

$$24 + 5x - 5 = 4$$

$$5x + 19 = 4$$

$$5x = -15$$

$$x = -3$$

Check: $\dfrac{4}{x-1} + \dfrac{5}{6} = \dfrac{2}{3x-3}$

$$\begin{array}{c|c}
\dfrac{4}{-3-1} + \dfrac{5}{6} & \dfrac{2}{3(-3)-3} \\
-1 + \dfrac{5}{6} & \dfrac{2}{-12} \\
-\dfrac{1}{6} \overset{?}{=} -\dfrac{1}{6} & \text{TRUE}
\end{array}$$

The solution is $-3$.

**21.** $\dfrac{2}{6} + \dfrac{1}{2x} = \dfrac{1}{3}$

Because $\dfrac{1}{2x}$ is undefined when $x$ is 0, at the outset we state the restriction that $x \neq 0$. Observe that $\dfrac{2}{6}$ is equivalent to $\dfrac{1}{3}$. This means that $\dfrac{1}{2x}$ must be 0 in order for the equation to be true. But there is no value of $x$ for which $\dfrac{1}{2x} = 0$, so the equation has no solution.

**23.** $y + \dfrac{4}{y} = -5$

Because $\dfrac{4}{y}$ is undefined when $y$ is 0, we note at the outset that $y \neq 0$. Then we multiply both sides by the LCD, $y$.

$$y\left(y + \dfrac{4}{y}\right) = y(-5)$$
$$y \cdot y + y \cdot \dfrac{4}{y} = -5y$$
$$y^2 + 4 = -5y$$
$$y^2 + 5y + 4 = 0$$
$$(y+1)(y+4) = 0$$
$$y + 1 = 0 \quad or \quad y + 4 = 0$$
$$y = -1 \quad or \quad y = -4$$

Both values check. The solutions are $-1$ and $-4$.

**25.** Because $\dfrac{12}{x}$ is undefined when $x$ is 0, at the outset we state the restriction that $x \neq 0$.

$$x - \dfrac{12}{x} = 4, \text{ LCD is } x$$
$$x\left(x - \dfrac{12}{x}\right) = x \cdot 4$$
$$x \cdot x - x \cdot \dfrac{12}{x} = x \cdot 4$$
$$x^2 - 12 = 4x$$
$$x^2 - 4x - 12 = 0$$
$$(x-6)(x+2) = 0$$
$$x = 6 \text{ or } x = -2$$

Both numbers check. The solutions are $-2$ and 6.

**27.** $\dfrac{t-1}{t-3} = \dfrac{2}{t-3}$

To ensure that neither denominator is 0, we note at the outset that $t \neq 3$. Then we multiply both sides by the LCD, $t - 3$.

$$(t-3) \cdot \dfrac{t-1}{t-3} = (t-3) \cdot \dfrac{2}{t-3}$$
$$t - 1 = 2$$
$$t = 3$$

Recall that, because of the restriction above, 3 cannot be a solution. A check confirms this.

Check: $\dfrac{t-1}{t-3} = \dfrac{2}{t-3}$

$$\begin{array}{c|c}
\dfrac{3-1}{3-3} & \dfrac{2}{3-3} \\
\dfrac{2}{0} \overset{?}{=} \dfrac{2}{0} & \text{UNDEFINED}
\end{array}$$

The equation has no solution.

**29.** $\dfrac{x}{x-5} = \dfrac{25}{x^2 - 5x}$

$$\dfrac{x}{x-5} = \dfrac{25}{x(x-5)}$$

To ensure that neither denominator is 0, we note at the outset that $x \neq 0$ and $x \neq 5$. Then we multiply both sides by the LCD, $x(x-5)$.

$$x(x-5) \cdot \dfrac{x}{x-5} = x(x-5) \cdot \dfrac{25}{x(x-5)}$$
$$x^2 = 25$$
$$x^2 - 25 = 0$$
$$(x+5)(x-5) = 0$$
$$x = -5 \text{ or } x = 5$$

Recall that, because of the restrictions above, 5 cannot be a solution. The number $-5$ checks and is the solution.

**31.** $\dfrac{5}{4t} = \dfrac{7}{5t-2}$

To ensure that neither denominator is 0, we note at the outset that $t \neq 0$ and $t \neq \dfrac{2}{5}$. Then we multiply both sides by the LCD, $4t(5t-2)$.

$$4t(5t-2) \cdot \dfrac{5}{4t} = 4t(5t-2) \cdot \dfrac{7}{5t-2}$$
$$5(5t-2) = 4t \cdot 7$$
$$25t - 10 = 28t$$
$$-10 = 3t$$
$$-\dfrac{10}{3} = t$$

This value checks. The solution is $-\dfrac{10}{3}$.

**33.** $\dfrac{x^2+4}{x-1} = \dfrac{5}{x-1}$

To ensure that neither denominator is 0, we note at the outset that $x \neq 1$. Then we multiply both sides by the LCD, $x - 1$.

$$(x-1) \cdot \dfrac{x^2+4}{x-1} = (x-1) \cdot \dfrac{5}{x-1}$$
$$x^2 + 4 = 5$$
$$x^2 - 1 = 0$$
$$(x+1)(x-1) = 0$$

$$x + 1 = 0 \quad or \quad x - 1 = 0$$
$$x = -1 \quad or \quad x = 1$$

Recall that, because of the restriction above, 1 cannot be a solution. The number $-1$ checks and is the solution.

We might also observe that since the denominators are the same, the numerators must be the same. Solving $x^2 + 4 = 5$, we get $x = -1$ or $x = 1$ as shown above. Again, because of the restriction $x \neq 1$, only $-1$ is a solution of the equation.

**35.** $\dfrac{6}{a+1} = \dfrac{a}{a-1}$

To ensure that neither denominator is 0, we note at the outset that $a \neq -1$ and $a \neq 1$. Then we multiply both sides by the LCD, $(a+1)(a-1)$.

$$(a+1)(a-1) \cdot \frac{6}{a+1} = (a+1)(a-1) \cdot \frac{a}{a-1}$$
$$6(a-1) = a(a+1)$$
$$6a - 6 = a^2 + a$$
$$0 = a^2 - 5a + 6$$
$$0 = (a-2)(a-3)$$
$$a - 2 = 0 \quad or \quad a - 3 = 0$$
$$a = 2 \quad or \quad a = 3$$

Both values check. The solutions are 2 and 3.

**37.** $\dfrac{60}{t-5} - \dfrac{18}{t} = \dfrac{40}{t}$

To ensure that none of the denominators is 0, we note at the outset that $t \neq 5$ and $t \neq 0$. Then we multiply both sides by the LCD, $t(t-5)$.

$$t(t-5)\left(\frac{60}{t-5} - \frac{18}{t}\right) = t(t-5) \cdot \frac{40}{t}$$
$$60t - 18(t-5) = 40(t-5)$$
$$60t - 18t + 90 = 40t - 200$$
$$2t = -290$$
$$t = -145$$

This value checks. The solution is $-145$.

**39.** $\dfrac{3}{x-3} + \dfrac{5}{x+2} = \dfrac{5x}{x^2 - x - 6}$

$$\frac{3}{x-3} + \frac{5}{x+2} = \frac{5x}{(x-3)(x+2)}$$

To ensure that none of the denominators is 0, we note at the outset that $x \neq 3$ and $x \neq -2$. Then we multiply both sides by the LCD, $(x-3)(x+2)$.

$$(x-3)(x+2)\left(\frac{3}{x-3} + \frac{5}{x+2}\right) = (x-3)(x+2) \cdot \frac{5x}{(x-3)(x+2)}$$
$$3(x+2) + 5(x-3) = 5x$$
$$3x + 6 + 5x - 15 = 5x$$
$$8x - 9 = 5x$$
$$-9 = -3x$$
$$3 = x$$

Recall that, because of the restriction above, 3 cannot be a solution. Thus, the equation has no solution.

**41.** $\dfrac{3}{x} + \dfrac{x}{x+2} = \dfrac{4}{x^2 + 2x}$

$$\frac{3}{x} + \frac{x}{x+2} = \frac{4}{x(x+2)}$$

To ensure that none of the denominators is 0, we note at the outset that $x \neq 0$ and $x \neq -2$. Then we multiply both sides by the LCD, $x(x+2)$.

$$x(x+2)\left(\frac{3}{x} + \frac{x}{x+2}\right) = x(x+2) \cdot \frac{4}{x(x+2)}$$
$$3(x+2) + x \cdot x = 4$$
$$3x + 6 + x^2 = 4$$
$$x^2 + 3x + 2 = 0$$
$$(x+1)(x+2) = 0$$
$$x + 1 = 0 \quad or \quad x + 2 = 0$$
$$x = -1 \quad or \quad x = -2$$

Recall that, because of the restrictions above, $-2$ cannot be a solution. The number $-1$ checks. The solution is $-1$.

**43.** $\dfrac{5}{x+2} - \dfrac{3}{x-2} = \dfrac{2x}{4-x^2}$

$$\frac{5}{x+2} - \frac{3}{x-2} = \frac{2x}{(2+x)(2-x)}$$
$$\frac{5}{x+2} + \frac{3}{2-x} = \frac{2x}{(2+x)(2-x)} \quad \left(-\frac{3}{x-2} = \frac{3}{2-x}\right)$$

First note that $x \neq -2$ and $x \neq 2$. Then multiply both sides by the LCD, $(2+x)(2-x)$.

$$(2+x)(2-x)\left(\frac{5}{x+2} + \frac{3}{2-x}\right) =$$
$$(2+x)(2-x) \cdot \frac{2x}{(2+x)(2-x)}$$
$$5(2-x) + 3(2+x) = 2x$$
$$10 - 5x + 6 + 3x = 2x$$
$$16 - 2x = 2x$$
$$16 = 4x$$
$$4 = x$$

This value checks. The solution is 4.

**45.** $\dfrac{3}{x^2 - 6x + 9} + \dfrac{x-2}{3x-9} = \dfrac{x}{2x-6}$

$$\frac{3}{(x-3)(x-3)} + \frac{x-2}{3(x-3)} = \frac{x}{2(x-3)}$$

Note that $x \neq 3$. Then multiply both sides by the LCD, $6(x-3)(x-3)$.

$$6(x-3)(x-3)\left(\frac{3}{(x-3)(x-3)} + \frac{x-2}{3(x-3)}\right) =$$
$$6(x-3)(x-3) \cdot \frac{x}{2(x-3)}$$
$$6 \cdot 3 + 2(x-3)(x-2) = 3x(x-3)$$
$$18 + 2x^2 - 10x + 12 = 3x^2 - 9x$$
$$0 = x^2 + x - 30$$
$$0 = (x+6)(x-5)$$
$$x = -6 \ or \ x = 5$$

Both values check. The solutions are $-6$ and 5.

**47.** We find all values of $a$ for which $2a - \dfrac{15}{a} = 7$. First note that $a \neq 0$. Then multiply both sides by the LCD, $a$.

$$a\left(2a - \frac{15}{a}\right) = a \cdot 7$$

$$a \cdot 2a - a \cdot \frac{15}{a} = 7a$$

$$2a^2 - 15 = 7a$$

$$2a^2 - 7a - 15 = 0$$

$$(2a + 3)(a - 5) = 0$$

$$a = -\frac{3}{2} \ or \ a = 5$$

Both values check. The solutions are $-\dfrac{3}{2}$ and 5.

**49.** We find all values of $a$ for which $\dfrac{a-5}{a+1} = \dfrac{3}{5}$. First note that $a \neq -1$. Then multiply both sides by the LCD, $5(a + 1)$.

$$5(a+1) \cdot \frac{a-5}{a+1} = 5(a+1) \cdot \frac{3}{5}$$

$$5(a-5) = 3(a+1)$$

$$5a - 25 = 3a + 3$$

$$2a = 28$$

$$a = 14$$

This value checks. The solution is 14.

**51.** We find all values of $a$ for which $\dfrac{12}{a} - \dfrac{12}{2a} = 8$. First note that $a \neq 0$. Then multiply both sides by the LCD, $2a$.

$$2a\left(\frac{12}{a} - \frac{12}{2a}\right) = 2a \cdot 8$$

$$2a \cdot \frac{12}{a} - 2a \cdot \frac{12}{2a} = 16a$$

$$24 - 12 = 16a$$

$$12 = 16a$$

$$\frac{3}{4} = a$$

This value checks. The solution is $\dfrac{3}{4}$.

**53.**
$$f(a) = g(a)$$

$$\frac{3a-1}{a^2 - 7a + 10} = \frac{a-1}{a^2-4} + \frac{2a+1}{a^2-3a-10}$$

$$\frac{3a-1}{(a-2)(a-5)} = \frac{a-1}{(a+2)(a-2)} + \frac{2a+1}{(a-5)(a+2)}$$

First note that $a \neq 2$, $a \neq 5$, and $a \neq -2$. Then multiply both sides by the LCD, $(a - 2)(a - 5)(a + 2)$.

$$(a-2)(a-5)(a+2) \cdot \frac{3a-1}{(a-2)(a-5)} =$$

$$(a-2)(a-5)(a+2)\left(\frac{a-1}{(a+2)(a-2)} + \frac{2a+1}{(a-5)(a+2)}\right)$$

$$(a+2)(3a-1) = (a-5)(a-1) + (a-2)(2a+1)$$

$$3a^2 + 5a - 2 = a^2 - 6a + 5 + 2a^2 - 3a - 2$$

$$3a^2 + 5a - 2 = 3a^2 - 9a + 3$$

$$5a - 2 = -9a + 3$$

$$14a - 2 = 3$$

$$14a = 5$$

$$a = \frac{5}{14}$$

This number checks. Then $f(a) = g(a)$ for $a = \dfrac{5}{14}$.

**55.**
$$f(a) = g(a)$$

$$\frac{2}{a^2 - 8a + 7} = \frac{3}{a^2 - 2a - 3} - \frac{1}{a^2 - 1}$$

$$\frac{2}{(a-1)(a-7)} = \frac{3}{(a+1)(a-3)} - \frac{1}{(a+1)(a-1)}$$

First note that $a \neq 1$, $a \neq 7$, $a \neq -1$, and $a \neq 3$. Then multiply both sides by the LCD, $(a-1)(a-7)(a+1)(a-3)$.

$$(a-1)(a-7)(a+1)(a-3) \cdot \frac{2}{(a-1)(a-7)} =$$

$$(a-1)(a-7)(a+1)(a-3)\left(\frac{3}{(a+1)(a-3)} - \frac{1}{(a+1)(a-1)}\right)$$

$$2(a+1)(a-3) = 3(a-1)(a-7) - (a-7)(a-3)$$

$$2(a^2 - 2a - 3) = 3(a^2 - 8a + 7) - (a^2 - 10a + 21)$$

$$2a^2 - 4a - 6 = 3a^2 - 24a + 21 - a^2 + 10a - 21$$

$$2a^2 - 4a - 6 = 2a^2 - 14a$$

$$-4a - 6 = -14a$$

$$-6 = -10a$$

$$\frac{3}{5} = a$$

This number checks. Then $f(a) = g(a)$ for $a = \dfrac{3}{5}$.

**57.** *Writing Exercise*

**59. Familiarize.** Let $x$, $y$, and $z$ represent the number of multiple-choice, true-false and fill-in questions, respectively.

**Translate.** The total number of questions is 70.

$$x + y + z = 70$$

The number of true-false is twice the number of fill-ins.

$$y = 2z$$

The number of multiple-choice is 5 less than the number of true-false.

$$x = y - 5$$

**Carry out.** Solving the system of three equations we get $(25, 30, 15)$.

**Check.** The sum of 25, 30, and 15 is 70. The number of true-false, 30, is twice the number of fill-ins, 15. The number of multiple-choice, 25, is 5 less than the number of true-false, 30.

**State**. On the test there are 25 multiple-choice, 30 true-false, and 15 fill-in questions.

**61.** *Familiarize*. Let $l$ represent the length and $w$ represent the width, in meters. Recall that perimeter $P = 2l + 2w$, and area $A = lw$.

*Translate*.

$$\underbrace{\text{The perimeter}}_{2l + 2w} \text{ } \underbrace{\text{is}}_{=} \text{ } \underbrace{628 \text{ m.}}_{628}$$

$$\underbrace{\text{The length}}_{l} \text{ } \underbrace{\text{is}}_{=} \text{ } \underbrace{6 \text{ m}}_{6} \text{ } \underbrace{\text{more than}}_{+} \text{ } \underbrace{\text{the width.}}_{w}$$

We have a system of equations:

$$2l + 2w = 628, \quad (1)$$
$$l = 6 + w \qquad (2)$$

*Carry out*. Use the substitution method to solve the system of equations and then compute the area. Substitute $6 + w$ for $l$ in (1).

$$2(6 + w) + 2w = 628$$
$$12 + 2w + 2w = 628$$
$$12 + 4w = 628$$
$$4w = 616$$
$$w = 154$$

Substitute 154 for $w$ in (2).

$$l = 6 + 154 = 160$$
$$A = lw = 160(154) = 24,640$$

**Check**. $2l + 2w = 2 \cdot 160 + 2 \cdot 154 = 320 + 308 = 628$. 160 is 6 more than 154. The numbers check.

**State**. The area is 24,640 m².

**63.** a) $\quad 2x - 3y = 4, \quad (1)$
$\qquad 4x - 6y = 7 \quad (2)$

Multiply Equation (1) by $-2$ and add.

$$-4x + 6y = -8$$
$$\underline{4x - 6y = \phantom{-}7}$$
$$0 = -1$$

We get a false equation, so the system has no solution. It is inconsistent.

b) $\quad x + 3y = 2, \quad (1)$
$\qquad 2x - 3y = 1 \quad (2)$

First we add the equations.

$$x + 3y = 2$$
$$\underline{2x - 3y = 1}$$
$$3x \qquad = 3$$
$$x = 1$$

Now substitute 1 for $x$ in one of the equations and solve for $y$.

$$x + 3y = 2 \quad (1)$$
$$1 + 3y = 2$$
$$3y = 1$$
$$y = \frac{1}{3}$$

The solution is $\left(1, \dfrac{1}{3}\right)$. The system is consistent.

**65.** *Writing Exercise*

**67.** $\qquad f(a) = g(a)$

$$\frac{a - \dfrac{2}{3}}{a + \dfrac{1}{2}} = \frac{a + \dfrac{2}{3}}{a - \dfrac{3}{2}}$$

$$\frac{a - \dfrac{2}{3}}{a + \dfrac{1}{2}} \cdot \frac{6}{6} = \frac{a + \dfrac{2}{3}}{a - \dfrac{3}{2}} \cdot \frac{6}{6}$$

$$\frac{6a - \dfrac{2}{3} \cdot 6}{6a + \dfrac{1}{2} \cdot 6} = \frac{6a + \dfrac{2}{3} \cdot 6}{6a - \dfrac{3}{2} \cdot 6}$$

$$\frac{6a - 4}{6a + 3} = \frac{6a + 4}{6a - 9}$$

$$\frac{6a - 4}{3(2a + 1)} = \frac{6a + 4}{3(2a - 3)}$$

To ensure that neither denominator is 0, we note at the outset that $a \neq -\dfrac{1}{2}$ and $a \neq \dfrac{3}{2}$. Then we multiply both sides by the LCD, $3(2a + 1)(2a - 3)$.

$$3(2a + 1)(2a - 3) \cdot \frac{6a - 4}{3(2a + 1)} =$$
$$\qquad\qquad 3(2a + 1)(2a - 3) \cdot \frac{6a + 4}{3(2a - 3)}$$
$$(2a - 3)(6a - 4) = (2a + 1)(6a + 4)$$
$$12a^2 - 26a + 12 = 12a^2 + 14a + 4$$
$$-26a + 12 = 14a + 4$$
$$-40a + 12 = 4$$
$$-40a = -8$$
$$a = \frac{1}{5}$$

This number checks. For $a = \dfrac{1}{5}$, $f(a) = g(a)$.

**69.** $\dfrac{a + 3}{a + 2} - \dfrac{a + 4}{a + 3} = \dfrac{a + 5}{a + 4} - \dfrac{a + 6}{a + 5}$

Note that $a \neq -2$ and $a \neq -3$ and $a \neq -4$ and $a \neq -5$.

$$(a + 2)(a + 3)(a + 4)(a + 5)\left(\frac{a + 3}{a + 2} - \frac{a + 4}{a + 3}\right) =$$
$$(a + 2)(a + 3)(a + 4)(a + 5)\left(\frac{a + 5}{a + 4} - \frac{a + 6}{a + 5}\right)$$

$$(a+3)(a+4)(a+5)(a+3) - (a+2)(a+4)(a+5)(a+4) =$$
$$(a+2)(a+3)(a+5)(a+5) - (a+2)(a+3)(a+4)(a+6)$$

$a^4+15a^3+83a^2+201a+180-$

$\quad(a^4+15a^3+82a^2+192a+160)=$

$a^4+15a^3+81a^2+185a+150-$

$\quad(a^4+15a^3+80a^2+180a+144)$

$a^2+9a+20=a^2+5a+6$

$4a=-14$

$a=-\dfrac{7}{2}$

This value checks. When $a=-\dfrac{7}{2}$, $f(a)=g(a)$.

**71.** Set $f(a)$ equal to $g(a)$ and solve for $a$.

$$\frac{0.793}{a}+18.15=\frac{6.034}{a}-43.17$$

Note that $a\neq 0$. Then multiply on both sides by the LCD, $a$.

$$a\left(\frac{0.793}{a}+18.15\right)=a\left(\frac{6.034}{a}-43.17\right)$$

$$0.793+18.15a=6.034-43.17a$$

$$61.32a=5.241$$

$$a\approx 0.0854697$$

This value checks. When $a\approx 0.0854697$, $f(a)=g(a)$.

**73.** $\dfrac{x^2+6x-16}{x-2}=x+8, x\neq 2$

$\dfrac{(x+8)(x-2)}{x-2}=x+8$

$\dfrac{(x+8)(x-2)}{x-2}=x+8$

$x+8=x+8$

$8=8$

Since $8=8$ is true for all values of $x$, the original equation is true for any possible replacements of the variable. It is an identity.

**75.**

**77.** Let $y_1=\dfrac{x^2-4}{x-2}$ and observe that for $x=2$ the entry in the Y1-column of the table is "ERROR."

---

## Exercise Set 6.5

**1.** *Familiarize.* Let $x=$ the number.

*Translate.*

| The reciprocal of 3 | plus | the reciprocal of 6 | is | the reciprocal of the number. |
|---|---|---|---|---|
| ↓ | ↓ | ↓ | ↓ | ↓ |
| $\dfrac{1}{3}$ | $+$ | $\dfrac{1}{6}$ | $=$ | $\dfrac{1}{x}$ |

*Carry out.* We solve the equation.

$\dfrac{1}{3}+\dfrac{1}{6}=\dfrac{1}{x}$, LCD is $6x$

$6x\left(\dfrac{1}{3}+\dfrac{1}{6}\right)=6x\cdot\dfrac{1}{x}$

$2x+x=6$

$3x=6$

$x=2$

*Check.* $\dfrac{1}{3}+\dfrac{1}{6}=\dfrac{2}{6}+\dfrac{1}{6}=\dfrac{3}{6}=\dfrac{1}{2}$. This is the reciprocal of 2, so the result checks.

*State.* The number is 2.

**3.** *Familiarize.* We let $x=$ the number.

*Translate.*

| A number | plus | 6 | times | its reciprocal | is | $-5$. |
|---|---|---|---|---|---|---|
| ↓ | ↓ | ↓ | ↓ | ↓ | ↓ | ↓ |
| $x$ | $+$ | $6$ | $\cdot$ | $\dfrac{1}{x}$ | $=$ | $-5$ |

*Carry out.* We solve the equation.

$x+\dfrac{6}{x}=-5$, LCD is $x$

$x\left(x+\dfrac{6}{x}\right)=x(-5)$

$x^2+6=-5x$

$x^2+5x+6=0$

$(x+3)(x+2)=0$

$x=-3$ *or* $x=-2$

*Check.* The possible solutions are $-3$ and $-2$. We check $-3$ in the conditions of the problem.

| Number: | $-3$ |
|---|---|
| 6 times the reciprocal of the number: | $6\left(-\dfrac{1}{3}\right)=-2$ |
| Sum of the number and 6 times its reciprocal: | $-3+(-2)=-5$ |

The number $-3$ checks.

Now we check $-2$:

| Number: | $-2$ |
|---|---|
| 6 times the reciprocal of the number: | $6\left(-\dfrac{1}{2}\right)=-3$ |
| Sum of the number and 6 times its reciprocal: | $-2+(-3)=-5$ |

The number $-2$ also checks.

*State.* The number is $-3$ or $-2$.

**5.** *Familiarize.* We let $x=$ the first integer. Then $x+1=$ the second, and their product $=x(x+1)$.

*Translate.*

| Reciprocal of the product | is | $\dfrac{1}{42}$. |
|---|---|---|
| ↓ | ↓ | ↓ |
| $\dfrac{1}{x(x+1)}$ | $=$ | $\dfrac{1}{42}$ |

*Carry out.* We solve the equation.

$$\frac{1}{x(x+1)} = \frac{1}{42}, \text{ LCD is } 42x(x+1)$$

$$42x(x+1) \cdot \frac{1}{x(x+1)} = 42x(x+1) \cdot \frac{1}{42}$$

$$42 = x(x+1)$$

$$42 = x^2 + x$$

$$0 = x^2 + x - 42$$

$$0 = (x+7)(x-6)$$

$$x = -7 \text{ or } x = 6$$

**Check**. When $x = -7$, then $x+1 = -6$ and $-7(-6) = 42$. The reciprocal of this product is $\frac{1}{42}$.

When $x = 6$, then $x+1 = 7$ and $6 \cdot 7 = 42$. The reciprocal of this product is also $\frac{1}{42}$. Both possible solutions check.

**State**. The integers are $-7$ and $-6$ or $6$ and $7$.

7. **Familiarize**. The job takes Nadine 5 hours working alone and Willy 9 hours working alone. Then in 1 hour, Nadine does $\frac{1}{5}$ of the job and Willy does $\frac{1}{9}$ of the job. Working together, they can do $\frac{1}{5} + \frac{1}{9}$ of the job in 1 hour. Let $t$ represent the number of hours required for Nadine and Willy, working together, to do the job.

**Translate**. We want to find $t$ such that
$$t\left(\frac{1}{5}\right) + t\left(\frac{1}{9}\right) = 1, \text{ or } \frac{t}{5} + \frac{t}{9} = 1,$$
where 1 represents one entire job.

**Carry out**. We solve the equation.
$$\frac{t}{5} + \frac{t}{9} = 1, \text{ LCD is } 45$$

$$45\left(\frac{t}{5} + \frac{t}{9}\right) = 45 \cdot 1$$

$$9t + 5t = 45$$

$$14t = 45$$

$$t = \frac{45}{14}$$

**Check**. In $\frac{45}{14}$ hours, Nadine will do $\frac{1}{5} \cdot \frac{45}{14}$, or $\frac{9}{14}$ of the job and Willy will do $\frac{1}{9} \cdot \frac{45}{14}$, or $\frac{5}{14}$ of the job. Together, they do $\frac{9}{14} + \frac{5}{14}$, or 1 entire job. The answer checks.

**State**. It will take $\frac{45}{14}$ hr, or $3\frac{3}{14}$ hr, for Nadine and Willy, together, to refinish the floor.

9. **Familiarize**. The pool can be filled in 12 hours with only the pipe and in 30 hours with only the hose. Then in 1 hour, the pipe fills $\frac{1}{12}$ of the pool, and the hose fills $\frac{1}{30}$ of the pool. Using both the pipe and the hose, $\frac{1}{12} + \frac{1}{30}$ of the pool can be filled in 1 hour.

Suppose that it takes $t$ hours to fill the pool using both the pipe and hose.

**Translate**. We want to find $t$ such that

$$t\left(\frac{1}{12}\right) + t\left(\frac{1}{30}\right) = 1, \text{ or } \frac{t}{12} + \frac{t}{30} = 1,$$
where 1 represents one entire job.

**Carry out**. We solve the equation. We multiply on both sides by the LCD, $60t$.
$$60\left(\frac{t}{12} + \frac{t}{30}\right) = 60 \cdot 1$$

$$5t + 2t = 60$$

$$7t = 60$$

$$t = \frac{60}{7}$$

**Check**. The possible solution is $\frac{60}{7}$ hours. If the pipe is used $\frac{60}{7}$ hours, it fills $\frac{1}{12} \cdot \frac{60}{7}$, or $\frac{5}{7}$ of the pool. If the hose is used $\frac{60}{7}$ hours, it fills $\frac{1}{30} \cdot \frac{60}{7}$, or $\frac{2}{7}$ of the pool. Using both, $\frac{5}{7} + \frac{2}{7}$ of the pool, or all of it, will be filled in $\frac{60}{7}$ hours.

**State**. Using both the pipe and the hose, it will take $\frac{60}{7}$ hours, or $8\frac{4}{7}$ hours, to fill the pool.

11. **Familiarize**. In 1 minute the Craftsman pump does $\frac{1}{44}$ of the job and the Simer pump does $\frac{1}{36}$ of the job. Working together, they do $\frac{1}{44} + \frac{1}{36}$ of the job in 1 minute. Suppose it takes $t$ minutes to do the job working together.

**Translate**. We find $t$ such that
$$t\left(\frac{1}{44}\right) + t\left(\frac{1}{36}\right) = 1, \text{ or } \frac{t}{44} + \frac{t}{36} = 1.$$

**Carry out**. We solve the equation. We multiply both sides by the LCD, 396.
$$396\left(\frac{t}{44} + \frac{t}{36}\right) = 396 \cdot 1$$

$$9t + 11t = 396$$

$$20t = 396$$

$$t = \frac{99}{5}$$

**Check**. In $\frac{99}{5}$ min the Craftsman pump does $\frac{99}{5} \cdot \frac{1}{44}$, or $\frac{9}{20}$ of the job, and the Simer pump does $\frac{99}{5} \cdot \frac{1}{36}$, or $\frac{11}{20}$ of the job. Together they do $\frac{9}{20} + \frac{11}{20}$, or 1 entire job. The answer checks.

**State**. The two pumps can pump out the basement in $\frac{99}{5}$ min, or $19\frac{4}{5}$ min, working together.

13. **Familiarize**. Let $t$ represent the time, in minutes, that it takes the Canon copier to copy the brochures, working alone. Then $2t$ represents the time it takes the HP copier to do the job, working alone. In 1 minute the Canon copier does $\frac{1}{t}$ and the HP copier does $\frac{1}{2t}$ of the job.

*Translate*. Working together, they can do the entire job in 24 min, so we want to find $t$ such that

$$24\left(\frac{1}{t}\right) + 24\left(\frac{1}{2t}\right) = 1, \text{ or } \frac{24}{t} + \frac{12}{t} = 1.$$

*Carry out*. We solve the equation.

$$\frac{24}{t} + \frac{12}{t} = 1, \text{ LCD is } t$$

$$t\left(\frac{24}{t} + \frac{12}{t}\right) = t \cdot 1$$

$$24 + 12 = t$$

$$36 = t$$

*Check*. If the Canon copier can do the job in 36 min, then in 24 min it does $24 \cdot \frac{1}{36}$, or $\frac{2}{3}$ of the job. If it takes the HP copier $2 \cdot 36$, or 72 min, to do the job, then in 24 min it does $24 \cdot \frac{1}{72}$, or $\frac{1}{3}$ of the job. Working together the two machines do $\frac{2}{3} + \frac{1}{3}$, or 1 entire job, in 24 min.

*State*. Working alone, it takes the Canon copier 36 min and the HP copier 72 min to copy the brochure.

15. *Familiarize*. Let $t$ represent the number of minutes it takes the Panasonic machine to purify the air working alone. Then $t - 10$ represents the time it takes the Blueair machine to purify the air, working alone. In 1 minute the Panasonic does $\frac{1}{t}$ of the job and the Blueair does $\frac{1}{t - 10}$ of the job.

*Translate*. Working together, the two machines can purify the air in $\frac{120}{7}$ min, so we want to find $t$ such that

$$\frac{120}{7}\left(\frac{1}{t}\right) + \frac{120}{7}\left(\frac{1}{t - 10}\right) = 1.$$

*Carry out*. We solve the equation. First we multiply both sides by the LCD, $7t(t - 10)$.

$$7t(t-10)\left(\frac{120}{7}\left(\frac{1}{t}\right) + \frac{120}{7}\left(\frac{1}{t-10}\right)\right) = 7t(t-10) \cdot 1$$

$$120(t - 10) + 120t = 7t(t - 10)$$

$$120t - 1200 + 120t = 7t^2 - 70t$$

$$240t - 1200 = 7t^2 - 70t$$

$$0 = 7t^2 - 310t + 1200$$

$$0 = (7t - 30)(t - 40)$$

$$t = \frac{30}{7} \text{ or } t - 40$$

*Check*. If $t = \frac{30}{7}$, then $t - 10 = \frac{30}{7} - 10 = -\frac{40}{7}$. Since negative time has no meaning in this application, $\frac{30}{7}$ cannot be a solution. If $t = 40$, then $t - 10 = 40 - 10 = 30$. In $\frac{120}{7}$ min the Panasonic machine does $\frac{120}{7} \cdot \frac{1}{40}$, or $\frac{3}{7}$ of the job and the Blueair machine does $\frac{120}{7} \cdot \frac{1}{30}$, or $\frac{4}{7}$ of the job. Together they do $\frac{3}{7} + \frac{4}{7}$, or 1 entire job. The answer checks.

*State*. Working alone, the Panasonic machine can purify the air in 40 min and the Blueair machine can purify the air in 30 min.

17. *Familiarize*. Let $t$ represent the number of hours it takes the Erickson helicopter to douse the fire, working alone. Then $4t$ represents the time it takes the S-58T helicopter to douse the fire. In 1 hour the Erickson does $\frac{1}{t}$ of the job and the S-58T does $\frac{1}{4t}$ of the job.

*Translate*. Working together, the two helicopters can douse the fire in 8 hours, so we want to find $t$ such that

$$8\left(\frac{1}{t}\right) + 8\left(\frac{1}{4t}\right) = 1.$$

*Carry out*. We solve the equation.

$$8\left(\frac{1}{t}\right) + 8\left(\frac{1}{4t}\right) = 1$$

$$\frac{8}{t} + \frac{2}{t} = 1$$

$$\frac{10}{t} = 1, \text{ LCD is } t$$

$$t\left(\frac{10}{t}\right) = t \cdot 1$$

$$10 = t$$

Then $4t = 4 \cdot 10 = 40$.

*Check*. In 8 hr the Erickson does $8 \cdot \frac{1}{10}$, or $\frac{4}{5}$ of the job, working alone, and the S-58T does $8 \cdot \frac{1}{40}$, or $\frac{1}{5}$ of the job. Working together, they do $\frac{4}{5} + \frac{1}{5}$, or 1 entire job. The answer checks.

*State*. The Erickson helicopter can douse the fire in 10 hr working alone and the S-58T helicopter can do the same job in 40 hr.

19. *Familiarize*. Let $t$ represent the number of hours it takes Zsuzanna to deliver the papers alone. Then $3t$ represents the number of hours it takes Stan to deliver the papers alone.

*Translate*. In 1 hr Zsuzanna and Stan will do one entire job, so we have

$$1\left(\frac{1}{t}\right) + 1\left(\frac{1}{3t}\right) = 1, \text{ or } \frac{1}{t} + \frac{1}{3t} = 1.$$

*Carry out*. We solve the equation. Multiply on both sides by the LCD, $3t$.

$$3t\left(\frac{1}{t} + \frac{1}{3t}\right) = 3t \cdot 1$$

$$3 + 1 = 3t$$

$$4 = 3t$$

$$\frac{4}{3} = t$$

*Check*. If Zsuzanna does the job alone in $\frac{4}{3}$ hr, then in 1 hr she does $\frac{1}{4/3}$, or $\frac{3}{4}$ of the job. If Stan does the job alone in $3 \cdot \frac{4}{3}$, or 4 hr, then in 1 hr he does $\frac{1}{4}$ of the job. Together, they do $\frac{3}{4} + \frac{1}{4}$, or 1 entire job, in 1 hr. The result checks.

***State***. It would take Zsuzanna $\frac{4}{3}$ hours and it would take Stan 4 hours to deliver the papers alone.

**21.** ***Familiarize***. Let $t$ represent the number of hours it would take Mo to pave the driveway, working alone. Then $t + 4$ represents the time it would take Larry to do the job. In 1 hr Mo does $\frac{1}{t}$ of the job and Larry does $\frac{1}{t+4}$.

We convert minutes to hours:

$$48 \text{ min} = 48 \text{ min} \cdot \frac{1 \text{ hr}}{60 \text{ min}} = \frac{48}{60} \text{ hr} = 0.8 \text{ hr}$$

Then 4 hr 48 min is 4.8 hr.

***Translate***. In 4.8 hr they do 1 entire job working together, so we have

$$4.8\left(\frac{1}{t}\right) + 4.8\left(\frac{1}{t+4}\right) = 1, \text{ or } \frac{4.8}{t} + \frac{4.8}{t+4} = 1.$$

***Carry out***. We solve the equation. First we multiply both sides by the LCD, $t(t+4)$.

$$t(t+4)\left(\frac{4.8}{t} + \frac{4.8}{t+4}\right) = t(t+4) \cdot 1$$
$$4.8(t+4) + 4.8t = t(t+4)$$
$$4.8t + 19.2 + 4.8t = t^2 + 4t$$
$$9.6t + 19.2 = t^2 + 4t$$
$$0 = t^2 - 5.6t - 19.2$$
$$0 = 10t^2 - 56t - 192$$

Multiplying by 10

$$0 = 2(5t^2 - 28t - 96)$$
$$0 = 2(t-8)(5t+12)$$

$$t = 8 \ \text{ or } \ t = -\frac{12}{5}$$

***Check***. Since negative time has no meaning in this application, $-\frac{12}{5}$ cannot be a solution. In 8 hr, Mo does $4.8\left(\frac{1}{8}\right)$, or 0.6 of the job working alone, and Larry does $4.8\left(\frac{1}{8+4}\right)$, or $4.8\left(\frac{1}{12}\right)$, or 0.4 of the job. Together they do $0.6 + 0.4$, or 1 entire job. The answer checks.

***State***. It takes Mo 8 hr to pave the driveway working alone.

**23.** ***Familiarize***. We first make a drawing. Let $r =$ the kayak's speed in still water in mph. Then $r - 3 =$ the speed upstream and $r + 3 =$ the speed downstream.

$$\text{Upstream} \quad 4 \text{ miles} \quad r - 3 \text{ mph}$$
$$10 \text{ miles} \quad r + 3 \text{ mph} \quad \text{Downstream}$$

We organize the information in a table. The time is the same both upstream and downstream so we use $t$ for each time.

|  | Distance | Speed | Time |
|---|---|---|---|
| Upstream | 4 | $r - 3$ | $t$ |
| Downstream | 10 | $r + 3$ | $t$ |

***Translate***. Using the formula Time = Distance/Rate in each row of the table and the fact that the times are the same, we can write an equation.

$$\frac{4}{r-3} = \frac{10}{r+3}$$

***Carry out***. We solve the equation.

$$\frac{4}{r-3} = \frac{10}{r+3}, \text{ LCD is } (r-3)(r+3)$$
$$(r-3)(r+3) \cdot \frac{4}{r-3} = (r-3)(r+3) \cdot \frac{10}{r+3}$$
$$4(r+3) = 10(r-3)$$
$$4r + 12 = 10r - 30$$
$$42 = 6r$$
$$7 = r$$

***Check***. If $r = 7$ mph, then $r - 3$ is 4 mph and $r + 3$ is 10 mph. The time upstream is $\frac{4}{4}$, or 1 hour. The time downstream is $\frac{10}{10}$, or 1 hour. Since the times are the same, the answer checks.

***State***. The speed of the kayak in still water is 7 mph.

**25.** ***Familiarize***. We first make a drawing. Let $r =$ Benny's speed on a nonmoving sidewalk in ft/sec. Then his speed moving forward on the moving sidewalk is $r + 1.7$, and his speed in the opposite direction is $r - 1.7$.

$$\text{Forward} \quad r + 1.7 \quad 120 \text{ ft}$$
$$52 \text{ ft} \quad r - 1.7 \quad \text{Opposite direction}$$

We organize the information in a table. The time is the same both forward and in the opposite direction so we use $t$ for each time.

|  | Distance | Speed | Time |
|---|---|---|---|
| Forward | 120 | $r + 1.7$ | $t$ |
| Opposite direction | 52 | $r - 1.7$ | $t$ |

***Translate***. Using the formula Time = Distance/Rate in each row of the table and the fact that the times are the same, we can write an equation.

$$\frac{120}{r+1.7} = \frac{52}{r-1.7}$$

***Carry out***. We solve the equation.

$$\frac{120}{r+1.7} = \frac{52}{r-1.7},$$
$$\text{LCD is } (r+1.7)(r-1.7)$$
$$(r+1.7)(r-1.7) \cdot \frac{120}{r+1.7} = (r+1.7)(r-1.7) \cdot \frac{52}{r-1.7}$$
$$120(r - 1.7) = 52(r + 1.7)$$
$$120r - 204 = 52r + 88.4$$
$$68r = 292.4$$
$$r = 4.3$$

***Check***. If Benny's speed on a nonmoving sidewalk is 4.3 ft/sec, then his speed moving forward on the moving sidewalk is $4.3 + 1.7$, or 6 ft/sec, and his speed moving in the opposite direction on the sidewalk is $4.3 - 1.7$, or 2.6 ft/sec. Moving 120 ft at 6 ft/sec takes $\frac{120}{6} = 20$ sec.

Moving 52 ft at 2.6 ft/sec takes $\frac{52}{2.6} = 20$ sec. Since the times are the same, the answer checks.

**State**. Benny would be walking 4.3 ft/sec on a nonmoving sidewalk.

**27. Familiarize**. Let $r$ = the speed of the passenger train in mph. Then $r - 14$ = the speed of the freight train in mph. We organize the information in a table. The time is the same for both trains so we use $t$ for each time.

|  | Distance | Speed | Time |
|---|---|---|---|
| Passenger train | 400 | $r$ | $t$ |
| Freight train | 330 | $r - 14$ | $t$ |

**Translate**. Using the formula Time = Distance/Rate in each row of the table and the fact that the times are the same, we can write an equation.

$$\frac{400}{r} = \frac{330}{r - 14}$$

**Carry out**. We solve the equation.

$$\frac{400}{r} = \frac{330}{r - 14}, \text{ LCD is } r(r - 14)$$

$$r(r - 14) \cdot \frac{400}{r} = r(r - 14) \cdot \frac{330}{r - 14}$$

$$400(r - 14) = 330r$$

$$400r - 5600 = 330r$$

$$-5600 = -70r$$

$$80 = r$$

**Check**. If the passenger train's speed is 80 mph, then the freight train's speed is $80 - 14$, or 66 mph. Traveling 400 mi at 80 mph takes $\frac{400}{80} = 5$ hr. Traveling 330 mi at 66 mph takes $\frac{330}{66} = 5$ hr. Since the times are the same, the answer checks.

**State**. The speed of the passenger train is 80 mph; the speed of the freight train is 66 mph.

**29.** Note that 38 mi is 7 mi less than 45 mi and that the local bus travels 7 mph slower than the express. Then the express travels 45 mi in one hr, or 45 mph, and the local bus travels 38 mi in one hr, or 38 mph.

**31. Familiarize**. We let $r$ = the speed of the river. Then $15 + r$ = Suzie's speed downstream in km/h and $15 - r$ = her speed upstream in km/h. The times are the same. Let $t$ represent the time. We organize the information in a table.

|  | Distance | Speed | Time |
|---|---|---|---|
| Downstream | 140 | $15 + r$ | $t$ |
| Upstream | 35 | $15 - r$ | $t$ |

**Translate**. Using the formula Time = Distance/Rate in each row of the table and the fact that the times are the same, we can write an equation.

$$\frac{140}{15 + r} = \frac{35}{15 - r}$$

**Carry out**. We solve the equation.

$$\frac{140}{15 + r} = \frac{35}{15 - r},$$

$$\text{LCD is } (15 + r)(15 - r)$$

$$(15 + r)(15 - r) \cdot \frac{140}{15 + r} = (15 + r)(15 - r) \cdot \frac{35}{15 - r}$$

$$140(15 - r) = 35(15 + r)$$

$$2100 - 140r = 525 + 35r$$

$$1575 = 175r$$

$$9 = r$$

**Check**. If $r = 9$, then the speed downstream is $15 + 9$, or 24 km/h and the speed upstream is $15 - 9$, or 6 km/h. The time for the trip is downstream is $\frac{140}{24}$, or $5\frac{5}{6}$ hours. The time for the trip upstream is $\frac{35}{6}$, or $5\frac{5}{6}$ hours. The times are the same. The values check.

**State**. The speed of the river is 9 km/h.

**33. Familiarize**. Let $c$ = the speed of the current, in km/h. Then $7 + c$ = the speed downriver and $7 - c$ = the speed upriver. We organize the information in a table.

|  | Distance | Speed | Time |
|---|---|---|---|
| Downriver | 45 | $7 + c$ | $t_1$ |
| Upriver | 45 | $7 - c$ | $t_2$ |

**Translate**. Using the formula Time = Distance/Rate we see that $t_1 = \frac{45}{7 + c}$ and $t_2 = \frac{45}{7 - c}$. The total time upriver and back is 14 hr, so $t_1 + t_2 = 14$, or

$$\frac{45}{7 + c} + \frac{45}{7 - c} = 14.$$

**Carry out**. We solve the equation. Multiply both sides by the LCD, $(7 + c)(7 - c)$.

$$(7 + c)(7 - c)\left(\frac{45}{7 + c} + \frac{45}{7 - c}\right) = (7 + c)(7 - c)14$$

$$45(7 - c) + 45(7 + c) = 14(49 - c^2)$$

$$315 - 45c + 315 + 45c = 686 - 14c^2$$

$$14c^2 - 56 = 0$$

$$14(c + 2)(c - 2) = 0$$

$$c + 2 = 0 \quad \text{or} \quad c - 2 = 0$$

$$c = -2 \quad \text{or} \quad c = 2$$

**Check**. Since speed cannot be negative in this problem, $-2$ cannot be a solution of the original problem. If the speed of the current is 2 km/h, the barge travels upriver at $7 - 2$, or 5 km/h. At this rate it takes $\frac{45}{5}$, or 9 hr, to travel 45 km. The barge travels downriver at $7 + 2$, or 9 km/h. At this rate it takes $\frac{45}{9}$, or 5 hr, to travel 45 km. The total travel time is $9 + 5$, or 14 hr. The answer checks.

**State**. The speed of the current is 2 km/h.

**35. *Familiarize*.** Let $w$ = the wind speed, in mph. Then the speed into the wind is $350 - w$, and the speed with the wind is $350 + w$. We organize the information in a table.

|                | Distance | Speed     | Time  |
|----------------|----------|-----------|-------|
| Into the wind  | 487.5    | $350 - w$ | $t_1$ |
| With the wind  | 487.5    | $350 + w$ | $t_2$ |

***Translate***. Using the formula Time = Distance/Rate we see that $t_1 = \dfrac{487.5}{350 - w}$ and $t_2 = \dfrac{487.5}{350 + w}$. The total time upstream and back is 2.8 hr, so $t_1 + t_2 = 2.8$, or

$$\frac{487.5}{350 - w} + \frac{487.5}{350 + w} = 2.8.$$

***Carry out***. We solve the equation. Multiply on both sides by the LCD, $(350 - w)(350 + w)$.

$$(350 - w)(350 + w)\left(\frac{487.5}{350 - w} + \frac{487.5}{350 + w}\right) =$$
$$(350 - w)(350 + w)(2.8)$$
$$487.5(350 + w) + 487.5(350 - w) =$$
$$2.8(122,500 - w^2)$$
$$170,625 + 487.5w + 170,625 - 487.5w =$$
$$343,000 - 2.8w^2$$
$$341,250 =$$
$$343,000 - 2.8w^2$$
$$2.8w^2 - 1750 = 0$$
$$2.8(w^2 - 625) = 0$$
$$2.8(w + 25)(w - 25) = 0$$
$$w = -25 \ or \ w = 25$$

***Check***. We check only 25 since the wind speed cannot be negative. If the wind speed is 25 mph, then the plane's speed into the wind is $350 - 25$, or 325 mph, and the speed with the wind is $350 + 25$, or 375 mph. Flying 487.5 mi into the wind takes $\dfrac{478.5}{325}$, or 1.5 hr. Flying 487.5 mi with the wind takes $\dfrac{487.5}{375}$, or 1.3 hr. The total time is $1.5 + 1.3$, or 2.8 hr. The answer checks.

***State***. The wind speed is 25 mph.

**37. *Familiarize*.** Let $r$ = the speed at which the train actually traveled in mph, and let $t$ = the actual travel time in hours. We organize the information in a table.

|               | Distance | Speed    | Time    |
|---------------|----------|----------|---------|
| Actual speed  | 120      | $r$      | $t$     |
| Faster speed  | 120      | $r + 10$ | $t - 2$ |

***Translate***. From the first row of the table we have $120 = rt$, and from the second row we have $120 = (r+10)(t-2)$. Solving the first equation for $t$, we have $t = \dfrac{120}{r}$. Substituting for $t$ in the second equation, we have

$$120 = (r+10)\left(\frac{120}{r} - 2\right).$$

***Carry out***. We solve the equation.

$$120 = (r + 10)\left(\frac{120}{r} - 2\right)$$
$$120 = 120 - 2r + \frac{1200}{r} - 20$$
$$20 = -2r + \frac{1200}{r}$$
$$r \cdot 20 = r\left(-2r + \frac{1200}{r}\right)$$
$$20r = -2r^2 + 1200$$
$$2r^2 + 20r - 1200 = 0$$
$$2(r^2 + 10r - 600) = 0$$
$$2(r + 30)(r - 20) = 0$$
$$r = -30 \ or \ r = 20$$

***Check***. Since speed cannot be negative in this problem, $-30$ cannot be a solution of the original problem. If the speed is 20 mph, it takes $\dfrac{120}{20}$, or 6 hr, to travel 120 mi. If the speed is 10 mph faster, or 30 mph, it takes $\dfrac{120}{30}$, or 4 hr, to travel 120 mi. Since 4 hr is 2 hr less time than 6 hr, the answer checks.

***State***. The speed was 20 mph.

**39. *Writing Exercise***

**41.** $\dfrac{35a^6b^8}{7a^2b^2} = \dfrac{35}{7}a^{6-2}b^{8-2} = 5a^4b^6$

**43.** $\dfrac{36s^{15}t^{10}}{9s^5t^2} = \dfrac{36}{9}s^{15-5}t^{10-2}$
$$= 4s^{10}t^8$$

**45.** $2(x^3 + 4x^2 - 5x + 7) - 5(2x^3 - 4x^2 + 3x - 1)$
$$= 2x^3 + 8x^2 - 10x + 14 - 10x^3 + 20x^2 - 15x + 5$$
$$= -8x^3 + 28x^2 - 25x + 19$$

**47. *Writing Exercise***

**49. *Familiarize*.** If the drainage gate is closed, $\dfrac{1}{9}$ of the bog is filled in 1 hr. If the bog is not being filled, $\dfrac{1}{11}$ of the bog is drained in 1 hr. If the bog is being filled with the drainage gate left open, $\dfrac{1}{9} - \dfrac{1}{11}$ of the bog is filled in 1 hr. Let $t$ = the time it takes to fill the bog with the drainage gate left open.

***Translate***. We want to find $t$ such that
$$t\left(\frac{1}{9} - \frac{1}{11}\right) = 1, \text{ or } \frac{t}{9} - \frac{t}{11} = 1.$$

***Carry out***. We solve the equation. First we multiply by the LCD, 99.
$$99\left(\frac{t}{9} - \frac{t}{11}\right) = 99 \cdot 1$$
$$11t - 9t = 99$$
$$2t = 99$$
$$t = \frac{99}{2}$$

***Check***. In $\dfrac{99}{2}$ hr, we have $\dfrac{99}{2}\left(\dfrac{1}{9} - \dfrac{1}{11}\right) = \dfrac{11}{2} - \dfrac{9}{2} = \dfrac{2}{2} = $ 1 full bog.

***State.*** It will take $\frac{99}{2}$, or $49\frac{1}{2}$ hr, to fill the bog.

51. Monica's speed downstream is $12 + 4$, or $16$ mph. Using Time = Distance/Rate, we find that the time it will take Monica to motor 3 mi downstream is 3/16 hr. We can convert this time to minutes:

$$\frac{3}{16} \text{ hr} = \frac{3}{16} \times 1 \text{ hr} = \frac{3}{16} \times 60 \text{ min} = 11.25 \text{ min}$$

53. ***Familiarize.*** Let $p = $ the number of people per hour moved by the 60 cm-wide escalator. Then $2p = $ the number of people per hour moved by the 100 cm-wide escalator. We convert 1575 people per 14 minutes to people per hour:

$$\frac{1575 \text{ people}}{14 \text{ min}} \cdot \frac{60 \text{ min}}{1 \text{ hr}} = 6750 \text{ people/hr}$$

***Translate.*** We use the information that together the escalators move 6750 people per hour to write an equation.

$$p + 2p = 6750$$

***Carry out.*** We solve the equation.

$$p + 2p = 6750$$
$$3p = 6750$$
$$p = 2250$$

***Check.*** If the 60 cm-wide escalator moves 2250 people per hour, then the 100 cm-wide escalator moves $2 \cdot 2250$, or 4500 people per hour. Together, they move $2250 + 4500$, or 6750 people per hour. The answer checks.

***State.*** The 60 cm-wide escalator moves 2250 people per hour.

55. ***Familiarize.*** Let $d = $ the distance, in miles, the paddle-boat can cruise upriver before it is time to turn around. The boat's speed upriver is $12 - 5$, or 7 mph, and its speed downriver is $12 + 5$, or 17 mph. We organize the information in a table.

|  | Distance | Speed | Time |
|---|---|---|---|
| Upriver | $d$ | 7 | $t_1$ |
| Downriver | $d$ | 17 | $t_2$ |

***Translate.*** Using the formula Time = Distance/Rate we see that $t_1 = \frac{d}{7}$ and $t_2 = \frac{d}{17}$. The time upriver and back is 3 hr, so $t_1 + t_2 = 3$, or

$$\frac{d}{7} + \frac{d}{17} = 3.$$

***Carry out.*** We solve the equation.

$$7 \cdot 17\left(\frac{d}{7} + \frac{d}{17}\right) = 7 \cdot 17 \cdot 3$$
$$17d + 7d = 357$$
$$24d = 357$$
$$d = \frac{119}{8}$$

***Check.*** Traveling $\frac{119}{8}$ mi upriver at a speed of 7 mph takes $\frac{119/8}{7} = \frac{17}{8}$ hr. Traveling $\frac{119}{8}$ mi downriver at a

speed of 17 mph takes $\frac{119/8}{17} = \frac{7}{8}$ hr. The total time is $\frac{17}{8} + \frac{7}{8} = \frac{24}{8} = 3$ hr. The answer checks.

***State.*** The pilot can go $\frac{119}{8}$, or $14\frac{7}{8}$ mi upriver before it is time to turn around.

57. ***Familiarize.*** Let $t$ represent the time it takes the printers to print 500 pages working together.

***Translate.*** The faster machine can print 500 pages in 40 min, and it takes the slower printer 50 min to do the same job. Then we have

$$\frac{t}{40} + \frac{t}{50} = 1.$$

***Carry out.*** We solve the equation.

$$\frac{t}{40} + \frac{t}{50} = 1, \text{ LCD is } 200$$
$$200\left(\frac{t}{40} + \frac{t}{50}\right) = 200 \cdot 1$$
$$5t + 4t = 200 \cdot 1$$
$$9t = 200$$
$$t = \frac{200}{9}$$

In $\frac{200}{9}$ min, the faster printer does $\frac{200/9}{40}$, or $\frac{200}{9} \cdot \frac{1}{40}$, or $\frac{5}{9}$ of the job. Then starting at page 1, it would print $\frac{5}{9} \cdot 500$, or $277\frac{7}{9}$ pages. Thus, in $\frac{200}{9}$ min, the two machines will meet on page 278.

***Check.*** We can check to see that the slower machine is also printing page 278 after $\frac{200}{9}$ min. In $\frac{200}{9}$ min, the slower machine does $\frac{200/9}{50}$, or $\frac{200}{9} \cdot \frac{1}{50}$, or $\frac{4}{9}$ of the job. Then it would print $\frac{4}{9} \cdot 500$, or $222\frac{2}{9}$ pages. Working backward from page 500, this machine would be on page $500 - 222\frac{2}{9}$, or $277\frac{2}{9}$. Thus, both machines are printing page 278 after $\frac{200}{9}$ min. The answer checks.

***State.*** The two machines will meet on page 278.

59. ***Familiarize*** Express the position of the hands in terms of minute units on the face of the clock. At 10:30 the hour hand is at $\frac{10.5}{12}$ hr$\times \frac{60 \text{ min}}{1 \text{ hr}}$, or 52.5 minutes, and the minute hand is at 30 minutes. The rate of the minute hand is 12 times the rate of the hour hand. (When the minute hand moves 60 minutes, the hour hand moves 5 minutes.) Let $t = $ the number of minutes after 10:30 that the hands will first be perpendicular. After $t$ minutes the minute hand has moved $t$ units, and the hour hand has moved $\frac{t}{12}$ units. The position of the hour hand will be 15 units "ahead" of the position of the minute hand when they are first perpendicular.

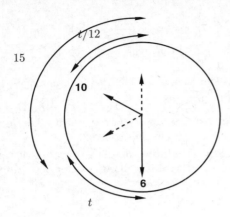

**Translate**.

Position of        position of
hour hand    is   minute hand   plus 15 min.
after $t$ min      after $t$ min

$$52.5 + \frac{t}{12} = 30 + t + 15$$

**Solve**. We solve the equation.

$$52.5 + \frac{t}{12} = 30 + t + 15$$

$$52.5 + \frac{t}{12} = 45 + t, \text{ LCM is } 12$$

$$12\left(52.5 + \frac{t}{12}\right) = 12(45 + t)$$

$$630 + t = 540 + 12t$$

$$90 = 11t$$

$$\frac{90}{11} = t, \text{ or}$$

$$8\frac{2}{11} = t$$

**Check**. At $\frac{90}{11}$ min after 10:30, the position of the hour hand is at $52.5 + \frac{90/11}{12}$, or $53\frac{2}{11}$ min. The minute hand is at $30 + \frac{90}{11}$, or $38\frac{2}{11}$ min. The hour hand is 15 minutes ahead of the minute hand so the hands are perpendicular. The answer checks.

**State**. After 10:30 the hands of a clock will first be perpendicular in $8\frac{2}{11}$ min. The time is $10:38\frac{2}{11}$, or $21\frac{9}{11}$ min before 11:00.

**61. Familiarize**. Let $r = $ the speed in mph Chip would have to travel for the last half of the trip in order to average a speed of 45 mph for the entire trip. We organize the information in a table.

|            | Distance | Speed | Time  |
|------------|----------|-------|-------|
| First half | 50       | 40    | $t_1$ |
| Last half  | 50       | $r$   | $t_2$ |

The total distance is $50 + 50$, or 100 mi.

The total time is $t_1 + t_2$, or $\frac{50}{40} + \frac{50}{r}$, or $\frac{5}{4} + \frac{50}{r}$. The average speed is 45 mph.

**Translate**.

$$\text{Average speed} = \frac{\text{Total distance}}{\text{Total time}}$$

$$45 = \frac{100}{\frac{5}{4} + \frac{50}{r}}$$

**Carry out**. We solve the equation.

$$45 = \frac{100}{\frac{5}{4} + \frac{50}{r}}$$

$$45 = \frac{100}{\frac{5r + 200}{4r}}$$

$$45 = 100 \cdot \frac{4r}{5r + 200}$$

$$45 = \frac{400r}{5r + 200}$$

$$(5r + 200)(45) = (5r + 200) \cdot \frac{400r}{5r + 200}$$

$$225r + 9000 = 400r$$

$$9000 = 175r$$

$$\frac{360}{7} = r$$

**Check**. Traveling 50 mi at 40 mph takes $\frac{50}{40}$, or $\frac{5}{4}$ hr. Traveling 50 mi at $\frac{360}{7}$ mph takes $\frac{50}{360/7}$, or $\frac{35}{36}$ hr. Then the total time is $\frac{5}{4} + \frac{35}{36} = \frac{80}{36} = \frac{20}{9}$ hr. The average speed when traveling 100 mi for $\frac{20}{9}$ hr is $\frac{100}{20/9} = 45$ mph. The answer checks.

**State**. Chip would have to travel at a speed of $\frac{360}{7}$, or $51\frac{3}{7}$ mph for the last half of the trip so that the average speed for the entire trip would be 45 mph.

## Exercise Set 6.6

**1.** True; see page 402 in the text.

**3.** True; see page 402 in the text.

**5.** False; see pages 403 and 404 in the text.

**7.** $\dfrac{32x^6 + 18x^5 - 27x^2}{6x^2}$

$$= \frac{32x^6}{6x^2} + \frac{18x^5}{6x^2} - \frac{27x^2}{6x^2}$$

$$= \frac{16}{3}x^4 + 3x^3 - \frac{9}{2}$$

**9.** $\dfrac{21a^3 + 7a^2 - 3a - 14}{7a}$

$= \dfrac{21a^3}{7a} + \dfrac{7a^2}{7a} - \dfrac{3a}{7a} - \dfrac{14}{7a}$

$= 3a^2 + a - \dfrac{3}{7} - \dfrac{2}{a}$

**11.** $\dfrac{18t^4 - 15t + 21}{-3t}$

$= \dfrac{18t^4}{-3t} - \dfrac{15t}{-3t} + \dfrac{21}{-3t}$

$= -6t^3 + 5 - \dfrac{7}{t}$

**13.** $\dfrac{16y^4z^2 - 8y^6z^4 + 12y^8z^3}{-4y^4z}$

$= \dfrac{16y^4z^2}{-4y^4z} - \dfrac{8y^6z^4}{-4y^4z} + \dfrac{12y^8z^3}{-4y^4z}$

$= -4z + 2y^2z^3 - 3y^4z^2$

**15.** $\dfrac{16y^3 - 9y^2 - 8y}{2y^2}$

$= \dfrac{16y^3}{2y^2} - \dfrac{9y^2}{2y^2} - \dfrac{8y}{2y^2}$

$= 8y - \dfrac{9}{2} - \dfrac{4}{y}$

**17.** $\dfrac{15x^7 - 21x^4 - 3x^2}{-3x^2}$

$= \dfrac{15x^7}{-3x^2} + \dfrac{-21x^4}{-3x^2} + \dfrac{-3x^2}{-3x^2}$

$= -5x^5 + 7x^2 + 1$

**19.** $(a^2b - a^3b^3 - a^5b^5) \div (a^2b)$

$= \dfrac{a^2b}{a^2b} - \dfrac{a^3b^3}{a^2b} - \dfrac{a^5b^5}{a^2b}$

$= 1 - ab^2 - a^3b^4$

**21.** $(x^2 + 10x + 21) \div (x + 7)$

$= \dfrac{(x+7)(x+3)}{x+7}$

$= \dfrac{(\cancel{x+7})(x+3)}{\cancel{x+7}}$

$= x + 3$

The answer is $x + 3$.

**23.**

$$\begin{array}{r} a - 12 \\ a + 4 \overline{)\, a^2 - 8a - 16} \\ \underline{a^2 + 4a} \\ -12a - 16 \\ \underline{-12a - 48} \\ 32 \end{array}$$

$(a^2 - 8a) - (a^2 + 4a) = -12a$

$(-12a - 16) - (-12a - 48) = 32$

The answer is $a - 12$, R 32, or $a - 12 + \dfrac{32}{a+4}$.

**25.**

$$\begin{array}{r} x - 5 \\ x - 4 \overline{)\, x^2 - 9x + 21} \\ \underline{x^2 - 4x} \\ -5x + 21 \\ \underline{-5x + 20} \\ 1 \end{array}$$

The answer is $x - 5$, R 1, or $x - 5 + \dfrac{1}{x-4}$.

**27.** $(y^2 - 25) \div (y + 5) = \dfrac{y^2 - 25}{y + 5}$

$\qquad\qquad = \dfrac{(y+5)(y-5)}{y+5}$

$\qquad\qquad = \dfrac{(\cancel{y+5})(y-5)}{\cancel{y+5}}$

$\qquad\qquad = y - 5$

We could also find this quotient as follows.

$$\begin{array}{r} y - 5 \\ y + 5 \overline{)\, y^2 + 0y - 25} \\ \underline{y^2 + 5y} \\ -5y - 25 \\ \underline{-5y - 25} \\ 0 \end{array}$$

Writing in the missing term

The answer is $y - 5$.

**29.**

$$\begin{array}{r} y^2 - 2y - 1 \\ y - 2 \overline{)\, y^3 - 4y^2 + 3y - 6} \\ \underline{y^3 - 2y^2} \\ -2y^2 + 3y \\ \underline{-2y^2 + 4y} \\ -y - 6 \\ \underline{-y + 2} \\ -8 \end{array}$$

The answer is $y^2 - 2y - 1$, R $-8$, or

$y^2 - 2y - 1 + \dfrac{-8}{y-2}$.

**31.**

$$\begin{array}{r} 2x^2 - x + 1 \\ x + 2 \overline{)\, 2x^3 + 3x^2 - x - 3} \\ \underline{2x^3 + 4x^2} \\ -x^2 - x \\ \underline{-x^2 - 2x} \\ x - 3 \\ \underline{x + 2} \\ -5 \end{array}$$

The answer is $2x^2 - x + 1$, R $-5$, or

$2x^2 - x + 1 + \dfrac{-5}{x+2}$.

**33.**

$$\begin{array}{r} a^2 + 4a + 15 \\ a - 4 \overline{)\, a^3 + 0a^2 - a + 10} \\ \underline{a^3 - 4a^2} \\ 4a^2 - a \\ \underline{4a^2 - 16a} \\ 15a + 10 \\ \underline{15a - 60} \\ 70 \end{array}$$

The answer is $a^2 + 4a + 15$, R 70, or

$a^2 + 4a + 15 + \dfrac{70}{a-4}$.

**35.**

$$
\begin{array}{r}
2y^2 + \phantom{0}2y - \phantom{0}1 \\
5y - 2 \overline{\smash)10y^3 + 6y^2 - 9y + 10} \\
\underline{10y^3 - 4y^2} \\
10y^2 - 9y \\
\underline{10y^2 - 4y} \\
-5y + 10 \\
\underline{-5y + \phantom{0}2} \\
8
\end{array}
$$

The answer is $2y^2 + 2y - 1$, R 8, or

$2y^2 + 2y - 1 + \dfrac{8}{5y - 2}$.

**37.**

$$
\begin{array}{r}
2x^2 - \phantom{0}x - \phantom{0}9 \\
x^2 + 2 \overline{\smash)2x^4 - x^3 - 5x^2 + \phantom{0}x - \phantom{0}6} \\
\underline{2x^4 \phantom{00000} + 4x^2} \\
-x^3 - 9x^2 + \phantom{0}x \\
\underline{-x^3 \phantom{0000000} - 2x} \\
-9x^2 + 3x - \phantom{0}6 \\
\underline{-9x^2 \phantom{0000} - 18} \\
3x + 12
\end{array}
$$

The answer is $2x^2 - x - 9$, R $3x + 12$, or

$2x^2 - x - 9 + \dfrac{3x + 12}{x^2 + 2}$.

**39.** $F(x) = \dfrac{f(x)}{g(x)} = \dfrac{8x^3 - 27}{2x - 3}$

$$
\begin{array}{r}
4x^2 + \phantom{0}6x + \phantom{0}9 \\
2x - 3 \overline{\smash)8x^3 \phantom{00000000} - 27} \\
\underline{8x^3 - 12x^2} \\
12x^2 + \phantom{0}0x \\
\underline{12x^2 - 18x} \\
18x - 27 \\
\underline{18x - 27} \\
0
\end{array}
$$

Since $g(x)$ is 0 for $x = \dfrac{3}{2}$, we have

$F(x) = 4x^2 + 6x + 9$, provided $x \neq \dfrac{3}{2}$.

**41.** $F(x) = \dfrac{f(x)}{g(x)} = \dfrac{6x^2 - 11x - 10}{3x + 2}$

$$
\begin{array}{r}
2x - \phantom{0}5 \\
3x + 2 \overline{\smash)6x^2 - 11x - 10} \\
\underline{6x^2 + \phantom{0}4x} \\
-15x - 10 \\
\underline{-15x - 10} \\
0
\end{array}
$$

Since $g(x)$ is 0 for $x = -\dfrac{2}{3}$, we have

$F(x) = 2x - 5$, provided $x \neq -\dfrac{2}{3}$.

**43.** $F(x) = \dfrac{f(x)}{g(x)} = \dfrac{x^4 - 24x^2 - 25}{x^2 - 25}$

$$
\begin{array}{r}
x^2 + \phantom{0}1 \\
x^2 - 25 \overline{\smash)x^4 - 24x^2 - 25} \\
\underline{x^4 - 25x^2} \\
x^2 - 25 \\
\underline{x^2 - 25} \\
0
\end{array}
$$

Since $g(x)$ is 0 for $x = -5$ or $x = 5$, we have $F(x) = x^2 + 1$, provided $x \neq -5$ and $x \neq 5$.

**45.** We rewrite $f(x)$ in descending order.

$$F(x) = \dfrac{f(x)}{g(x)} = \dfrac{2x^5 - 3x^4 - 2x^3 + 8x^2 - 5}{x^2 - 1}$$

$$
\begin{array}{r}
2x^3 - 3x^2 + \phantom{0}5 \\
x^2 - 1 \overline{\smash)2x^5 - 3x^4 - 2x^3 + 8x^2 - 5} \\
\underline{2x^5 \phantom{00000} - 2x^3} \\
-3x^4 \phantom{000000} + 8x^2 \\
\underline{-3x^4 \phantom{000000} + 3x^2} \\
5x^2 - 5 \\
\underline{5x^2 - 5} \\
0
\end{array}
$$

Since $g(x)$ is 0 for $x = -1$ or $x = 1$, we have $F(x) = 2x^3 - 3x^2 + 5$, provided $x \neq -1$ and $x \neq 1$.

**47.** *Writing Exercise*

**49.** $ab - cd = k$

$-cd = k - ab$

$c = \dfrac{k - ab}{-d}$, or $\dfrac{ab - k}{d}$

**51.** **Familiarize.** Let $x$, $x + 1$, and $x + 2$ represent the three consecutive positive integers.

**Translate.** Rewording, we write an equation.

| Product of first and second | is | product of second and third | | less | 26. |
|---|---|---|---|---|---|
| $\downarrow$ | $\downarrow$ | $\downarrow$ | | $\downarrow$ | $\downarrow$ |
| $x(x + 1)$ | $=$ | $(x + 1)(x + 2)$ | $-$ | | $26$ |

**Carry out.** We solve the equation.

$$x^2 + x = x^2 + 3x + 2 - 26$$
$$x = 3x - 24$$
$$24 = 2x$$
$$12 = x$$

If the first integer is 12, the next two are 13 and 14.

**Check.** The product of 12 and 13 is 156. The product of 13 and 14 is 182, and $182 - 26 = 156$. The numbers check.

**State.** The three consecutive positive integers are 12, 13, and 14.

**53.** $|2x - 3| > 7$

$2x - 3 < -7 \ \text{ or } \ 2x - 3 > 7$

$2x < -4 \ \text{ or } \phantom{0000} 2x > 10$

$x < -2 \ \text{ or } \phantom{0000} x > 5$

The solution set is $\{x | x < -2 \text{ or } x > 5\}$, or $(-\infty, -2) \cup (5, \infty)$.

**55.** *Writing Exercise*

**57.**
$$
a^2 + 3ab + 2b^2 \overline{)\ a^4 + 4a^3b + 5a^2b^2 + 2ab^3}
$$
with quotient $a^2 + ab$

$$
\begin{array}{r}
a^2 + \phantom{0}ab \\
a^2 + 3ab + 2b^2 \overline{)\ a^4 + 4a^3b + 5a^2b^2 + 2ab^3} \\
\underline{a^4 + 3a^3b + \phantom{0}2a^2b^2} \\
a^3b + 3a^2b^2 + 2ab^3 \\
\underline{a^3b + 3a^2b^2 + 2ab^3} \\
0
\end{array}
$$

The answer is $a^2 + ab$.

**59.**
$$
\begin{array}{r}
a^6 - a^5b + a^4b^2 - a^3b^3 + a^2b^4 - \phantom{0}ab^5 + \phantom{0}b^6 \\
a + b \overline{)\ a^7 \phantom{aaaaaaaaaaaaaaaaaaaaaaaaaaaaaa} + b^7} \\
\underline{a^7 + a^6b} \\
-a^6b \\
\underline{-a^6b - a^5b^2} \\
a^5b^2 \\
\underline{a^5b^2 + a^4b^3} \\
-a^4b^3 \\
\underline{-a^4b^3 - a^3b^4} \\
a^3b^4 \\
\underline{a^3b^4 + a^2b^5} \\
-a^2b^5 \\
\underline{-a^2b^5 - ab^6} \\
ab^6 + b^7 \\
\underline{ab^6 + b^7} \\
0
\end{array}
$$

The answer is $a^6 - a^5b + a^4b^2 - a^3b^3 + a^2b^4 - ab^5 + b^6$.

**61.**
$$
\begin{array}{r}
x - \phantom{00}5 \\
x + 2 \overline{)\ x^2 - \phantom{0}3x + 2k} \\
\underline{x^2 + \phantom{0}2x} \\
-5x + 2k \\
\underline{-5x - 10} \\
2k + 10
\end{array}
$$

The remainder is 7. Thus, we solve the following equation for $k$.
$$
\begin{aligned}
2k + 10 &= 7 \\
2k &= -3 \\
k &= -\frac{3}{2}
\end{aligned}
$$

**63.** *Writing Exercise*

**65.**

# Exercise Set 6.7

**1.** True; see page 410 in the text.

**3.** True

**5.** True; see page 409 in the text.

**7.** True by substitution and the remainder theorem

**9.**
$$
\begin{array}{r|rrrr}
1 & 1 & -2 & 2 & -7 \\
  &   & 1 & -1 & 1 \\
\hline
  & 1 & -1 & 1 & -6
\end{array}
$$

$x^2 - x + 1$, R $-6$, or $x^2 - x + 1 + \dfrac{-6}{x - 1}$

**11.** $(a^2 + 8a + 11) \div (a + 3) =$
$(a^2 + 8a + 11) \div [a - (-3)]$

$$
\begin{array}{r|rrr}
-3 & 1 & 8 & 11 \\
   &   & -3 & -15 \\
\hline
   & 1 & 5 & -4
\end{array}
$$

The answer is $a + 5$, R $-4$, or $a + 5 + \dfrac{-4}{a + 3}$.

**13.** $(x^3 - 7x^2 - 13x + 3) \div (x + 2) =$
$(x^3 - 7x^2 - 13x + 3) \div [x - (-2)]$

$$
\begin{array}{r|rrrr}
-2 & 1 & -7 & -13 & 3 \\
   &   & -2 & 18 & -10 \\
\hline
   & 1 & -9 & 5 & -7
\end{array}
$$

$x^2 - 9x + 5$, R $-7$, or $x^2 - 9x + 5 + \dfrac{-7}{x + 2}$

**15.** $(3x^3 + 7x^2 - 4x + 3) \div (x + 3) =$
$(3x^3 + 7x^2 - 4x + 3) \div [x - (-3)]$

$$
\begin{array}{r|rrrr}
-3 & 3 & 7 & -4 & 3 \\
   &   & -9 & 6 & -6 \\
\hline
   & 3 & -2 & 2 & -3
\end{array}
$$

The answer is $3x^2 - 2x + 2$, R $-3$, or
$3x^2 - 2x + 2 + \dfrac{-3}{x + 3}$.

**17.** $(y^3 - 3y + 10) \div (y - 2) =$
$(y^3 + 0y^2 - 3y + 10) \div (y - 2)$

$$
\begin{array}{r|rrrr}
2 & 1 & 0 & -3 & 10 \\
  &   & 2 & 4 & 2 \\
\hline
  & 1 & 2 & 1 & 12
\end{array}
$$

The answer is $y^2 + 2y + 1$, R 12, or
$y^2 + 2y + 1 + \dfrac{12}{y - 2}$.

**19.** $(x^5 - 32) \div (x - 2) =$
$(x^5 + 0x^4 + 0x^3 + 0x^2 + 0x - 32) \div (x - 2)$

$$
\begin{array}{r|rrrrrr}
2 & 1 & 0 & 0 & 0 & 0 & -32 \\
  &   & 2 & 4 & 8 & 16 & 32 \\
\hline
  & 1 & 2 & 4 & 8 & 16 & 0
\end{array}
$$

The answer is $x^4 + 2x^3 + 4x^2 + 8x + 16$.

**21.** $(3x^3 + 1 - x + 7x^2) \div \left(x + \dfrac{1}{3}\right) =$

$(3x^3 + 7x^2 - x + 1) \div \left[x - \left(-\dfrac{1}{3}\right)\right]$

$$
\begin{array}{r|rrrr}
-\frac{1}{3} & 3 & 7 & -1 & 1 \\
             &   & -1 & -2 & 1 \\
\hline
             & 3 & 6 & -3 & 2
\end{array}
$$

The answer is $3x^2 + 6x - 3$ R 2, or
$3x^2 + 6x - 3 + \dfrac{2}{x + \dfrac{1}{3}}$.

**23.**
$$
\begin{array}{r|rrrrr}
-3 & 5 & 12 & 0 & 28 & 9 \\
   &   & -15 & 9 & -27 & -3 \\
\hline
   & 5 & -3 & 9 & 1 & 6
\end{array}
$$

The remainder tells us that $f(-3) = 6$.

**25.**

$$\begin{array}{r|rrrrr} -1 & 6 & -1 & -7 & 1 & 2 \\ & & -6 & 7 & -1 & -1 \\ \hline & 6 & -7 & 0 & 0\, | & 1 \end{array}$$

The remainder tells us that $P(-1) = 1$.

**27.**

$$\begin{array}{r|rrrrr} 4 & 1 & -1 & -19 & 49 & -30 \\ & & 4 & 12 & -28 & 84 \\ \hline & 1 & 3 & -7 & 21\, | & 54 \end{array}$$

The remainder tells us that $f(4) = 54$.

**29.** *Writing Exercise*

**31.**
$$9 + cb = a - b$$
$$9 + cb + b = a$$
$$cb + b = a - 9$$
$$b(c + 1) = a - 9$$
$$b = \frac{a - 9}{c + 1}$$

**33.** $f(x) = \dfrac{5}{3x^2 - 75}$

$f(x)$ cannot be calculated for any $x$-value for which the denominator is 0. We solve an equation to find those values.

$$3x^2 - 75 = 0$$
$$3(x^2 - 25) = 0$$
$$3(x + 5)(x - 5) = 0$$
$$x + 5 = 0 \quad or \quad x - 5 = 0$$
$$x = -5 \quad or \qquad x = 5$$

The domain of $f$ is $\{x | x$ is a real number *and* $x \neq -5$ *and* $x \neq 5\}$.

**35.** Graph: $y - 2 = \dfrac{3}{4}(x + 1)$

This is the equation of the line with slope $\dfrac{3}{4}$ and passing through $(-1, 2)$. To graph this equation, start at $(-1, 2)$ and count off a slope of $\dfrac{3}{4}$ by going up 3 units and to the right 4 units (or down 3 units and to the left 4 units). Then draw the line.

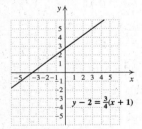

**37.** *Writing Exercise*

**39.** a) The degree of the remainder must be less than the degree of the divisor. Thus, the degree of the remainder must be 0, so $R$ must be a constant.

    b) $P(x) = (x - r) \cdot Q(x) + R$
$$P(r) = (r - r) \cdot Q(r) + R = 0 \cdot Q(r) + R = R$$

**41.**

$$\begin{array}{r|rrrr} -3 & 4 & 16 & -3 & -45 \\ & & -12 & -12 & 45 \\ \hline & 4 & 4 & -15\, | & 0 \end{array}$$

The remainder tells us that $f(-3) = 0$.

$$f(x) = (x + 3)(4x^2 + 4x - 15) = (x + 3)(2x + 5)(2x - 3)$$

Solve $f(x) = 0$:

$$(x + 3)(2x + 5)(2x - 3) = 0$$
$$x + 3 = 0 \quad or \quad 2x + 5 = 0 \quad or \quad 2x - 3 = 0$$
$$x = -3 \quad or \qquad x = -\frac{5}{2} \quad or \qquad x = \frac{3}{2}$$

The solutions are $-3$, $-\dfrac{5}{2}$, and $\dfrac{3}{2}$.

**43.** ▮

**45.** $f(x) = 4x^3 + 16x^2 - 3x - 45$
$$= x(4x^2 + 16x - 3) - 45$$
$$= x(x(4x + 16) - 3) - 45$$
$$f(-3) = -3(-3(4(-3) + 16) - 3) - 45$$
$$= -3(-3(-12 + 16) - 3) - 45$$
$$= -3(-3 \cdot 4 - 3) - 45$$
$$= -3(-12 - 3) - 45$$
$$= -3(-15) - 45$$
$$= 45 - 45$$
$$= 0$$

---

## Exercise Set 6.8

**1.** LCD; see Examples 1, 2, and 3.

**3.** factor; see Examples 1 and 3.

**5.** As the number of painters increases, the time required to scrape the house decreases, so we have inverse variation.

**7.** As the number of laps increases, the time required to swim them increases, so we have direct variation.

**9.** As the number of volunteers increases, the time required to wrap the toys decreases, so we have inverse variation.

**11.** $f = \dfrac{L}{d}$

$df = L$    Multiplying by $d$

$d = \dfrac{L}{f}$    Dividing by $f$

**13.**
$$s = \frac{(v_1 + v_2)t}{2}$$
$$2s = (v_1 + v_2)t \quad \text{Multiplying by 2}$$
$$\frac{2s}{t} = v_1 + v_2 \quad \text{Dividing by } t$$
$$\frac{2s}{t} - v_2 = v_1$$

This result can also be expressed as $v_1 = \dfrac{2s - tv_2}{t}$.

**15.**
$$\frac{t}{a} + \frac{t}{b} = 1$$

$$ab\left(\frac{t}{a} + \frac{t}{b}\right) = ab \cdot 1 \quad \text{Multiplying by the LCD}$$

$$ab \cdot \frac{t}{a} + ab \cdot \frac{t}{b} = ab$$

$$bt + at = ab$$

$$at = ab - bt$$

$$at = b(a - t) \quad \text{Factoring}$$

$$\frac{at}{a - t} = b$$

**17.**
$$I = \frac{2V}{R + 2r}$$

$$I(R + 2r) = \frac{2V}{R + 2r} \cdot (R + 2r) \quad \begin{array}{l}\text{Multiplying}\\\text{by the LCD}\end{array}$$

$$I(R + 2r) = 2V$$

$$R + 2r = \frac{2V}{I}$$

$$R = \frac{2V}{I} - 2r, \text{ or } \frac{2V - 2Ir}{I}$$

**19.**
$$R = \frac{gs}{g + s}$$

$$(g + s) \cdot R = (g + s) \cdot \frac{gs}{g + s} \quad \begin{array}{l}\text{Multiplying}\\\text{by the LCD}\end{array}$$

$$Rg + Rs = gs$$

$$Rs = gs - Rg$$

$$Rs = g(s - R) \quad \text{Factoring out } g$$

$$\frac{Rs}{s - R} = g \quad \text{Multiplying by } \frac{1}{s - R}$$

**21.**
$$I = \frac{nE}{R + nr}$$

$$I(R + nr) = \frac{nE}{R + nr} \cdot (R + nr) \quad \begin{array}{l}\text{Multiplying}\\\text{by the LCD}\end{array}$$

$$IR + Inr = nE$$

$$IR = nE - Inr$$

$$IR = n(E - Ir)$$

$$\frac{IR}{E - Ir} = n$$

**23.**
$$\frac{1}{p} + \frac{1}{q} = \frac{1}{f}$$

$$pqf\left(\frac{1}{p} + \frac{1}{q}\right) = pqf \cdot \frac{1}{f} \quad \begin{array}{l}\text{Multiplying by}\\\text{the LCD}\end{array}$$

$$qf + pf = pq$$

$$pf = pq - qf$$

$$pf = q(p - f)$$

$$\frac{pf}{p - f} = q$$

**25.**
$$S = \frac{H}{m(t_1 - t_2)}$$

$$(t_1 - t_2)S = \frac{H}{m} \quad \text{Multiplying by } t_1 - t_2$$

$$t_1 - t_2 = \frac{H}{Sm} \quad \text{Dividing by } S$$

$$t_1 = \frac{H}{Sm} + t_2, \text{ or } \frac{H + Smt_2}{Sm}$$

**27.**
$$\frac{E}{e} = \frac{R + r}{r}$$

$$er \cdot \frac{E}{e} = er \cdot \frac{R + r}{r} \quad \text{Multiplying by the LCD}$$

$$Er = e(R + r)$$

$$Er = eR + er$$

$$Er - er = eR$$

$$r(E - e) = eR$$

$$r = \frac{er}{E - e}$$

**29.**
$$S = \frac{a}{1 - r}$$

$$(1 - r)S = a \quad \text{Multiplying by the LCD, } 1 - r$$

$$1 - r = \frac{a}{S} \quad \text{Dividing by } S$$

$$1 - \frac{a}{S} = r \quad \text{Adding } r \text{ and } -\frac{a}{S}$$

This result can also be expressed as $r = \dfrac{S - a}{S}$.

**31.**
$$c = \frac{f}{(a + b)c}$$

$$\frac{a + b}{c} \cdot c = \frac{a + b}{c} \cdot \frac{f}{(a + b)c}$$

$$a + b = \frac{f}{c^2}$$

**33.**
$$P = \frac{A}{1 + r}$$

$$P(1 + r) = \frac{A}{1 + r} \cdot (1 + r)$$

$$P(1 + r) = A$$

$$1 + r = \frac{A}{P}$$

$$r = \frac{A}{P} - 1, \text{ or } \frac{A - P}{P}$$

**35.**
$$v = \frac{d_2 - d_1}{t_2 - t_1}$$

$$(t_2 - t_1)v = (t_2 - t_1) \cdot \frac{d_2 - d_1}{t_2 - t_1}$$

$$(t_2 - t_1)v = d_2 - d_1$$

$$t_2 - t_1 = \frac{d_2 - d_1}{v}$$

$$t_2 = \frac{d_2 - d_1}{v} + t_1, \text{ or } \frac{d_2 - d_1 + t_1 v}{v}$$

**37.**
$$\frac{x^2}{a^2} + \frac{y^2}{b^2} = 1$$

$$a^2 b^2 \left( \frac{x^2}{a^2} + \frac{y^2}{b^2} \right) = a^2 b^2 \cdot 1$$

$$a^2 b^2 \cdot \frac{x^2}{a^2} + a^2 b^2 \cdot \frac{y^2}{b^2} = a^2 b^2$$

$$b^2 x^2 + a^2 y^2 = a^2 b^2$$

$$a^2 y^2 = a^2 b^2 - b^2 x^2$$

$$a^2 y^2 = b^2 (a^2 - x^2)$$

$$\frac{a^2 y^2}{a^2 - x^2} = b^2$$

**39.**
$$A = \frac{2Tt + Qq}{2T + Q}$$

$$(2T + Q) \cdot A = (2T + Q) \cdot \frac{2Tt + Qq}{2T + Q}$$

$$2AT + AQ = 2Tt + Qq$$

$$AQ - Qq = 2Tt - 2AT \quad \text{Adding } -2AT$$
$$\text{and } -Qq$$

$$Q(A - q) = 2Tt - 2AT$$

$$Q = \frac{2Tt - 2AT}{A - q}$$

**41.**    $y = kx$

$28 = k \cdot 4$   Substituting

$7 = k$

The variation constant is 7.
The equation of variation is $y = 7x$.

**43.**    $y = kx$

$3.4 = k \cdot 2$   Substituting

$1.7 = k$

The variation constant is 1.7.
The equation of variation is $y = 1.7x$.

**45.**   $y = kx$

$2 = k \cdot \dfrac{1}{3}$   Substituting

$6 = k$        Multiplying by 3

The variation constant is 6.

The equation of variation is $y = 6x$.

**47.**   $y = \dfrac{k}{x}$

$3 = \dfrac{k}{20}$   Substituting

$60 = k$

The variation constant is 60.

The equation of variation is $y = \dfrac{60}{x}$.

**49.**    $y = \dfrac{k}{x}$

$28 = \dfrac{k}{4}$   Substituting

$112 = k$

The variation constant is 112.

The equation of variation is $y = \dfrac{112}{x}$.

**51.**    $y = \dfrac{k}{x}$

$27 = \dfrac{k}{\dfrac{1}{3}}$   Substituting

$9 = k$

The variation constant is 9.

The equation of variation is $y = \dfrac{9}{x}$.

**53. *Familiarize*.** Because $N$ varies directly as the number of people $P$ using the cans, we write $N$ as a function of $P$: $N(P) = kP$. We know that $N(250) = 60,000$.

***Translate*.**

$N(P) = kP$

$N(250) = k \cdot 250$   Replacing $P$ with 250

$60,000 = k \cdot 250$   Replacing $N(250)$ with 60,000

$\dfrac{60,000}{250} = k$

$240 = k$        Variation constant

$N(P) = 240P$    Equation of variation

***Carry out*.** Find $N(1,008,000)$.

$$N(P) = 240P$$
$$N(1,008,000) = 240 \cdot 1,008,000$$
$$= 241,920,000$$

***Check*.** Reexamine the calculation.

***State*.** 241,920,000 aluminum cans are used each year in Dallas.

**55. *Familiarize*.** Because of the phrase "$I$ ... varies directly as ... $V$," we express the current as a function of the voltage. Thus we have $I(V) = kV$. We know that $I(15) = 5$.

***Translate*.** We find the variation constant and then find the equation of variation.

$I(V) = kV$

$I(15) = k \cdot 15$   Replacing $V$ with 15

$5 = k \cdot 15$   Replacing $I(15)$ with 5

$\dfrac{5}{15} = k$

$\dfrac{1}{3} = k$        Variation constant

The equation of variation is $I(V) = \dfrac{1}{3}V$.

**Carry out**. We compute $I(18)$.

$$I(V) = \frac{1}{3}V$$

$$I(18) = \frac{1}{3} \cdot 18 \quad \text{Replacing } V \text{ with } 18$$

$$= 6$$

**Check**. Reexamine the calculations. Note that the answer seems reasonable since $15/5 = 18/6$.

**State**. The current is 6 amperes when 18 volts is applied.

**57. Familiarize**. Because $T$ varies inversely as $P$, we write $T(p) = k/p$. We know that $T(7) = 5$.

**Translate**. We find the variation constant and the equation of variation.

$$T(P) = \frac{k}{p}$$

$$T(7) = \frac{k}{7} \quad \text{Replacing } P \text{ with } 7$$

$$5 = \frac{k}{7} \quad \text{Replacing } T(P) \text{ with } 5$$

$$35 = k \quad \text{Variation constant}$$

$$T(P) = \frac{35}{P} \quad \text{Equation of variation}$$

**Carry out**. We find $T(10)$.

$$T(10) = \frac{35}{10}$$

$$= 3.5$$

**Check**. Reexamine the calculations.

**State**. It would take 3.5 hr for 10 volunteers to complete the job.

**59.** Since we have direct variation and $48 = \frac{1}{2} \cdot 96$, then the result is $\frac{1}{2} \cdot 64$ kg, or 32 kg. We could also do this problem as follows.

**Familiarize**. Because $W$ varies directly as the total mass, we write $W(m) = km$. We know that $W(96) = 64$.

**Translate**.

$$W(m) = km$$

$$W(96) = k \cdot 96 \quad \text{Replacing } m \text{ with } 96$$

$$64 = k \cdot 96 \quad \text{Replacing } W(96) \text{ with } 64$$

$$\frac{2}{3} = k \quad \text{Variation constant}$$

$$W(m) = \frac{2}{3}m \quad \text{Equation of variation}$$

**Carry out**. Find $W(48)$.

$$W(m) = \frac{2}{3}m$$

$$W(48) = \frac{2}{3} \cdot 48$$

$$= 32$$

**Check**. Reexamine the calculations.

**State**. There are 32 kg of water in a 64 kg person.

**61. Familiarize**. Because the number of calories burned $C$ varies directly as the time $t$ spent bicycling, we write $C(t) = kt$. We know that $C(30) = 150$.

**Translate**.

$$C(t) = kt$$

$$C(30) = k \cdot 30 \quad \text{Replacing } t \text{ with } 30$$

$$150 = k \cdot 30 \quad \text{Replacing } C(30) \text{ with } 150$$

$$5 = k \quad \text{Variation constant}$$

$$C(t) = 5t \quad \text{Equation of variation}$$

**Carry out**. We find $t$ when $C(t)$ is 250.

$$C(t) = 5t$$

$$250 = 5t$$

$$50 = t$$

**Check**. Reexamine the calculations.

**State**. It would take 50 min to burn 250 calories when biking 10 mph.

**63. Familiarize**. Because of the phrase "$t$ varies inversely as $\ldots u$," we write $t(u) = k/u$. We know that $t(4) = 75$.

**Translate**. We find the variation constant and then we find the equation of variation.

$$t(u) = \frac{k}{u}$$

$$t(4) = \frac{k}{4} \quad \text{Replacing } u \text{ with } 4$$

$$75 = \frac{k}{4} \quad \text{Replacing } t(4) \text{ with } 70$$

$$300 = k \quad \text{Variation constant}$$

$$t(u) = \frac{300}{u} \quad \text{Equation of variation}$$

**Carry out**. We find $t(14)$.

$$t(14) = \frac{300}{14} \approx 21$$

**Check**. Reexamine the calculations. Note that, as expected, as the UV rating increases, the time it takes to burn goes down.

**State**. It will take about 21 min to burn when the UV rating is 14.

**65. Familiarize**. The amount $A$ of carbon monoxide released, in tons, varies directly as the population $P$. We write $A$ as a function of $P$: $A(P) = kP$. We know that $A(2.6) = 1.1$.

**Translate**.

$$A(P) = kP$$

$$A(2.6) = k \cdot 2.6 \quad \text{Replacing } P \text{ with } 2.6$$

$$1.1 = k \cdot 2.6 \quad \text{Replacing } A(2.6) \text{ with } 1.1$$

$$\frac{11}{26} = k \quad \text{Variation constant}$$

$$A(P) = \frac{11}{26}P \quad \text{Equation of variation}$$

*Carry out*. Find $A(289,000,000)$.

$$A(P) = \frac{11}{26}P$$

$$A(289,000,000) = \frac{11}{26}(289,000,000)$$

$$\approx 122,269,231$$

*Check*. Reexamine the calculations. Answers may vary slightly due to rounding differences.

*State*. About 122,269,231 tons of carbon monoxide were released nationally.

**67.** $y = kx^2$

$\quad 6 = k \cdot 3^2 \qquad$ Substituting

$\quad 6 = 9k$

$\quad \dfrac{6}{9} = k$

$\quad \dfrac{2}{3} = k \qquad$ Variation constant

The equation of variation is $y = \dfrac{2}{3}x^2$.

**69.** $y = \dfrac{k}{x^2}$

$\quad 6 = \dfrac{k}{3^2} \qquad$ Substituting

$\quad 6 = \dfrac{k}{9}$

$\quad 6 \cdot 9 = k$

$\quad 54 = k \qquad$ Variation constant

The equation of variation is $y = \dfrac{54}{x^2}$.

**71.** $y = kxz^2$

$\quad 105 = k \cdot 14 \cdot 5^2 \qquad$ Substituting 105 for $y$,
$\qquad\qquad\qquad\qquad$ 14 for $x$, and 5 for $z$

$\quad 105 = 350k$

$\quad \dfrac{105}{350} = k$

$\quad 0.3 = k$

The equation of variation is $y = 0.3xz^2$.

**73.** $y = k \cdot \dfrac{wx^2}{z}$

$\quad 49 = k \cdot \dfrac{3 \cdot 7^2}{12} \qquad$ Substituting

$\quad 4 = k \qquad$ Variation constant

The equation of variation is $y = \dfrac{4wx^2}{z}$.

**75.** *Familiarize*. Because time $t$, in seconds, varies inversely as square of the current $c$, in amperes, we write $t = k/c^2$. We know that $t = 3.4$ when $c = 0.089$.

*Translate*. Find $k$ and the equation of variation.

$$t = \frac{k}{c^2}$$

$$3.4 = \frac{k}{(0.089)^2}$$

$$3.4(0.089)^2 = k$$

$$0.027 \approx k$$

$$t = \frac{0.027}{c^2}$$

*Carry out*. We find the value of $t$ when $c$ is 0.096.

$$t = \frac{0.027}{(0.096)^2} \approx 2.9$$

*Check*. Reexamine the calculations.

*State*. It would take about 2.9 sec for a 0.096-amp current to stop a 150-lb person's heart from beating.

**77.** *Familiarize*. Because $V$ varies directly as $T$ and inversely as $P$, we write $V = kT/P$. We know that $V = 231$ when $T = 300$ and $P = 20$.

*Translate*. Find $k$ and the equation of variation.

$$V = \frac{kT}{P}$$

$$231 = \frac{k \cdot 300}{20}$$

$$\frac{20}{300} \cdot 231 = k$$

$$15.4 = k$$

$$V = \frac{15.4T}{P} \qquad \text{Equation of variation}$$

*Carry out*. Substitute 320 for $T$ and 16 for $P$ and find $V$.

$$V = \frac{15.4(320)}{16} = 308$$

*Check*. Reexamine the calculations.

*State*. The volume is 308 cm$^3$ when $T = 320°$K and $P = 16$ lb/cm$^2$.

**79.** *Familiarize*. The drag $W$ varies jointly as the surface area $A$ and velocity $v$, so we write $W = kAv$. We know that $W = 222$ when $A = 37.8$ and $v = 40$.

*Translate*. Find $k$.

$$W = kAv$$

$$222 = k(37.8)(40)$$

$$\frac{222}{37.8(40)} = k$$

$$\frac{37}{252} = k$$

$$W = \frac{37}{252}Av \qquad \text{Equation of variation}$$

*Carry out*. Substitute 51 for $A$ and 430 for $W$ and solve for $v$.

$$430 = \frac{37}{252} \cdot 51 \cdot v$$

$$57.42 \text{ mph} \approx v$$

(If we had used the rounded value 0.1468 for $k$, the resulting speed would have been approximately 57.43 mph.)

**Check.** Reexamine the calculations.

**State.** The car must travel about 57.42 mph.

**81.** *Writing Exercise*

**83.** $f(x) = \dfrac{2x - 1}{x^2 + 1}$

$x^2 + 1 > 0$ for all real numbers $x$, so the domain of $f$ is $\{x | x \text{ is a real number}\}$.

**85.** Graph: $6x - y < 6$

First graph the line $6x - y = 6$. The intercepts are $(0, -6)$ and $(1, 0)$. We draw the line dashed since the inequality is $<$. Since the ordered pair $(0, 0)$ is a solution of the inequality ($6 \cdot 0 - 0 < 6$ is true), we shade the half-plane containing $(0, 0)$.

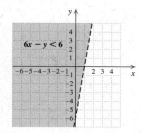

**87.** $t^3 + 8b^3 = t^3 + (2b)^3 = (t + 2b)(t^2 - 2tb + 4b^2)$

**89.** *Writing Exercise*

**91.** Use the result of Example 2.

$$h = \frac{2R^2 g}{V^2} - R$$

We have $V = 6.5$ mi/sec, $R = 3960$ mi, and $g = 32.2$ ft/sec$^2$. We must convert 32.2 ft/sec$^2$ to mi/sec$^2$ so all units of length are the same.

$$32.2 \frac{\cancel{\text{ft}}}{\text{sec}^2} \cdot \frac{1 \text{ mi}}{5280 \cancel{\text{ft}}} \approx 0.0060984 \frac{\text{mi}}{\text{sec}^2}$$

Now we substitute and compute.

$$h = \frac{2(3960)^2(0.0060984)}{(6.5)^2} - 3960$$

$$h \approx 567$$

The satellite is about 567 mi from the surface of the earth.

**93.** $c = \dfrac{a}{a + 12} \cdot d$

$c = \dfrac{2a}{2a + 12} \cdot d$  Doubling $a$

$= \dfrac{\cancel{2}a}{\cancel{2}(a + 6)} \cdot d$

$= \dfrac{a}{a + 6} \cdot d$  Simplifying

The ratio of the larger dose to the smaller dose is

$$\frac{\dfrac{a}{a + 6} \cdot d}{\dfrac{a}{a + 12} \cdot d} = \frac{\dfrac{ad}{a + 6}}{\dfrac{ad}{a + 12}}$$

$$= \frac{ad}{a + 6} \cdot \frac{a + 12}{ad}$$

$$= \frac{\cancel{ad}(a + 12)}{(a + 6)\cancel{ad}}$$

$$= \frac{a + 12}{a + 6}.$$

The amount by which the dosage increases is

$$\frac{a}{a + 6} \cdot d - \frac{a}{a + 12} \cdot d$$

$$\frac{ad}{a + 6} - \frac{ad}{a + 12}$$

$$= \frac{ad}{a + 6} \cdot \frac{a + 12}{a + 12} - \frac{ad}{a + 12} \cdot \frac{a + 6}{a + 6}$$

$$= \frac{ad(a + 12) - ad(a + 6)}{(a + 6)(a + 12)}$$

$$= \frac{a^2 d + 12ad - a^2 d - 6ad}{(a + 6)(a + 12)}$$

$$= \frac{6ad}{(a + 6)(a + 12)}.$$

Then the percent by which the dosage increases is

$$\frac{\dfrac{6ad}{(a + 6)(a + 12)}}{\dfrac{a}{a + 12} \cdot d} = \frac{\dfrac{6ad}{(a + 6)(a + 12)}}{\dfrac{ad}{a + 12}}$$

$$= \frac{6ad}{(a + 6)(a + 12)} \cdot \frac{a + 12}{ad}$$

$$= \frac{6 \cdot \cancel{ad} \cdot (a \cancel{+ 12})}{(a + 6)(a \cancel{+ 12}) \cdot \cancel{ad}}$$

$$= \frac{6}{a + 6}.$$

This is a decimal representation for the percent of increase. To give the result in percent notation we multiply by 100 and use a percent symbol. We have

$$\frac{6}{a + 6} \cdot 100\%, \text{ or } \frac{600}{a + 6}\%.$$

**95.**

$$a = \frac{\dfrac{d_4 - d_3}{t_4 - t_3} - \dfrac{d_2 - d_1}{t_2 - t_1}}{t_4 - t_2}$$

$$a(t_4 - t_2) = \frac{d_4 - d_3}{t_4 - t_3} - \frac{d_2 - d_1}{t_2 - t_1} \quad \begin{array}{l}\text{Multiplying} \\ \text{by } t_4 - t_2\end{array}$$

$$a(t_4 - t_2)(t_4 - t_3)(t_2 - t_1) = (d_4 - d_3)(t_2 - t_1) - (d_2 - d_1)(t_4 - t_3)$$
$$\text{Multiplying by } (t_4 - t_3)(t_2 - t_1)$$
$$a(t_4 - t_2)(t_4 - t_3)(t_2 - t_1) - (d_4 - d_3)(t_2 - t_1) = -(d_2 - d_1)(t_4 - t_3)$$
$$(t_2 - t_1)[a(t_4 - t_2)(t_4 - t_3) - (d_4 - d_3)] = -(d_2 - d_1)(t_4 - t_3)$$
$$t_2 - t_1 = \frac{-(d_2 - d_1)(t_4 - t_3)}{a(t_4 - t_2)(t_4 - t_3) - (d_4 - d_3)}$$
$$t_2 + \frac{(d_2 - d_1)(t_4 - t_3)}{a(t_4 - t_2)(t_4 - t_3) + d_3 - d_4} = t_1$$

190                            **Chapter 6: Rational Expressions, Equations, and Functions**

**97.** Let $w$ = the wattage of the bulb. Then we have $I = \dfrac{kw}{d^2}$.

Now substitute $2w$ for $w$ and $2d$ for $d$.

$$I = \frac{k(2w)}{(2d)^2} = \frac{2kw}{4d^2} = \frac{kw}{2d^2} = \frac{1}{2} \cdot \frac{kw}{d^2}$$

We see that the intensity is halved.

**99. *Familiarize*.** We write $T = kml^2 f^2$. We know that $T = 100$ when $m = 5$, $l = 2$, and $f = 80$.

***Translate*.** Find $k$.

$$T = kml^2 f^2$$
$$100 = k(5)(2)^2(80)^2$$
$$0.00078125 = k$$
$$T = 0.00078125 ml^2 f^2$$

***Carry out*.** Substitute 72 for $T$, 5 for $m$, and 80 for $f$ and solve for $l$.

$$72 = 0.00078125(5)(l^2)(80)^2$$
$$2.88 = l^2$$
$$1.697 \approx l$$

***Check*.** Recheck the calculations.

***State*.** The string should be about 1.697 m long.

**101. *Familiarize*.** Because $d$ varies inversely as $s$, we write $d(s) = k/s$. We know that $d(0.56) = 50$.

***Translate*.**

$$d(s) = \frac{k}{s}$$
$$d(0.56) = \frac{k}{0.56} \quad \text{Replacing } s \text{ with } 0.56$$
$$50 = \frac{k}{0.56} \quad \text{Replacing } d(0.56) \text{ with } 50$$
$$28 = k$$
$$d(s) = \frac{28}{s} \quad \text{Equation of variation}$$

***Carry out*.** Find $d(0.40)$.

$$d(0.40) = \frac{28}{0.40}$$
$$= 70$$

***Check*.** Reexamine the calculations. Also observe that, as expected, when $d$ decreases, then $s$ increases.

***State*.** The equation of variation is $d(s) = \dfrac{28}{s}$. The distance is 70 yd.

# Chapter 7

# Exponents and Radicals

1. two; see page 436 in the text.

3. positive; see Example 4.

5. irrational; see page 437 in the text.

7. nonnegative; see page 441 in the text.

9. The square roots of 49 are 7 and $-7$, because $7^2 = 49$ and $(-7)^2 = 49$.

11. The square roots of 144 are 12 and $-12$, because $12^2 = 144$ and $(-12)^2 = 144$.

13. The square roots of 400 are 20 and $-20$, because $20^2 = 400$ and $(-20)^2 = 400$.

15. The square roots of 900 are 30 and $-30$, because $30^2 = 900$ and $(-30)^2 = 900$.

17. $-\sqrt{\dfrac{36}{49}} = -\dfrac{6}{7}$   Since $\sqrt{\dfrac{36}{49}} = \dfrac{6}{7}$, $-\sqrt{\dfrac{36}{49}} = -\dfrac{6}{7}$.

19. $\sqrt{441} = 21$   Remember, $\sqrt{\phantom{-}}$ indicates the principle square root.

21. $-\sqrt{\dfrac{16}{81}} = -\dfrac{4}{9}$   Since $\sqrt{\dfrac{16}{81}} = \dfrac{4}{9}$, $-\sqrt{\dfrac{16}{81}} = -\dfrac{4}{9}$.

23. $\sqrt{0.04} = 0.2$

25. $-\sqrt{-0.0025} = -0.05$

27. $5\sqrt{p^2 + 4}$

The radicand is the expression written under the radical sign, $p^2 + 4$.

Since the index is not written, we know it is 2.

29. $x^2 y^2 \sqrt{\dfrac{x}{y+4}}$

The radicand is the expression written under the radical sign, $\dfrac{x}{y+4}$.

The index is 3.

31. $f(t) = \sqrt{5t - 10}$

$f(6) = \sqrt{5 \cdot 6 - 10} = \sqrt{20}$

$f(2) = \sqrt{5 \cdot 2 - 10} = \sqrt{0} = 0$

$f(1) = \sqrt{5 \cdot 1 - 10} = \sqrt{-5}$

Since negative numbers do not have real-number square roots, $f(1)$ does not exist.

$f(-1) = \sqrt{5(-1) - 10} = \sqrt{-15}$

Since negative numbers do not have real-number square roots, $f(-1)$ does not exist.

33. $t(x) = -\sqrt{2x + 1}$

$t(4) = -\sqrt{2 \cdot 4 + 1} = -\sqrt{9} = -3$

$t(0) = -\sqrt{2 \cdot 0 + 1} = -\sqrt{1} = -1$

$t(-1) = -\sqrt{2(-1) + 1} = -\sqrt{-1}$;

$t(-1)$ does not exist.

$t\left(-\dfrac{1}{2}\right) = -\sqrt{2\left(-\dfrac{1}{2}\right) + 1} = -\sqrt{0} = 0$

35. $f(t) = \sqrt{t^2 + 1}$

$f(0) = \sqrt{0^2 + 1} = \sqrt{1} = 1$

$f(-1) = \sqrt{(-1)^2 + 1} = \sqrt{2}$

$f(-10) = \sqrt{(-10)^2 + 1} = \sqrt{101}$

37. $g(x) = \sqrt{x^3 + 9}$

$g(-2) = \sqrt{(-2)^3 + 9} = \sqrt{1} = 1$

$g(-3) = \sqrt{(-3)^3 + 9} = \sqrt{-18}$;

$g(-3)$ does not exist.

$g(3) = \sqrt{3^3 + 9} = \sqrt{36} = 6$

39. $\sqrt{36x^2} = \sqrt{(6x)^2} = |6x| = 6|x|$

Since $x$ might be negative, absolute-value notation is necessary.

41. $\sqrt{(-6b)^2} = |-6b| = |-6| \cdot |b| = 6|b|$

Since $b$ might be negative, absolute-value notation is necessary.

43. $\sqrt{(8-t)^2} = |8 - t|$

Since $8 - t$ might be negative, absolute-value notation is necessary.

45. $\sqrt{y^2 + 16y + 64} = \sqrt{(y+8)^2} = |y + 8|$

Since $y + 8$ might be negative, absolute-value notation is necessary.

47. $\sqrt{4x^2 + 28x + 49} = \sqrt{(2x+7)^2} = |2x + 7|$

Since $2x + 7$ might be negative, absolute-value notation is necessary.

49. $\sqrt[4]{256} = 4$   Since $4^4 = 256$

51. $\sqrt[5]{-1} = -1$   Since $(-1)^5 = -1$

53. $\sqrt[5]{-\dfrac{32}{243}} = -\dfrac{2}{3}$   Since $\left(-\dfrac{2}{3}\right)^5 = -\dfrac{32}{243}$

**55.** $\sqrt[6]{x^6} = |x|$

The index is even. Use absolute-value notation since $x$ could have a negative value.

**57.** $\sqrt[4]{(6a)^4} = |6a| = 6|a|$

The index is even. Use absolute-value notation since $a$ could have a negative value.

**59.** $\sqrt[10]{(-6)^{10}} = |-6| = 6$

**61.** $\sqrt[414]{(a+b)^{414}} = |a+b|$

The index is even. Use absolute-value notation since $a+b$ could have a negative value.

**63.** $\sqrt{a^{22}} = |a^{11}|$  Note that $(a^{11})^2 = a^{22}$; $a^{11}$ could have a negative value.

**65.** $\sqrt{-25}$ is not a real number, so $\sqrt{-25}$ cannot be simplified.

**67.** $\sqrt{16x^2} = \sqrt{(4x)^2} = 4x$  Assuming $x$ is nonnegative

**69.** $\sqrt{(3c)^2} = 3c$  Assuming $3c$ is nonnegative

**71.** $\sqrt{(a+1)^2} = a+1$  Assuming $a+1$ is nonnegative

**73.** $\sqrt{4x^2 + 8x + 4} = \sqrt{4(x^2 + 2x + 1)} =$
$\sqrt{[2(x+1)]^2} = 2(x+1)$, or $2x+2$

**75.** $\sqrt{9t^2 - 12t + 4} = \sqrt{(3t-2)^2} = 3t - 2$

**77.** $\sqrt[3]{27} = 3$      $(3^3 = 27)$

**79.** $\sqrt[4]{16x^4} = \sqrt[4]{(2x)^4} = 2x$

**81.** $\sqrt[3]{-216} = -6$      $[(-6)^3 = -216]$

**83.** $-\sqrt[3]{-125y^3} = -(-5y)$   $[(-5y)^3 = -125y^3]$
$= 5y$

**85.** $\sqrt{t^{18}} = \sqrt{(t^9)^2} = t^9$

**87.** $\sqrt{(x-2)^8} = \sqrt{[(x-2)^4]^2} = (x-2)^4$

**89.**      $f(x) = \sqrt[3]{x+1}$
      $f(7) = \sqrt[3]{7+1} = \sqrt[3]{8} = 2$
      $f(26) = \sqrt[3]{26+1} = \sqrt[3]{27} = 3$
      $f(-9) = \sqrt[3]{-9+1} = \sqrt[3]{-8} = -2$
      $f(-65) = \sqrt[3]{-65+1} = \sqrt[3]{-64} = -4$

**91.**      $g(t) = \sqrt[4]{t-3}$
      $g(19) = \sqrt[4]{19-3} = \sqrt[4]{16} = 2$
      $g(-13) = \sqrt[4]{-13-3} = \sqrt[4]{-16};$
            $g(-13)$ does not exist.
      $g(1) = \sqrt[4]{1-3} = \sqrt[4]{-2};$
            $g(1)$ does not exist.
      $g(84) = \sqrt[4]{84-3} = \sqrt[4]{81} = 3$

**93.** $f(x) = \sqrt{x-6}$

Since the index is even, the radicand, $x - 6$, must be nonnegative. We solve the inequality:
$$x - 6 \geq 0$$
$$x \geq 6$$
Domain of $f = \{x | x \geq 6\}$, or $[6, \infty)$

**95.** $g(t) = \sqrt[4]{t+8}$

Since the index is even, the radicand, $t + 8$, must be nonnegative. We solve the inequality:
$$t + 8 \geq 0$$
$$t \geq -8$$
Domain of $g = \{t | t \geq -8\}$, or $[-8, \infty)$

**97.** $g(x) = \sqrt[4]{2x - 10}$

Since the index is even, the radicand, $2x - 10$, must be nonnegative. We solve the inequality:
$$2x - 10 \geq 0$$
$$2x \geq 10$$
$$x \geq 5$$
Domain of $g = \{x | x \geq 5\}$, or $[5, \infty)$

**99.** $f(t) = \sqrt[5]{8 - 3t}$

Since the index is odd, the radicand can be any real number.

Domain of $f = \{t | t$ is a real number$\}$, or $(-\infty, \infty)$

**101.** $h(z) = -\sqrt[6]{5z + 2}$

Since the index is even, the radicand, $5z + 2$, must be nonnegative. We solve the inequality:
$$5z + 2 \geq 0$$
$$5z \geq -2$$
$$z \geq -\frac{2}{5}$$
Domain of $h = \left\{ z | z \geq -\frac{2}{5} \right\}$, or $\left[ -\frac{2}{5}, \infty \right)$

**103.** $f(t) = 7 + \sqrt[8]{t^8}$

Since we can compute $7 + \sqrt[8]{t^8}$ for any real number $t$, the domain is the set of real numbers, or $\{x | x$ is a real number$\}$, or $(-\infty, \infty)$.

**105.** *Writing Exercise*

**107.** $(a^3 b^2 c^5)^3 = a^{3 \cdot 3} b^{2 \cdot 3} c^{5 \cdot 3} = a^9 b^6 c^{15}$

**109.** $(2a^{-2} b^3 c^{-4})^{-3} = 2^{-3} a^{-2(-3)} b^{3(-3)} c^{-4(-3)} =$
$\frac{1}{2^3} a^6 b^{-9} c^{12} = \frac{a^6 c^{12}}{8b^9}$

**111.** $\frac{8x^{-2} y^5}{4x^{-6} z^{-2}} = \frac{8}{4} x^{-2-(-6)} y^5 z^2 = 2x^4 y^5 z^2$

**113.** *Writing Exercise*

**115.** $M = -5 + \sqrt{6.7x - 444}$

    a) Substitute 300 for $x$.
$$M = -5 + \sqrt{6.7(300) - 444}$$
$$= -5 + \sqrt{1566}$$
$$\approx 34.6 \text{ lb}$$

    b) Substitute 100 for $x$.
$$M = -5 + \sqrt{6.7(100) - 444}$$
$$= -5 + \sqrt{226}$$
$$\approx 10.0 \text{ lb}$$

    c) Substitute 200 for $x$.
$$M = -5 + \sqrt{6.7(200) - 444}$$
$$= -5 + \sqrt{896}$$
$$\approx 24.9 \text{ lb}$$

    d) Substitute 400 for $x$.
$$M = -5 + \sqrt{6.7(400) - 444}$$
$$= -5 + \sqrt{2236}$$
$$\approx 42.3 \text{ lb}$$

**117.** $f(x) = \sqrt{x + 5}$

Since the index is even, the radicand, $x + 5$, must be non-negative. Solve:
$$x + 5 \geq 0$$
$$x \geq -5$$

Domain of $f = \{x | x \geq -5\}$, or $[-5, \infty)$

Make a table of values, keeping in mind that $x$ must be $-5$ or greater. Plot these points and draw the graph.

| $x$ | $f(x)$ |
|-----|--------|
| $-5$ | $0$ |
| $-4$ | $1$ |
| $-1$ | $2$ |
| $1$ | $2.4$ |
| $3$ | $2.8$ |
| $4$ | $3$ |

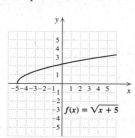

$f(x) = \sqrt{x + 5}$

**119.** $g(x) = \sqrt{x} - 2$

Since the index is even, the radicand, $x$, must be nonnegative, so we have $x \geq 0$.

Domain of $g = \{x | x \geq 0\}$, or $[0, \infty)$

Make a table of values, keeping in mind that $x$ must be nonnegative. Plot these points and draw the graph.

| $x$ | $g(x)$ |
|-----|--------|
| $0$ | $-2$ |
| $1$ | $-1$ |
| $4$ | $0$ |
| $6$ | $0.5$ |
| $8$ | $0.8$ |

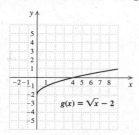

$g(x) = \sqrt{x} - 2$

**121.** $f(x) = \dfrac{\sqrt{x + 3}}{\sqrt[4]{2 - x}}$

In the numerator we must have $x + 3 \geq 0$, or $x \geq -3$, and in the denominator we must have $2 - x > 0$, or $x < 2$. Thus, we have $x \geq -3$ *and* $x < 2$, so

Domain of $f = \{x | -3 \leq x < 2\}$, or $[-3, 2)$.

**123.** ■

---

## Exercise Set 7.2

**1.** Choice (g) is correct because $a^{m/n} = \sqrt[n]{a^m}$.

**3.** $x^{-5/2} = \dfrac{1}{x^{5/2}} = \dfrac{1}{(\sqrt{x})^5}$, so choice (e) is correct.

**5.** $x^{1/5} \cdot x^{2/5} = x^{1/5 + 2/5} = x^{3/5}$, so choice (a) is correct.

**7.** Choice (b) is correct because $\sqrt[n]{a^m}$ and $(\sqrt[n]{a})^m$ are equivalent.

**9.** $x^{1/6} = \sqrt[6]{x}$

**11.** $(16)^{1/2} = \sqrt{16} = 4$

**13.** $81^{1/4} = \sqrt[4]{81} = 3$

**15.** $9^{1/2} = \sqrt{9} = 3$

**17.** $(xyz)^{1/3} = \sqrt[3]{xyz}$

**19.** $(a^2b^2)^{1/5} = \sqrt[5]{a^2b^2}$

**21.** $t^{2/5} = \sqrt[5]{t^2}$

**23.** $16^{3/4} = \sqrt[4]{16^3} = (\sqrt[4]{16})^3 = 2^3 = 8$

**25.** $27^{4/3} = \sqrt[3]{27^4} = (\sqrt[3]{27})^4 = 3^4 = 81$

**27.** $(81x)^{3/4} = \sqrt[4]{(81x)^3} = \sqrt[4]{81^3 x^3}$, or $\sqrt[4]{81^3} \cdot \sqrt[4]{x^3} = (\sqrt[4]{81})^3 \cdot \sqrt[4]{x^3} = 3^3 \sqrt[4]{x^3} = 27\sqrt[4]{x^3}$

**29.** $(25x^4)^{3/2} = \sqrt{(25x^4)^3} = \sqrt{25^3 \cdot x^{12}} = \sqrt{25^3} \cdot \sqrt{x^{12}} = (\sqrt{25})^3 x^6 = 5^3 x^6 = 125x^6$

**31.** $\sqrt[3]{20} = 20^{1/3}$

**33.** $\sqrt{17} = 17^{1/2}$

**35.** $\sqrt{x^3} = x^{3/2}$

**37.** $\sqrt[5]{m^2} = m^{2/5}$

**39.** $\sqrt[4]{cd} = (cd)^{1/4}$    Parentheses are required.

**41.** $\sqrt[5]{xy^2z} = (xy^2z)^{1/5}$

**43.** $(\sqrt{3mn})^3 = (3mn)^{3/2}$

**45.** $(\sqrt[7]{8x^2y})^5 = (8x^2y)^{5/7}$

**47.** $\dfrac{2x}{\sqrt[3]{z^2}} = \dfrac{2x}{z^{2/3}}$

**49.** $x^{-1/3} = \dfrac{1}{x^{1/3}}$

**51.** $(2rs)^{-3/4} = \dfrac{1}{(2rs)^{3/4}}$

**53.** $\left(\dfrac{1}{16}\right)^{-3/4} = \left(\dfrac{16}{1}\right)^{3/4} = (2^4)^{3/4} = 2^{4(3/4)} = 2^3 = 8$

**55.** $\dfrac{2c}{a^{-3/5}} = 2a^{3/5}c$

**57.** $5x^{-2/3}y^{4/5}z = 5 \cdot \dfrac{1}{x^{2/3}} \cdot y^{4/5} \cdot z = \dfrac{5y^{4/5}z}{x^{2/3}}$

**59.** $3^{-5/2}a^3b^{-7/3} = \dfrac{1}{3^{5/2}} \cdot a^3 \cdot \dfrac{1}{b^{7/3}} = \dfrac{a^3}{3^{5/2}b^{7/3}}$

**61.** $\left(\dfrac{2ab}{3c}\right)^{-5/6} = \left(\dfrac{3c}{2ab}\right)^{5/6}$  Finding the reciprocal of the base and changing the sign of the exponent

**63.** $\dfrac{6a}{\sqrt[4]{b}} = \dfrac{6a}{b^{1/4}}$

**65.** $7^{3/4} \cdot 7^{1/8} = 7^{3/4+1/8} = 7^{6/8+1/8} = 7^{7/8}$

We added exponents after finding a common denominator.

**67.** $\dfrac{3^{5/8}}{3^{-1/8}} = 3^{5/8-(-1/8)} = 3^{5/8+1/8} = 3^{6/8} = 3^{3/4}$

We subtracted exponents and simplified.

**69.** $\dfrac{5.2^{-1/6}}{5.2^{-2/3}} = 5.2^{-1/6-(-2/3)} = 5.2^{-1/6+2/3} =$
$5.2^{-1/6+4/6} = 5.2^{3/6} = 5.2^{1/2}$

We subtracted exponents after finding a common denominator. Then we simplified.

**71.** $(10^{3/5})^{2/5} = 10^{3/5 \cdot 2/5} = 10^{6/25}$

We multiplied exponents.

**73.** $a^{2/3} \cdot a^{5/4} = a^{2/3+5/4} = a^{8/12+15/12} = a^{23/12}$

We added exponents after finding a common denominator.

**75.** $(64^{3/4})^{4/3} = 64^{\frac{3}{4}\cdot\frac{4}{3}} = 64^1 = 64$

**77.** $(m^{2/3}n^{-1/4})^{1/2} = m^{2/3 \cdot 1/2}n^{-1/4 \cdot 1/2} = m^{1/3}n^{-1/8} =$
$m^{1/3} \cdot \dfrac{1}{n^{1/8}} = \dfrac{m^{1/3}}{n^{1/8}}$

**79.** $\sqrt[6]{x^4} = x^{4/6}$    Converting to exponential notation
$\phantom{\sqrt[6]{x^4}} = x^{2/3}$    Simplifying the exponent
$\phantom{\sqrt[6]{x^4}} = \sqrt[3]{x^2}$    Returning to radical notation

**81.** $\sqrt[4]{a^{12}} = a^{12/4}$    Converting to exponential notation
$\phantom{\sqrt[4]{a^{12}}} = a^3$    Simplifying

**83.** $\sqrt[5]{a^{10}} = a^{10/5}$    Converting to exponential notation
$\phantom{\sqrt[5]{a^{10}}} = a^2$    Simplifying

**85.** $\left(\sqrt[7]{xy}\right)^{14} = (xy)^{14/7}$    Converting to exponential notation
$\phantom{\left(\sqrt[7]{xy}\right)^{14}} = (xy)^2$    Simplifying the exponent
$\phantom{\left(\sqrt[7]{xy}\right)^{14}} = x^2y^2$    Using the laws of exponents

**87.** $\sqrt[4]{(7a)^2} = (7a)^{2/4}$    Converting to exponential notation
$\phantom{\sqrt[4]{(7a)^2}} = (7a)^{1/2}$    Simplifying the exponent
$\phantom{\sqrt[4]{(7a)^2}} = \sqrt{7a}$    Returning to radical notation

**89.** $(\sqrt[8]{2x})^6 = (2x)^{6/8}$    Converting to exponential notation
$\phantom{(\sqrt[8]{2x})^6} = (2x)^{3/4}$    Simplifying the exponent
$\phantom{(\sqrt[8]{2x})^6} = \sqrt[4]{(2x)^3}$    Returning to radical notation
$\phantom{(\sqrt[8]{2x})^6} = \sqrt[4]{8x^3}$    Using the laws of exponents

**91.** $\sqrt[3]{\sqrt[6]{a}} = \sqrt[3]{a^{1/6}}$    Converting to
$\phantom{\sqrt[3]{\sqrt[6]{a}}} = (a^{1/6})^{1/3}$    exponential notation
$\phantom{\sqrt[3]{\sqrt[6]{a}}} = a^{1/18}$    Using the laws of exponents
$\phantom{\sqrt[3]{\sqrt[6]{a}}} = \sqrt[18]{a}$    Returning to radical notation

**93.** $\sqrt[4]{(xy)^{12}} = (xy)^{12/4}$    Converting to exponential notation
$\phantom{\sqrt[4]{(xy)^{12}}} = (xy)^3$    Simplifying the exponent
$\phantom{\sqrt[4]{(xy)^{12}}} = x^3y^3$    Using the laws of exponents

**95.** $(\sqrt[5]{a^2b^4})^{15} = (a^2b^4)^{15/5}$    Converting to exponential notation
$\phantom{(\sqrt[5]{a^2b^4})^{15}} = (a^2b^4)^3$    Simplifying the exponent
$\phantom{(\sqrt[5]{a^2b^4})^{15}} = a^6b^{12}$    Using the laws of exponents

**97.** $\sqrt[3]{\sqrt[4]{xy}} = \sqrt[3]{(xy)^{1/4}}$    Converting to
$\phantom{\sqrt[3]{\sqrt[4]{xy}}} = [(xy)^{1/4}]^{1/3}$    exponential notation
$\phantom{\sqrt[3]{\sqrt[4]{xy}}} = (xy)^{1/12}$    Using a law of exponents
$\phantom{\sqrt[3]{\sqrt[4]{xy}}} = \sqrt[12]{xy}$    Returning to radical notation

**99.** *Writing Exercise*

**101.**    $3x(x^3 - 2x^2) + 4x^2(2x^2 + 5x)$
$= 3x^4 - 6x^3 + 8x^4 + 20x^3$
$= 11x^4 + 14x^3$

**103.**    $(3a - 4b)(5a + 3b)$
$= 3a \cdot 5a + 3a \cdot 3b - 4b \cdot 5a - 4b \cdot 3b$
$= 15a^2 + 9ab - 20ab - 12b^2$
$= 15a^2 - 11ab - 12b^2$

**105.** *Familiarize*. Let $p =$ the selling price of the home.
*Translate*.
$\underbrace{0.5\% \text{ of the selling price}}_{0.005p}$ $\underset{=}{\text{ is }}$ $\underset{467.50}{\$467.50}$

*Carry out*. We solve the equation.
$0.005p = 467.50$
$\phantom{0.005}p = 93,500$   Dividing by 0.005

*Check*. 0.5% of \$93,500 is 0.005(\$93,500), or \$467.50. The answer checks.

*State*. The selling price of the home was \$93,500.

**107.** *Writing Exercise*

**109.** $\sqrt{x\sqrt[3]{x^2}} = \sqrt{x \cdot x^{2/3}} = (x^{5/3})^{1/2} = x^{5/6} = \sqrt[6]{x^5}$

**111.** $\sqrt[12]{p^2 + 2pq + q^2} = \sqrt[12]{(p+q)^2} = [(p+q)^2]^{1/12} =$
$(p+q)^{2/12} = (p+q)^{1/6} = \sqrt[6]{p+q}$

**113.** $2^{7/12} \approx 1.498 \approx 1.5$ so the G that is 7 half steps above middle C has a frequency that is about 1.5 times that of middle C.

**115.** $P = \dfrac{r^{1.83}}{r^{1.83} + \sigma^{1.83}}$

$= \dfrac{799^{1.83}}{799^{1.83} + 749^{1.83}}$

$\approx 53.0\%$    Using a calculator

**117.** $r(1900) = 10^{-12}2^{-1900/5700} \approx 10^{-12}(0.7937)$, or $7.937 \times 10^{-13}$. The ratio is about $7.937 \times 10^{-13}$ to 1.

**119.**   $y_1 = x^{1/2}$, $y_2 = 3x^{2/5}$,
$y_3 = x^{4/7}$, $y_4 = \frac{1}{5}x^{3/4}$

---

## Exercise Set 7.3

**1.** True; see page 453 in the text.

**3.** False; see page 454 in the text.

**5.** True; see page 454 in the text.

**7.** $\sqrt{5}\sqrt{7} = \sqrt{5 \cdot 7} = \sqrt{35}$

**9.** $\sqrt[3]{7}\sqrt[3]{2} = \sqrt[3]{7 \cdot 2} = \sqrt[3]{14}$

**11.** $\sqrt[4]{6}\sqrt[4]{3} = \sqrt[4]{6 \cdot 3} = \sqrt[4]{18}$

**13.** $\sqrt{2x}\sqrt{13y} = \sqrt{2x \cdot 13y} = \sqrt{26xy}$

**15.** $\sqrt[5]{8y^3}\sqrt[5]{10y} = \sqrt[5]{8y^3 \cdot 10y} = \sqrt[5]{80y^4}$

**17.** $\sqrt{y-b}\sqrt{y+b} = \sqrt{(y-b)(y+b)} = \sqrt{y^2 - b^2}$

**19.** $\sqrt[3]{0.7y}\sqrt[3]{0.3y} = \sqrt[3]{0.7y \cdot 0.3y} = \sqrt[3]{0.21y^2}$

**21.** $\sqrt[5]{x-2}\sqrt[5]{(x-2)^2} = \sqrt[5]{(x-2)(x-2)^2} = \sqrt[5]{(x-2)^3}$

**23.** $\sqrt{\dfrac{7}{t}}\sqrt{\dfrac{s}{11}} = \sqrt{\dfrac{7}{t} \cdot \dfrac{s}{11}} = \sqrt{\dfrac{7s}{11t}}$

**25.** $\sqrt[7]{\dfrac{x-3}{4}}\sqrt[7]{\dfrac{5}{x+2}} = \sqrt[7]{\dfrac{x-3}{4} \cdot \dfrac{5}{x+2}} = \sqrt[7]{\dfrac{5x-15}{4x+8}}$

**27.**   $\sqrt{18}$

$= \sqrt{9 \cdot 2}$    9 is the largest perfect square factor of 18.

$= \sqrt{9} \cdot \sqrt{2}$

$= 3\sqrt{2}$

**29.**   $\sqrt{27}$

$= \sqrt{9 \cdot 3}$    9 is the largest perfect square factor of 27.

$= \sqrt{9} \cdot \sqrt{3}$

$= 3\sqrt{3}$

**31.** $\sqrt{8} = \sqrt{4 \cdot 2} = \sqrt{4} \cdot \sqrt{2} = 2\sqrt{2}$

**33.** $\sqrt{198} = \sqrt{9 \cdot 22} = \sqrt{9} \cdot \sqrt{22} = 3\sqrt{22}$

**35.**   $\sqrt{36a^4b}$

$= \sqrt{36a^4 \cdot b}$    $36a^4$ is a perfect square.

$= \sqrt{36a^4} \cdot \sqrt{b}$    Factoring into two radicals

$= 6a^2\sqrt{b}$    Taking the square root of $36a^4$

**37.**   $\sqrt[3]{8x^3y^2}$

$= \sqrt[3]{8x^3 \cdot y^2}$    $8x^3$ is a perfect cube.

$= \sqrt[3]{8x^3} \cdot \sqrt[3]{y^2}$    Factoring into two radicals

$= 2x\sqrt[3]{y^2}$    Taking the cube root of $8x^3$

**39.**   $\sqrt[3]{-16x^6}$

$= \sqrt[3]{-8x^6 \cdot 2}$    $-8x^6$ is a perfect cube.

$= \sqrt[3]{-8x^6} \cdot \sqrt[3]{2}$

$= -2x^2\sqrt[3]{2}$    Taking the cube root of $-8x^6$

**41.** $f(x) = \sqrt[3]{125x^5}$

$= \sqrt[3]{125x^3 \cdot x^2}$

$= \sqrt[3]{125x^3} \cdot \sqrt[3]{x^2}$

$= 5x\sqrt[3]{x^2}$

**43.** $f(x) = \sqrt{49(x-3)^2}$    $49(x-3)^2$ is a perfect square.

$= |7(x-3)|$, or $7|x-3|$

**45.** $f(x) = \sqrt{5x^2 - 10x + 5}$

$= \sqrt{5(x^2 - 2x + 1)}$

$= \sqrt{5(x-1)^2}$

$= \sqrt{(x-1)^2} \cdot \sqrt{5}$

$= |x-1|\sqrt{5}$

**47.**   $\sqrt{a^6b^7}$

$= \sqrt{a^6 \cdot b^6 \cdot b}$    Identifying the largest even powers of $a$ and $b$

$= \sqrt{a^6}\sqrt{b^6}\sqrt{b}$    Factoring into several radicals

$= a^3b^3\sqrt{b}$

**49.**   $\sqrt[3]{x^5y^6z^{10}}$

$= \sqrt[3]{x^3 \cdot x^2 \cdot y^6 \cdot z^9 \cdot z}$    Identifying the largest perfect-cube powers of $x$, $y$, and $z$

$= \sqrt[3]{x^3} \cdot \sqrt[3]{y^6} \cdot \sqrt[3]{z^9} \cdot \sqrt[3]{x^2z}$    Factoring into several radicals

$= xy^2z^3\sqrt[3]{x^2z}$

**51.** $\sqrt[5]{-32a^7b^{11}} = \sqrt[5]{-32 \cdot a^5 \cdot a^2 \cdot b^{10} \cdot b} =$
$\sqrt[5]{-32}\sqrt[5]{a^5}\sqrt[5]{b^{10}}\sqrt[5]{a^2b} = -2ab^2\sqrt[5]{a^2b}$

**53.** $\sqrt[5]{x^{13}y^8z^{17}} = \sqrt[5]{x^{10} \cdot x^3 \cdot y^5 \cdot y^3 \cdot z^{15} \cdot z^2} =$
$\sqrt[5]{x^{10}}\sqrt[5]{y^5}\sqrt[5]{z^{15}}\sqrt[5]{x^3y^3z^2} = x^2yz^3\sqrt[5]{x^3y^3z^2}$

**55.** $\sqrt[3]{-80a^{14}} = \sqrt[3]{-8 \cdot 10 \cdot a^{12} \cdot a^2} =$
$\sqrt[3]{-8} \cdot \sqrt[3]{a^{12}} \cdot \sqrt[3]{10a^2} = -2a^4\sqrt[3]{10a^2}$

**57.** $\sqrt{6}\sqrt{3} = \sqrt{6 \cdot 3} = \sqrt{18} = \sqrt{9 \cdot 2} = 3\sqrt{2}$

**59.** $\sqrt{15}\sqrt{21} = \sqrt{15 \cdot 21} = \sqrt{315} = \sqrt{9 \cdot 35} = 3\sqrt{35}$

**61.** $\sqrt[3]{9}\sqrt[3]{3} = \sqrt[3]{9 \cdot 3} = \sqrt[3]{27} = 3$

**63.** $\sqrt{18a^3}\sqrt{18a^3} = \sqrt{(18a^3)^2} = 18a^3$

**65.** $\sqrt[3]{5a^2}\sqrt[3]{2a} = \sqrt[3]{5a^2 \cdot 2a} = \sqrt[3]{10a^3} = \sqrt[3]{a^3 \cdot 10} = a\sqrt[3]{10}$

**67.** $\sqrt{2x^5}\sqrt{10x^2} = \sqrt{20x^7} = \sqrt{4x^6 \cdot 5x} = 2x^3\sqrt{5x}$

**69.** $\sqrt[3]{s^2t^4}\sqrt[3]{s^4t^6} = \sqrt[3]{s^6t^{10}} = \sqrt[3]{s^6t^9 \cdot t} = s^2t^3\sqrt[3]{t}$

**71.** $\sqrt[3]{(x+5)^2}\sqrt[3]{(x+5)^4} = \sqrt[3]{(x+5)^6} = (x+5)^2$

**73.** $\sqrt[4]{20a^3b^7}\sqrt[4]{4a^2b^5} = \sqrt[4]{80a^5b^{12}} = \sqrt[4]{16a^4b^{12} \cdot 5a} =$
$2ab^3\sqrt[4]{5a}$

**75.** $\sqrt[5]{x^3(y+z)^6}\sqrt[5]{x^3(y+z)^4} = \sqrt[5]{x^6(y+z)^{10}} =$
$\sqrt[5]{x^5(y+z)^{10} \cdot x} = x(y+z)^2\sqrt[5]{x}$

**77.** *Writing Exercise*

**79.** $\dfrac{3x}{16y} + \dfrac{5y}{64x}$, LCD is $64xy$

$= \dfrac{3x}{16y} \cdot \dfrac{4x}{4x} + \dfrac{5y}{64x} \cdot \dfrac{y}{y}$

$= \dfrac{12x^2}{64xy} + \dfrac{5y^2}{64xy}$

$= \dfrac{12x^2 + 5y^2}{64xy}$

**81.** $\dfrac{4}{x^2-9} - \dfrac{7}{2x-6}$

$= \dfrac{4}{(x+3)(x-3)} - \dfrac{7}{2(x-3)}$, LCD is $2(x+3)(x-3)$

$= \dfrac{4}{(x+3)(x-3)} \cdot \dfrac{2}{2} - \dfrac{7}{2(x-3)} \cdot \dfrac{x+3}{x+3}$

$= \dfrac{8}{2(x+3)(x-3)} - \dfrac{7(x+3)}{2(x+3)(x-3)}$

$= \dfrac{8 - 7(x+3)}{2(x+3)(x-3)}$

$= \dfrac{8 - 7x - 21}{2(x+3)(x-3)}$

$= \dfrac{-7x - 13}{2(x+3)(x-3)}$

**83.** $\dfrac{9a^4b^7}{3a^2b^5} = \dfrac{9}{3}a^{4-2}b^{7-5} = 3a^2b^2$

**85.** *Writing Exercise*

**87.** $R(x) = \dfrac{1}{2}\sqrt[4]{\dfrac{x \cdot 3.0 \times 10^6}{\pi^2}}$

$R(5 \times 10^4) = \dfrac{1}{2}\sqrt[4]{\dfrac{5 \times 10^4 \cdot 3.0 \times 10^6}{\pi^2}}$

$= \dfrac{1}{2}\sqrt[4]{\dfrac{15 \times 10^{10}}{\pi^2}}$

$\approx 175.6 \text{ mi}$

**89.** a) $T_w = 33 - \dfrac{(10.45 + 10\sqrt{8} - 8)(33-7)}{22}$

     $\approx -3.3°$ C

b) $T_w = 33 - \dfrac{(10.45 + 10\sqrt{12} - 12)(33-0)}{22}$

     $\approx -16.6°$ C

c) $T_w = 33 - \dfrac{(10.45 + 10\sqrt{14} - 14)[33-(-5)]}{22}$

     $\approx -25.5°$ C

d) $T_w = 33 - \dfrac{(10.45 + 10\sqrt{15} - 15)[33-(-23)]}{22}$

     $\approx -54.0°$ C

**91.** $(\sqrt[3]{25x^4})^4 = \sqrt[3]{(25x^4)^4} = \sqrt[3]{25^4x^{16}} =$
$\sqrt[3]{25^3 \cdot 25 \cdot x^{15} \cdot x} = \sqrt[3]{25^3}\sqrt[3]{x^{15}}\sqrt[3]{25x} =$
$25x^5\sqrt[3]{25x}$

**93.** $(\sqrt{a^3b^5})^7 = \sqrt{(a^3b^5)^7} = \sqrt{a^{21}b^{35}} =$
$\sqrt{a^{20} \cdot a \cdot b^{34} \cdot b} = \sqrt{a^{20}}\sqrt{b^{34}}\sqrt{ab} = a^{10}b^{17}\sqrt{ab}$

**95.**

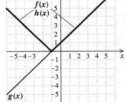

We see that $f(x) = h(x)$ and $f(x) \neq g(x)$.

**97.** $g(x) = x^2 - 6x + 8$

We must have $x^2 - 6x + 8 \geq 0$, or $(x-2)(x-4) \geq 0$. We graph $y = x^2 - 6x + 8$.

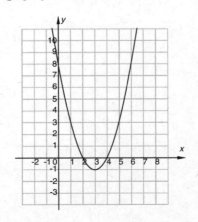

From the graph we see that $y \geq 0$ for $x \leq 2$ or $x \geq 4$, so the domain of $g$ is $\{x | x \leq 2 \ or \ x \geq 4\}$, or $(-\infty, 2] \cup [4, \infty)$.

**99.** $\sqrt[5]{4a^{3k+2}} \sqrt[5]{8a^{6-k}} = 2a^4$

$$\sqrt[5]{32a^{2k+8}} = 2a^4$$

$$2\sqrt[5]{a^{2k+8}} = 2a^4$$

$$\sqrt[5]{a^{2k+8}} = a^4$$

$$a^{\frac{2k+8}{5}} = a^4$$

Since the base is the same, the exponents must be equal. We have:

$$\frac{2k+8}{5} = 4$$

$$2k + 8 = 20$$

$$2k = 12$$

$$k = 6$$

**101.** *Writing Exercise*

## Exercise Set 7.4

**1.** $\sqrt[3]{\dfrac{a^2}{b^6}} = \dfrac{\sqrt[3]{a^2}}{\sqrt[3]{b^6}} = \dfrac{\sqrt[3]{a^2}}{b^2}$, so choice (e) is correct.

**3.** $\sqrt[5]{\dfrac{a^6}{b^4}} = \sqrt[5]{\dfrac{a^6}{b^4} \cdot \dfrac{b}{b}} = \sqrt[5]{\dfrac{a^6 b}{b^4 \cdot b}}$, so choice (f) is correct.

**5.** $\dfrac{\sqrt[5]{a^2}}{\sqrt[5]{b^2}} = \dfrac{\sqrt[5]{a^2}}{\sqrt[5]{b^2}} \cdot \dfrac{\sqrt[5]{b^3}}{\sqrt[5]{b^3}} = \dfrac{\sqrt[5]{a^2 b^3}}{\sqrt[5]{b^5}}$, so choice (h) is the correct.

**7.** $\dfrac{\sqrt[5]{a^2}}{\sqrt[5]{b^3}} = \dfrac{\sqrt[5]{a^2}}{\sqrt[5]{b^3}} \cdot \dfrac{\sqrt[5]{b^2}}{\sqrt[5]{b^2}} = \dfrac{\sqrt[5]{a^2} \sqrt[5]{b^2}}{\sqrt[5]{b^5}}$, so choice (a) is correct.

**9.** $\sqrt{\dfrac{36}{25}} = \dfrac{\sqrt{36}}{\sqrt{25}} = \dfrac{6}{5}$

**11.** $\sqrt[3]{\dfrac{64}{27}} = \dfrac{\sqrt[3]{64}}{\sqrt[3]{27}} = \dfrac{4}{3}$

**13.** $\sqrt{\dfrac{49}{y^2}} = \dfrac{\sqrt{49}}{\sqrt{y^2}} = \dfrac{7}{y}$

**15.** $\sqrt{\dfrac{36y^3}{x^4}} = \dfrac{\sqrt{36y^3}}{\sqrt{x^4}} = \dfrac{\sqrt{36y^2 \cdot y}}{\sqrt{x^4}} = \dfrac{\sqrt{36y^2} \sqrt{y}}{\sqrt{x^4}} = \dfrac{6y\sqrt{y}}{x^2}$

**17.** $\sqrt[3]{\dfrac{27a^4}{8b^3}} = \dfrac{\sqrt[3]{27a^4}}{\sqrt[3]{8b^3}} = \dfrac{\sqrt[3]{27a^3 \cdot a}}{\sqrt[3]{8b^3}} = \dfrac{\sqrt[3]{27a^3} \sqrt[3]{a}}{\sqrt[3]{8b^3}} = \dfrac{3a\sqrt[3]{a}}{2b}$

**19.** $\sqrt[4]{\dfrac{16a^4}{b^4 c^8}} = \dfrac{\sqrt[4]{16a^4}}{\sqrt[4]{b^4 c^8}} = \dfrac{2a}{bc^2}$

**21.** $\sqrt[4]{\dfrac{a^5 b^8}{c^{10}}} = \dfrac{\sqrt[4]{a^5 b^8}}{\sqrt[4]{c^{10}}} = \dfrac{\sqrt[4]{a^4 b^8 \cdot a}}{\sqrt[4]{c^8 \cdot c^2}} = \dfrac{\sqrt[4]{a^4 b^8} \sqrt[4]{a}}{\sqrt[4]{c^8} \sqrt[4]{c^2}} = \dfrac{ab^2 \sqrt[4]{a}}{c^2 \sqrt[4]{c^2}}$, or $\dfrac{ab^2}{c^2} \sqrt[4]{\dfrac{a}{c^2}}$

**23.** $\sqrt[5]{\dfrac{32x^6}{y^{11}}} = \dfrac{\sqrt[5]{32x^6}}{\sqrt[5]{y^{11}}} = \dfrac{\sqrt[5]{32x^5 \cdot x}}{\sqrt[5]{y^{10} \cdot y}} = \dfrac{\sqrt[5]{32x^5} \cdot \sqrt[5]{x}}{\sqrt[5]{y^{10}} \sqrt[5]{y}} =$

$\dfrac{2x \sqrt[5]{x}}{y^2 \sqrt[5]{y}}$, or $\dfrac{2x}{y^2} \sqrt[5]{\dfrac{x}{y}}$

**25.** $\sqrt[6]{\dfrac{x^6 y^8}{z^{15}}} = \dfrac{\sqrt[6]{x^6 y^8}}{\sqrt[6]{z^{15}}} = \dfrac{\sqrt[6]{x^6 y^6 \cdot y^2}}{\sqrt[6]{z^{12} \cdot z^3}} = \dfrac{\sqrt[6]{x^6 y^6} \sqrt[6]{y^2}}{\sqrt[6]{z^{12}} \sqrt[6]{z^3}} =$

$\dfrac{xy \sqrt[6]{y^2}}{z^2 \sqrt[6]{z^3}}$, or $\dfrac{xy}{z^2} \sqrt[6]{\dfrac{y^2}{z^3}}$

**27.** $\dfrac{\sqrt{35x}}{\sqrt{7x}} = \sqrt{\dfrac{35x}{7x}} = \sqrt{5}$

**29.** $\dfrac{\sqrt[3]{270}}{\sqrt[3]{10}} = \sqrt[3]{\dfrac{270}{10}} = \sqrt[3]{27} = 3$

**31.** $\dfrac{\sqrt{40xy^3}}{\sqrt{8x}} = \sqrt{\dfrac{40xy^3}{8x}} = \sqrt{5y^3} = \sqrt{y^2 \cdot 5y} =$

$\sqrt{y^2} \sqrt{5y} = y\sqrt{5y}$

**33.** $\dfrac{\sqrt[3]{96a^4 b^2}}{\sqrt[3]{12a^2 b}} = \sqrt[3]{\dfrac{96a^4 b^2}{12a^2 b}} = \sqrt[3]{8a^2 b} = \sqrt[3]{8} \sqrt[3]{a^2 b} =$

$2\sqrt[3]{a^2 b}$

**35.** $\dfrac{\sqrt{100ab}}{5\sqrt{2}} = \dfrac{1}{5} \dfrac{\sqrt{100ab}}{\sqrt{2}} = \dfrac{1}{5} \sqrt{\dfrac{100ab}{2}} = \dfrac{1}{5} \sqrt{50ab} =$

$\dfrac{1}{5} \sqrt{25 \cdot 2ab} = \dfrac{1}{5} \cdot 5\sqrt{2ab} = \sqrt{2ab}$

**37.** $\dfrac{\sqrt[4]{48x^9 y^{13}}}{\sqrt[4]{3xy^{-2}}} = \sqrt[4]{\dfrac{48x^9 y^{13}}{3xy^{-2}}} = \sqrt[4]{16x^8 y^{15}} =$

$\sqrt[4]{16x^8 y^{12}} \sqrt[4]{y^3} = 2x^2 y^3 \sqrt[4]{y^3}$

**39.** $\dfrac{\sqrt[3]{x^3 - y^3}}{\sqrt[3]{x - y}} = \sqrt[3]{\dfrac{x^3 - y^3}{x - y}} =$

$\sqrt[3]{\dfrac{(x - y)(x^2 + xy + y^2)}{x - y}} =$

$\sqrt[3]{\dfrac{\cancel{(x - y)}(x^2 + xy + y^2)}{\cancel{x - y}}} = \sqrt[3]{x^2 + xy + y^2}$

**41.** $\sqrt{\dfrac{3}{2}} = \sqrt{\dfrac{3}{2} \cdot \dfrac{2}{2}} = \sqrt{\dfrac{6}{4}} = \dfrac{\sqrt{6}}{\sqrt{4}} = \dfrac{\sqrt{6}}{2}$

**43.** $\dfrac{6\sqrt{5}}{5\sqrt{3}} = \dfrac{6\sqrt{5}}{5\sqrt{3}} \cdot \dfrac{\sqrt{3}}{\sqrt{3}} = \dfrac{6\sqrt{15}}{5 \cdot 3} = \dfrac{2\sqrt{15}}{5}$

**45.** $\sqrt[3]{\dfrac{16}{9}} = \sqrt[3]{\dfrac{16}{9} \cdot \dfrac{3}{3}} = \sqrt[3]{\dfrac{48}{27}} = \dfrac{\sqrt[3]{8 \cdot 6}}{\sqrt[3]{27}} = \dfrac{2\sqrt[3]{6}}{3}$

**47.** $\dfrac{\sqrt[3]{3a}}{\sqrt[3]{5c}} = \dfrac{\sqrt[3]{3a}}{\sqrt[3]{5c}} \cdot \dfrac{\sqrt[3]{5^2 c^2}}{\sqrt[3]{5^2 c^2}} = \dfrac{\sqrt[3]{75ac^2}}{\sqrt[3]{5^3 c^3}} = \dfrac{\sqrt[3]{75ac^2}}{5c}$

**49.** $\dfrac{\sqrt[3]{5y^4}}{\sqrt[3]{6x^4}} = \dfrac{\sqrt[3]{5y^4}}{\sqrt[3]{6x^4}} \cdot \dfrac{\sqrt[3]{36x^2}}{\sqrt[3]{36x^2}} = \dfrac{\sqrt[3]{y^3 \cdot 180x^2 y}}{\sqrt[3]{216x^6}} =$

$\dfrac{y\sqrt[3]{180x^2 y}}{6x^2}$

**51.** $\sqrt[3]{\dfrac{2}{x^2y}} = \sqrt[3]{\dfrac{2}{x^2y} \cdot \dfrac{xy^2}{xy^2}} = \sqrt[3]{\dfrac{2xy^2}{x^3y^3}} = \dfrac{\sqrt[3]{2xy^2}}{\sqrt[3]{x^3y^3}} =$

$\dfrac{\sqrt[3]{2xy^2}}{xy}$

**53.** $\sqrt{\dfrac{7a}{18}} = \sqrt{\dfrac{7a}{18} \cdot \dfrac{2}{2}} = \sqrt{\dfrac{14a}{36}} = \dfrac{\sqrt{14a}}{\sqrt{36}} = \dfrac{\sqrt{14a}}{6}$

**55.** $\sqrt{\dfrac{9}{20x^2y}} = \sqrt{\dfrac{9}{20x^2y} \cdot \dfrac{5y}{5y}} = \sqrt{\dfrac{9 \cdot 5y}{100x^2y^2}} =$

$\dfrac{\sqrt{9 \cdot 5y}}{\sqrt{100x^2y^2}} = \dfrac{3\sqrt{5y}}{10xy}$

**57.** $\sqrt{\dfrac{10ab^2}{72a^3b}} = \sqrt{\dfrac{5b}{36a^2}} = \dfrac{\sqrt{5b}}{6a}$

**59.** $\dfrac{\sqrt{5}}{\sqrt{7x}} = \dfrac{\sqrt{5}}{\sqrt{7x}} \cdot \dfrac{\sqrt{5}}{\sqrt{5}} = \dfrac{\sqrt{25}}{\sqrt{35x}} = \dfrac{5}{\sqrt{35x}}$

**61.** $\sqrt{\dfrac{14}{21}} = \sqrt{\dfrac{2}{3}} = \sqrt{\dfrac{2}{3} \cdot \dfrac{2}{2}} = \sqrt{\dfrac{4}{6}} = \dfrac{\sqrt{4}}{\sqrt{6}} = \dfrac{2}{\sqrt{6}}$

**63.** $\dfrac{4\sqrt{13}}{3\sqrt{7}} = \dfrac{4\sqrt{13}}{3\sqrt{7}} \cdot \dfrac{\sqrt{13}}{\sqrt{13}} = \dfrac{4\sqrt{169}}{3\sqrt{91}} = \dfrac{4 \cdot 13}{3\sqrt{91}} = \dfrac{52}{3\sqrt{91}}$

**65.** $\dfrac{\sqrt[3]{7}}{\sqrt[3]{2}} = \dfrac{\sqrt[3]{7}}{\sqrt[3]{2}} \cdot \dfrac{\sqrt[3]{7^2}}{\sqrt[3]{7^2}} = \dfrac{\sqrt[3]{7^3}}{\sqrt[3]{98}} = \dfrac{7}{\sqrt[3]{98}}$

**67.** $\sqrt{\dfrac{7x}{3y}} = \sqrt{\dfrac{7x}{3y} \cdot \dfrac{7x}{7x}} = \dfrac{\sqrt{(7x)^2}}{\sqrt{21xy}} = \dfrac{7x}{\sqrt{21xy}}$

**69.** $\sqrt[3]{\dfrac{2a^5}{5b}} = \sqrt[3]{\dfrac{2a^5}{5b} \cdot \dfrac{4a}{4a}} = \sqrt[3]{\dfrac{8a^6}{20ab}} = \dfrac{2a^2}{\sqrt[3]{20ab}}$

**71.** $\sqrt{\dfrac{x^3y}{2}} = \sqrt{\dfrac{x^3y}{2} \cdot \dfrac{xy}{xy}} = \sqrt{\dfrac{x^4y^2}{2xy}} = \dfrac{\sqrt{x^4y^2}}{\sqrt{2xy}} = \dfrac{x^2y}{\sqrt{2xy}}$

**73.** *Writing Exercise*

**75.** $\dfrac{3}{x-5} \cdot \dfrac{x-1}{x+5} = \dfrac{3(x-1)}{(x-5)(x+5)}$

**77.** $\dfrac{a^2 - 8a + 7}{a^2 - 49} = \dfrac{(a-1)(a-7)}{(a+7)(a-7)}$

$= \dfrac{(a-1)(a\!-\!7)}{(a+7)(a\!-\!7)}$

$= \dfrac{a-1}{a+7}$

**79.** $(5a^3b^4)^3 = 5^3(a^3)^3(b^4)^3 = 125a^{3\cdot3}b^{4\cdot3} = 125a^9b^{12}$

**81.** *Writing Exercise*

**83.** a) $T = 2\pi\sqrt{\dfrac{65}{980}} \approx 1.62 \text{ sec}$

b) $T = 2\pi\sqrt{\dfrac{98}{980}} \approx 1.99 \text{ sec}$

c) $T = 2\pi\sqrt{\dfrac{120}{980}} \approx 2.20 \text{ sec}$

**85.** $\dfrac{(\sqrt[3]{81mn^2})^2}{(\sqrt[3]{mn})^2} = \dfrac{\sqrt[3]{(81mn^2)^2}}{\sqrt[3]{(mn)^2}}$

$= \dfrac{\sqrt[3]{6561m^2n^4}}{\sqrt[3]{m^2n^2}}$

$= \sqrt[3]{\dfrac{6561m^2n^4}{m^2n^2}}$

$= \sqrt[3]{6561n^2}$

$= \sqrt[3]{729 \cdot 9n^2}$

$= \sqrt[3]{729}\,\sqrt[3]{9n^2}$

$= 9\sqrt[3]{9n^2}$

**87.** $\sqrt{a^2 - 3} - \dfrac{a^2}{\sqrt{a^2 - 3}}$

$= \sqrt{a^2 - 3} - \dfrac{a^2}{\sqrt{a^2 - 3}} \cdot \dfrac{\sqrt{a^2 - 3}}{\sqrt{a^2 - 3}}$

$= \sqrt{a^2 - 3} - \dfrac{a^2\sqrt{a^2 - 3}}{a^2 - 3}$

$= \sqrt{a^2 - 3} \cdot \dfrac{a^2 - 3}{a^2 - 3} - \dfrac{a^2\sqrt{a^2 - 3}}{a^2 - 3}$

$= \dfrac{a^2\sqrt{a^2 - 3} - 3\sqrt{a^2 - 3} - a^2\sqrt{a^2 - 3}}{a^2 - 3}$

$= \dfrac{-3\sqrt{a^2 - 3}}{a^2 - 3}, \text{ or } \dfrac{-3}{\sqrt{a^2 - 3}}$

**89.** Step 1: $\sqrt[n]{x} = x^{1/n}$, by definition;

Step 2: $\left(\dfrac{x}{y}\right)^n = \dfrac{x^n}{y^n}$, raising a quotient to a power;

Step 3: $x^{1/n} = \sqrt[n]{x}$, by definition

**91.** $f(x) = \sqrt{18x^3},\; g(x) = \sqrt{2x}$

$(f/g)(x) = \dfrac{f(x)}{g(x)} = \dfrac{\sqrt{18x^3}}{\sqrt{2x}} = \sqrt{\dfrac{18x^3}{2x}} = \sqrt{9x^2} = 3x$

$\sqrt{2x}$ is defined for $2x \geq 0$, or $x \geq 0$. To avoid division by 0, we must exclude 0 from the domain. Thus, the domain of $f/g = \{x|x$ is a real number and $x > 0\}$, or $(0, \infty)$.

**93.** $f(x) = \sqrt{x^2 - 9},\; g(x) = \sqrt{x - 3}$

$(f/g)(x) = \dfrac{f(x)}{g(x)} = \dfrac{\sqrt{x^2 - 9}}{\sqrt{x - 3}} = \sqrt{\dfrac{x^2 - 9}{x - 3}} =$

$\sqrt{\dfrac{(x+3)(x-3)}{x-3}} = \sqrt{x + 3}$

$\sqrt{x - 3}$ is defined for $x - 3 \geq 0$, or $x \geq 3$. To avoid division by 0 we must exclude 3 from the domain. Thus, the domain of $f/g = \{x|x$ is a real number and $x > 3\}$, or $(3, \infty)$.

## Exercise Set 7.5

**1.** To add radical expressions, the <u>indices</u> and the <u>radicands</u> must be the same.

**3.** To find a product by adding exponents, the <u>bases</u> must be the same.

**5.** To rationalize the <u>numerator</u> of $\dfrac{\sqrt{a+0.1}-\sqrt{a}}{0.1}$, we multiply by a form of 1, using the <u>conjugate of $\sqrt{a+0.1}-\sqrt{a}$</u>, or $\sqrt{a+0.1}+\sqrt{a}$, to write 1.

**7.** $2\sqrt{5}+7\sqrt{5}=(2+7)\sqrt{5}=9\sqrt{5}$

**9.** $7\sqrt[3]{4}-5\sqrt[3]{4}=(7-5)\sqrt[3]{4}=2\sqrt[3]{4}$

**11.** $\sqrt[3]{y}+9\sqrt[3]{y}=(1+9)\sqrt[3]{y}=10\sqrt[3]{y}$

**13.** $8\sqrt{2}-6\sqrt{2}+5\sqrt{2}=(8-6+5)\sqrt{2}=7\sqrt{2}$

**15.** $9\sqrt[3]{7}-\sqrt{3}+4\sqrt[3]{7}+2\sqrt{3}=$
$(9+4)\sqrt[3]{7}+(-1+2)\sqrt{3}=13\sqrt[3]{7}+\sqrt{3}$

**17.** $\quad 4\sqrt{27}-3\sqrt{3}$
$\quad = 4\sqrt{9\cdot3}-3\sqrt{3}\quad$ Factoring the
$\quad = 4\sqrt{9}\cdot\sqrt{3}-3\sqrt{3}\quad$ first radical
$\quad = 4\cdot3\sqrt{3}-3\sqrt{3}\quad$ Taking the square root of 9
$\quad = 12\sqrt{3}-3\sqrt{3}$
$\quad = 9\sqrt{3}\quad\quad\quad$ Combining like radicals

**19.** $\quad 3\sqrt{45}+7\sqrt{20}$
$\quad = 3\sqrt{9\cdot5}+7\sqrt{4\cdot5}\quad$ Factoring the
$\quad = 3\sqrt{9}\cdot\sqrt{5}+7\sqrt{4}\cdot\sqrt{5}\quad$ radicals
$\quad = 3\cdot3\sqrt{5}+7\cdot2\sqrt{5}\quad$ Taking the square roots
$\quad = 9\sqrt{5}+14\sqrt{5}$
$\quad = 23\sqrt{5}\quad\quad\quad$ Combining like radicals

**21.** $3\sqrt[3]{16}+\sqrt[3]{54}=3\sqrt[3]{8\cdot2}+\sqrt[3]{27\cdot2}=$
$3\sqrt[3]{8}\cdot\sqrt[3]{2}+\sqrt[3]{27}\cdot\sqrt[3]{2}=3\cdot2\sqrt[3]{2}+3\sqrt[3]{2}=$
$6\sqrt[3]{2}+3\sqrt[3]{2}=9\sqrt[3]{2}$

**23.** $\sqrt{5a}+2\sqrt{45a^3}=\sqrt{5a}+2\sqrt{9a^2\cdot5a}=$
$\sqrt{5a}+2\sqrt{9a^2}\cdot\sqrt{5a}=\sqrt{5a}+2\cdot3a\sqrt{5a}=$
$\sqrt{5a}+6a\sqrt{5a}=(1+6a)\sqrt{5a}$

**25.** $\sqrt[3]{6x^4}+\sqrt[3]{48x}=\sqrt[3]{x^3\cdot6x}+\sqrt[3]{8\cdot6x}=$
$\sqrt[3]{x^3}\cdot\sqrt[3]{6x}+\sqrt[3]{8}\cdot\sqrt[3]{6x}=x\sqrt[3]{6x}+2\sqrt[3]{6x}=$
$(x+2)\sqrt[3]{6x}$

**27.** $\sqrt{4a-4}+\sqrt{a-1}=\sqrt{4(a-4)}+\sqrt{a-1}=$
$\sqrt{4}\sqrt{a-1}+\sqrt{a-1}=2\sqrt{a-1}+\sqrt{a-1}=3\sqrt{a-1}$

**29.** $\sqrt{x^3-x^2}+\sqrt{9x-9}=\sqrt{x^2(x-1)}+\sqrt{9(x-1)}=$
$\sqrt{x^2}\cdot\sqrt{x-1}+\sqrt{9}\cdot\sqrt{x-1}=$
$x\sqrt{x-1}+3\sqrt{x-1}=(x+3)\sqrt{x-1}$

**31.** $\sqrt{3}(4+\sqrt{3})=\sqrt{3}\cdot4+\sqrt{3}\cdot\sqrt{3}=4\sqrt{3}+3$

**33.** $3\sqrt{5}(\sqrt{5}-\sqrt{2})=3\sqrt{5}\cdot\sqrt{5}-3\sqrt{5}\cdot\sqrt{2}=$
$3\cdot5-3\sqrt{10}=15-3\sqrt{10}$

**35.** $\sqrt{2}(3\sqrt{10}-2\sqrt{2})=\sqrt{2}\cdot3\sqrt{10}-\sqrt{2}\cdot2\sqrt{2}=$
$3\sqrt{20}-2\cdot2=3\sqrt{4\cdot5}-4=6\sqrt{5}-4$

**37.** $\sqrt[3]{3}(\sqrt[3]{9}-4\sqrt[3]{21})=\sqrt[3]{3}\cdot\sqrt[3]{9}-\sqrt[3]{3}\cdot4\sqrt[3]{21}=$
$\sqrt[3]{27}-4\sqrt[3]{63}=3-4\sqrt[3]{63}$

**39.** $\sqrt[3]{a}(\sqrt[3]{a^2}+\sqrt[3]{24a^2})=\sqrt[3]{a}\cdot\sqrt[3]{a^2}+\sqrt[3]{a}\sqrt[3]{24a^2}=$
$\sqrt[3]{a^3}+\sqrt[3]{24a^3}=\sqrt[3]{a^3}+\sqrt[3]{8a^3\cdot3}=$
$a+2a\sqrt[3]{3}$

**41.** $(2+\sqrt{6})(5-\sqrt{6})=2\cdot5-2\sqrt{6}+5\sqrt{6}-\sqrt{6}\cdot\sqrt{6}=$
$10+3\sqrt{6}-6=4+3\sqrt{6}$

**43.** $(\sqrt{2}+\sqrt{7})(\sqrt{3}-\sqrt{7})=\sqrt{2}\cdot\sqrt{3}-\sqrt{2}\cdot\sqrt{7}+\sqrt{7}\cdot\sqrt{3}-\sqrt{7}\cdot\sqrt{7}=$
$\sqrt{6}-\sqrt{14}+\sqrt{21}-7$

**45.** $(3-\sqrt{5})(3+\sqrt{5})=3^2-(\sqrt{5})^2=9-5=4$

**47.** $(\sqrt{6}+\sqrt{8})(\sqrt{6}-\sqrt{8})=(\sqrt{6})^2-(\sqrt{8})^2=6-8=-2$

**49.** $(3\sqrt{7}+2\sqrt{5})(2\sqrt{7}-4\sqrt{5})=$
$3\sqrt{7}\cdot2\sqrt{7}-3\sqrt{7}\cdot4\sqrt{5}+2\sqrt{5}\cdot2\sqrt{7}-2\sqrt{5}\cdot4\sqrt{5}=$
$6\cdot7-12\sqrt{35}+4\sqrt{35}-8\cdot5=42-8\sqrt{35}-40=$
$2-8\sqrt{35}$

**51.** $(2+\sqrt{3})^2=2^2+2\cdot2\cdot\sqrt{3}+(\sqrt{3})^2=4+4\sqrt{3}+3=7+4\sqrt{3}$

**53.** $(\sqrt{3}-\sqrt{2})^2=(\sqrt{3})^2-2\cdot\sqrt{3}\cdot\sqrt{2}+(\sqrt{2})^2=$
$3-2\sqrt{6}+2=5-2\sqrt{6}$

**55.** $(\sqrt{2t}+\sqrt{5})^2=(\sqrt{2t})^2+2\cdot\sqrt{2t}\cdot\sqrt{5}+(\sqrt{5})^2=$
$2t+2\sqrt{10t}+5$

**57.** $(3-\sqrt{x+5})^2=3^2-2\cdot3\cdot\sqrt{x+5}+(\sqrt{x+5})^2=$
$9-6\sqrt{x+5}+x+5=14-6\sqrt{x+5}+x$

**59.** $(2\sqrt[4]{7}-\sqrt[4]{6})(3\sqrt[4]{9}+2\sqrt[4]{5})=$
$2\sqrt[4]{7}\cdot3\sqrt[4]{9}+2\sqrt[4]{7}\cdot2\sqrt[4]{5}-\sqrt[4]{6}\cdot3\sqrt[4]{9}-\sqrt[4]{6}\cdot2\sqrt[4]{5}=$
$6\sqrt[4]{63}+4\sqrt[4]{35}-3\sqrt[4]{54}-2\sqrt[4]{30}$

**61.** $\dfrac{5}{4-\sqrt{3}}=\dfrac{5}{4-\sqrt{3}}\cdot\dfrac{4+\sqrt{3}}{4+\sqrt{3}}=\dfrac{5(4+\sqrt{3})}{(4-\sqrt{3})(4+\sqrt{3})}=$
$\dfrac{20+5\sqrt{3}}{4^2-(\sqrt{3})^2}=\dfrac{20+5\sqrt{3}}{16-3}=\dfrac{20+5\sqrt{3}}{13}$

**63.** $\dfrac{2+\sqrt{5}}{6+\sqrt{3}}=\dfrac{2+\sqrt{5}}{6+\sqrt{3}}\cdot\dfrac{6-\sqrt{3}}{6-\sqrt{3}}=$
$\dfrac{(2+\sqrt{5})(6-\sqrt{3})}{(6+\sqrt{3})(6-\sqrt{3})}=\dfrac{12-2\sqrt{3}+6\sqrt{5}-\sqrt{15}}{36-3}=$
$\dfrac{12-2\sqrt{3}+6\sqrt{5}-\sqrt{15}}{33}$

**65.** $\dfrac{\sqrt{a}}{\sqrt{a}+\sqrt{b}}=\dfrac{\sqrt{a}}{\sqrt{a}+\sqrt{b}}\cdot\dfrac{\sqrt{a}-\sqrt{b}}{\sqrt{a}-\sqrt{b}}=$
$\dfrac{\sqrt{a}(\sqrt{a}-\sqrt{b})}{(\sqrt{a}+\sqrt{b})(\sqrt{a}-\sqrt{b})}=\dfrac{a-\sqrt{ab}}{a-b}$

**67.** $\dfrac{\sqrt{7}-\sqrt{3}}{\sqrt{3}-\sqrt{7}}=\dfrac{-1(\sqrt{3}-\sqrt{7})}{\sqrt{3}-\sqrt{7}}=-1\cdot\dfrac{\sqrt{3}-\sqrt{7}}{\sqrt{3}-\sqrt{7}}=$
$-1\cdot1=-1$

**69.** $\dfrac{3\sqrt{2}-\sqrt{7}}{4\sqrt{2}+2\sqrt{5}} = \dfrac{3\sqrt{2}-\sqrt{7}}{4\sqrt{2}+2\sqrt{5}}\cdot\dfrac{4\sqrt{2}-2\sqrt{5}}{4\sqrt{2}-2\sqrt{5}} =$

$\dfrac{(3\sqrt{2}-\sqrt{7})(4\sqrt{2}-2\sqrt{5})}{(4\sqrt{2}+2\sqrt{5})(4\sqrt{2}-2\sqrt{5})} =$

$\dfrac{12\cdot 2 - 6\sqrt{10} - 4\sqrt{14} + 2\sqrt{35}}{16\cdot 2 - 4\cdot 5} =$

$\dfrac{24-6\sqrt{10}-4\sqrt{14}+2\sqrt{35}}{32-20} = \dfrac{24-6\sqrt{10}-4\sqrt{14}+2\sqrt{35}}{12} =$

$\dfrac{2(12-3\sqrt{10}-2\sqrt{14}+\sqrt{35})}{2\cdot 6} = \dfrac{12-3\sqrt{10}-2\sqrt{14}+\sqrt{35}}{6}$

**71.** $\dfrac{\sqrt{7}+2}{5} = \dfrac{\sqrt{7}+2}{5}\cdot\dfrac{\sqrt{7}-2}{\sqrt{7}-2} =$

$\dfrac{(\sqrt{7}+2)(\sqrt{7}-2)}{5(\sqrt{7}-2)} = \dfrac{(\sqrt{7})^2 - 2^2}{5\sqrt{7}-10} =$

$\dfrac{7-4}{5\sqrt{7}-10} = \dfrac{3}{5\sqrt{7}-10}$

**73.** $\dfrac{\sqrt{6}-2}{\sqrt{3}+7} = \dfrac{\sqrt{6}-2}{\sqrt{3}+7}\cdot\dfrac{\sqrt{6}+2}{\sqrt{6}+2} =$

$\dfrac{(\sqrt{6}-2)(\sqrt{6}+2)}{(\sqrt{3}+7)(\sqrt{6}+2)} = \dfrac{6-4}{\sqrt{18}+2\sqrt{3}+7\sqrt{6}+14} =$

$\dfrac{2}{3\sqrt{2}+2\sqrt{3}+7\sqrt{6}+14}$

**75.** $\dfrac{\sqrt{x}-\sqrt{y}}{\sqrt{x}+\sqrt{y}} = \dfrac{\sqrt{x}-\sqrt{y}}{\sqrt{x}+\sqrt{y}}\cdot\dfrac{\sqrt{x}+\sqrt{y}}{\sqrt{x}+\sqrt{y}} =$

$\dfrac{(\sqrt{x}-\sqrt{y})(\sqrt{x}+\sqrt{y})}{(\sqrt{x}+\sqrt{y})(\sqrt{x}+\sqrt{y})} = \dfrac{x-y}{x+2\sqrt{xy}+y}$

**77.** $\dfrac{\sqrt{a+h}-\sqrt{a}}{h} = \dfrac{\sqrt{a+h}-\sqrt{a}}{h}\cdot\dfrac{\sqrt{a+h}+\sqrt{a}}{\sqrt{a+h}+\sqrt{a}} =$

$\dfrac{(\sqrt{a+h}-\sqrt{a})(\sqrt{a+h}+\sqrt{a})}{h(\sqrt{a+h}+\sqrt{a})} = \dfrac{a+h-a}{h(\sqrt{a+h}+\sqrt{a})} =$

$\dfrac{h}{h(\sqrt{a+h}+\sqrt{a})} = \dfrac{1}{\sqrt{a+h}+\sqrt{a}}$

**79.**    $\sqrt{a}\sqrt[4]{a^3}$

$= a^{1/2}\cdot a^{3/4}$    Converting to exponential notation

$= a^{5/4}$    Adding exponents

$= a^{1+1/4}$    Writing 5/4 as a mixed number

$= a\cdot a^{1/4}$    Factoring

$= a\sqrt[4]{a}$    Returning to radical notation

**81.**    $\sqrt[5]{b^2}\sqrt{b^3}$

$= b^{2/5}\cdot b^{3/2}$    Converting to exponential notation

$= b^{19/10}$    Adding exponents

$= b^{1+9/10}$    Writing 19/10 as a mixed number

$= b\cdot b^{9/10}$    Factoring

$= b\sqrt[10]{b^9}$    Returning to radical notation

**83.** $\sqrt{xy^3}\sqrt[3]{x^2y} = (xy^3)^{1/2}(x^2y)^{1/3}$

$= (xy^3)^{3/6}(x^2y)^{2/6}$

$= [(xy^3)^3(x^2y)^2]^{1/6}$

$= \sqrt[6]{x^3y^9\cdot x^4y^2}$

$= \sqrt[6]{x^7y^{11}}$

$= \sqrt[6]{x^6y^6\cdot xy^5}$

$= xy\sqrt[6]{xy^5}$

**85.** $\sqrt[4]{9ab^3}\sqrt{3a^4b} = (9ab^3)^{1/4}(3a^4b)^{1/2}$

$= (9ab^3)^{1/4}(3a^4b)^{2/4}$

$= [(9ab^3)(3a^4b)^2]^{1/4}$

$= \sqrt[4]{9ab^3\cdot 9a^8b^2}$

$= \sqrt[4]{81a^9b^5}$

$= \sqrt[4]{81a^8b^4\cdot ab}$

$= 3a^2b\sqrt[4]{ab}$

**87.** $\sqrt{a^4b^3c^4}\sqrt[3]{ab^2c} = (a^4b^3c^4)^{1/2}(ab^2c)^{1/3}$

$= (a^4b^3c^4)^{3/6}(ab^2c)^{2/6}$

$= [(a^4b^3c^4)^3(ab^2c)^2]^{1/6}$

$= \sqrt[6]{a^{12}b^9c^{12}\cdot a^2b^4c^2}$

$= \sqrt[6]{a^{14}b^{13}c^{14}}$

$= \sqrt[6]{a^{12}b^{12}c^{12}\cdot a^2bc^2}$

$= a^2b^2c^2\sqrt[6]{a^2bc^2}$

**89.**    $\dfrac{\sqrt[3]{a^2}}{\sqrt[4]{a}}$

$= \dfrac{a^{2/3}}{a^{1/4}}$    Converting to exponential notation

$= a^{2/3-1/4}$    Subtracting exponents

$= a^{5/12}$    Converting back

$= \sqrt[12]{a^5}$    to radical notation

**91.**    $\dfrac{\sqrt[4]{x^2y^3}}{\sqrt[3]{xy}}$

$= \dfrac{(x^2y^3)^{1/4}}{(xy)^{1/3}}$    Converting to exponential notation

$= \dfrac{x^{2/4}y^{3/4}}{x^{1/3}y^{1/3}}$    Using the power and product rules

$= x^{2/4-1/3}y^{3/4-1/3}$    Subtracting exponents

$= x^{2/12}y^{5/12}$

$= (x^2y^5)^{1/2}$    Converting back to

$= \sqrt[12]{x^2y^5}$    radical notation

**93.** $\dfrac{\sqrt{ab^3}}{\sqrt[5]{a^2b^3}}$

$= \dfrac{(ab^3)^{1/2}}{(a^2b^3)^{1/5}}$    Converting to exponential notation

$= \dfrac{a^{1/2}b^{3/2}}{a^{2/5}b^{3/5}}$

$= a^{1/10}b^{9/10}$    Subtracting exponents

$= (ab^9)^{1/10}$    Converting back to

$= \sqrt[10]{ab^9}$     radical notation

**95.** $\dfrac{\sqrt[4]{(3x-1)^3}}{\sqrt[5]{(3x-1)^3}}$

$= \dfrac{(3x-1)^{3/4}}{(3x-1)^{3/5}}$    Converting to exponential notation

$= (3x-1)^{3/4-3/5}$    Subtracting exponents

$= (3x-1)^{3/20}$    Converting back

$= \sqrt[20]{(3x-1)^3}$    to radical notation

**97.** $\dfrac{\sqrt[3]{(2x+1)^2}}{\sqrt[5]{(2x+1)^2}}$

$= \dfrac{(2x+1)^{2/3}}{(2x+1)^{2/5}}$    Converting to exponential notation

$= (2x+1)^{2/3-2/5}$    Subtracting exponents

$= (2x+1)^{4/15}$    Converting back to

$= \sqrt[15]{(2x+1)^4}$    radical notation

**99.** $\sqrt[3]{x^2y}(\sqrt{xy} - \sqrt[5]{xy^3})$

$= (x^2y)^{1/3}[(xy)^{1/2} - (xy^3)^{1/5}]$

$= x^{2/3}y^{1/3}(x^{1/2}y^{1/2} - x^{1/5}y^{3/5})$

$= x^{2/3}y^{1/3}x^{1/2}y^{1/2} - x^{2/3}y^{1/3}x^{1/5}y^{3/5}$

$= x^{2/3+1/2}y^{1/3+1/2} - x^{2/3+1/5}y^{1/3+3/5}$

$= x^{7/6}y^{5/6} - x^{13/15}y^{14/15}$

$= x^{1\frac{1}{6}}y^{\frac{5}{6}} - x^{13/15}y^{14/15}$

        Writing a mixed numeral

$= x \cdot x^{1/6}y^{5/6} - x^{13/15}y^{14/15}$

$= x(xy^5)^{1/6} - (x^{13}y^{14})^{1/15}$

$= x\sqrt[6]{xy^5} - \sqrt[15]{x^{13}y^{14}}$

**101.** $(m + \sqrt[3]{n^2})(2m + \sqrt[4]{n})$

$= (m + n^{2/3})(2m + n^{1/4})$    Converting to exponential notation

$= 2m^2 + mn^{1/4} + 2mn^{2/3} + n^{2/3}n^{1/4}$    Using FOIL

$= 2m^2 + mn^{1/4} + 2mn^{2/3} + n^{2/3+1/4}$    Adding exponents

$= 2m^2 + mn^{1/4} + 2mn^{2/3} + n^{11/12}$

$= 2m^2 + m\sqrt[4]{n} + 2m\sqrt[3]{n^2} + \sqrt[12]{n^{11}}$    Converting back to radical notation

**103.** $f(x) = \sqrt[4]{x}, \; g(x) = \sqrt[4]{2x} - \sqrt[4]{x^{11}}$

$(f \cdot g)(x) = \sqrt[4]{x}(\sqrt[4]{2x} - \sqrt[4]{x^{11}})$

$= \sqrt[4]{2x^2} - \sqrt[4]{x^{12}}$

$= \sqrt[4]{2x^2} - x^3$

**105.** $f(x) = x + \sqrt{7}, \; g(x) = x - \sqrt{7}$

$(f \cdot g)(x) = (x + \sqrt{7})(x - \sqrt{7})$

$= x^2 - (\sqrt{7})^2$

$= x^2 - 7$

**107.** $f(x) = x^2$

$f(5 + \sqrt{2}) = (5 + \sqrt{2})^2 = 25 + 10\sqrt{2} + (\sqrt{2})^2 =$

$25 + 10\sqrt{2} + 2 = 27 + 10\sqrt{2}$

**109.** $f(x) = x^2$

$f(\sqrt{3} - \sqrt{5}) = (\sqrt{3} - \sqrt{5})^2 =$

$(\sqrt{3})^2 - 2 \cdot \sqrt{3} \cdot \sqrt{5} + (\sqrt{5})^2 =$

$3 - 2\sqrt{15} + 5 = 8 - 2\sqrt{15}$

**111.** *Writing Exercise*

**113.** $\dfrac{12x}{x-4} - \dfrac{3x^2}{x+4} = \dfrac{384}{x^2-16}$

$\dfrac{12x}{x-4} - \dfrac{3x^2}{x+4} = \dfrac{384}{(x+4)(x-4)}$,

        LCM is $(x+4)(x-4)$.

Note that $x \neq -4$ and $x \neq 4$.

$(x+4)(x-4)\left[\dfrac{12x}{x-4} - \dfrac{3x^2}{x+4}\right] =$

        $(x+4)(x-4) \cdot \dfrac{384}{(x+4)(x-4)}$

$12x(x+4) - 3x^2(x-4) = 384$

$12x^2 + 48x - 3x^3 + 12x^2 = 384$

$-3x^3 + 24x^2 + 48x - 384 = 0$

$-3(x^3 - 8x^2 - 16x + 128) = 0$

$-3[x^2(x-8) - 16(x-8)] = 0$

$-3(x-8)(x^2-16) = 0$

$-3(x-8)(x+4)(x-4) = 0$

$x - 8 = 0 \; or \; x + 4 = 0 \;\; or \; x - 4 = 0$

$x = 8 \; or \;\;\;\;\; x = -4 \; or \;\;\;\;\; x = 4$

Check: For 8:

$$\dfrac{\dfrac{12x}{x-4} - \dfrac{3x^2}{x+4} = \dfrac{384}{x^2-16}}{\begin{array}{c|c} \dfrac{12 \cdot 8}{8-4} - \dfrac{3 \cdot 8^2}{8+4} & \dfrac{384}{8^2-16} \\[2mm] \dfrac{96}{4} - \dfrac{192}{12} & \dfrac{384}{48} \\[2mm] 24 - 16 & 8 \end{array}}$$

           $8 \overset{?}{=} 8$      TRUE

8 is a solution.

For $-4$:

$$\frac{\dfrac{12x}{x-4} - \dfrac{3x^2}{x+4} = \dfrac{384}{x^2-16}}{\dfrac{12(-4)}{-4-4} - \dfrac{3(-4)^2}{-4+4} \;\Big|\; \dfrac{384}{(-4)^2-16}}$$

$$\frac{-48}{-8} - \frac{48}{0} \overset{?}{=} \frac{384}{16-16} \qquad \text{UNDEFINED}$$

$-4$ is not a solution.

For $4$:

$$\frac{\dfrac{12x}{x-4} - \dfrac{3x^2}{x+4} = \dfrac{384}{x^2-16}}{\dfrac{12\cdot 4}{4-4} - \dfrac{3\cdot 4^2}{4+4} \;\Big|\; \dfrac{384}{4^2-16}}$$

$$\frac{48}{0} - \frac{48}{8} \overset{?}{=} \frac{384}{16-16} \qquad \text{UNDEFINED}$$

$4$ is not a solution.

The checks confirm that $-4$ and $4$ are not solutions. The solution is 8.

**115.** $\quad 5x^2 - 6x + 1 = 0$

$(5x-1)(x-1) = 0$

$5x - 1 = 0 \quad or \quad x - 1 = 0$

$5x = 1 \quad or \qquad x = 1$

$x = \dfrac{1}{5} \quad or \qquad x = 1$

The solutions are $\dfrac{1}{5}$ and 1.

**117.** *Familiarize.* Let $x = $ the number.

*Translate.*

$$\underbrace{\text{A number}}_{\downarrow \atop x} \underset{\downarrow \atop +}{\text{plus}} \underbrace{\text{its square}}_{\downarrow \atop x^2} \underset{\downarrow \atop =}{\text{is}} \underset{\downarrow \atop 20}{20.}$$

*Carry out.* We solve the equation.

$$x + x^2 = 20$$
$$x^2 + x - 20 = 0$$
$$(x+5)(x-4) = 0$$
$$x + 5 = 0 \quad or \quad x - 4 = 0$$
$$x = -5 \quad or \qquad x = 4$$

*Check.* $-5 + (-5)^2 = 20$ and $4 + 4^2 = 20$, so both numbers check.

*State.* The number is $-5$ or 4.

**119.** *Writing Exercise*

**121.** $f(x) = \sqrt{20x^2 + 4x^3} - 3x\sqrt{45 + 9x} + \sqrt{5x^2 + x^3}$

$\quad = \sqrt{4x^2(5+x)} - 3x\sqrt{9(5+x)} + \sqrt{x^2(5+x)}$

$\quad = \sqrt{4x^2}\sqrt{5+x} - 3x\sqrt{9}\sqrt{5+x} + \sqrt{x^2}\sqrt{5+x}$

$\quad = 2x\sqrt{5+x} - 3x\cdot 3\sqrt{5+x} + x\sqrt{5+x}$

$\quad = 2x\sqrt{5+x} - 9x\sqrt{5+x} + x\sqrt{5+x}$

$\quad = -6x\sqrt{5+x}$

**123.** $f(x) = \sqrt[4]{x^5 - x^4} + 3\sqrt[4]{x^9 - x^8}$

$\quad = \sqrt[4]{x^4(x-1)} + 3\sqrt[4]{x^8(x-1)}$

$\quad = \sqrt[4]{x^4}\cdot\sqrt[4]{x-1} + 3\sqrt[4]{x^8}\sqrt[4]{x-1}$

$\quad = x\sqrt[4]{x-1} + 3x^2\sqrt[4]{x-1}$

$\quad = (x + 3x^2)\sqrt[4]{x-1}$

**125.** $\dfrac{1}{2}\sqrt{36a^5bc^4} - \dfrac{1}{2}\sqrt[3]{64a^4bc^6} + \dfrac{1}{6}\sqrt{144a^3bc^6} =$

$\dfrac{1}{2}\sqrt{36a^4c^4\cdot ab} - \dfrac{1}{2}\sqrt[3]{64a^3c^6\cdot ab} + \dfrac{1}{6}\sqrt{144a^2c^6\cdot ab} =$

$\dfrac{1}{2}(6a^2c^2)\sqrt{ab} - \dfrac{1}{2}(4ac^2)\sqrt[3]{ab} + \dfrac{1}{6}(12ac^3)\sqrt{ab} =$

$3a^2c^2\sqrt{ab} - 2ac^2\sqrt[3]{ab} + 2ac^3\sqrt{ab}$

$(3a^2c^2 + 2ac^3)\sqrt{ab} - 2ac^2\sqrt[3]{ab}$, or

$ac^2[(3a + 2c)\sqrt{ab} - 2\sqrt[3]{ab}]$

**127.** $\quad \sqrt{27a^5(b+1)}\,\sqrt[3]{81a(b+1)^4}$

$= [27a^5(b+1)]^{1/2}[81a(b+1)^4]^{1/3}$

$= [27a^5(b+1)]^{3/6}[81a(b+1)^4]^{2/6}$

$= \{[3^3a^5(b+1)]^3[3^4a(b+1)^4]^2\}^{1/6}$

$= \sqrt[6]{3^9a^{15}(b+1)^3\cdot 3^8a^2(b+1)^8}$

$= \sqrt[6]{3^{17}a^{17}(b+1)^{11}}$

$= \sqrt[6]{3^{12}a^{12}(b+1)^6\cdot 3^5a^5(b+1)^5}$

$= 3^2a^2(b+1)\sqrt[6]{3^5a^5(b+1)^5}$, or

$\quad 9a^2(b+1)\sqrt[6]{243a^5(b+1)^5}$

**129.** $\dfrac{\dfrac{1}{\sqrt{w}} - \sqrt{w}}{\dfrac{\sqrt{w}+1}{\sqrt{w}}} = \dfrac{\dfrac{1}{\sqrt{w}} - \sqrt{w}}{\dfrac{\sqrt{w}+1}{\sqrt{w}}}\cdot\dfrac{\sqrt{w}}{\sqrt{w}} = \dfrac{1-w}{\sqrt{w}+1} =$

$\dfrac{1-w}{\sqrt{w}+1}\cdot\dfrac{\sqrt{w}-1}{\sqrt{w}-1} = \dfrac{\sqrt{w}-1-w\sqrt{w}+w}{w-1} =$

$\dfrac{(w-1) - \sqrt{w}(w-1)}{w-1} = \dfrac{(w-1)(1-\sqrt{w})}{w-1} =$

$1 - \sqrt{w}$

**131.** $x - 5 = (\sqrt{x})^2 - (\sqrt{5})^2 = (\sqrt{x} + \sqrt{5})(\sqrt{x} - \sqrt{5})$

**133.** $x - a = (\sqrt{x})^2 - (\sqrt{a})^2 = (\sqrt{x} + \sqrt{a})(\sqrt{x} - \sqrt{a})$

**135.** $(\sqrt{x+2} - \sqrt{x-2})^2 =$

$x + 2 - 2\sqrt{(x+2)(x-2)} + x - 2 =$

$x + 2 - 2\sqrt{x^2-4} + x - 2 = 2x - 2\sqrt{x^2-4}$s

## Exercise Set 7.6

**1.** False; if $x^2 = 25$, then $x = 5$ or $x = -5$.

**3.** True by the principle of powers

**5.** If we add 8 to both sides of $\sqrt{x} - 8 = 7$, we get $\sqrt{x} = 15$, so the statement is true.

**7.** $\sqrt{5x - 2} = 7$

$(\sqrt{5x - 2})^2 = 7^2$    Principle of powers (squaring)

$5x - 2 = 49$

$5x = 51$

$x = \dfrac{51}{5}$

Check:    $\dfrac{\sqrt{5x - 2} = 7}{}$

$$\sqrt{5 \cdot \dfrac{51}{5} - 2} \; \bigg| \; 7$$

$$\sqrt{49} \; \bigg|$$

$$7 \overset{?}{=} 7 \qquad \text{TRUE}$$

The solution is $\dfrac{51}{5}$.

**9.** $\sqrt{3x} + 1 = 6$

$\sqrt{3x} = 5$    Adding to isolate the radical

$(\sqrt{3x})^2 = 5^2$    Principle of powers (squaring)

$3x = 25$

$x = \dfrac{25}{3}$

Check:    $\dfrac{\sqrt{3x} + 1 = 6}{}$

$$\sqrt{3 \cdot \dfrac{25}{3}} + 1 \; \bigg| \; 6$$

$$5 + 1 \; \bigg|$$

$$6 \overset{?}{=} 6 \qquad \text{TRUE}$$

The solution is $\dfrac{25}{3}$.

**11.** $\sqrt{y + 1} - 5 = 8$

$\sqrt{y + 1} = 13$    Adding to isolate the radical

$(\sqrt{y + 1})^2 = 13^2$    Principle of powers (squaring)

$y + 1 = 169$

$y = 168$

Check:    $\dfrac{\sqrt{y + 1} - 5 = 8}{}$

$$\sqrt{168 + 1} - 5 \; \bigg| \; 8$$

$$13 - 5 \; \bigg|$$

$$8 \overset{?}{=} 8 \qquad \text{TRUE}$$

The solution is 168.

**13.** $\sqrt{x - 7} + 3 = 10$

$\sqrt{x - 7} = 7$    Adding to isolate the radical

$(\sqrt{x - 7})^2 = 7^2$    Principle of powers (squaring)

$x - 7 = 49$

$x = 56$

Check:    $\dfrac{\sqrt{x - 7} + 3 = 10}{}$

$$\sqrt{56 - 7} + 3 \; \bigg| \; 10$$

$$\sqrt{49} + 3 \; \bigg|$$

$$7 + 3 \; \bigg|$$

$$10 \overset{?}{=} 10 \qquad \text{TRUE}$$

The solution is 56.

**15.** $\sqrt[3]{x + 5} = 2$

$(\sqrt[3]{x + 5})^3 = 2^3$

$x + 5 = 8$

$x = 3$

Check:    $\dfrac{\sqrt[3]{x + 5} = 2}{}$

$$\sqrt[3]{3 + 5} \; \bigg| \; 2$$

$$\sqrt[3]{8} \; \bigg|$$

$$2 \overset{?}{=} 2 \qquad \text{TRUE}$$

The solution is 3.

**17.** $\sqrt[4]{y - 1} = 3$

$(\sqrt[4]{y - 1})^4 = 3^4$

$y - 1 = 81$

$y = 82$

Check:    $\dfrac{\sqrt[4]{y - 1} = 3}{}$

$$\sqrt[4]{82 - 1} \; \bigg| \; 3$$

$$\sqrt[4]{81} \; \bigg|$$

$$3 \overset{?}{=} 3 \qquad \text{TRUE}$$

The solution is 82.

**19.** $3\sqrt{x} = x$

$(3\sqrt{x})^2 = x^2$

$9x = x^2$

$0 = x^2 - 9x$

$0 = x(x - 9)$

$x = 0 \;\; or \;\; x = 9$

Check:

For 0:    $\dfrac{3\sqrt{x} = x}{}$

$$3\sqrt{0} \; \bigg| \; 0$$

$$3 \cdot 0 \; \bigg|$$

$$0 \overset{?}{=} 0 \qquad \text{TRUE}$$

For 9:    $\dfrac{3\sqrt{x} = x}{}$

$$3\sqrt{9} \; \bigg| \; 9$$

$$3 \cdot 3 \; \bigg|$$

$$9 \overset{?}{=} 9 \qquad \text{TRUE}$$

The solutions are 0 and 9.

**21.** $2y^{1/2} - 7 = 9$

$2\sqrt{y} - 7 = 9$

$2\sqrt{y} = 16$

$\sqrt{y} = 8$

$(\sqrt{y})^2 = 8^2$

$y = 64$

Check: $\quad$ $2y^{1/2} - 7 = 9$

$$\begin{array}{c|c} 2 \cdot 64^{1/2} - 7 & 9 \\ 2 \cdot 8 - 7 & \\ & \overset{?}{9 = 9} \quad \text{TRUE} \end{array}$$

The solution is 64.

**23.** $\sqrt[3]{x} = -3$

$(\sqrt[3]{x})^3 = (-3)^3$

$x = -27$

Check: $\quad$ $\sqrt[3]{x} = -3$

$$\begin{array}{c|c} \sqrt{-27} & -3 \\ \overset{?}{-3 = -3} & \text{TRUE} \end{array}$$

The solution is $-27$.

**25.** $t^{1/3} - 2 = 3$

$t^{1/3} = 5$

$(t^{1/3})^3 = 5^3$ $\quad$ Principle of powers

$t = 125$

Check: $\quad$ $t^{1/3} - 2 = 3$

$$\begin{array}{c|c} 125^{1/3} - 2 & 3 \\ 5 - 2 & \\ & \overset{?}{3 = 3} \quad \text{TRUE} \end{array}$$

The solution is 125.

**27.** $(y - 3)^{1/2} = -2$

$\sqrt{y - 3} = -2$

This equation has no solution, since the principal square root is never negative.

**29.** $\sqrt[4]{3x + 1} - 4 = -1$

$\sqrt[4]{3x + 1} = 3$

$(\sqrt[4]{3x + 1})^4 = 3^4$

$3x + 1 = 81$

$3x = 80$

$x = \dfrac{80}{3}$

Check: $\quad$ $\sqrt[4]{3x + 1} - 4 = -1$

$$\begin{array}{c|c} \sqrt[4]{3 \cdot \dfrac{80}{3} - 4} & -1 \\ \sqrt[4]{81} - 4 & \\ 3 - 4 & \\ & \overset{?}{-1 = -1} \quad \text{TRUE} \end{array}$$

The solution is $\dfrac{80}{3}$.

**31.** $(x + 7)^{1/3} = 4$

$[(x + 7)^{1/3}]^3 = 4^3$

$x + 7 = 64$

$x = 57$

Check: $\quad$ $(x + 7)^{1/3} = 4$

$$\begin{array}{c|c} (57 + 7)^{1/3} & 4 \\ 64^{1/3} & \\ & \overset{?}{4 = 4} \quad \text{TRUE} \end{array}$$

The solution is 57.

**33.** $\sqrt[3]{3y + 6} + 2 = 3$

$\sqrt[3]{3y + 6} = 1$

$(\sqrt[3]{3y + 6})^3 = 1^3$

$3y + 6 = 1$

$3y = -5$

$y = -\dfrac{5}{3}$

Check: $\quad$ $\sqrt[3]{3y + 6} + 2 = 3$

$$\begin{array}{c|c} \sqrt[3]{3\left(-\dfrac{5}{3}\right) + 6} + 2 & 3 \\ \sqrt[3]{1} + 2 & \\ 1 + 2 & \\ & \overset{?}{3 = 3} \quad \text{TRUE} \end{array}$$

The solution is $-\dfrac{5}{3}$.

**35.** $\sqrt{3t + 4} = \sqrt{4t + 3}$

$(\sqrt{3t + 4})^2 = (\sqrt{4t + 3})^2$

$3t + 4 = 4t + 3$

$4 = t + 3$

$1 = t$

Check: $\quad$ $\sqrt{3t + 4} = \sqrt{4t + 3}$

$$\begin{array}{c|c} \sqrt{3 \cdot 1 + 4} & \sqrt{4 \cdot 1 + 3} \\ \overset{?}{\sqrt{7} = \sqrt{7}} & \text{TRUE} \end{array}$$

The solution is 1.

**37.** $\quad 3(4 - t)^{1/4} = 6^{1/4}$

$[3(4 - t)^{1/4}]^4 = (6^{1/4})^4$

$81(4 - t) = 6$

$324 - 81t = 6$

$-81t = -318$

$t = \dfrac{106}{27}$

The number $\dfrac{106}{27}$ checks and is the solution.

**39.** $\quad 3 + \sqrt{5 - x} = x$

$\sqrt{5 - x} = x - 3$

$(\sqrt{5 - x})^2 = (x - 3)^2$

$5 - x = x^2 - 6x + 9$

$0 = x^2 - 5x + 4$

$0 = (x - 1)(x - 4)$

$x - 1 = 0 \ or \ x - 4 = 0$

$x = 1 \ or \ \quad x = 4$

Check:

For 1: $\quad 3 + \sqrt{5 - x} = x$

$\begin{array}{r|l} 3 + \sqrt{5 - 1} & 1 \\ 3 + \sqrt{4} & \\ 3 + 2 & \end{array}$

$\qquad \qquad \overset{?}{5 = 1} \qquad$ FALSE

For 4: $\quad 3 + \sqrt{5 - x} = x$

$\begin{array}{r|l} 3 + \sqrt{5 - 4} & 4 \\ 3 + \sqrt{1} & \\ 3 + 1 & \end{array}$

$\qquad \qquad \overset{?}{4 = 4} \qquad$ TRUE

Since 4 checks but 1 does not, the solution is 4.

**41.** $\quad \sqrt{4x - 3} = 2 + \sqrt{2x - 5} \qquad$ One radical is already isolated.

$(\sqrt{4x - 3})^2 = (2 + \sqrt{2x - 5})^2 \qquad$ Squaring both sides

$4x - 3 = 4 + 4\sqrt{2x - 5} + 2x - 5$

$2x - 2 = 4\sqrt{2x - 5}$

$x - 1 = 2\sqrt{2x - 5}$

$x^2 - 2x + 1 = 8x - 20$

$x^2 - 10x + 21 = 0$

$(x - 7)(x - 3) = 0$

$x - 7 = 0 \ or \ x - 3 = 0$

$x = 7 \ or \ \quad x = 3$

Both numbers check. The solutions are 7 and 3.

**43.** $\quad \sqrt{20 - x} + 8 = \sqrt{9 - x} + 11$

$\sqrt{20 - x} = \sqrt{9 - x} + 3 \quad$ Isolating one radical

$(\sqrt{20 - x})^2 = (\sqrt{9 - x} + 3)^2 \quad$ Squaring both sides

$20 - x = 9 - x + 6\sqrt{9 - x} + 9$

$2 = 6\sqrt{9 - x} \qquad$ Isolating the remaining radical

$1 = 3\sqrt{9 - x} \qquad$ Multiplying by $\dfrac{1}{2}$

$1^2 = (3\sqrt{9 - x})^2 \quad$ Squaring both sides

$1 = 9(9 - x)$

$1 = 81 - 9x$

$-80 = -9x$

$\dfrac{80}{9} = x$

The number $\dfrac{80}{9}$ checks and is the solution.

**45.** $\quad \sqrt{x + 2} + \sqrt{3x + 4} = 2$

$\sqrt{x + 2} = 2 - \sqrt{3x + 4} \qquad$ Isolating one radical

$(\sqrt{x + 2})^2 = (2 - \sqrt{3x + 4})^2$

$x + 2 = 4 - 4\sqrt{3x + 4} + 3x + 4$

$-2x - 6 = -4\sqrt{3x + 4} \qquad$ Isolating the remaining radical

$x + 3 = 2\sqrt{3x + 4} \qquad$ Multiplying by $-\dfrac{1}{2}$

$(x + 3)^2 = (2\sqrt{3x + 4})^2$

$x^2 + 6x + 9 = 4(3x + 4)$

$x^2 + 6x + 9 = 12x + 16$

$x^2 - 6x - 7 = 0$

$(x - 7)(x + 1) = 0$

$x - 7 = 0 \ or \ x + 1 = 0$

$x = 7 \ or \ x = -1$

Check:

For 7:

$\sqrt{x + 2} + \sqrt{3x + 4} = 2$

$\begin{array}{r|l} \sqrt{7 + 2} + \sqrt{3 \cdot 7 + 4} & 2 \\ \sqrt{9} + \sqrt{25} & \end{array}$

$\qquad \qquad \overset{?}{8 = 2} \qquad$ FALSE

For $-1$:

$\sqrt{x + 2} + \sqrt{3x + 4} = 2$

$\begin{array}{r|l} \sqrt{-1 + 2} + \sqrt{3 \cdot (-1) + 4} & 2 \\ \sqrt{1} + \sqrt{1} & \end{array}$

$\qquad \qquad \overset{?}{2 = 2} \qquad$ TRUE

Since $-1$ checks but 7 does not, the solution is $-1$.

**47.** We must have $f(x) = 2$, or $\sqrt{x} + \sqrt{x-9} = 1$.

$$\sqrt{x} + \sqrt{x-9} = 1$$

$\sqrt{x-9} = 1 - \sqrt{x}$     Isolating one radical term

$(\sqrt{x-9})^2 = (1 - \sqrt{x})^2$

$x - 9 = 1 - 2\sqrt{x} + x$

$-10 = -2\sqrt{x}$     Isolating the remaining radical term

$5 = \sqrt{x}$

$25 = x$

This value does not check. There is no solution, so there is no value of $x$ for which $f(x) = 1$.

**49.** $\sqrt{t-2} - \sqrt{4t+1} = -3$

$\sqrt{t-2} = \sqrt{4t+1} - 3$

$(\sqrt{t-2})^2 = (\sqrt{4t+1} - 3)^2$

$t - 2 = 4t + 1 - 6\sqrt{4t+1} + 9$

$-3t - 12 = -6\sqrt{4t+1}$

$t + 4 = 2\sqrt{4t+1}$

$(t+4)^2 = (2\sqrt{4t+1})^2$

$t^2 + 8t + 16 = 4(4t+1)$

$t^2 + 8t + 16 = 16t + 4$

$t^2 - 8t + 12 = 0$

$(t-2)(t-6) = 0$

$t - 2 = 0 \ \ or \ \ t - 6 = 0$

$t = 2 \ \ or \qquad t = 6$

Both numbers check, so we have $f(t) = -3$ when $t = 2$ and when $t = 6$.

**51.** We must have $\sqrt{2x-3} = \sqrt{x+7} - 2$.

$\sqrt{2x-3} = \sqrt{x+7} - 2$

$(\sqrt{2x-3})^2 = (\sqrt{x+7} - 2)^2$

$2x - 3 = x + 7 - 4\sqrt{x+7} + 4$

$x - 14 = -4\sqrt{x+7}$

$(x-14)^2 = (-4\sqrt{x+7})^2$

$x^2 - 28x + 196 = 16(x+7)$

$x^2 - 28x + 196 = 16x + 112$

$x^2 - 44x + 84 = 0$

$(x-2)(x-42) = 0$

$x = 2 \ \ or \ \ x = 42$

Since 2 checks but 42 does not, we have $f(x) = g(x)$ when $x = 2$.

**53.** We must have $4 - \sqrt{t-3} = (t+5)^{1/2}$.

$4 - \sqrt{t-3} = (t+5)^{1/2}$

$(4 - \sqrt{t-3})^2 = [(t+5)^{1/2}]^2$

$16 - 8\sqrt{t-3} + t - 3 = t + 5$

$-8\sqrt{t-3} = -8$

$\sqrt{t-3} = 1$

$(\sqrt{t-3})^2 = 1^2$

$t - 3 = 1$

$t = 4$

The number 4 checks, so we have $f(t) = g(t)$ when $t = 4$.

**55.** *Writing Exercise*

**57.** **Familiarize.** Let $h = $ the height of the triangle, in inches. Then $h + 2 = $ the base. Recall that the formula for the area of a triangle with base $b$ and height $h$ is $A = \frac{1}{2}bh$.

**Translate.** Substitute in the formula.

$$31\frac{1}{2} = \frac{1}{2}(h+2)(h)$$

**Carry out.** We solve the equation.

$31\frac{1}{2} = \frac{1}{2}(h+2)(h)$

$\frac{63}{2} = \frac{1}{2}(h+2)(h)$

$63 = (h+2)(h)$   Multiplying by 2

$63 = h^2 + 2h$

$0 = h^2 + 2h - 63$

$0 = (h+9)(h-7)$

$h + 9 = 0 \ \ or \ \ h - 7 = 0$

$h = -9 \ \ or \qquad h = 7$

**Check.** Since the height of the triangle cannot be negative we check only 7. If the height is 7 in., then the base is $7 + 2$, or 9 in., and the area is $\frac{1}{2} \cdot 9 \cdot 7 = \frac{63}{2} = 31\frac{1}{2}$ in². The answer checks.

**State.** The height of the triangle is 7 in., and the base is 9 in.

**59.** Graph $f(x) = \frac{2}{3}x - 5$.

The $y$-intercept is $(0, -5)$. The slope is $\frac{2}{3}$. We plot $(0, -5)$. From that point we go up 2 units and right 3 units and plot a second point at $(3, -3)$. Then we draw the graph.

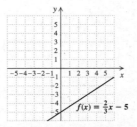

To check that the line is drawn correctly, we calculate the coordinates of another point on the line. For $x = 6$, we have

$$f(6) = \frac{2}{3} \cdot 6 - 5 = 4 - 5 = -1.$$

Since the point $(6, -1)$ appears to lie on the line, we have a check.

**61.** Graph $F(x) < -2x + 4$

First graph $F(x) = -2x + 4$ using a dashed line since the inequality symbol is $<$. Because the inequality is of the form $y < mx + b$, we shade below the line.

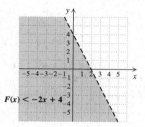

$F(x) < -2x + 4$

**63.** *Writing Exercise*

**65.** Substitute 1880 for $S(t)$ and solve for $t$.

$$1880 = 1087.7\sqrt{\frac{9t + 2617}{2457}}$$

$$1.7284 \approx \sqrt{\frac{9t + 2617}{2457}} \quad \text{Dividing by 1087.7}$$

$$(1.7284)^2 \approx \left(\sqrt{\frac{9t + 2617}{2457}}\right)^2$$

$$2.9874 \approx \frac{9t + 2617}{2457}$$

$$7340.0418 \approx 9t + 2617$$

$$4723.0418 \approx 9t$$

$$524.7824 \approx t$$

The temperature is about $524.8°C$.

**67.**
$$S = 1087.7\sqrt{\frac{9t + 2617}{2457}}$$

$$\frac{S}{1087.7} = \sqrt{\frac{9t + 2617}{2457}}$$

$$\left(\frac{S}{1087.7}\right)^2 = \left(\sqrt{\frac{9t + 2617}{2457}}\right)^2$$

$$\frac{S^2}{1087.7^2} = \frac{9t + 2617}{2457}$$

$$\frac{2457S^2}{1087.7^2} = 9t + 2617$$

$$\frac{2457S^2}{1087.7^2} - 2617 = 9t$$

$$\frac{1}{9}\left(\frac{2457S^2}{1087.7^2} - 2617\right) = t$$

**69.** $d(n) = 0.75\sqrt{2.8n}$

Substitute 84 for $d(n)$ and solve for $n$.

$$84 = 0.75\sqrt{2.8n}$$

$$112 = \sqrt{2.8n}$$

$$(112)^2 = (\sqrt{2.8n})^2$$

$$12,544 = 2.8n$$

$$4480 = n$$

About 4480 rpm will produce peak performance.

**71.**
$$v = \sqrt{2gr}\sqrt{\frac{h}{r + h}}$$

$$v^2 = 2gr \cdot \frac{h}{r + h} \quad \text{Squaring both sides}$$

$$v^2(r + h) = 2grh \quad \text{Multiplying by } r+h$$

$$v^2r + v^2h = 2grh$$

$$v^2h = 2grh - v^2r$$

$$v^2h = r(2gh - v^2)$$

$$\frac{v^2h}{2gh - v^2} = r$$

**73.** $D(h) = 1.2\sqrt{h}$

$$10.2 = 1.2\sqrt{h}$$

$$8.5 = \sqrt{h}$$

$$(8.5)^2 = (\sqrt{h})^2$$

$$72.25 = h$$

The sailor must climb 72.25 ft above sea level.

**75.**
$$\frac{x + \sqrt{x + 1}}{x - \sqrt{x + 1}} = \frac{5}{11}$$

$$11(x + \sqrt{x + 1}) = 5(x - \sqrt{x + 1})$$

$$11x + 11\sqrt{x + 1} = 5x - 5\sqrt{x + 1}$$

$$16\sqrt{x + 1} = -6x$$

$$8\sqrt{x + 1} = -3x$$

$$(8\sqrt{x + 1})^2 = (-3x)^2$$

$$64(x + 1) = 9x^2$$

$$64x + 64 = 9x^2$$

$$0 = 9x^2 - 64x - 64$$

$$0 = (9x + 8)(x - 8)$$

$$9x + 8 = 0 \quad \text{or} \quad x - 8 = 0$$

$$9x = -8 \quad \text{or} \quad x = 8$$

$$x = -\frac{8}{9} \quad \text{or} \quad x = 8$$

Since $-\frac{8}{9}$ checks but 8 does not, the solution is $-\frac{8}{9}$.

**77.**
$$(z^2 + 17)^{3/4} = 27$$

$$[(z^2 + 17)^{3/4}]^{4/3} = (3^3)^{4/3}$$

$$z^2 + 17 = 3^4$$

$$z^2 + 17 = 81$$

$$z^2 - 64 = 0$$

$$(z + 8)(z - 8) = 0$$

$z = -8 \ \ or \ \ z = 8$

Both $-8$ and $8$ check. They are the solutions.

**79.**
$$\sqrt{8-b} = b\sqrt{8-b}$$
$$(\sqrt{8-b})^2 = (b\sqrt{8-b})^2$$
$$(8-b) = b^2(8-b)$$
$$0 = b^2(8-b) - (8-b)$$
$$0 = (8-b)(b^2-1)$$
$$0 = (8-b)(b+1)(b-1)$$
$$8-b = 0 \ \ or \ \ b+1 = 0 \ or \ b-1 = 0$$
$$8 = b \ \ or \ \ \ \ \ \ \ b = -1 \ or \ \ \ b = 1$$

Since the numbers 8 and 1 check but $-1$ does not, 8 and 1 are the solutions.

**81.** We find the values of $x$ for which $g(x) = 0$.
$$6x^{1/2} + 6x^{-1/2} - 37 = 0$$
$$6\sqrt{x} + \frac{6}{\sqrt{x}} = 37$$
$$\left(6\sqrt{x} + \frac{6}{\sqrt{x}}\right)^2 = 37^2$$
$$36x + 72 + \frac{36}{x} = 1369$$
$$36x^2 + 72x + 36 = 1369x \quad \text{Multiplying by } x$$
$$36x^2 - 1297x + 36 = 0$$
$$(36x - 1)(x - 36) = 0$$
$$36x - 1 = 0 \ \ \ or \ \ x - 36 = 0$$
$$36x = 1 \ \ \ or \ \ \ \ \ \ \ x = 36$$
$$x = \frac{1}{36} \ \ or \ \ \ \ \ \ \ x = 36$$

Both numbers check. The $x$-intercepts are $\left(\frac{1}{36}, 0\right)$ and $(36, 0)$.

**83.**

**85.**

## Exercise Set 7.7

**1.** right; hypotenuse; see page 483 in the text.

**3.** square roots; see page 483 in the text.

**5.** 30°-60°-90°; leg; see page 486 in the text.

**7.** $a = 5, \ \ b = 3$

Find $c$.
$$c^2 = a^2 + b^2 \quad \text{Pythagorean theorem}$$
$$c^2 = 5^2 + 3^2 \quad \text{Substituting}$$
$$c^2 = 25 + 9$$
$$c^2 = 34$$
$$c = \sqrt{34} \quad \quad \text{Exact answer}$$
$$c \approx 5.831 \quad \text{Approximation}$$

**9.** $a = 9, \ \ b = 9$

Observe that the legs have the same length, so this is an isosceles right triangle. Then we know that the length of the hypotenuse is the length of a leg times $\sqrt{2}$, or $9\sqrt{2}$, or approximately 12.728.

**11.** $b = 12, \ \ c = 13$

Find $a$.
$$a^2 + b^2 = c^2 \quad \text{Pythagorean theorem}$$
$$a^2 + 12^2 = 13^2 \quad \text{Substituting}$$
$$a^2 + 144 = 169$$
$$a^2 = 25$$
$$a = 5$$

**13.**
$$a^2 + b^2 = c^2 \quad \text{Pythagorean theorem}$$
$$(4\sqrt{3})^2 + b^2 = 8^2$$
$$16 \cdot 3 + b^2 = 64$$
$$48 + b^2 = 64$$
$$b^2 = 16$$
$$b = 4$$

The other leg is 4 m long.

**15.**
$$a^2 + b^2 = c^2 \quad \quad \text{Pythagorean theorem}$$
$$1^2 + b^2 = (\sqrt{20})^2 \quad \text{Substituting}$$
$$1 + b^2 = 20$$
$$b^2 = 19$$
$$b = \sqrt{19}$$
$$b \approx 4.359$$

The length of the other leg is $\sqrt{19}$ in., or about 4.359 in.

**17.** Observe that the length of the hypotenuse, $\sqrt{2}$, is $\sqrt{2}$ times the length of the given leg, 1. Thus, we have an isosceles right triangle and the length of the other leg is also 1.

**19.** From the drawing in the text we see that we have a right triangle with legs of 150 ft and 200 ft. Let $d =$ the length of the diagonal, in feet. We use the Pythagorean theorem to find $d$.
$$150^2 + 200^2 = d^2$$
$$22,500 + 40,000 = d^2$$
$$62,500 = d^2$$
$$250 = d$$

Clare travels 250 ft across the parking lot.

**21.** We make a drawing and let $d =$ the distance from home plate to second base.

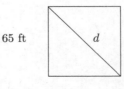

65 ft

Note that we have an isosceles right triangle. Then the length of the hypotenuse is the length of a leg times $\sqrt{2}$, or $65\sqrt{2}$ ft. This is about 91.924 ft.

(We could also have used the Pythagorean theorem, solving $65^2 + 65^2 = d^2$.)

**23.** We make a drawing similar to the one in the text.

We use the Pythagorean theorem to find $h$.

$$16^2 + h^2 = 20^2$$
$$256 + h^2 = 400$$
$$h^2 = 144$$
$$h = 12$$

The height of the screen is 12 in.

**25.** First we will find the diagonal distance, $d$, in feet, across the room. We make a drawing.

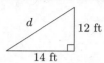

Now we use the Pythagorean theorem.

$$12^2 + 14^2 = d^2$$
$$144 + 196 = d^2$$
$$340 = d^2$$
$$\sqrt{340} = d$$
$$18.439 \approx d$$

Recall that 4 ft of slack is required on each end. Thus, $\sqrt{340} + 2 \cdot 4$, or $(\sqrt{340} + 8)$ ft, of wire should be purchased. This is about 26.439 ft.

**27.** The diagonal is the hypotenuse of a right triangle with legs of 70 paces and 40 paces. First we use the Pythagorean theorem to find the length $d$ of the diagonal, in paces.

$$70^2 + 40^2 = d^2$$
$$4900 + 1600 = d^2$$
$$6500 = d^2$$
$$\sqrt{6500} = d$$
$$80.623 \approx d$$

If Marissa walks along two sides of the quad she takes $70 + 40$, or 110 paces. Then by using the diagonal she saves $(110 - \sqrt{6500})$ paces. This is approximately $110 - 80.623$, or 29.377 paces.

**29.** Since one acute angle is $45°$, this is an isosceles right triangle with one leg $= 5$. Then the other leg $= 5$ also. And the hypotenuse is the length of the a leg times $\sqrt{2}$, or $5\sqrt{2}$.

Exact answer: Leg $= 5$, hypotenuse $= 5\sqrt{2}$

Approximation: hypotenuse $\approx 7.071$

**31.** This is a $30°$-$60°$-$90°$ right triangle with hypotenuse 14. We find the legs:

$$2a = 14, \text{ so } a = 7 \text{ and } a\sqrt{3} = 7\sqrt{3}$$

Exact answer: shorter leg $= 7$; longer leg $= 7\sqrt{3}$

Approximation: longer leg $\approx 12.124$

**33.** This is a $30°$-$60°$-$90°$ right triangle with one leg $= 15$. We substitute to find the length of the other leg, $a$, and the hypotenuse, $c$.

$$b = a\sqrt{3}$$
$$15 = a\sqrt{3}$$
$$\frac{15}{\sqrt{3}} = a$$
$$\frac{15\sqrt{3}}{3} = a \qquad \text{Rationalizing the denominator}$$
$$5\sqrt{3} = a \qquad \text{Simplifying}$$
$$c = 2a$$
$$c = 2 \cdot 5\sqrt{3}$$
$$c = 10\sqrt{3}$$

Exact answer: $a = 5\sqrt{3}$, $c = 10\sqrt{3}$

Approximations: $a \approx 8.660$, $c \approx 17.321$

**35.** This is an isosceles right triangle with hypotenuse 13. The two legs have the same length, $a$.

$$a\sqrt{2} = 13$$
$$a = \frac{13}{\sqrt{2}} = \frac{13\sqrt{2}}{2}$$

Exact answer: $\dfrac{13\sqrt{2}}{2}$

Approximation: 9.192

**37.** This is a $30°$-$60°$-$90°$ triangle with the shorter leg $= 14$. We find the longer leg and the hypotenuse.

$$a\sqrt{3} = 14\sqrt{3}, \text{ and } 2a = 2 \cdot 14 = 28.$$

Exact answer: longer leg $= 14\sqrt{3}$, hypotenuse $= 28$

Approximation: longer leg $\approx 24.249$

**39.**

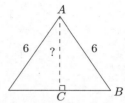

This is an equilateral triangle, so all the angles are $60°$. The altitude bisects one angle and one side. Then triangle $ABC$ is a 30-60-90 right triangle with the shorter leg of length $6/2$, or 3, and hypotenuse of length 6. Then the length of the other leg is the length of the shorter leg times $\sqrt{3}$:

Exact answer: $3\sqrt{3}$

Approximation: 5.196

**41.**

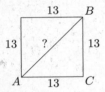

Triangle $ABC$ is an isosceles right triangle with legs of length 13. Then the length of the hypotenuse is the length of a leg times $\sqrt{2}$.

Exact answer: $13\sqrt{2}$

Approximation: 18.385

**43.**

Triangle $ABC$ is an isosceles right triangle with hypotenuse 19. Then the length of a side of the square $a$ is the length of the legs of the triangle. We have:

$$a\sqrt{2} = 19$$
$$a = \frac{19}{\sqrt{2}}, \text{ or}$$
$$a = \frac{19\sqrt{2}}{2}$$

Exact answer: $a = \dfrac{19\sqrt{2}}{2}$

Approximation: $a \approx 13.435$

**45.** We will express all distances in feet. Recall that 1 mi = 5280 ft.

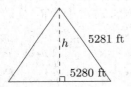

We use the Pythagorean theorem to find $h$.
$$h^2 + (5280)^2 = (5281)^2$$
$$h^2 + 27,878,400 = 27,888,961$$
$$h^2 = 10,561$$
$$h = \sqrt{10,561}$$
$$h \approx 102.767$$

The height of the bulge is $\sqrt{10,561}$ ft, or about 102.767 ft.

**47.**

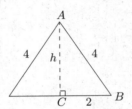

The entrance is an equilateral triangle, so all the angles are 60°. The altitude bisects one angle and one side. Then triangle $ABC$ is a 30-60-90 right triangle with the shorter leg of length 4/2, or 2, and hypotenuse of length 4. Then the height of the tent is the length of the shorter leg times $\sqrt{3}$.

Exact answer: $h = 2\sqrt{3}$ ft

Approximation: $h \approx 3.464$ ft

**49.**

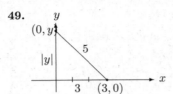

$$|y|^2 + 3^2 = 5^2$$
$$y^2 + 9 = 25$$
$$y^2 = 16$$
$$y = \pm 4$$

The points are $(0, -4)$ and $(0, 4)$.

**51.** *Writing Exercise*

**53.** $47(-1)^{19} = 47(-1) = -47$

**55.** $x^3 - 9x = x \cdot x^2 - 9 \cdot x = x(x^2 - 9) = x(x + 3)(x - 3)$

**57.** $|3x - 5| = 7$

$3x - 5 = 7 \quad or \quad 3x - 5 = -7$

$3x = 12 \quad or \quad 3x = -2$

$x = 4 \quad or \quad x = -\dfrac{2}{3}$

The solution set is $\left\{ 4, -\dfrac{2}{3} \right\}$.

**59.** *Writing Exercise*

**61.** The length of a side of the hexagon is 72/6, or 12 cm. Then the shaded region is a triangle with base 12 cm. To find the height of the triangle, note that it is the longer leg of a 30°-60°-90° right triangle. Thus its length is the length of the length of the shorter leg times $\sqrt{3}$. The length of the shorter leg is half the length of the base, $\dfrac{1}{2} \cdot 12$ cm, or 6 cm, so the length of the longer leg is $6\sqrt{3}$ cm. Now we find the area of the triangle.

$$A = \frac{1}{2}bh$$

$$= \frac{1}{2}(12 \text{ cm})(6\sqrt{3} \text{ cm})$$

$$= 36\sqrt{3} \text{ cm}^2$$

$$\approx 62.354 \text{ cm}^2$$

**63.**

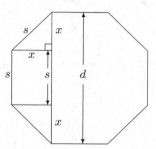

$d = s + 2x$

Use the Pythagoran theorem to find $x$.

$$x^2 + x^2 = s^2$$

$$2x^2 = s^2$$

$$x^2 = \frac{s^2}{2}$$

$$x = \frac{s}{\sqrt{2}} = \frac{s}{\sqrt{2}} \cdot \frac{\sqrt{2}}{2} = \frac{s\sqrt{2}}{2}$$

Then $d = s + 2x = s + 2\left(\frac{s\sqrt{2}}{2}\right) = s + s\sqrt{2}$.

**65.**

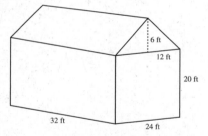

The area to be painted consists of two 20 ft by 24 ft rectangles, two 20 ft by 32 ft rectangles, and two triangles with height 6 ft and base 24 ft. The area of the two 20 ft by 24 ft rectangles is $2 \cdot 20 \text{ ft} \cdot 24 \text{ ft} = 960 \text{ ft}^2$. The area of the two 20 ft by 32 ft rectangles is $2 \cdot 20 \text{ ft} \cdot 32 \text{ ft} = 1280 \text{ ft}^2$. The area of the two triangles is $2 \cdot \frac{1}{2} \cdot 24 \text{ ft} \cdot 6 \text{ ft} = 144 \text{ ft}^2$. Thus, the total area to be painted is $960 \text{ ft}^2 + 1280 \text{ ft}^2 + 144 \text{ ft}^2 = 2384 \text{ ft}^2$.

One gallon of paint covers a minimum of 450 ft², so we divide to determine how many gallons of paint are required: $\frac{2384}{450} \approx 5.3$. Thus, 5 gallons of paint should be bought to paint the house. This answer assumes that the total area of the doors and windows is 134 ft² or more. ($5 \cdot 450 = 2250$ and $2384 = 2250 + 134$)

**67.** First we find the radius of a circle with an area of 6160 ft². This is the length of the hose.

$$A = \pi r^2$$

$$6160 = \pi r^2$$

$$\frac{6160}{\pi} = r^2$$

$$\sqrt{\frac{6160}{\pi}} = r$$

$$44.28 \approx r$$

Now we make a drawing of the room.

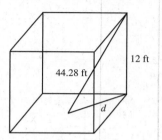

We use the Pythagorean theorem to find $d$.

$$d^2 + 12^2 = 44.28^2$$

$$d^2 + 144 = 1960.7184$$

$$d^2 = 1816.7184$$

$$d \approx 42.623$$

Now we make a drawing of the floor of the room.

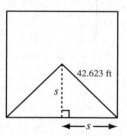

We have an isosceles right triangle with hypotenuse 42.623 ft. We find the length of a side $s$.

$$s\sqrt{2} = 42.623$$

$$s = \frac{42.623}{\sqrt{2}} \approx 30.14 \text{ ft}$$

Then the length of a side of the room is $2s = 2(30.14 \text{ ft}) = 60.28 \text{ ft}$; so the dimensions of the largest square room that meets the given conditions is 60.28 ft by 60.28 ft.

## Exercise Set 7.8

**1.** False; see page 491 in the text.

**3.** True; see page 492 in the text.

**5.** True; see page 492 in the text.

**7.** False; see Exercises 59-64.

**9.** $\sqrt{-36} = \sqrt{-1 \cdot 36} = \sqrt{-1} \cdot \sqrt{36} = i \cdot 6 = 6i$

**11.** $\sqrt{-13} = \sqrt{-1 \cdot 13} = \sqrt{-1} \cdot \sqrt{13} = i\sqrt{13}$, or $\sqrt{13}i$

**13.** $\sqrt{-18} = \sqrt{-1} \cdot \sqrt{9} \cdot \sqrt{2} = i \cdot 3 \cdot \sqrt{2} = 3i\sqrt{2}$, or $3\sqrt{2}i$

**15.** $\sqrt{-3} = \sqrt{-1 \cdot 3} = \sqrt{-1} \cdot \sqrt{3} = i\sqrt{3}$, or $\sqrt{3}i$

**17.** $\sqrt{-81} = \sqrt{-1 \cdot 81} = \sqrt{-1} \cdot \sqrt{81} = i \cdot 9 = 9i$

**19.** $\sqrt{-300} = \sqrt{-1} \cdot \sqrt{100} \cdot \sqrt{3} = i \cdot 10 \cdot \sqrt{3} = 10i\sqrt{3}$, or $10\sqrt{3}i$

**21.** $6 - \sqrt{-84} = 6 - \sqrt{-1 \cdot 4 \cdot 21} = 6 - i \cdot 2\sqrt{21} = 6 - 2i\sqrt{21}$, or $6 - 2\sqrt{21}i$

**23.** $-\sqrt{-76} + \sqrt{-125} = -\sqrt{-1 \cdot 4 \cdot 19} + \sqrt{-1 \cdot 25 \cdot 5} =$
$-i \cdot 2\sqrt{19} + i \cdot 5\sqrt{5} = -2i\sqrt{19} + 5i\sqrt{5} = (-2\sqrt{19} + 5\sqrt{5})i$

**25.** $\sqrt{-18} - \sqrt{-100} = \sqrt{-1 \cdot 9 \cdot 2} - \sqrt{-1 \cdot 100} =$
$i \cdot 3\sqrt{2} - i \cdot 10 = 3i\sqrt{2} - 10i = (3\sqrt{2} - 10)i$

**27.** $\quad (6 + 7i) + (5 + 3i)$
$\quad = (6 + 5) + (7 + 3)i \qquad$ Combining the real and
$\qquad\qquad\qquad\qquad\qquad$ the imaginary parts
$\quad = 11 + 10i$

**29.** $(9 + 8i) - (5 + 3i) = (9 - 5) + (8 - 3)i$
$\qquad\qquad\qquad\qquad = 4 + 5i$

**31.** $(7 - 4i) - (5 - 3i) = (7 - 5) + [-4 - (-3)]i = 2 - i$

**33.** $(-5 - i) - (7 + 4i) = (-5 - 7) + (-1 - 4)i = -12 - 5i$

**35.** $7i \cdot 6i = 42 \cdot i^2 = 42(-1) = -42$

**37.** $(-4i)(-6i) = 24 \cdot i^2 = 24(-1) = -24$

**39.** $\sqrt{-36}\sqrt{-9} = \sqrt{-1} \cdot \sqrt{36} \cdot \sqrt{-1} \cdot \sqrt{9} = i \cdot 6 \cdot i \cdot 3 =$
$i^2 \cdot 18 = -1 \cdot 18 = -18$

**41.** $\sqrt{-5}\sqrt{-2} = \sqrt{-1} \cdot \sqrt{5} \cdot \sqrt{-1} \cdot \sqrt{2} =$
$i \cdot \sqrt{5} \cdot i \cdot \sqrt{2} = i^2 \cdot \sqrt{10} = -1 \cdot \sqrt{10} = -\sqrt{10}$

**43.** $\sqrt{-6}\sqrt{-21} = \sqrt{-1} \cdot \sqrt{6} \cdot \sqrt{-1} \cdot \sqrt{21} =$
$i \cdot \sqrt{6} \cdot i \cdot \sqrt{21} = i^2\sqrt{126} = -1 \cdot \sqrt{9 \cdot 14} = -3\sqrt{14}$

**45.** $5i(2 + 6i) = 5i \cdot 2 + 5i \cdot 6i = 10i + 30i^2 =$
$10i - 30 = -30 + 10i$

**47.** $-7i(3 - 4i) = -7i \cdot 3 - 7i(-4i) = -21i + 28i^2 =$
$-21i - 28 = -28 - 21i$

**49.** $(1 + i)(3 + 2i) = 3 + 2i + 3i + 2i^2 =$
$3 + 2i + 3i - 2 = 1 + 5i$

**51.** $(6 - 5i)(3 + 4i) = 18 + 24i - 15i - 20i^2 =$
$18 + 24i - 15i + 20 = 38 + 9i$

**53.** $(7 - 2i)(2 - 6i) = 14 - 42i - 4i + 12i^2 =$
$14 - 42i - 4i - 12 = 2 - 46i$

**55.** $(-2 + 3i)(-2 + 5i) = 4 - 10i - 6i + 15i^2 =$
$4 - 10i - 6i - 15 = -11 - 16i$

**57.** $(-5 - 4i)(3 + 7i) = -15 - 35i - 12i - 28i^2 =$
$-15 - 35i - 12i + 28 = 13 - 47i$

**59.** $(4 - 2i)^2 = 4^2 - 2 \cdot 4 \cdot 2i + (2i)^2 = 16 - 16i + 4i^2 =$
$16 - 16i - 4 = 12 - 16i$

**61.** $(2 + 3i)^2 = 2^2 + 2 \cdot 2 \cdot 3i + (3i)^2 = 4 + 12i + 9i^2 =$
$4 + 12i - 9 = -5 + 12i$

**63.** $(-2 + 3i)^2 = (-2)^2 + 2(-2)(3i) + (3i)^2 =$
$4 - 12i + 9i^2 = 4 - 12i - 9 = -5 - 12i$

**65.**
$\dfrac{7}{4 + i}$
$= \dfrac{7}{4 + i} \cdot \dfrac{4 - i}{4 - i} \qquad$ Multiplying by 1, using
$\qquad\qquad\qquad\qquad$ the conjugate
$= \dfrac{28 - 7i}{16 - i^2}$
$= \dfrac{28 - 7i}{16 - (-1)} \qquad i^2 = -1$
$= \dfrac{28 - 7i}{17}$
$= \dfrac{28}{17} - \dfrac{7}{17}i$

**67.** $\dfrac{2}{3 - 2i} = \dfrac{2}{3 - 2i} \cdot \dfrac{3 + 2i}{3 + 2i} \qquad$ Multiplying by 1,
$\qquad\qquad\qquad\qquad\qquad$ using the conjugate
$\qquad = \dfrac{6 + 4i}{9 - 4i^2}$
$\qquad = \dfrac{6 + 4i}{9 - 4(-1)}$
$\qquad = \dfrac{6 + 4i}{13}$
$\qquad = \dfrac{6}{13} + \dfrac{4}{13}i$

**69.** $\dfrac{2i}{5 + 3i} = \dfrac{2i}{5 + 3i} \cdot \dfrac{5 - 3i}{5 - 3i} = \dfrac{10i - 6i^2}{25 - 9i^2} = \dfrac{10i + 6}{25 + 9} =$
$\dfrac{10i + 6}{34} = \dfrac{6}{34} + \dfrac{10}{34}i = \dfrac{3}{17} + \dfrac{5}{17}i$

**71.** $\dfrac{5}{6i} = \dfrac{5}{6i} \cdot \dfrac{i}{i} = \dfrac{5i}{6i^2} = \dfrac{5i}{-6} = -\dfrac{5}{6}i$

**73.** $\dfrac{5 - 3i}{4i} = \dfrac{5 - 3i}{4i} \cdot \dfrac{i}{i} = \dfrac{5i - 3i^2}{4i^2} = \dfrac{5i + 3}{-4} =$
$-\dfrac{3}{4} - \dfrac{5}{4}i$

**75.** $\dfrac{7i + 14}{7i} = \dfrac{7i}{7i} + \dfrac{14}{7i} = 1 + \dfrac{2}{i} = 1 + \dfrac{2}{i} \cdot \dfrac{i}{i} =$
$1 + \dfrac{2i}{i^2} = 1 + \dfrac{2i}{-1} = 1 - 2i$

**77.** $\dfrac{4 + 5i}{3 - 7i} = \dfrac{4 + 5i}{3 - 7i} \cdot \dfrac{3 + 7i}{3 + 7i} = \dfrac{12 + 28i + 15i + 35i^2}{9 - 49i^2} =$
$\dfrac{12 + 28i + 15i - 35}{9 + 49} = \dfrac{-23 + 43i}{58} = -\dfrac{23}{58} + \dfrac{43}{58}i$

**79.** $\dfrac{2 + 3i}{2 + 5i} = \dfrac{2 + 3i}{2 + 5i} \cdot \dfrac{2 - 5i}{2 - 5i} = \dfrac{4 - 10i + 6i - 15i^2}{4 - 25i^2} =$
$\dfrac{4 - 10i + 6i + 15}{4 + 25} = \dfrac{19 - 4i}{29} = \dfrac{19}{29} - \dfrac{4}{29}i$

**81.** $\dfrac{3-2i}{4+3i} = \dfrac{3-2i}{4+3i} \cdot \dfrac{4-3i}{4-3i} = \dfrac{12-9i-8i+6i^2}{16-9i^2} =$

$\dfrac{12-9i-8i-6}{16+9} = \dfrac{6-17i}{25} = \dfrac{6}{25} - \dfrac{17}{25}i$

**83.** $i^7 = i^6 \cdot i = (i^2)^3 \cdot i = (-1)^3 \cdot i = -1 \cdot i = -i$

**85.** $i^{24} = (i^2)^{12} = (-1)^{12} = 1$

**87.** $i^{42} = (i^2)^{21} = (-1)^{21} = -1$

**89.** $i^9 = (i^2)^4 \cdot i = (-1)^4 \cdot i = 1 \cdot i = i$

**91.** $(-i)^6 = (-1 \cdot i)^6 = (-1)^6 \cdot i^6 = 1 \cdot i^6 = (i^2)^3 = (-1)^3 = -1$

**93.** $(5i)^3 = 5^3 \cdot i^3 = 125 \cdot i^2 \cdot i = 125(-1)(i) = -125i$

**95.** $i^2 + i^4 = -1 + (i^2)^2 = -1 + (-1)^2 = -1 + 1 = 0$

**97.** *Writing Exercise*

**99.** Graph $f(x) = 3x - 5$.

The $y$-intercept is $(0, -5)$ and the slope is 3, or $\dfrac{3}{1}$. First plot the $y$-intercept and, from there, go up 3 units and right 1 unit to $(1, -2)$. Plot this point and draw the graph.

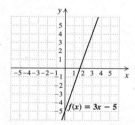

To check that the line is drawn correctly, we find the coordinates of another point on the line. When $x = 2$, we have:

$f(2) = 3 \cdot 2 - 5 = 6 - 5 = 1$.

Since the point $(2, 1)$ appears to be on the line, we have a check.

**101.** Graph $F(x) = x^2$.

We select numbers for $x$ and find the corresponding values of $y$. Then we plot these ordered pairs and draw the graph.

| $x$ | $F(x)$ |
|-----|--------|
| $-2$ | 4 |
| $-1$ | 1 |
| 0 | 0 |
| 1 | 1 |
| 2 | 4 |

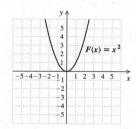

**103.** $28 = 3x^2 - 17x$

$0 = 3x^2 - 17x - 28$

$0 = (3x + 4)(x - 7)$

$3x + 4 = 0 \quad or \quad x - 7 = 0$

$3x = -4 \quad or \quad x = 7$

$x = -\dfrac{4}{3} \quad or \quad x = 7$

Both values check. The solutions are $-\dfrac{4}{3}$ and 7.

**105.** *Writing Exercise*

**107.** $g(3i) = \dfrac{(3i)^4 - (3i)^2}{3i - 1} = \dfrac{81i^4 - 9i^2}{-1 + 3i} = \dfrac{81 + 9}{-1 + 3i} =$

$\dfrac{90}{-1 + 3i} = \dfrac{90}{-1 + 3i} \cdot \dfrac{-1 - 3i}{-1 - 3i} = \dfrac{90(-1 - 3i)}{1 - 9i^2} =$

$\dfrac{90(-1 - 3i)}{1 + 9} = \dfrac{90(-1 - 3i)}{10} = \dfrac{9 \cdot \cancel{10}(-1 - 3i)}{\cancel{10}} =$

$9(-1 - 3i) = -9 - 27i$

**109.** First we simplify $g(z)$.

$g(z) = \dfrac{z^4 - z^2}{z - 1} = \dfrac{z^2(z^2 - 1)}{z - 1} = \dfrac{z^2(z + 1)(z - 1)}{z - 1} =$

$\dfrac{z^2(z + 1)\cancel{(z - 1)}}{\cancel{z - 1}} = z^2(z + 1)$

Now we substitute.

$g(5i - 1) = (5i - 1)^2(5i - 1 + 1) =$

$(25i^2 - 10i + 1)(5i) =$

$(-25 - 10i + 1)(5i) = (-24 - 10i)(5i) =$

$-120i - 50i^2 = 50 - 120i$

**111.** $\dfrac{1}{\dfrac{1-i}{10} - \left(\dfrac{1-i}{10}\right)^2} = \dfrac{1}{\dfrac{1-i}{10} - \left(\dfrac{-2i}{100}\right)} =$

$\dfrac{1}{\dfrac{1-i}{10} + \dfrac{i}{50}} = \dfrac{1}{\dfrac{1-i}{10} + \dfrac{i}{50}} \cdot \dfrac{50}{50} = \dfrac{50}{5 - 5i + i} =$

$\dfrac{50}{5 - 4i} = \dfrac{50}{5 - 4i} \cdot \dfrac{5 + 4i}{5 + 4i} = \dfrac{250 + 200i}{41} = \dfrac{250}{41} + \dfrac{200}{41}i$

**113.** $(1 - i)^3(1 + i)^3 =$

$(1 - i)(1 + i) \cdot (1 - i)(1 + i) \cdot (1 - i)(1 + i) =$

$(1 - i^2)(1 - i^2)(1 - i^2) = (1 + 1)(1 + 1)(1 + 1) =$

$2 \cdot 2 \cdot 2 = 8$

**115.** $\dfrac{6}{1 + \dfrac{3}{i}} = \dfrac{6}{\dfrac{i + 3}{i}} = \dfrac{6i}{i + 3} = \dfrac{6i}{i + 3} \cdot \dfrac{-i + 3}{-i + 3} =$

$\dfrac{-6i^2 + 18i}{-i^2 + 9} = \dfrac{6 + 18i}{10} = \dfrac{6}{10} + \dfrac{18}{10}i = \dfrac{3}{5} + \dfrac{9}{5}i$

**117.** $\dfrac{i - i^{38}}{1 + i} = \dfrac{i - (i^2)^{19}}{1 + i} = \dfrac{i - (-1)^{19}}{1 + i} = \dfrac{i - (-1)}{1 + i} =$

$\dfrac{i + 1}{1 + i} = 1$

# Chapter 8

# Quadratic Functions and Equations

## Exercise Set 8.1

**1.** $\sqrt{k}$; $-\sqrt{k}$

**3.** $t+3$; $t+3$

**5.** 25; 5

**7.** $4x^2 = 20$

$\qquad x^2 = 5 \qquad$ Multiplying by $\frac{1}{4}$

$\qquad x = \sqrt{5}$, or $x = -\sqrt{5} \quad$ Using the principle
of square roots

The solutions are $\sqrt{5}$ and $-\sqrt{5}$, or $\pm\sqrt{5}$.

**9.** $9x^2 + 16 = 0$

$\qquad x^2 = -\frac{16}{9} \quad$ Isolating $x^2$

$\qquad x = \sqrt{-\frac{16}{9}} \qquad or \quad x = -\sqrt{-\frac{16}{9}}$

$\qquad x = \sqrt{\frac{16}{9}}\sqrt{-1} \quad or \quad x = -\sqrt{\frac{16}{9}}\sqrt{-1}$

$\qquad x = \frac{4}{3}i \qquad\qquad or \quad x = -\frac{4}{3}i$

The solutions are $\frac{4}{3}i$ and $-\frac{4}{3}i$, or $\pm\frac{4}{3}i$.

**11.** $5t^2 - 7 = 0$

$\qquad t^2 = \frac{7}{5}$

$\qquad t = \sqrt{\frac{7}{5}} \qquad or \quad t = -\sqrt{\frac{7}{5}} \qquad$ Principle of
square roots

$\qquad t = \sqrt{\frac{7}{5}\cdot\frac{5}{5}} \quad or \quad t = -\sqrt{\frac{7}{5}\cdot\frac{5}{5}} \quad$ Rationalizing
denominators

$\qquad t = \frac{\sqrt{35}}{5} \qquad or \quad t = -\frac{\sqrt{35}}{5}$

The solutions are $\sqrt{\frac{7}{5}}$ and $-\sqrt{\frac{7}{5}}$. This can also be written

as $\pm\sqrt{\frac{7}{5}}$ or, if we rationalize the denominator, $\pm\frac{\sqrt{35}}{5}$.

**13.** $(x-1)^2 = 49$

$\qquad x - 1 = 7 \;\; or \;\; x - 1 = -7 \quad$ Principle of square
roots

$\qquad\quad x = 8 \;\; or \qquad\quad x = -6$

The solutions are 8 and $-6$.

**15.** $(a-13)^2 = 18$

$\qquad a - 13 = \sqrt{18} \qquad or \;\; a - 13 = -\sqrt{18} \quad$ Principle
of square roots

$\qquad a - 13 = 3\sqrt{2} \qquad or \;\; a - 13 = -3\sqrt{2}$

$\qquad\quad a = 13 + 3\sqrt{2} \;\; or \qquad\quad a = 13 - 3\sqrt{2}$

The solutions are $13 + 3\sqrt{2}$ and $13 - 3\sqrt{2}$, or $13 \pm 3\sqrt{2}$.

**17.** $(x+1)^2 = -9$

$\qquad x + 1 = \sqrt{-9} \qquad or \;\; x + 1 = -\sqrt{-9}$

$\qquad x + 1 = 3i \qquad\quad or \;\; x + 1 = -3i$

$\qquad\quad x = -1 + 3i \;\; or \qquad\quad x = -1 - 3i$

The solutions are $-1 + 3i$ and $-1 - 3i$, or $-1 \pm 3i$.

**19.** $\left(y + \frac{3}{4}\right)^2 = \frac{17}{16}$

$\qquad y + \frac{3}{4} = \pm\frac{\sqrt{17}}{4}$

$\qquad y = -\frac{3}{4} \pm \frac{\sqrt{17}}{4}$, or $\frac{-3 \pm \sqrt{17}}{4}$

The solutions are $-\frac{3}{4} \pm \frac{\sqrt{17}}{4}$, or $\frac{-3 \pm \sqrt{17}}{4}$.

**21.** $x^2 - 10x + 25 = 64$

$\qquad (x-5)^2 = 64$

$\qquad\quad x - 5 = \pm 8$

$\qquad\qquad x = 5 \pm 8$

$\qquad\qquad x = 13 \; or \; x = -3$

The solutions are 13 and $-13$.

**23.** $\qquad f(x) = 16$

$\qquad (x-5)^2 = 16 \qquad\qquad$ Substituting

$\qquad x - 5 = 4 \; or \; x - 5 = -4$

$\qquad x = 9 \; or \quad x = 1$

The solutions are 9 and 1.

**25.** $\qquad F(t) = 13$

$\qquad (t+4)^2 = 13 \quad$ Substituting

$\qquad t + 4 = \sqrt{13} \qquad or \;\; t + 4 = -\sqrt{13}$

$\qquad\quad t = -4 + \sqrt{13} \;\; or \qquad t = -4 - \sqrt{13}$

The solutions are $-4 + \sqrt{13}$ and $-4 - \sqrt{13}$, or $-4 \pm \sqrt{13}$.

**27.** $g(x) = x^2 + 14x + 49$

Observe first that $g(0) = 49$. Also observe that when $x = -14$, then $x^2 + 14x = (-14)^2 - (14)(14) = (14)^2 - (14)^2 = 0$, so $g(-14) = 49$ as well. Thus, we have $x = 0$ or $x = 14$.

We can also do this problem as follows.

$$g(x) = 49$$
$$x^2 + 14x + 49 = 49 \quad \text{Substituting}$$
$$(x+7)^2 = 49$$
$$x + 7 = 7 \;\; or \;\; x + 7 = -7$$
$$x = 0 \;\; or \quad\quad x = -14$$

The solutions are 0 and $-14$.

**29.** $x^2 + 16x$

We take half the coefficient of $x$ and square it: Half of 16 is 8, and $8^2 = 64$. We add 64.

$$x^2 + 16x + 64 = (x+8)^2$$

**31.** $t^2 - 10t$

We take half the coefficient of $t$ and square it:

Half of $-10$ is $-5$, and $(-5)^2 = 25$. We add 25.

$$t^2 - 10t + 25 = (t-5)^2$$

**33.** $x^2 + 3x$

We take half the coefficient of $x$ and square it:

$\dfrac{1}{2}(3) = \dfrac{3}{2}$ and $\left(\dfrac{3}{2}\right)^2 = \dfrac{9}{4}$. We add $\dfrac{9}{4}$.

$$x^2 + 3x + \frac{9}{4} = \left(x + \frac{3}{2}\right)^2$$

**35.** $t^2 - 9t$

We take half the coefficient of $t$ and square it:

$\dfrac{1}{2}(-9) = -\dfrac{9}{2}$, and $\left(-\dfrac{9}{2}\right)^2 = \dfrac{81}{4}$. We add $\dfrac{81}{4}$.

$$t^2 - 9t + \frac{81}{4} = \left(t - \frac{9}{2}\right)^2$$

**37.** $x^2 + \dfrac{2}{5}x$

$\dfrac{1}{2} \cdot \dfrac{2}{5} = \dfrac{1}{5}$, and $\left(\dfrac{1}{5}\right)^2 = \dfrac{1}{25}$. We add $\dfrac{1}{25}$.

$$x^2 + \frac{2}{5}x + \frac{1}{25} = \left(x + \frac{1}{5}\right)^2$$

**39.** $t^2 - \dfrac{5}{6}t$

$\dfrac{1}{2}\left(-\dfrac{5}{6}\right) = -\dfrac{5}{12}$, and $\left(-\dfrac{5}{12}\right)^2 = \dfrac{25}{144}$. We add $\dfrac{25}{144}$.

$$t^2 - \frac{5}{6}t + \frac{25}{144} = \left(t - \frac{5}{12}\right)^2$$

**41.**
$$x^2 + 6x = 7$$
$$x^2 + 6x + 9 = 7 + 9 \quad \text{Adding 9 to both sides}$$
$$\qquad\qquad\qquad\qquad \text{to complete the square}$$
$$(x+3)^2 = 16 \qquad \text{Factoring}$$
$$x + 3 = \pm 4 \qquad \text{Principle of square roots}$$
$$x = -3 \pm 4$$
$$x = -3 + 4 \;\; or \;\; x = -3 - 4$$
$$x = 1 \qquad\;\; or \;\; x = -7$$

The solutions are 1 and $-7$.

**43.**
$$t^2 - 10t = -24$$
$$t^2 - 10t + 25 = -24 + 25 \quad \text{Adding 25 to both sides}$$
$$\qquad\qquad\qquad\qquad\qquad \text{to complete the square}$$
$$(t-5)^2 = 1 \qquad\qquad \text{Factoring}$$
$$t - 5 = \pm 1 \qquad\quad \text{Principle of square roots}$$
$$t = 5 \pm 1$$
$$t = 5 + 1 \;\; or \;\; x = 5 - 1$$
$$t = 6 \quad\;\; or \;\; x = 4$$

The solutions are 6 and 4.

**45.**
$$x^2 + 10x + 9 = 0$$
$$x^2 + 10x = -9 \quad \text{Adding } -9 \text{ to both sides}$$
$$x^2 + 10x + 25 = -9 + 25 \quad \text{Completing the square}$$
$$(x+5)^2 = 16$$
$$x + 5 = \pm 4$$
$$x = -5 \pm 4$$
$$x = -5 + 4 \;\; or \;\; x = -5 - 4$$
$$x = -1 \quad\;\; or \;\; x = -9$$

The solutions are $-1$ and $-9$.

**47.**
$$t^2 + 8t - 3 = 0$$
$$t^2 + 8t = 3$$
$$t^2 + 8t + 16 = 3 + 16$$
$$(t+4)^2 = 19$$
$$t + 4 = \pm\sqrt{19}$$
$$t = -4 \pm \sqrt{19}$$

The solutions are $-4 \pm \sqrt{19}$.

**49.** The value of $f(x)$ must be 0 at any $x$-intercepts.
$$f(x) = 0$$
$$x^2 + 6x + 7 = 0$$
$$x^2 + 6x = -7$$
$$x^2 + 6x + 9 + -7 + 9$$
$$(x+3)^2 = 2$$
$$x + 3 = \pm\sqrt{2}$$
$$x = -3 \pm \sqrt{2}$$

The $x$-intercepts are $(-3 - \sqrt{2}, 0)$ and $(-3 + \sqrt{2}, 0)$.

**51.** The value of $g(x)$ must be 0 at any $x$-intercepts.
$$g(x) = 0$$
$$x^2 + 12x + 25 = 0$$
$$x^2 + 12x = -25$$
$$x^2 + 12x + 36 = -25 + 36$$
$$(x+6)^2 = 11$$
$$x + 6 = \pm\sqrt{11}$$
$$x = -6 \pm \sqrt{11}$$

The $x$-intercepts are $(-6 - \sqrt{11}, 0)$ and $(-6 + \sqrt{11}, 0)$.

**53.** The value of $f(x)$ must be 0 at any $x$-intercepts.
$$f(x) = 0$$
$$x^2 - 10x - 22 = 0$$
$$x^2 - 10x = 22$$
$$x^2 - 10x + 25 = 22 + 25$$
$$(x-5)^2 = 47$$
$$x - 5 = \pm\sqrt{47}$$
$$x = 5 \pm \sqrt{47}$$
The $x$-intercepts are $(5 - \sqrt{47}, 0)$ and $(5 + \sqrt{47}, 0)$.

**55.**
$$9x^2 + 18x = -8$$
$$x^2 + 2x = -\frac{8}{9} \quad \text{Dividing both sides by 9}$$
$$x^2 + 2x + 1 = -\frac{8}{9} + 1$$
$$(x+1)^2 = \frac{1}{9}$$
$$x + 1 = \pm\frac{1}{3}$$
$$x = -1 \pm \frac{1}{3}$$
$$x = -1 - \frac{1}{3} \quad or \quad x = -1 + \frac{1}{3}$$
$$x = -\frac{4}{3} \quad\quad or \quad x = -\frac{2}{3}$$
The solutions are $-\frac{4}{3}$ and $-\frac{2}{3}$.

**57.**
$$3x^2 - 5x - 2 = 0$$
$$3x^2 - 5x = 2$$
$$x^2 - \frac{5}{3}x = \frac{2}{3} \quad \text{Dividing both sides by 3}$$
$$x^2 - \frac{5}{3}x + \frac{25}{36} = \frac{2}{3} + \frac{25}{36}$$
$$\left(x - \frac{5}{6}\right)^2 = \frac{49}{36}$$
$$x - \frac{5}{6} = \pm\frac{7}{6}$$
$$x = \frac{5}{6} \pm \frac{7}{6}$$
$$x = \frac{5}{6} - \frac{7}{6} \quad or \quad x = \frac{5}{6} + \frac{7}{6}$$
$$x = -\frac{1}{3} \quad\quad or \quad x = 2$$
The solutions are $-\frac{1}{3}$ and 2.

**59.**
$$5x^2 + 4x - 3 = 0$$
$$5x^2 + 4x = 3$$
$$x^2 + \frac{4}{5}x = \frac{3}{5} \quad \text{Dividing both sides by 5}$$
$$x^2 + \frac{4}{5}x + \frac{4}{25} = \frac{3}{5} + \frac{4}{25}$$
$$\left(x + \frac{2}{5}\right)^2 = \frac{19}{25}$$
$$x + \frac{2}{5} = \pm\frac{\sqrt{19}}{5}$$
$$x = -\frac{2}{5} \pm \frac{\sqrt{19}}{5}, \text{ or } \frac{-2 \pm \sqrt{19}}{5}$$
The solutions are $-\frac{2}{5} \pm \frac{\sqrt{19}}{5}$, or $\frac{-2 \pm \sqrt{19}}{5}$.

**61.** The value of $f(x)$ must be 0 at any $x$-intercepts.
$$f(x) = 0$$
$$4x^2 + 2x - 3 = 0$$
$$4x^2 + 2x = 3$$
$$x^2 + \frac{1}{2}x = \frac{3}{4} \quad \text{Dividing both sides by 4}$$
$$x^2 + \frac{1}{2}x + \frac{1}{16} = \frac{3}{4} + \frac{1}{16}$$
$$\left(x + \frac{1}{4}\right)^2 = \frac{13}{16}$$
$$x + \frac{1}{4} = \pm\frac{\sqrt{13}}{4}$$
$$x = -\frac{1}{4} \pm \frac{\sqrt{13}}{4}, \text{ or } \frac{-1 \pm \sqrt{13}}{4}$$
The $x$-intercepts are $\left(-\frac{1}{4} - \frac{\sqrt{13}}{4}, 0\right)$ and $\left(-\frac{1}{4} + \frac{\sqrt{13}}{4}, 0\right)$, or $\left(\frac{-1 - \sqrt{13}}{4}, 0\right)$ and $\left(\frac{-1 + \sqrt{13}}{4}, 0\right)$.

**63.** The value of $g(x)$ must be 0 at any $x$-intercepts.
$$g(x) = 0$$
$$2x^2 - 3x - 1 = 0$$
$$2x^2 - 3x = 1$$
$$x^2 - \frac{3}{2}x = \frac{1}{2} \quad \text{Dividing both sides by 2}$$
$$x^2 - \frac{3}{2}x + \frac{9}{16} = \frac{1}{2} + \frac{9}{16}$$
$$\left(x - \frac{3}{4}\right)^2 = \frac{17}{16}$$
$$x - \frac{3}{4} = \pm\frac{\sqrt{17}}{4}$$
$$x = \frac{3}{4} \pm \frac{\sqrt{17}}{4}, \text{ or } \frac{3 \pm \sqrt{17}}{4}$$
The $x$-intercepts are $\left(\frac{3}{4} - \frac{\sqrt{17}}{4}, 0\right)$ and $\left(\frac{3}{4} + \frac{\sqrt{17}}{4}, 0\right)$, or $\left(\frac{3 - \sqrt{17}}{4}, 0\right)$ and $\left(\frac{3 + \sqrt{17}}{4}, 0\right)$.

**65. Familiarize**. We are already familiar with the compound-interest formula.

**Translate**. We substitute into the formula.

$$A = P(1+r)^t$$
$$2420 = 2000(1+r)^2$$

**Carry out**. We solve for $r$.

$$2420 = 2000(1+r)^2$$
$$\frac{2420}{2000} = (1+r)^2$$
$$\frac{121}{100} = (1+r)^2$$
$$\pm\sqrt{\frac{121}{100}} = 1+r$$
$$\pm\frac{11}{10} = 1+r$$
$$-\frac{10}{10} + \frac{11}{10} = r$$
$$\frac{1}{10} = r \; or \; -\frac{21}{10} = r$$

**Check**. Since the interest rate cannot be negative, we need only check $\frac{1}{10}$, or 10%. If \$2000 were invested at 10% interest, compounded annually, then in 2 years it would grow to \$2000$(1.1)^2$, or \$2420. The number 10% checks.

**State**. The interest rate is 10%.

**67. Familiarize**. We are already familiar with the compound-interest formula.

**Translate**. We substitute into the formula.

$$A = P(1+r)^t$$
$$1805 = 1280(1+r)^2$$

**Carry out**. We solve for $r$.

$$1805 = 1280(1+r)^2$$
$$\frac{1805}{1280} = (1+r)^2$$
$$\frac{361}{256} = (1+r)^2$$
$$\pm\frac{19}{16} = 1+r$$
$$-\frac{16}{16} \pm \frac{19}{16} = r$$
$$\frac{3}{16} = r \; or \; -\frac{35}{16} = r$$

**Check**. Since the interest rate cannot be negative, we need only check $\frac{3}{16}$ or 18.75%. If \$1280 were invested at 18.75% interest, compounded annually, then in 2 years it would grow to \$1280$(1.1875)^2$, or \$1805. The number 18.75% checks.

**State**. The interest rate is 18.75%.

**69. Familiarize**. We are already familiar with the compound-interest formula.

**Translate**. We substitute into the formula.

$$A = P(1+r)^t$$
$$6760 = 6250(1+r)^2$$

**Carry out**. We solve for $r$.

$$\frac{6760}{6250} = (1+r)^2$$
$$\frac{676}{625} = (1+r)^2$$
$$\pm\frac{26}{25} = 1+r$$
$$-\frac{25}{25} \pm \frac{26}{25} = r$$
$$\frac{1}{25} = r \; or \; -\frac{51}{25} = r$$

**Check**. Since the interest rate cannot be negative, we need only check $\frac{1}{25}$, or 4%. If \$6250 were invested at 4% interest, compounded annually, then in 2 years it would grow to \$6250$(1.04)^2$, or \$6760. The number 4% checks.

**State**. The interest rate is 4%.

**71. Familiarize**. We will use the formula $s = 16t^2$.

**Translate**. We substitute into the formula.

$$s = 16t^2$$
$$1053 = 16t^2$$

**Carry out**. We solve for $t$.

$$1053 = 16t^2$$
$$\frac{1053}{16} = t^2$$
$$\sqrt{\frac{1053}{16}} = t \quad \text{Principle of square roots; rejecting the negative square root}$$
$$8.1 \approx t$$

**Check**. Since $16(8.1)^2 = 1049.76 \approx 1053$, our answer checks.

**State**. It would take an object about 8.1 sec to fall freely from the Royal Gorge bridge.

**73. Familiarize**. We will use the formula $s = 16t^2$.

**Translate**. We substitute into the formula.

$$s = 16t^2$$
$$1454 = 16t^2$$

**Carry out**. We solve for $t$.

$$1454 = 16t^2$$
$$\frac{1454}{16} = t^2$$
$$\sqrt{\frac{1454}{16}} = t \quad \text{Principle of square roots; rejecting the negative square root}$$
$$9.5 \approx t$$

**Check**. Since $16(9.5)^2 = 1444 \approx 1454$, our answer checks.

**State**. It would take an object about 9.5 sec to fall freely from the top of the Sears Tower.

**75.** *Writing Exercise*

**77.** $at^2 - bt = 3 \cdot 4^2 - 5 \cdot 4$

$\qquad = 3 \cdot 16 - 5 \cdot 4$

$\qquad = 48 - 20$

$\qquad = 28$

**79.** $\sqrt[3]{270} = \sqrt[3]{27 \cdot 10} = \sqrt[3]{27}\sqrt[3]{10} = 3\sqrt[3]{10}$

**81.** $f(x) = \sqrt{3x - 5}$

$\quad f(10) = \sqrt{3 \cdot 10 - 5} = \sqrt{30 - 5} = \sqrt{25} = 5$

**83.** *Writing Exercise*

**85.** In order for $x^2 + bx + 81$ to be a square, the following must be true:

$$\left(\frac{b}{2}\right)^2 = 81$$

$$\frac{b^2}{4} = 81$$

$$b^2 = 324$$

$$b = 18 \text{ or } b = -18$$

**87.** We see that $x$ is a factor of each term, so $x$ is also a factor of $f(x)$. We have $f(x) = x(2x^4 - 9x^3 - 66x^2 + 45x + 280)$. Since $x^2 - 5$ is a factor of $f(x)$ it is also a factor of $2x^4 - 9x^3 - 66x^2 + 45x + 280$. We divide to find another factor.

$$
\begin{array}{r}
2x^2 - \phantom{0}9x - \phantom{0}56 \\
x^2 - 5 \overline{\smash{\big)}\, 2x^4 - \phantom{0}9x^3 - \phantom{0}66x^2 + 45x + 280} \\
\underline{2x^4 \phantom{00000} - 10x^2} \\
-9x^3 - \phantom{0}56x^2 + 45x \\
\underline{-9x^3 \phantom{0000000} + 45x} \\
-56x^2 \phantom{0000} + 280 \\
\underline{-56x^2 \phantom{0000} + 280} \\
0
\end{array}
$$

Then we have $f(x) = x(x^2 - 5)(2x^2 - 9x - 56)$, or $f(x) = x(x^2 - 5)(2x + 7)(x - 8)$. Now we find the values of $a$ for which $f(a) = 0$.

$$f(a) = 0$$

$$a(a^2 - 5)(2a + 7)(a - 8) = 0$$

$a = 0 \ \text{ or } \ a^2 - 5 = 0 \quad \text{ or } \ 2a + 7 = 0 \quad \text{ or } \ a - 8 = 0$

$a = 0 \ \text{ or } \quad a^2 = 5 \quad \text{ or } \quad 2a = -7 \ \text{ or } \quad a = 8$

$a = 0 \ \text{ or } \quad a = \pm\sqrt{5} \ \text{ or } \quad a = -\dfrac{7}{2} \ \text{ or } \quad a = 8$

The solutions are $0$, $\sqrt{5}$, $-\sqrt{5}$, $-\dfrac{7}{2}$, and $8$.

**89.** *Familiarize*. It is helpful to list information in a chart and make a drawing. Let $r$ represent the speed of the fishing boat. Then $r - 7$ represents the speed of the barge.

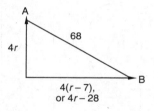

| Boat | $r$ | $t$ | $d$ |
|------|-----|-----|-----|
| Fishing | $r$ | 4 | $4r$ |
| Barge | $r - 7$ | 4 | $4(r-7)$ |

*Translate*. We use the Pythagorean equation:

$$a^2 + b^2 = c^2$$

$$(4r - 28)^2 + (4r)^2 = 68^2$$

*Carry out*.

$$(4r - 28)^2 + (4r)^2 = 68^2$$

$$16r^2 - 224r + 784 + 16r^2 = 4624$$

$$32r^2 - 224r - 3840 = 0$$

$$r^2 - 7r - 120 = 0$$

$$(r + 8)(r - 15) = 0$$

$$r + 8 = 0 \quad \text{ or } \ r - 15 = 0$$

$$r = -8 \ \text{ or } \qquad r = 15$$

*Check*. We check only 15 since the speeds of the boats cannot be negative. If the speed of the fishing boat is 15 km/h, then the speed of the barge is $15 - 7$, or 8 km/h, and the distances they travel are $4 \cdot 15$ (or 60) and $4 \cdot 8$ (or 32).

$$60^2 + 32^2 = 3600 + 1024 = 4624 = 68^2$$

The values check.

*State*. The speed of the fishing boat is 15 km/h, and the speed of the barge is 8 km/h.

**91.** ▮

**93.** *Writing Exercise*

## Exercise Set 8.2

**1.** True; see page 519 in the text.

**3.** False; see page 519 in the text.

**5.** False; the quadratic formula yields at most two solutions.

**7.** $x^2 + 7x - 3 = 0$

$\quad a = 1,\ b = 7,\ c = -3$

$$x = \frac{-b \pm \sqrt{b^2 - 4ac}}{2a}$$

$$x = \frac{-7 \pm \sqrt{7^2 - 4 \cdot 1 \cdot (-3)}}{2 \cdot 1} = \frac{-7 \pm \sqrt{49 + 12}}{2}$$

$$x = \frac{-7 \pm \sqrt{61}}{2} = -\frac{7}{2} \pm \frac{\sqrt{61}}{2}$$

The solutions are $-\dfrac{7}{2} - \dfrac{\sqrt{61}}{2}$ and $-\dfrac{7}{2} + \dfrac{\sqrt{61}}{2}$.

**9.** $\qquad 3p^2 = 18p - 6$

$\quad 3p^2 - 18p + 6 = 0$

$\quad\ p^2 - 6p + 2 = 0 \qquad$ Dividing by 3

$a = 1,\ b = -6,\ c = 2$

$$p = \frac{-b \pm \sqrt{b^2 - 4ac}}{2a}$$

$$p = \frac{-(-6) \pm \sqrt{(-6)^2 - 4 \cdot 1 \cdot 2}}{2 \cdot 1} = \frac{6 \pm \sqrt{36 - 8}}{2}$$

$$p = \frac{6 \pm \sqrt{28}}{2} = \frac{6 \pm 2\sqrt{7}}{2}$$

$$p = \frac{6}{2} \pm \frac{2\sqrt{7}}{2} = 3 \pm \sqrt{7}$$

The solutions are $3 + \sqrt{7}$ and $3 - \sqrt{7}$.

**11.** $x^2 - x + 1 = 0$

$a = 1, b = -1, c = 1$

$$x = \frac{-b \pm \sqrt{b^2 - 4ac}}{2a}$$

$$x = \frac{-(-1) \pm \sqrt{(-1)^2 - 4 \cdot 1 \cdot 1}}{2 \cdot 1} = \frac{1 \pm \sqrt{1 - 4}}{2}$$

$$x = \frac{1 \pm \sqrt{-3}}{2} = \frac{1 \pm i\sqrt{3}}{2} = \frac{1}{2} \pm \frac{\sqrt{3}}{2}i$$

The solutions are $\frac{1}{2} + \frac{\sqrt{3}}{2}i$ and $\frac{1}{2} - \frac{\sqrt{3}}{2}i$.

**13.** $\qquad x^2 + 13 = 4x$

$x^2 - 4x + 13 = 0$

$a = 1, b = -4, c = 13$

$$x = \frac{-b \pm \sqrt{b^2 - 4ac}}{2a}$$

$$x = \frac{-(-4) \pm \sqrt{(-4)^2 - 4 \cdot 1 \cdot 13}}{2 \cdot 1} = \frac{4 \pm \sqrt{16 - 52}}{2}$$

$$x = \frac{4 \pm \sqrt{-36}}{2} = \frac{4 \pm 6i}{2}$$

$$x = \frac{4}{2} \pm \frac{6}{2}i = 2 \pm 3i$$

The solutions are $2 + 3i$ and $2 - 3i$.

**15.** $\qquad h^2 + 4 = 6h$

$h^2 - 6h + 4 = 0$

$a = 1, b = -6, c = 4$

$$x = \frac{-(-6) \pm \sqrt{(-6)^2 - 4 \cdot 1 \cdot 4}}{2 \cdot 1} = \frac{6 \pm \sqrt{36 - 16}}{2}$$

$$x = \frac{6 \pm \sqrt{20}}{2} = \frac{6 \pm \sqrt{4 \cdot 5}}{2} = \frac{6 \pm 2\sqrt{5}}{2} = \frac{6}{2} \pm \frac{2\sqrt{5}}{2}$$

$$x = 3 \pm \sqrt{5}$$

The solutions are $3 + \sqrt{5}$ and $3 - \sqrt{5}$.

**17.** $\qquad \frac{1}{x^2} - 3 = \frac{8}{x}$, LCD is $x^2$

$$x^2 \left( \frac{1}{x^2} - 3 \right) = x^2 \cdot \frac{8}{x}$$

$$x^2 \cdot \frac{1}{x^2} - x^2 \cdot 3 = 8x$$

$$1 - 3x^2 = 8x$$

$$0 = 3x^2 + 8x - 1$$

$a = 3, b = 8, c = -1$

$$x = \frac{-8 \pm \sqrt{8^2 - 4 \cdot 3 \cdot (-1)}}{2 \cdot 3} = \frac{-8 \pm \sqrt{64 + 12}}{6}$$

$$x = \frac{-8 \pm \sqrt{76}}{6} = \frac{-8 \pm \sqrt{4 \cdot 19}}{6} = \frac{-8 \pm 2\sqrt{19}}{6}$$

$$x = \frac{-4 \pm \sqrt{19}}{3} = -\frac{4}{3} \pm \frac{\sqrt{19}}{3}$$

The solutions are $-\frac{4}{3} - \frac{\sqrt{19}}{3}$ and $-\frac{4}{3} + \frac{\sqrt{19}}{3}$.

**19.** $3x + x(x - 2) = 4$

$3x + x^2 - 2x = 4$

$\qquad x^2 + x = 4$

$\qquad x^2 + x - 4 = 0$

$a = 1, b = 1, c = -4$

$$x = \frac{-1 \pm \sqrt{1^2 - 4 \cdot 1 \cdot (-4)}}{2 \cdot 1} = \frac{-1 \pm \sqrt{1 + 16}}{2}$$

$$x = \frac{-1 \pm \sqrt{17}}{2} = -\frac{1}{2} \pm \frac{\sqrt{17}}{2}$$

The solutions are $-\frac{1}{2} - \frac{\sqrt{17}}{2}$ and $-\frac{1}{2} + \frac{\sqrt{17}}{2}$.

**21.** $\qquad 12x^2 + 9t = 1$

$12t^2 + 9t - 1 = 0$

$a = 12, b = 9, c = -1$

$$t = \frac{-9 \pm \sqrt{9^2 - 4 \cdot 12 \cdot (-1)}}{2 \cdot 12} = \frac{-9 \pm \sqrt{81 + 48}}{24}$$

$$t = \frac{-9 \pm \sqrt{129}}{24} = -\frac{9}{24} \pm \frac{\sqrt{129}}{24} = -\frac{3}{8} \pm \frac{\sqrt{129}}{24}$$

The solutions are $-\frac{3}{8} - \frac{\sqrt{129}}{24}$ and $-\frac{3}{8} + \frac{\sqrt{129}}{24}$.

**23.** $\qquad 25x^2 - 20x + 4 = 0$

$\qquad (5x - 2)(5x - 2) = 0$

$\qquad 5x - 2 = 0 \quad or \quad 5x - 2 = 0$

$\qquad\qquad 5x = 2 \quad or \qquad 5x = 2$

$$\qquad\qquad x = \frac{2}{5} \quad or \qquad x = \frac{2}{5}$$

The solution is $\frac{2}{5}$.

**25.** $7x(x + 2) + 5 = 3x(x + 1)$

$7x^2 + 14x + 5 = 3x^2 + 3x$

$\qquad 4x^2 + 11x + 5 = 0$

$a = 4, b = 11, c = 5$

$$x = \frac{-11 \pm \sqrt{11^2 - 4 \cdot 4 \cdot 5}}{2 \cdot 4} = \frac{-11 \pm \sqrt{121 - 80}}{8}$$

$$x = \frac{-11 \pm \sqrt{41}}{8} = -\frac{11}{8} \pm \frac{\sqrt{41}}{8}$$

The solutions are $-\frac{11}{8} - \frac{\sqrt{41}}{8}$ and $-\frac{11}{8} + \frac{\sqrt{41}}{8}$.

**27.** $14(x-4)-(x+2)=(x+2)(x-4)$

$14x-56-x-2=x^2-2x-8$    Removing parentheses

$13x-58=x^2-2x-8$

$0=x^2-15x+50$

$0=(x-10)(x-5)$

$x-10=0$   or   $x-5=0$

$x=10$   or     $x=5$

The solutions are 10 and 5.

**29.**        $5x^2=13x+17$

$5x^2-13x-17=0$

$a=5,\ b=-13,\ c=-17$

$x=\dfrac{-(-13)\pm\sqrt{(-13)^2-4(5)(-17)}}{2\cdot 5}$

$x=\dfrac{13\pm\sqrt{169+340}}{10}=\dfrac{13\pm\sqrt{509}}{10}$

$x=\dfrac{13}{10}\pm\dfrac{\sqrt{509}}{10}$

The solutions are $\dfrac{13}{10}-\dfrac{\sqrt{509}}{10}$ and $\dfrac{13}{10}+\dfrac{\sqrt{509}}{10}$.

**31.**       $x^2+9=4x$

$x^2-4x+9=0$

$a=1,\ b=-4,\ c=9$

$x=\dfrac{-(-4)\pm\sqrt{(-4)^2-4\cdot 1\cdot 9}}{2\cdot 1}=\dfrac{4\pm\sqrt{16-36}}{2}$

$x=\dfrac{4\pm\sqrt{-20}}{2}=\dfrac{4\pm\sqrt{-4\cdot 5}}{2}$

$x=\dfrac{4\pm 2i\sqrt{5}}{2}=\dfrac{4}{2}\pm\dfrac{2\sqrt{5}}{2}i=2\pm i\sqrt{5}$

The solutions are $2+\sqrt{5}i$ and $2-\sqrt{5}i$.

**33.**         $x^3-8=0$

$x^3-2^3=0$

$(x-2)(x^2+2x+4)=0$

$x-2=0$   or   $x^2+2x+4=0$

$x=2$   or   $x=\dfrac{-2\pm\sqrt{2^2-4\cdot 1\cdot 4}}{2\cdot 1}$

$x=2$   or   $x=\dfrac{-2\pm\sqrt{-12}}{2}=\dfrac{-2\pm 2i\sqrt{3}}{2}$

$x=2$   or   $x=-\dfrac{2}{2}\pm\dfrac{2\sqrt{3}}{2}i$

$x=2$   or   $x=-1\pm\sqrt{3}i$

The solutions are $2,\ -1+\sqrt{3}i$, and $-1-\sqrt{3}i$.

**35.**        $f(x)=0$

$3x^2-5x+2=0$

$(3x-2)(x-1)=0$

$3x-2=0$   or   $x-1=0$

$3x=2$   or     $x=1$

$x=\dfrac{2}{3}$   or     $x=1$

$f(x)=0$ for $x=\dfrac{2}{3}$ and $x=1$.

**37.**          $f(x)=1$

$\dfrac{7}{x}+\dfrac{7}{x+4}=1$     Substituting

$x(x+4)\left(\dfrac{7}{x}+\dfrac{7}{x+4}\right)=x(x+4)\cdot 1$

               Multiplying by the LCD

$7(x+4)+7x=x^2+4x$

$7x+28+7x=x^2+4x$

$14x+28=x^2+4x$

$0=x^2-10x-28$

$a=1,\ b=-10,\ c=-28$

$x=\dfrac{-(-10)\pm\sqrt{(-10)^2-4\cdot 1\cdot(-28)}}{2\cdot 1}$

$x=\dfrac{10\pm\sqrt{100+112}}{2}=\dfrac{10\pm\sqrt{212}}{2}$

$x=\dfrac{10\pm\sqrt{4\cdot 53}}{2}=\dfrac{10\pm 2\sqrt{53}}{2}$

$x=5\pm\sqrt{53}$

$f(x)=1$ for $x=5+\sqrt{53}$ and $x=5-\sqrt{53}$.

**39.**        $F(x)=G(x)$

$\dfrac{x+3}{x}=\dfrac{x-4}{3}$     Substituting

$3x\left(\dfrac{x+3}{x}\right)=3x\left(\dfrac{x-4}{3}\right)$    Multiplying by the LCD

$3x+9=x^2-4x$

$0=x^2-7x-9$

$a=1,\ b=-7,\ c=-9$

$x=\dfrac{-(-7)\pm\sqrt{(-7)^2-4\cdot 1\cdot(-9)}}{2\cdot 1}$

$x=\dfrac{7\pm\sqrt{49+36}}{2}=\dfrac{7\pm\sqrt{85}}{2}$

$x=\dfrac{7}{2}\pm\dfrac{\sqrt{85}}{2}$

$F(x)=G(x)$ for $x=\dfrac{7}{2}-\dfrac{\sqrt{85}}{2}$ and $x=\dfrac{7}{2}+\dfrac{\sqrt{85}}{2}$.

**41.**        $f(x)=g(x)$

$\dfrac{15-2x}{6}=\dfrac{3}{x}$,   LCD is $6x$

$6x\cdot\dfrac{15-2x}{6}=6x\cdot\dfrac{3}{x}$

$x(15-2x)=6\cdot 3$

$15x-2x^2=18$

$0=2x^2-15x+18$

$0=(2x-3)(x-6)$

$2x-3=0$   or   $x-6=0$

$2x=3$   or     $x=6$

$x=\dfrac{3}{2}$   or     $x=6$

$f(x)=g(x)$ for $x=\dfrac{3}{2}$ and $x=6$.

**43.** $x^2 + 4x - 7 = 0$

$a = 1, b = 4, c = -7$

$x = \dfrac{-4 \pm \sqrt{4^2 - 4 \cdot 1 \cdot (-7)}}{2 \cdot 1} = \dfrac{-4 \pm \sqrt{16 + 28}}{2}$

$x = \dfrac{-4 \pm \sqrt{44}}{2}$

Using a calculator we find that $\dfrac{-4 + \sqrt{44}}{2} \approx$

1.31662479 and $\dfrac{-4 - \sqrt{44}}{2} \approx -5.31662479$.

The solutions are approximately 1.31662479 and $-5.31662479$.

**45.** $x^2 - 6x + 4 = 0$

$a = 1, b = -6, c = 4$

$x = \dfrac{-(-6) \pm \sqrt{(-6)^2 - 4 \cdot 1 \cdot 4}}{2 \cdot 1} = \dfrac{6 \pm \sqrt{36 - 16}}{2}$

$x = \dfrac{6 \pm \sqrt{20}}{2}$

Using a calculator we find that $\dfrac{6 + \sqrt{20}}{2} \approx$

5.236067978 and $\dfrac{6 - \sqrt{20}}{2} \approx 0.7639320225$.

The solutions are approximately 5.236067978 and 0.7639320225.

**47.** $2x^2 - 3x - 7 = 0$

$a = 2, b = -3, c = -7$

$x = \dfrac{-(-3) \pm \sqrt{(-3)^2 - 4 \cdot 2 \cdot (-7)}}{2 \cdot 2}$

$x = \dfrac{3 \pm \sqrt{9 + 56}}{4} = \dfrac{3 \pm \sqrt{65}}{4}$

Using a calculator we find that $\dfrac{3 + \sqrt{65}}{4} \approx$

2.765564437 and $\dfrac{3 - \sqrt{65}}{4} \approx -1.265564437$.

The solutions are approximately 2.765564437 and $-1.265564437$.

**49.** *Writing Exercise*

**51.** **Familiarize**. Let $x =$ the number of pounds of Kenyan coffee and $y =$ the number of pounds of Kona coffee in the mixture. We organize the information in a table.

| Type of Coffee | Kenyan | Kona | Mixture |
|---|---|---|---|
| Price per pound | $6.75 | $11.25 | $8.55 |
| Number of pounds | $x$ | $y$ | 50 |
| Total cost | $6.75x$ | $11.25y$ | $8.55 \times 50$, or $427.50 |

**Translate**. From the last two rows of the table we get a system of equations.

$x + y = 50,$

$6.75x + 11.25y = 427.50$

**Solve**. Solving the system of equations, we get $(30, 20)$.

**Check**. The total number of pounds in the mixture is $30 + 20$, or 50. The total cost of the mixture is $6.75(30) + 11.25(20) = 427.50$. The values check.

**State**. The mixture should consist of 30 lb of Kenyan coffee and 20 lb of Kona coffee.

**53.** $\sqrt{27a^2 b^5} \cdot \sqrt{6a^3 b} = \sqrt{27a^2 b^5 \cdot 6a^3 b} =$

$\sqrt{162a^5 b^6} = \sqrt{81a^4 b^6 \cdot 2a} = \sqrt{81a^4 b^6}\sqrt{2a} =$

$9a^2 b^3 \sqrt{2a}$

**55.**

$\dfrac{\dfrac{3}{x-1}}{\dfrac{1}{x+1} + \dfrac{2}{x-1}}$

$= \dfrac{\dfrac{3}{x-1}}{\dfrac{1}{x+1} + \dfrac{2}{x-1}} \cdot \dfrac{(x-1)(x+1)}{(x-1)(x+1)}$

$= \dfrac{3(x+1)}{x-1 + 2(x+1)}$

$= \dfrac{3x + 3}{x - 1 + 2x + 2}$

$= \dfrac{3x + 3}{3x + 1}, \text{ or } \dfrac{3(x+1)}{3x + 1}$

**57.** *Writing Exercise*

**59.** $f(x) = \dfrac{x^2}{x - 2} + 1$

To find the $x$-coordinates of the $x$-intercepts of the graph of $f$, we solve $f(x) = 0$.

$\dfrac{x^2}{x - 2} + 1 = 0$

$x^2 + x - 2 = 0 \quad \text{Multiplying by } x - 2$

$(x + 2)(x - 1) = 0$

$x = -2 \ \text{ or } \ x = 1$

The $x$-intercepts are $(-2, 0)$ and $(1, 0)$.

**61.**
$$f(x) = g(x)$$

$$\dfrac{x^2}{x - 2} + 1 = \dfrac{4x - 2}{x - 2} + \dfrac{x + 4}{2}$$

Substituting

$$2(x-2)\left(\dfrac{x^2}{x-2} + 1\right) = 2(x-2)\left(\dfrac{4x-2}{x-2} + \dfrac{x+4}{2}\right)$$

Multiplying by the LCD

$$2x^2 + 2(x - 2) = 2(4x - 2) + (x - 2)(x + 4)$$

$$2x^2 + 2x - 4 = 8x - 4 + x^2 + 2x - 8$$

$$2x^2 + 2x - 4 = x^2 + 10x - 12$$

$$x^2 - 8x + 8 = 0$$

$a = 1, b = -8, c = 8$

$$x = \frac{-(-8) \pm \sqrt{(-8)^2 - 4 \cdot 1 \cdot 8}}{2 \cdot 1} = \frac{8 \pm \sqrt{64 - 32}}{2}$$

$$x = \frac{8 \pm \sqrt{32}}{2} = \frac{8 \pm \sqrt{16 \cdot 2}}{2} = \frac{8 \pm 4\sqrt{2}}{2}$$

$$x = \frac{8}{2} \pm \frac{4\sqrt{2}}{2} = 4 \pm 2\sqrt{2}$$

The solutions are $4 + 2\sqrt{2}$ and $4 - 2\sqrt{2}$.

**63.** $z^2 + 0.84z - 0.4 = 0$

$a = 1$, $b = 0.84$, $c = -0.4$

$$z = \frac{-0.84 \pm \sqrt{(0.84)^2 - 4 \cdot 1 \cdot (-0.4)}}{2 \cdot 1}$$

$$z = \frac{-0.84 \pm \sqrt{2.3056}}{2}$$

$$z = \frac{-0.84 + \sqrt{2.3056}}{2} \approx 0.3392101158$$

$$z = \frac{-0.84 - \sqrt{2.3056}}{2} \approx -1.179210116$$

The solutions are approximately $0.3392101158$ and $-1.179210116$.

**65.** $\sqrt{2}x^2 + 5x + \sqrt{2} = 0$

$$x = \frac{-5 \pm \sqrt{5^2 - 4 \cdot \sqrt{2} \cdot \sqrt{2}}}{2\sqrt{2}} = \frac{-5 \pm \sqrt{17}}{2\sqrt{2}}, \text{ or}$$

$$x = \frac{-5 \pm \sqrt{17}}{2\sqrt{2}} \cdot \frac{\sqrt{2}}{\sqrt{2}} = \frac{-5\sqrt{2} \pm \sqrt{34}}{4}$$

The solutions are $\dfrac{-5\sqrt{2} \pm \sqrt{34}}{4}$.

**67.**
$$kx^2 + 3x - k = 0$$
$$k(-2)^2 + 3(-2) - k = 0 \quad \text{Substituting } -2 \text{ for } x$$
$$4k - 6 - k = 0$$
$$3k = 6$$
$$k = 2$$
$$2x^2 + 3x - 2 = 0 \quad \text{Substituting 2 for } k$$
$$(2x - 1)(x + 2) = 0$$
$$2x - 1 = 0 \quad or \quad x + 2 = 0$$
$$x = \frac{1}{2} \quad or \quad x = -2$$

The other solution is $\dfrac{1}{2}$.

**69.**

## Exercise Set 8.3

**1. Familiarize.** We first make a drawing, labeling it with the known and unknown information. We can also organize the information in a table. We let $r$ represent the speed and $t$ the time for the first part of the trip.

$$\underbrace{r \text{ mph} \quad t \text{ hr}}_{120 \text{ mi}} \cdot \underbrace{r - 10 \text{ mph} \quad 4 - t \text{ hr}}_{100 \text{ mi}}$$

| Trip | Distance | Speed | Time |
|------|----------|-------|------|
| 1st part | 120 | $r$ | $t$ |
| 2nd part | 100 | $r - 10$ | $4 - t$ |

**Translate.** Using $r = \dfrac{d}{t}$, we get two equations from the table, $r = \dfrac{120}{t}$ and $r - 10 = \dfrac{100}{4 - t}$.

**Carry out.** We substitute $\dfrac{120}{t}$ for $r$ in the second equation and solve for $t$.

$$\frac{120}{t} - 10 = \frac{100}{4 - t}, \text{ LCD is } t(4 - t)$$

$$t(4 - t)\left(\frac{120}{t} - 10\right) = t(4 - t) \cdot \frac{100}{4 - t}$$

$$120(4 - t) - 10t(4 - t) = 100t$$

$$480 - 120t - 40t + 10t^2 = 100t$$

$$10t^2 - 260t + 480 = 0 \quad \text{Standard form}$$

$$t^2 - 26t + 48 = 0 \quad \text{Multiplying by } \frac{1}{10}$$

$$(t - 2)(t - 24) = 0$$

$$t = 2 \quad or \quad t = 24$$

**Check.** Since the time cannot be negative (If $t = 24$, $4 - t = -20$.), we check only 2 hr. If $t = 2$, then $4 - t = 2$. The speed of the first part is $\dfrac{120}{2}$, or 60 mph. The speed of the second part is $\dfrac{100}{2}$, or 50 mph. The speed of the second part is 10 mph slower than the first part. The value checks.

**State.** The speed of the first part was 60 mph, and the speed of the second part was 50 mph.

**3. Familiarize.** We first make a drawing. We also organize the information in a table. We let $r =$ the speed and $t =$ the time of the slower trip.

$$\underbrace{200 \text{ mi} \qquad r \text{ mph} \qquad t \text{ hr}}$$
$$\underbrace{200 \text{ mi} \qquad r + 10 \text{ mph} \qquad t - 1 \text{ hr}}$$

| Trip | Distance | Speed | Time |
|------|----------|-------|------|
| Slower | 200 | $r$ | $t$ |
| Faster | 200 | $r + 10$ | $t - 1$ |

**Translate.** Using $t = d/r$, we get two equations from the table:

$$t = \frac{200}{r} \text{ and } t - 1 = \frac{200}{r + 10}$$

**Carry out.** We substitute $\dfrac{200}{r}$ for $t$ in the second equation and solve for $r$.

$$\frac{200}{r} - 1 = \frac{200}{r + 10}, \text{ LCD is } r(r + 10)$$

$$r(r + 10)\left(\frac{200}{r} - 1\right) = r(r + 10) \cdot \frac{200}{r + 10}$$

$$200(r + 10) - r(r + 10) = 200r$$

$$200r + 2000 - r^2 - 10r = 200r$$

$$0 = r^2 + 10r - 2000$$

$$0 = (r + 50)(r - 40)$$

$$r = -50 \ \ or \ \ r = 40$$

**Check.** Since negative speed has no meaning in this problem, we check only 40. If $r = 40$, then the time for the slower trip is $\frac{200}{40}$, or 5 hours. If $r = 40$, then $r + 10 = 50$ and the time for the faster trip is $\frac{200}{50}$, or 4 hours. This is 1 hour less time than the slower trip took, so we have an answer to the problem.

**State.** The speed is 40 mph.

5. **Familiarize.** We make a drawing and then organize the information in a table. We let $r =$ the speed and $t =$ the time of the Cessna.

600 mi          $r$ mph          $t$ hr

1000 mi          $r + 50$ mph          $t + 1$ hr

| Plane | Distance | Speed | Time |
|---|---|---|---|
| Cessna | 600 | $r$ | $t$ |
| Beechcraft | 1000 | $r + 50$ | $t + 1$ |

**Translate.** Using $t = d/r$, we get two equations from the table:

$$t = \frac{600}{r} \ \ and \ \ t + 1 = \frac{1000}{r + 50}$$

**Carry out.** We substitute $\frac{600}{r}$ for $t$ in the second equation and solve for $r$.

$$\frac{600}{r} + 1 = \frac{1000}{r + 50},$$

$$\text{LCD is } r(r + 50)$$

$$r(r + 50)\left(\frac{600}{r} + 1\right) = r(r + 50) \cdot \frac{1000}{r + 50}$$

$$600(r + 50) + r(r + 50) = 1000r$$

$$600r + 30,000 + r^2 + 50r = 1000r$$

$$r^2 - 350r + 30,000 = 0$$

$$(r - 150)(r - 200) = 0$$

$$r = 150 \ \ or \ \ r = 200$$

**Check.** If $r = 150$, then the Cessna's time is $\frac{600}{150}$, or 4 hr and the Beechcraft's time is $\frac{1000}{150 + 50}$, or $\frac{1000}{200}$, or 5 hr. If $r = 200$, then the Cessna's time is $\frac{600}{200}$, or 3 hr and the Beechcraft's time is $\frac{1000}{200 + 50}$, or $\frac{1000}{250}$, or 4 hr. Since the Beechcraft's time is 1 hr longer in each case, both values check. There are two solutions.

**State.** The speed of the Cessna is 150 mph and the speed of the Beechcraft is 200 mph; or the speed of the Cessna is 200 mph and the speed of the Beechcraft is 250 mph.

7. **Familiarize.** We make a drawing and then organize the information in a table. We let $r$ represent the speed and $t$ the time of the trip to Hillsboro.

                                              Hillsboro
40 mi          $r$ mph          $t$ hr

40 mi          $r - 6$ mph          $14 - t$ hr

| Trip | Distance | Speed | Time |
|---|---|---|---|
| To Hillsboro | 40 | $r$ | $t$ |
| Return | 40 | $r - 6$ | $14 - t$ |

**Translate.** Using $t = \frac{d}{r}$, we get two equations from the table,

$$t = \frac{40}{r} \ \ and \ \ 14 - t = \frac{40}{r - 6}.$$

**Carry out.** We substitute $\frac{40}{r}$ for $t$ in the second equation and solve for $r$.

$$14 - \frac{40}{r} = \frac{40}{r - 6},$$

$$\text{LCD is } r(r - 6)$$

$$r(r - 6)\left(14 - \frac{40}{r}\right) = r(r - 6) \cdot \frac{40}{r - 6}$$

$$14r(r - 6) - 40(r - 6) = 40r$$

$$14r^2 - 84r - 40r + 240 = 40r$$

$$14r^2 - 164r + 240 = 0$$

$$7r^2 - 82r + 120 = 0$$

$$(7r - 12)(r - 10) = 0$$

$$r = \frac{12}{7} \ \ or \ \ r = 10$$

**Check.** Since negative speed has no meaning in this problem (If $r = \frac{12}{7}$, then $r - 6 = -\frac{30}{7}$.), we check only 10 mph. If $r = 10$, then the time of the trip to Hillsboro is $\frac{40}{10}$, or 4 hr. The speed of the return trip is $10 - 6$, or 4 mph, and the time is $\frac{40}{4}$, or 10 hr. The total time for the round trip is 4 hr + 10 hr, or 14 hr. The value checks.

**State.** Naoki's speed on the trip to Hillsboro was 10 mph and it was 4 mph on the return trip.

9. **Familiarize.** We make a drawing and organize the information in a table. Let $r$ represent the speed of the boat in still water, and let $t$ represent the time of the trip upriver.

60 mi          $r - 3$ mph          $t$ hr
                                        Upriver

Downriver          60 mi          $r + 3$ mph          $9 - t$ hr

| Trip | Distance | Speed | Time |
|------|----------|-------|------|
| Upriver | 60 | $r - 3$ | $t$ |
| Downriver | 60 | $r + 3$ | $9 - t$ |

**Translate**. Using $t = \dfrac{d}{r}$, we get two equations from the table,

$$t = \frac{60}{r - 3} \text{ and } 9 - t = \frac{60}{r + 3}.$$

**Carry out**. We substitute $\dfrac{60}{r - 3}$ for $t$ in the second equation and solve for $r$.

$$9 - \frac{60}{r - 3} = \frac{60}{r + 3}$$

$$(r - 3)(r + 3)\left(9 - \frac{60}{r - 3}\right) = (r - 3)(r + 3) \cdot \frac{60}{r + 3}$$

$$9(r - 3)(r + 3) - 60(r + 3) = 60(r - 3)$$

$$9r^2 - 81 - 60r - 180 = 60r - 180$$

$$9r^2 - 120r - 81 = 0$$

$$3r^2 - 40r - 27 = 0 \quad \text{Dividing by 3}$$

We use the quadratic formula.

$$r = \frac{-(-40) \pm \sqrt{(-40)^2 - 4 \cdot 3 \cdot (-27)}}{2 \cdot 3}$$

$$r = \frac{40 \pm \sqrt{1924}}{6}$$

$$r \approx 14 \quad or \quad r \approx -0.6$$

**Check**. Since negative speed has no meaning in this problem, we check only 14 mph. If $r \approx 14$, then the speed upriver is about $14 - 3$, or 11 mph, and the time is about $\dfrac{60}{11}$, or 5.5 hr. The speed downriver is about $14 + 3$, or 17 mph, and the time is about $\dfrac{60}{17}$, or 3.5 hr. The total time of the round trip is $5.5 + 3.5$, or 9 hr. The value checks.

**State**. The speed of the boat in still water is about 14 mph.

**11.** **Familiarize**. Let $x$ represent the time it takes the spring to fill the pool. Then $x - 6$ represents the time it takes the well to fill the pool. It takes them 4 hr to fill the pool working together, so they can fill $\dfrac{1}{4}$ of the pool in 1 hr. The spring will fill $\dfrac{1}{x}$ of the pool in 1 hr, and the well will fill $\dfrac{1}{x - 6}$ of the pool in 1 hr.

**Translate**. We have an equation.

$$\frac{1}{x} + \frac{1}{x - 6} = \frac{1}{4}$$

**Carry out**. We solve the equation.
We multiply by the LCD, $4x(x - 6)$.

$$4x(x - 6)\left(\frac{1}{x} + \frac{1}{x - 6}\right) = 4x(x - 6) \cdot \frac{1}{4}$$

$$4(x - 6) + 4x = x(x - 6)$$

$$4x - 24 + 4x = x^2 - 6x$$

$$0 = x^2 - 14x + 24$$

$$0 = (x - 2)(x - 12)$$

$$x = 2 \quad or \quad x = 12$$

**Check**. Since negative time has no meaning in this problem, 2 is not a solution $(2 - 6 = -4)$. We check only 12 hr. This is the time it would take the spring working alone. Then the well would take $12 - 6$, or 6 hr working alone. The well would fill $4\left(\dfrac{1}{6}\right)$, or $\dfrac{2}{3}$, of the pool in 4 hr, and the spring would fill $4\left(\dfrac{1}{12}\right)$, or $\dfrac{1}{3}$, of the pool in 4 hr. Thus in 4 hr they would fill $\dfrac{2}{3} + \dfrac{1}{3}$ of the pool. This is all of it, so the numbers check.

**State**. It takes the spring, working alone, 12 hr to fill the pool.

**13.** We make a drawing and then organize the information in a table. We let $r$ represent Ellen's speed in still water. Then $r - 2$ is the speed upstream and $r + 2$ is the speed downstream. Using $t = \dfrac{d}{r}$, we let $\dfrac{1}{r - 2}$ represent the time upstream and $\dfrac{1}{r + 2}$ represent the time downstream.

$$\xrightarrow[\text{1 mi} \qquad\qquad r - 2 \text{ mph}]{} \text{Upstream}$$

$$\text{Downstream} \xleftarrow[\text{1 mi} \qquad\qquad r + 2 \text{ mph}]{}$$

| Trip | Distance | Speed | Time |
|------|----------|-------|------|
| Upstream | 1 | $r - 2$ | $\dfrac{1}{r - 2}$ |
| Downstream | 1 | $r + 2$ | $\dfrac{1}{r + 2}$ |

**Translate**. The time for the round trip is 1 hour. We now have an equation.

$$\frac{1}{r - 2} + \frac{1}{r + 2} = 1$$

**Carry out**. We solve the equation. We multiply by the LCD, $(r - 2)(r + 2)$.

$$(r - 2)(r + 2)\left(\frac{1}{r - 2} + \frac{1}{r + 2}\right) = (r - 2)(r + 2) \cdot 1$$

$$(r + 2) + (r - 2) = (r - 2)(r + 2)$$

$$2r = r^2 - 4$$

$$0 = r^2 - 2r - 4$$

$$a = 1,\ b = -2,\ c = -4$$

$$r = \frac{-(-2) \pm \sqrt{(-2)^2 - 4 \cdot 1(-4)}}{2 \cdot 1}$$

$$r = \frac{2 \pm \sqrt{4 + 16}}{2} = \frac{2 \pm \sqrt{20}}{2}$$

$$r = \frac{2 \pm 2\sqrt{5}}{2} = 1 \pm \sqrt{5}$$

$$1 + \sqrt{5} \approx 1 + 2.236 \approx 3.24$$

$$1 - \sqrt{5} \approx 1 - 2.236 \approx -1.24$$

**Check**. Since negative speed has no meaning in this problem, we check only 3.24 mph. If $r \approx 3.24$, then $r - 2 \approx 1.24$ and $r + 2 \approx 5.24$. The time it takes to travel upstream is approximately $\dfrac{1}{1.24}$, or 0.806 hr, and the time it takes to travel downstream is approximately $\dfrac{1}{5.24}$, or 0.191 hr. The total time is 0.997 which is approximately 1 hour. The value checks.

**State.** Ellen's speed in still water is approximately 3.24 mph.

**15.** $\quad A = 4\pi r^2$

$\dfrac{A}{4\pi} = r^2 \qquad$ Dividing by $4\pi$

$\dfrac{1}{2}\sqrt{\dfrac{A}{\pi}} = r \qquad$ Taking the positive square root

**17.** $\quad A = 2\pi r^2 + 2\pi rh$

$0 = 2\pi r^2 + 2\pi rh - A \qquad$ Standard form

$a = 2\pi,\ b = 2\pi h,\ c = -A$

$r = \dfrac{-2\pi h \pm \sqrt{(2\pi h)^2 - 4 \cdot 2\pi \cdot (-A)}}{2 \cdot 2\pi} \qquad$ Using the quadratic formula

$r = \dfrac{-2\pi h \pm \sqrt{4\pi^2 h^2 + 8\pi A}}{4\pi}$

$r = \dfrac{-2\pi h \pm 2\sqrt{\pi^2 h^2 + 2\pi A}}{4\pi}$

$r = \dfrac{-\pi h \pm \sqrt{\pi^2 h^2 + 2\pi A}}{2\pi}$

Since taking the negative square root would result in a negative answer, we take the positive one.

$r = \dfrac{-\pi h + \sqrt{\pi^2 h^2 + 2\pi A}}{2\pi}$

**19.** $\quad N = \dfrac{kQ_1 Q_2}{s^2}$

$Ns^2 = kQ_1 Q_2 \qquad$ Multiplying by $s^2$

$s^2 = \dfrac{kQ_1 Q_2}{N} \qquad$ Dividing by $N$

$s = \sqrt{\dfrac{kQ_1 Q_2}{N}} \qquad$ Taking the positive square root

**21.** $\quad T = 2\pi\sqrt{\dfrac{l}{g}}$

$\dfrac{T}{2\pi} = \sqrt{\dfrac{l}{g}} \qquad$ Multiplying by $\dfrac{1}{2\pi}$

$\dfrac{T^2}{4\pi^2} = \dfrac{l}{g} \qquad$ Squaring

$gT^2 = 4\pi^2 l \qquad$ Multiplying by $4\pi^2 g$

$g = \dfrac{4\pi^2 l}{T^2} \qquad$ Multiplying by $\dfrac{1}{T^2}$

**23.** $\quad a^2 + b^2 + c^2 = d^2$

$c^2 = d^2 - a^2 - b^2 \qquad$ Subtracting $a^2$ and $b^2$

$c = \sqrt{d^2 - a^2 - b^2} \qquad$ Taking the positive square root

**25.** $\quad s = v_0 t + \dfrac{gt^2}{2}$

$0 = \dfrac{gt^2}{2} + v_0 t - s \qquad$ Standard form

$a = \dfrac{g}{2},\ b = v_0,\ c = -s$

$t = \dfrac{-v_0 \pm \sqrt{v_0^2 - 4\left(\dfrac{g}{2}\right)(-s)}}{2\left(\dfrac{g}{2}\right)}$

$t = \dfrac{-v_0 \pm \sqrt{v_0^2 + 2gs}}{g}$

Since taking the negative square root would result in a negative answer, we take the positive one.

$t = \dfrac{-v_0 + \sqrt{v_0^2 + 2gs}}{g}$

**27.** $\quad N = \dfrac{1}{2}(n^2 - n)$

$N = \dfrac{1}{2}n^2 - \dfrac{1}{2}n$

$0 = \dfrac{1}{2}n^2 - \dfrac{1}{2}n - N$

$a = \dfrac{1}{2},\ b = -\dfrac{1}{2},\ c = -N$

$n = \dfrac{-\left(-\dfrac{1}{2}\right) \pm \sqrt{\left(-\dfrac{1}{2}\right)^2 - 4 \cdot \dfrac{1}{2} \cdot (-N)}}{2\left(\dfrac{1}{2}\right)}$

$n = \dfrac{1}{2} \pm \sqrt{\dfrac{1}{4} + 2N}$

$n = \dfrac{1}{2} \pm \sqrt{\dfrac{1 + 8N}{4}}$

$n = \dfrac{1}{2} \pm \dfrac{1}{2}\sqrt{1 + 8N}$

Since taking the negative square root would result in a negative answer, we take the positive one.

$n = \dfrac{1}{2} + \dfrac{1}{2}\sqrt{1 + 8N}, \text{ or } \dfrac{1 + \sqrt{1 + 8N}}{2}$

**29.** $\quad V = 3.5\sqrt{h}$

$V = 12.25h \qquad$ Squaring

$\dfrac{V^2}{12.25} = h$

**31.** $\quad at^2 + bt + c = 0$

The quadratic formula gives the result.

$t = \dfrac{-b \pm \sqrt{b^2 - 4ac}}{2a}$

**33.** a) **Familiarize and Translate.** From Example 4, we know

$t = \dfrac{-v_0 + \sqrt{v_0{}^2 + 19.6s}}{9.8}.$

**Carry out.** Substituting 500 for $s$ and 0 for $v_0$, we have

$t = \dfrac{0 + \sqrt{0^2 + 19.6(500)}}{9.8}$

$t \approx 10.1$

**Check**. Substitute 10.1 for $t$ and 0 for $v_0$ in the original formula. (See Example 4.)

$$s = 4.9t^2 + v_0 t = 4.9(10.1)^2 + 0 \cdot (10.1)^2$$
$$\approx 500$$

The answer checks.

**State**. It takes the bolt about 10.1 sec to reach the ground.

b) **Familiarize and Translate**. From Example 4, we know

$$t = \frac{-v_0 + \sqrt{v_0^2 + 19.6s}}{9.8}.$$

**Carry out**. Substitute 500 for $s$ and 30 for $v_0$.

$$t = \frac{-30 + \sqrt{30^2 + 19.6(500)}}{9.8}$$
$$t \approx 7.49$$

**Check**. Substitute 30 for $v_0$ and 7.49 for $t$ in the original formula. (See Example 4.)

$$s = 4.9t^2 + v_0 t = 4.9(7.49)^2 + (30)(7.49)$$
$$\approx 500$$

The answer checks.

**State**. It takes the ball about 7.49 sec to reach the ground.

c) **Familiarize and Translate**. We will use the formula in Example 4, $s = 4.9t^2 + v_0 t$.

**Carry out**. Substitute 5 for $t$ and 30 for $v_0$.

$$s = 4.9(5)^2 + 30(5) = 272.5$$

**Check**. We can substitute 30 for $v_0$ and 272.5 for $s$ in the form of the formula we used in part (b).

$$t = \frac{-v_0 + \sqrt{v_0^2 + 19.6s}}{9.8}$$
$$= \frac{-30 + \sqrt{(30)^2 + 19.6(272.5)}}{9.8} = 5$$

The answer checks.

**State**. The object will fall 272.5 m.

35. **Familiarize**. We will use the formula $4.9t^2 = s$.

**Translate**. Substitute 40 for $s$.

$$4.9t^2 = 40$$

**Carry out**. We solve the equation.

$$4.9t^2 = 40$$
$$t^2 = \frac{40}{4.9}$$
$$t = \sqrt{\frac{40}{4.9}}$$
$$t \approx 2.9$$

**Check**. Substitute 2.9 for $t$ in the formula.

$$s = 4.9(2.9)^2 = 41.209 \approx 40$$

The answer checks.

**State**. Jesse will fall for about 2.9 sec before the cord begins to stretch.

37. **Familiarize and Translate**. From Example 3, we know

$$T = \frac{\sqrt{3V}}{12}.$$

**Carry out**. Substituting 45 for $V$, we have

$$T = \frac{\sqrt{3 \cdot 45}}{12}$$
$$T \approx 0.968$$

**Check**. Substitute 0.968 for $T$ in the original formula. (See Example 3.)

$$48T^2 = V$$
$$48(0.968)^2 = V$$
$$45 \approx V$$

The answer checks.

**State**. Steve Francis' hang time is about 0.968 sec.

39. **Familiarize and Translate**. We will use the formula in Example 4, $s = 4.9t^2 + v_0 t$.

**Carry out**. Solve the formula for $v_0$.

$$s - 4.9t^2 = v_0 t$$
$$\frac{s - 4.9t^2}{t} = v_0$$

Now substitute 51.6 for $s$ and 3 for $t$.

$$\frac{51.6 - 4.9(3)^2}{3} = v_0$$
$$2.5 = v_0$$

**Check**. Substitute 3 for $t$ and 2.5 for $v_0$ in the original formula.

$$s = 4.9(3)^2 + 2.5(3) = 51.6$$

The solution checks.

**State**. The initial velocity is 2.5 m/sec.

41. **Familiarize and Translate**. From Exercise 32 we know that

$$r = -1 + \frac{-P_2 + \sqrt{P_2^2 + 4P_1 A}}{2P_1},$$

where $A$ is the total amount in the account after two years, $P_1$ is the amount of the original deposit, $P_2$ is deposited at the beginning of the second year, and $r$ is the annual interest rate.

**Carry out**. Substitute 3000 for $P_1$, 1700 for $P_2$, and 5253.70 for $A$.

$$r = -1 + \frac{-1700 + \sqrt{(1700)^2 + 4(3000)(5253.70)}}{2(3000)}$$

Using a calculator, we have $r = 0.07$.

**Check**. Substitute in the original formula in Exercise 32.

$$P_1(1 + r)^2 + P_2(1 + r) = A$$
$$3000(1.07)^2 + 1700(1.07) = A$$
$$5253.70 = A$$

The answer checks.

**State**. The annual interest rate is 0.07, or 7%.

43. *Writing Exercise*

**45.** $b^2 - 4ac = 6^2 - 4 \cdot 5 \cdot 7$

$\qquad\qquad\quad = 36 - 4 \cdot 5 \cdot 7$

$\qquad\qquad\quad = 36 - 140$

$\qquad\qquad\quad = -104$

**47.** $\dfrac{x^2 + xy}{2x} = \dfrac{x(x+y)}{2x}$

$\qquad\qquad = \dfrac{x(x+y)}{2 \cdot x}$

$\qquad\qquad = \dfrac{\cancel{x}(x+y)}{2 \cdot \cancel{x}}$

$\qquad\qquad = \dfrac{x+y}{2}$

**49.** $\dfrac{3 + \sqrt{45}}{6} = \dfrac{3 + \sqrt{9 \cdot 5}}{6} = \dfrac{3 + 3\sqrt{5}}{6} = \dfrac{\cancel{3}(1 + \sqrt{5})}{\cancel{3} \cdot 2} =$

$\dfrac{1 + \sqrt{5}}{2}$

**51.** *Writing Exercise*

**53.** $\qquad\qquad A = 6.5 - \dfrac{20.4t}{t^2 + 36}$

$(t^2 + 36)A = (t^2 + 36)\left(6.5 - \dfrac{20.4t}{t^2 + 36}\right)$

$At^2 + 36A = (t^2 + 36)(6.5) - (t^2 + 36)\left(\dfrac{20.4t}{t^2 + 36}\right)$

$At^2 + 36A = 6.5t^2 + 234 - 20.4t$

$At^2 - 6.5t^2 + 20.4t + 36A - 234 = 0$

$(A - 6.5)t^2 + 20.4t + (36A - 234) = 0$

$a = A - 6.5,\ b = 20.4,\ c = 36A - 234$

$t = \dfrac{-20.4 \pm \sqrt{(20.4)^2 - 4(A - 6.5)(36A - 234)}}{2(A - 6.5)}$

$t = \dfrac{-20.4 \pm \sqrt{416.16 - 144A^2 + 1872A - 6084}}{2(A - 6.5)}$

$t = \dfrac{-20.4 \pm \sqrt{-144A^2 + 1872A - 5667.84}}{2(A - 6.5)}$

$t = \dfrac{-20.4 \pm \sqrt{144(-A^2 + 13A - 39.36)}}{2(A - 6.5)}$

$t = \dfrac{-20.4 \pm 12\sqrt{-A^2 + 13A - 39.36}}{2(A - 6.5)}$

$t = \dfrac{2(-10.2 \pm 6\sqrt{-A^2 + 13A - 39.36})}{2(A - 6.5)}$

$t = \dfrac{-10.2 \pm 6\sqrt{-A^2 + 13A - 39.36}}{A - 6.5}$

**55. Familiarize**. Let $a =$ the number. Then $a - 1$ is 1 less than $a$ and the reciprocal of that number is $\dfrac{1}{a - 1}$. Also, 1 more than the number is $a + 1$.

**Translate**.

$$\underbrace{\text{The reciprocal of 1 less than a number}}_{\dfrac{1}{(a-1)}} \underset{=}{\underbrace{\text{is}}} \underbrace{\text{1 more than the number.}}_{a + 1}$$

**Carry out**. We solve the equation.

$\qquad \dfrac{1}{a - 1} = a + 1,\ \text{LCD is } a - 1$

$(a - 1) \cdot \dfrac{1}{a - 1} = (a - 1)(a + 1)$

$\qquad\qquad 1 = a^2 - 1$

$\qquad\qquad 2 = a^2$

$\qquad \pm\sqrt{2} = a$

**Check**. $\dfrac{1}{\sqrt{2} - 1} \approx 2.4142 \approx \sqrt{2} + 1$ and $\dfrac{1}{-\sqrt{2} - 1} \approx$ $-0.4142 \approx -\sqrt{2} + 1$. The answers check.

**State**. The numbers are $\sqrt{2}$ and $-\sqrt{2}$, or $\pm\sqrt{2}$.

**57.** $\qquad\qquad \dfrac{w}{l} = \dfrac{l}{w + l}$

$l(w + l) \cdot \dfrac{w}{l} = l(w + l) \cdot \dfrac{l}{w + l}$

$\qquad w(w + l) = l^2$

$\qquad w^2 + lw = l^2$

$\qquad\qquad 0 = l^2 - lw - w^2$

Use the quadratic formula with $a = 1$, $b = -w$, and $c = -w^2$.

$l = \dfrac{-(-w) \pm \sqrt{(-w)^2 - 4 \cdot 1 \cdot (-w^2)}}{2 \cdot 1}$

$l = \dfrac{w \pm \sqrt{w^2 + 4w^2}}{2} = \dfrac{w \pm \sqrt{5w^2}}{2}$

$l = \dfrac{w \pm w\sqrt{5}}{2}$

Since $\dfrac{w - w\sqrt{5}}{2}$ is negative we use the positive square root:

$l = \dfrac{w + w\sqrt{5}}{2}$ Then $L(A) = \sqrt{\dfrac{A}{2}}$.

**59.** $mn^4 - r^2pm^3 - r^2n^2 + p = 0$

Let $u = n^2$. Substitute and rearrange.

$mu^2 - r^2u - r^2pm^3 + p = 0$

$a = m,\ b = -r^2,\ c = -r^2pm^3 + p$

$u = \dfrac{-(-r^2) \pm \sqrt{(-r^2)^2 - 4 \cdot m(-r^2pm^3 + p)}}{2 \cdot m}$

$u = \dfrac{r^2 \pm \sqrt{r^4 + 4m^4r^2p - 4mp}}{2m}$

$n^2 = \dfrac{r^2 \pm \sqrt{r^4 + 4m^4r^2p - 4mp}}{2m}$

$n = \pm\sqrt{\dfrac{r^2 \pm \sqrt{r^4 + 4m^4r^2p - 4mp}}{2m}}$

**61.** Let $s$ represent a length of a side of the cube, let $S$ represent the surface area of the cube, and let $A$ represent the surface area of the sphere. Then the diameter of the sphere is $s$, so the radius $r$ is $s/2$. From Exercise 15, we know, $A = 4\pi r^2$, so when $r = s/2$ we have $A = 4\pi\left(\dfrac{s}{2}\right)^2 =$ $4\pi \cdot \dfrac{s^2}{4} = \pi s^2$. From the formula for the surface area of a cube (See Exercise 16.) we know that $S = 6s^2$, so $\dfrac{S}{6} = s^2$ and then $A = \pi \cdot \dfrac{S}{6}$, or $A(S) = \dfrac{\pi S}{6}$.

## Exercise Set 8.4

**1.** discriminant; see page 532 in the text.

**3.** two; see page 533 in the text.

**5.** rational; see page 533 in the text.

**7.** $x^2 - 7x + 5 = 0$

$a = 1,\ b = -7,\ c = 5$

We substitute and compute the discriminant.

$$b^2 - 4ac = (-7)^2 - 4 \cdot 1 \cdot 5$$
$$= 49 - 20$$
$$= 29$$

Since the discriminant is a positive number that is not a perfect square, there are two irrational solutions.

**9.** $x^2 + 3 = 0$

$a = 1,\ b = 0,\ c = 3$

We substitute and compute the discriminant.

$$b^2 - 4ac = 0^2 - 4 \cdot 1 \cdot 3$$
$$= -12$$

Since the discriminant is negative, there are two imaginary-number solutions.

**11.** $x^2 - 5 = 0$

$a = 1,\ b = 0,\ c = -5$

We substitute and compute the discriminant.

$$b^2 - 4ac = 0^2 - 4 \cdot 1 \cdot (-5)$$
$$= 20$$

Since the discriminant is a positive number that is not a perfect square, there are two irrational solutions.

**13.** $4x^2 + 8x - 5 = 0$

$a = 4,\ b = 8,\ c = -5$

We substitute and compute the discriminant.

$$b^2 - 4ac = 8^2 - 4 \cdot 4 \cdot (-5)$$
$$= 64 + 80$$
$$= 144$$

Since the discriminant is a positive number and a perfect square, there are two rational solutions.

**15.** $x^2 + 4x + 6 = 0$

$a = 1,\ b = 4,\ c = 6$

We substitute and compute the discriminant.

$$b^2 - 4ac = 4^2 - 4 \cdot 1 \cdot 6$$
$$= 16 - 24$$
$$= -8$$

Since the discriminant is negative, there are two imaginary-number solutions.

**17.** $9t^2 - 48t + 64 = 0$

$a = 9,\ b = -48,\ c = 64$

We substitute and compute the discriminant.

$$b^2 - 4ac = (-48)^2 - 4 \cdot 9 \cdot 64$$
$$= 2304 - 2304$$
$$= 0$$

Since the discriminant is 0, there is just one solution and it is a rational number.

**19.** $10x^2 - x - 2 = 0$

$a = 10,\ b = -1,\ c = -2$

We substitute and compute the discriminant.

$$b^2 - 4ac = (-1)^2 - 4 \cdot 10 \cdot (-2)$$
$$= 1 + 80 = 81$$

Since the discriminant is a positive number and a perfect square, there are two rational solutions.

**21.** $9t^2 - 3t = 0$

Observe that we can factor $9t^2 - 3t$. This tells us that there are two rational solutions. We could also do this problem as follows.

$a = 9,\ b = -3,\ c = 0$

We substitute and compute the discriminant.

$$b^2 - 4ac = (-3)^2 - 4 \cdot 9 \cdot 0$$
$$= 9 - 0$$
$$= 9$$

Since the discriminant is a positive number and a perfect square, there are two rational solutions.

**23.** $x^2 + 4x = 8$

$x^2 + 4x - 8 = 0$    Standard form

$a = 1,\ b = 4,\ c = -8$

We substitute and compute the discriminant.

$$b^2 - 4ac = 4^2 - 4 \cdot 1 \cdot (-8)$$
$$= 16 + 32 = 48$$

Since the discriminant is a positive number that is not a perfect square, there are two irrational solutions.

**25.** $2a^2 - 3a = -5$

$2a^2 - 3a + 5 = 0$    Standard form

$a = 2,\ b = -3,\ c = 5$

We substitute and compute the discriminant.

$$b^2 - 4ac = (-3)^2 - 4 \cdot 2 \cdot 5$$
$$= 9 - 40$$
$$= -31$$

Since the discriminant is negative, there are two imaginary-number solutions.

**27.**
$$y^2 + \frac{9}{4} = 4y$$

$$y^2 - 4y + \frac{9}{4} = 0 \quad \text{Standard form}$$

$$a = 1, \ b = -4, \ c = \frac{9}{4}$$

We substitute and compute the discriminant.

$$b^2 - 4ac = (-4)^2 - 4 \cdot 1 \cdot \frac{9}{4}$$
$$= 16 - 9$$
$$= 7$$

The discriminant is a positive number that is not a perfect square. There are two irrational solutions.

**29.** The solutions are $-7$ and $3$.
$$x = -7 \quad or \quad x = 3$$
$$x + 7 = 0 \quad or \quad x - 3 = 0$$
$$(x + 7)(x - 3) = 0 \quad \text{Principle of zero products}$$
$$x^2 + 4x - 21 = 0 \quad \text{FOIL}$$

**31.** The only solution is 3. It must be a repeated solution.
$$x = 3 \quad or \quad x = 3$$
$$x - 3 = 0 \quad or \quad x - 3 = 0$$
$$(x - 3)(x - 3) = 0 \quad \text{Principle of zero products}$$
$$x^2 - 6x + 9 = 0 \quad \text{FOIL}$$

**33.** The solutions are $-1$ and $3$.
$$x = -1 \quad or \quad x = -3$$
$$x + 1 = 0 \quad or \quad x + 3 = 0$$
$$(x + 1)(x + 3) = 0$$
$$x^2 + 4x + 3 = 0$$

**35.** The solutions are $5$ and $\frac{3}{4}$.
$$x = 5 \quad or \quad x = \frac{3}{4}$$
$$x - 5 = 0 \quad or \quad x - \frac{3}{4} = 0$$
$$(x - 5)\left(x - \frac{3}{4}\right) = 0$$
$$x^2 - \frac{3}{4}x - 5x + \frac{15}{4} = 0$$
$$x^2 - \frac{23}{4}x + \frac{15}{4} = 0$$
$$4x^2 - 23x + 15 = 0 \quad \text{Multiplying by 4}$$

**37.** The solutions are $-\frac{1}{4}$ and $-\frac{1}{2}$.
$$x = -\frac{1}{4} \quad or \quad x = -\frac{1}{2}$$
$$x + \frac{1}{4} = 0 \quad or \quad x + \frac{1}{2} = 0$$
$$\left(x + \frac{1}{4}\right)\left(x + \frac{1}{2}\right) = 0$$
$$x^2 + \frac{1}{2}x + \frac{1}{4}x + \frac{1}{8} = 0$$
$$x^2 + \frac{3}{4}x + \frac{1}{8} = 0$$
$$8x^2 + 6x + 1 = 0 \quad \text{Multiplying by 8}$$

**39.** The solutions are 2, 4 and $-0.4$.
$$x = 2.4 \quad or \quad x = -0.4$$
$$x - 2.4 = 0 \quad or \quad x + 0.4 = 0$$
$$(x - 2.4)(x + 0.4) = 0$$
$$x^2 + 0.4x - 2.4x - 0.96 = 0$$
$$x^2 - 2x - 0.96 = 0$$

**41.** The solutions are $-\sqrt{3}$ and $\sqrt{3}$.
$$x = -\sqrt{3} \quad or \quad x = \sqrt{3}$$
$$x + \sqrt{3} = 0 \quad or \quad x - \sqrt{3} = 0$$
$$(x + \sqrt{3})(x - \sqrt{3}) = 0$$
$$x^2 - 3 = 0$$

**43.** The solutions are $2\sqrt{5}$ and $-2\sqrt{5}$.
$$x = 2\sqrt{5} \quad or \quad x = -2\sqrt{5}$$
$$x - 2\sqrt{5} = 0 \quad or \quad x + 2\sqrt{5} = 0$$
$$(x - 2\sqrt{5})(x + 2\sqrt{5}) = 0$$
$$x^2 - (2\sqrt{5})^2 = 0$$
$$x^2 - 4 \cdot 5 = 0$$
$$x^2 - 20 = 0$$

**45.** The solutions are $4i$ and $-4i$.
$$x = 4i \quad or \quad x = -4i$$
$$x - 4i = 0 \quad or \quad x + 4i = 0$$
$$(x - 4i)(x + 4i) = 0$$
$$x^2 - (4i)^2 = 0$$
$$x^2 + 16 = 0$$

**47.** The solutions are $2 - 7i$ and $2 + 7i$.
$$x = 2 - 7i \quad or \quad x = 2 + 7i$$
$$x - 2 + 7i = 0 \quad or \quad x - 2 - 7i = 0$$
$$(x - 2) + 7i = 0 \quad or \quad (x - 2) - 7i = 0$$
$$[(x - 2) + 7i][(x - 2) - 7i] = 0$$
$$(x - 2)^2 - (7i)^2 = 0$$
$$x^2 - 4x + 4 - 49i^2 = 0$$
$$x^2 - 4x + 4 + 49 = 0$$
$$x^2 - 4x + 53 = 0$$

**49.** The solutions are $3 - \sqrt{14}$ and $3 + \sqrt{14}$.
$$x = 3 - \sqrt{14} \quad or \quad x = 3 + \sqrt{14}$$
$$x - 3 + \sqrt{14} = 0 \quad or \quad x - 3 - \sqrt{14} = 0$$
$$(x - 3) + \sqrt{14} = 0 \quad or \quad (x - 3) - \sqrt{14} = 0$$
$$[(x - 3) + \sqrt{14}][(x - 3) - \sqrt{14}] = 0$$
$$(x - 3)^2 - (\sqrt{14})^2 = 0$$
$$x^2 - 6x + 9 - 14 = 0$$
$$x^2 - 6x - 5 = 0$$

**51.** The solutions are $1 - \dfrac{\sqrt{21}}{3}$ and $1 + \dfrac{\sqrt{21}}{3}$.

$$x = 1 - \frac{\sqrt{21}}{3} \quad or \qquad\qquad x = 1 + \frac{\sqrt{21}}{3}$$

$$x - 1 + \frac{\sqrt{21}}{3} = 0 \quad or \quad x - 1 - \frac{\sqrt{21}}{3} = 0$$

$$(x-1) + \frac{\sqrt{21}}{3} = 0 \quad or \quad (x-1) - \frac{\sqrt{21}}{3} = 0$$

$$\left[(x-1) + \frac{\sqrt{21}}{3}\right]\left[(x-1) - \frac{\sqrt{21}}{3}\right] = 0$$

$$(x-1)^2 - \left(\frac{\sqrt{21}}{3}\right)^2 = 0$$

$$x^2 - 2x + 1 - \frac{21}{9} = 0$$

$$x^2 - 2x + 1 - \frac{7}{3} = 0$$

$$x^2 - 2x - \frac{4}{3} = 0$$

$$3x^2 - 6x - 4 = 0 \quad \text{Multiplying by 3}$$

**53.** The solutions are $-2$, $1$, and $5$.

$$x = -2 \quad or \quad x = 1 \quad or \quad x = 5$$

$$x + 2 = 0 \quad or \quad x - 1 = 0 \quad or \quad x - 5 = 0$$

$$(x+2)(x-1)(x-5) = 0$$

$$(x^2 + x - 2)(x-5) = 0$$

$$x^3 + x^2 - 2x - 5x^2 - 5x + 10 = 0$$

$$x^3 - 4x^2 - 7x + 10 = 0$$

**55.** The solutions are $-1$, $0$, and $3$.

$$x = -1 \quad or \quad x = 0 \quad or \quad x = 3$$

$$x + 1 = 0 \quad or \quad x = 0 \quad or \quad x - 3 = 0$$

$$(x+1)(x)(x-3) = 0$$

$$(x^2 + x)(x-3) = 0$$

$$x^3 - 3x^2 + x^2 - 3x = 0$$

$$x^3 - 2x^2 - 3x = 0$$

**57.** *Writing Exercise*

**59.** $(3a^2)^4 = 3^4(a^2)^4 = 81a^{2 \cdot 4} = 81a^8$

**61.** $f(x) = x^2 - 7x - 8$

We find the values of $x$ for which $f(x) = 0$.

$$x^2 - 7x - 8 = 0$$

$$(x - 8)(x + 1) = 0$$

$$x - 8 = 0 \quad or \quad x + 1 = 0$$

$$x = 8 \quad or \quad x = -1$$

The $x$-intercepts are $(8, 0)$ and $(-1, 0)$.

**63.** *Familiarize*. Let $x$ and $y$ represent the number of 30-sec and 60-sec commercials, respectively. Then the amount of time for the 30-sec commercials was $30x$ sec, or $\dfrac{30x}{60} =$

$\dfrac{x}{2}$ min. The amount of time for the 60-sec commercials was $60x$ sec, or $\dfrac{60x}{60} = x$ min.

*Translate*. Rewording, we write two equations. We will express time in minutes.

$$\underbrace{\text{Total number of commercials}}_{\downarrow} \quad \text{is} \quad 12.$$
$$x + y \qquad\qquad = 12$$

$$\underbrace{\begin{array}{c}\text{Time for}\\\text{30-sec}\\\text{commercials}\end{array}}_{\downarrow} \; \text{is} \; \underbrace{\begin{array}{c}\text{total}\\\text{commercial}\\\text{time}\end{array}}_{\downarrow} \; \text{less 6 min.}$$
$$\frac{x}{2} \qquad = \qquad \frac{x}{2} + x \quad - \quad 6$$

*Carry out*. Solving the system of equations we get $(6, 6)$.

*Check*. If there are six 30-sec and six 60-sec commercials, the total number of commercials is 12. The amount of time for six 30-sec commercials is 180 sec, or 3 min, and for six 60-sec commercials is 360 sec, or 6 min. The total commercial time is 9 min, and the amount of time for 30-sec commercials is 6 min less than this. The numbers check.

*State*. There were six 30-sec commercials.

**65.** *Writing Exercise*

**67.** The graph includes the points $(-3, 0)$, $(0, -3)$, and $(1, 0)$. Substituting in $y = ax^2 + bx + c$, we have three equations.

$$0 = 9a - 3b + c,$$
$$-3 = \phantom{9a - 3b + {}} c,$$
$$0 = a + b + c$$

The solution of this system of equations is $a = 1$, $b = 2$, $c = -3$.

**69.** a) $kx^2 - 2x + k = 0$; one solution is $-3$

We first find $k$ by substituting $-3$ for $x$.

$$k(-3)^2 - 2(-3) + k = 0$$

$$9k + 6 + k = 0$$

$$10k = -6$$

$$k = -\frac{6}{10}$$

$$k = -\frac{3}{5}$$

b) Now substitute $-\dfrac{3}{5}$ for $k$ in the original equation.

$$-\frac{3}{5}x^2 - 2x + \left(-\frac{3}{5}\right) = 0$$

$$3x^2 + 10x + 3 = 0 \quad \text{Multiplying by } -5$$

$$(3x + 1)(x + 3) = 0$$

$$x = -\frac{1}{3} \text{ or } x = -3$$

The other solution is $-\dfrac{1}{3}$.

**71.** a) $x^2 - (6 + 3i)x + k = 0$; one solution is 3.

We first find $k$ by substituting 3 for $x$.
$$3^2 - (6 + 3i)3 + k = 0$$
$$9 - 18 - 9i + k = 0$$
$$-9 - 9i + k = 0$$
$$k = 9 + 9i$$

b) Now we substitute $9 + 9i$ for $k$ in the original equation.
$$x^2 - (6 + 3i)x + (9 + 9i) = 0$$
$$x^2 - (6 + 3i)x + 3(3 + 3i) = 0$$
$$[x - (3 + 3i)][x - 3] = 0$$
$$x = 3 + 3i \quad or \quad x = 3$$

The other solution is $3 + 3i$.

**73.** The solutions of $ax^2 + bx + c = 0$ are $x = \dfrac{-b \pm \sqrt{b^2 - 4ac}}{2a}$.
When there is just one solution, $b^2 - 4ac = 0$, so
$$x = \frac{-b \pm 0}{2a} = -\frac{b}{2a}.$$

**75.** We substitute $(-3, 0)$, $\left(\dfrac{1}{2}, 0\right)$, and $(0, -12)$ in $f(x) = ax^2 + bx + c$ and get three equations.
$$0 = 9a - 3b + c,$$
$$0 = \frac{1}{4}a + \frac{1}{2}b + c,$$
$$-12 = c$$

The solution of this system of equations is $a = 8$, $b = 20$, $c = -12$.

**77.** If $1 - \sqrt{5}$ and $3 + 2i$ are two solutions, then $1 + \sqrt{5}$ and $3 - 2i$ are also solutions. The equation of lowest degree that has these solutions is found as follows.
$$[x-(1-\sqrt{5})][x-(1+\sqrt{5})][x-(3+2i)][x-(3-2i)] = 0$$
$$(x^2 - 2x - 4)(x^2 - 6x + 13) = 0$$
$$x^4 - 8x^3 + 21x^2 - 2x - 52 = 0$$

**79.** *Writing Exercise*

## Exercise Set 8.5

**1.** $x^6 = (x^3)^2$, so (f) is an appropriate choice.

**3.** $x^8 = (x^4)^2$, so (h) is an appropriate choice.

**5.** $x^{4/3} = (x^{2/3})^2$, so (g) is an appropriate choice.

**7.** $x^{-4/3} = (x^{-2/3})^2$, so (e) is an appropriate choice.

**9.** $x^4 - 5x^2 + 4 = 0$

Let $u = x^2$ and $u^2 = x^4$.
$$u^2 - 5u + 4 = 0 \quad \text{Substituting } u \text{ for } x^2$$
$$(u - 1)(u - 4) = 0$$
$$u - 1 = 0 \quad or \quad u - 4 = 0$$
$$u = 1 \quad or \qquad u = 4$$

Now replace $u$ with $x^2$ and solve these equations.
$$x^2 = 1 \quad or \quad x^2 = 4$$
$$x = \pm 1 \quad or \quad x = \pm 2$$

The numbers 1, $-1$, 2, and $-2$ check. They are the solutions.

**11.** $x^4 - 9x^2 + 20 = 0$

Let $u = x^2$ and $u^2 = x^4$.
$$u^2 - 9u + 20 = 0 \quad \text{Substituting}$$
$$(u - 4)(u - 5) = 0$$
$$u = 4 \quad or \quad u = 5$$

Now replace $u$ with $x^2$ and solve these equations:
$$x^2 = 4 \quad or \quad x^2 = 5$$
$$x = \pm 2 \quad or \quad x = \pm\sqrt{5}$$

The numbers 2, $-2$, $\sqrt{5}$, and $-\sqrt{5}$ check. They are the solutions.

**13.** $4t^4 - 19t^2 + 12 = 0$

Let $u = t^2$ and $u^2 = t^4$.
$$4u^2 - 19u + 12 = 0 \quad \text{Substituting}$$
$$(4u - 3)(u - 4) = 0$$
$$4u - 3 = 0 \quad or \quad u - 4 = 0$$
$$u = \frac{3}{4} \quad or \qquad u = 4$$

Now replace $u$ with $t^2$ and solve these equations:
$$t^2 = \frac{3}{4} \quad or \quad t^2 = 4$$
$$t = \pm\frac{\sqrt{3}}{2} \quad or \quad t = \pm 2$$

The numbers $\dfrac{\sqrt{3}}{2}$, $-\dfrac{\sqrt{3}}{2}$, 2, and $-2$ check. They are the solutions.

**15.** $r - 2\sqrt{r} - 6 = 0$

Let $u = \sqrt{r}$ and $u^2 = r$.
$$u^2 - 2u - 6 = 0$$
$$u = \frac{-(-2) \pm \sqrt{(-2)^2 - 4 \cdot 1 \cdot (-6)}}{2 \cdot 1}$$
$$u = \frac{2 \pm \sqrt{28}}{2} = \frac{2 + 2\sqrt{7}}{2}$$
$$u = 1 \pm \sqrt{7}$$

Replace $u$ with $\sqrt{r}$ and solve these equations:
$$\sqrt{r} = 1 + \sqrt{7} \qquad or \quad \sqrt{r} = 1 - \sqrt{7}$$
$$(\sqrt{r})^2 = (1 + \sqrt{7})^2$$
$$r = 1 + 2\sqrt{7} + 7 \qquad \text{No solution:}$$
$$\qquad\qquad\qquad\qquad 1 - \sqrt{7} \text{ is negative}$$
$$r = 8 + 2\sqrt{7}$$

The number $8 + 2\sqrt{7}$ checks. It is the solution.

**17.** $(x^2 - 7)^2 - 3(x^2 - 7) + 2 = 0$

Let $u = x^2 - 7$ and $u^2 = (x^2 - 7)^2$.
$$u^2 - 3u + 2 = 0 \quad \text{Substituting}$$
$$(u - 1)(u - 2) = 0$$

$$u = 1 \qquad or \qquad u = 2$$
$$x^2 - 7 = 1 \quad or \quad x^2 - 7 = 2 \quad \text{Replacing } u$$
$$\text{with } x^2 - 7$$
$$x^2 = 8 \qquad or \qquad x^2 = 9$$
$$x = \pm\sqrt{8} \quad or \qquad x = \pm 3$$
$$x = \pm 2\sqrt{2} \quad or \qquad x = \pm 3$$

The numbers $2\sqrt{2}$, $-2\sqrt{3}$, 3, and $-3$ check. They are the solutions.

**19.** $(1 + \sqrt{x})^2 + 5(1 + \sqrt{x}) + 6 = 0$

Let $u = 1 + \sqrt{x}$ and $u^2 = (1 + \sqrt{x})^2$.

$$u^2 + 5u + 6 = 0 \quad \text{Substituting}$$
$$(u + 3)(u + 2) = 0$$
$$u = -3 \quad or \qquad u = -2$$
$$1 + \sqrt{x} = -3 \quad or \quad 1 + \sqrt{x} = -2 \quad \text{Replacing } u$$
$$\text{with } 1 + \sqrt{x}$$
$$\sqrt{x} = -4 \quad or \qquad \sqrt{x} = -3$$

Since the principal square root cannot be negative, this equation has no solution.

**21.** $x^{-2} - x^{-1} - 6 = 0$

Let $u = x^{-1}$ and $u^2 = x^{-2}$.

$$u^2 - u - 6 = 0 \quad \text{Substituting}$$
$$(u - 3)(u + 2) = 0$$
$$u = 3 \quad or \quad u = -2$$

Now we replace $u$ with $x^{-1}$ and solve these equations:

$$x^{-1} = 3 \quad or \quad x^{-1} = -2$$
$$\frac{1}{x} = 3 \quad or \quad \frac{1}{x} = -2$$
$$\frac{1}{3} = x \quad or \quad -\frac{1}{2} = x$$

Both $\frac{1}{3}$ and $-\frac{1}{2}$ check. They are the solutions.

**23.** $4x^{-2} + x^{-1} - 5 = 0$

Let $u = x^{-1}$ and $u^2 = x^{-2}$.

$$4u^2 + u - 5 = 0 \quad \text{Substituting}$$
$$(4u + 5)(u - 1) = 0$$
$$u = -\frac{5}{4} \quad or \quad u = 1$$

Now we replace $u$ with $x^{-1}$ and solve these equations:

$$x^{-1} = -\frac{5}{4} \quad or \quad x^{-1} = 1$$
$$\frac{1}{x} = -\frac{5}{4} \quad or \quad \frac{1}{x} = 1$$
$$4 = -5x \quad or \quad 1 = x$$
$$-\frac{4}{5} = x \quad or \quad 1 = x$$

The numbers $-\frac{4}{5}$ and 1 check. They are the solutions.

**25.** $t^{2/3} + t^{1/3} - 6 = 0$

Let $u = t^{1/3}$ and $u^2 = t^{2/3}$.

$$u^2 + u - 6 = 0 \quad \text{Substituting}$$
$$(u + 3)(u - 2) = 0$$
$$u = -3 \quad or \quad u = 2$$

Now we replace $u$ with $t^{1/3}$ and solve these equations:

$$t^{1/3} = -3 \quad or \quad t^{1/3} = 2$$
$$t = (-3)^3 \quad or \quad t = 2^3 \quad \text{Raising to the}$$
$$\text{third power}$$
$$t = -27 \quad or \quad t = 8$$

Both $-27$ and 8 check. They are the solutions.

**27.** $y^{1/3} - y^{1/6} - 6 = 0$

Let $u = y^{1/6}$ and $u^2 = y^{2/3}$.

$$u^2 - u - 6 = 0 \quad \text{Substituting}$$
$$(u - 3)(u + 2) = 0$$
$$u = 3 \quad or \quad u = -2$$

Now we replace $u$ with $y^{1/6}$ and solve these equations:

$$y^{1/6} = 3 \quad or \quad y^{1/6} = -2$$
$$\sqrt[6]{y} = 3 \quad or \quad \sqrt[6]{y} = -2$$
$$y = 3^6 \qquad \text{This equation has no}$$
$$y = 729 \qquad \text{solution since principal}$$
$$\text{sixth roots are never negative.}$$

The number 729 checks and is the solution.

**29.** $\qquad t^{1/3} + 2t^{1/6} = 3$
$$t^{1/3} + 2t^{1/6} - 3 = 0$$

Let $u = t^{1/6}$ and $u^2 = t^{2/6} = t^{1/3}$.

$$u^2 + 2u - 3 = 0 \quad \text{Substituting}$$
$$(u + 3)(u - 1) = 0$$
$$u = -3 \quad or \quad u = 1$$
$$t^{1/6} = -3 \quad or \quad t^{1/6} = 1 \quad \text{Substituting } t^{1/6} \text{ for } u$$
$$\text{No solution} \qquad t = 1$$

The number 1 checks and is the solution.

**31.** $(3 - \sqrt{x})^2 - 10(3 - \sqrt{x}) + 23 = 0$

Let $u = 3 - \sqrt{x}$ and $u^2 = (3 - \sqrt{x})^2$.

$$u^2 - 10u + 23 = 0 \quad \text{Substituting}$$
$$u = \frac{-(-10) \pm \sqrt{(-10)^2 - 4 \cdot 1 \cdot 23}}{2 \cdot 1}$$
$$u = \frac{10 \pm \sqrt{8}}{2} = \frac{2 \cdot 5 \pm 2\sqrt{2}}{2}$$
$$u = 5 \pm \sqrt{2}$$
$$u = 5 + \sqrt{2} \quad or \quad u = 5 - \sqrt{2}$$

Now we replace $u$ with $3 - \sqrt{x}$ and solve these equations:

$$3 - \sqrt{x} = 5 + \sqrt{2} \quad or \quad 3 - \sqrt{x} = 5 - \sqrt{2}$$
$$-\sqrt{x} = 2 + \sqrt{2} \quad or \quad -\sqrt{x} = 2 - \sqrt{2}$$
$$\sqrt{x} = -2 - \sqrt{2} \quad or \quad \sqrt{x} = -2 + \sqrt{2}$$

Since both $-2 - \sqrt{2}$ and $-2 + \sqrt{2}$ are negative and principal square roots are never negative, the equation has no solution.

**33.** $16\left(\dfrac{x-1}{x-8}\right)^2 + 8\left(\dfrac{x-1}{x-8}\right) + 1 = 0$

Let $u = \dfrac{x-1}{x-8}$ and $u^2 = \left(\dfrac{x-1}{x-8}\right)^2$.

$\quad 16u^2 + 8u + 1 = 0 \quad$ Substituting

$\quad (4u+1)(4u+1) = 0$

$\qquad\qquad u = -\dfrac{1}{4}$

Now we replace $u$ with $\dfrac{x-1}{x-8}$ and solve this equation:

$\quad \dfrac{x-1}{x-8} = -\dfrac{1}{4}$

$\quad 4x - 4 = -x + 8 \quad$ Multiplying by $4(x-8)$

$\qquad 5x = 12$

$\qquad\ x = \dfrac{12}{5}$

The number $\dfrac{12}{5}$ checks and is the solution.

**35.** The $x$-intercepts occur where $f(x) = 0$. Thus, we must have $5x + 13\sqrt{x} - 6 = 0$.

Let $u = \sqrt{x}$ and $u^2 = x$.

$\quad 5u^2 + 13u - 6 = 0 \quad$ Substituting

$\quad (5u - 2)(u + 3) = 0$

$\quad u = \dfrac{2}{5} \ or \ u = -3$

Now replace $u$ with $\sqrt{x}$ and solve these equations:

$\quad \sqrt{x} = \dfrac{2}{5} \quad or \quad \sqrt{x} = -3$

$\qquad x = \dfrac{4}{25} \qquad\quad$ No solution

The number $\dfrac{4}{25}$ checks. Thus, the $x$-intercept is $\left(\dfrac{4}{25}, 0\right)$.

**37.** The $x$-intercepts occur where $f(x) = 0$. Thus, we must have $(x^2 - 3x)^2 - 10(x^2 - 3x) + 24 = 0$.

Let $u = x^2 - 3x$ and $u^2 = (x^2 - 3x)^2$.

$\quad u^2 - 10u + 24 = 0 \quad$ Substituting

$\quad (u - 6)(u - 4) = 0$

$\quad u = 6 \ or \ u = 4$

Now replace $u$ with $x^2 - 3x$ and solve these equations:

$\quad x^2 - 3x = 6 \quad or \qquad x^2 - 3x = 4$

$\quad x^2 - 3x - 6 = 0 \ or \ x^2 - 3x - 4 = 0$

$\quad x = \dfrac{-(-3) \pm \sqrt{(-3)^2 - 4(1)(-6)}}{2 \cdot 1} \ or$

$\qquad\qquad\qquad\qquad (x - 4)(x + 1) = 0$

$\quad x = \dfrac{3}{2} \pm \dfrac{\sqrt{33}}{2} \ or \ x = 4 \ or \ x = -1$

All four numbers check. Thus, the $x$-intercepts are $\left(\dfrac{3}{2} + \dfrac{\sqrt{33}}{2}, 0\right)$, $\left(\dfrac{3}{2} - \dfrac{\sqrt{33}}{2}, 0\right)$, $(4, 0)$, and $(-1, 0)$.

**39.** The $x$-intercepts occur where $f(x) = 0$. Thus, we must have $x^{2/5} + x^{1/5} - 6 = 0$.

Let $u = x^{1/5}$ and $u^2 = x^{2/5}$.

$\quad u^2 + u - 6 = 0 \quad$ Substituting

$\quad (u + 3)(u - 2) = 0$

$\quad u = -3 \quad or \quad u = 2$

$\quad x^{1/5} = -3 \quad or \ x^{1/5} = 2 \quad$ Replacing $u$
$\qquad\qquad\qquad\qquad\qquad\qquad$ with $x^{1/5}$

$\quad x = -243 \ or \qquad x = 32 \quad$ Raising to the fifth
$\qquad\qquad\qquad\qquad\qquad\qquad\qquad$ power

Both $-243$ and $32$ check. Thus, the $x$-intercepts are $(-243, 0)$ and $(32, 0)$.

**41.** $f(x) = \left(\dfrac{x^2 + 2}{x}\right)^4 + 7\left(\dfrac{x^2 + 2}{x}\right)^2 + 5$

Observe that, for all real numbers $x$, each term is positive. Thus, there are no real-number values of $x$ for which $f(x) = 0$ and hence no $x$-intercepts.

**43.** *Writing Exercise*

**45.** Graph $f(x) = \dfrac{3}{2}x$.

We find some ordered pairs, plot points, and draw the graph.

| $x$ | $y$ |
|-----|-----|
| $-4$ | $-6$ |
| $-2$ | $-3$ |
| $0$ | $0$ |
| $2$ | $3$ |
| $4$ | $6$ |

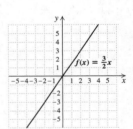

**47.** Graph $g(x) = \dfrac{2}{x}$.

We find some ordered pairs, plot points, and draw the graph. Note that we cannot use 0 as a first coordinate since division by 0 is undefined.

| $x$ | $y$ |
|-----|-----|
| $-4$ | $-\dfrac{1}{2}$ |
| $-2$ | $-1$ |
| $-\dfrac{1}{2}$ | $-4$ |
| $\dfrac{1}{2}$ | $4$ |
| $2$ | $1$ |
| $4$ | $\dfrac{1}{2}$ |

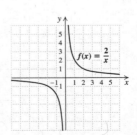

**49.** *Familiarize.* Let $a$ = the number of pounds of Hiker's Mix in the mixture and $b$ = the number of pounds of Trail Snax. We organize the information in a table.

| Solution | Hiker's Mix | Trail Snax | Mixture |
|---|---|---|---|
| Number of pounds | $a$ | $b$ | 12 |
| Percent of peanuts | 18% | 45% | 36% |
| Pounds of peanuts | $0.18a$ | $0.45b$ | 0.36(12), or 4.32 lb |

From the first row of the table we get one equation:

$a + b = 12$

We get a second equation from the last row of the table:

$0.18a + 0.45b = 4.32$

After clearing decimals, we have the following system of equations:

$$a + \ \ b = 12, \quad (1)$$
$$18a + 45b = 432 \quad (2)$$

**Carry out**. We use the elimination method. First we multiply equation (1) by $-18$ and then add.

$$
\begin{array}{r}
-18a - 18b = -216 \\
18a + 45b = 432 \\
\hline
27b = 216 \\
b = 8
\end{array}
$$

Now we substitute 8 for $b$ in one of the original equations and solve for $a$.

$$a + b = 12 \quad (1)$$
$$a + 8 = 12$$
$$a = 4$$

**Check**. If 4 lb of Hiker's Mix and 8 lb of Trail Snax are used, the mixture has $4 + 8$, or 12 lb. The amount of peanuts in 4 lb of Hiker's Mix is 0.18(4), or 0.72 lb. The amount of peanuts in 8 lb of Trail Snax is 0.45(8), or 3.6 lb. Then the amount of peanuts in the mixture is $0.72 + 3.6$, or 4.32 lb. The answer checks.

**State**. The mixture should contain 4 lb of Hiker's Mix and 8 lb of Trail Snax.

**51.** *Writing Exercise*

**53.** $5x^4 - 7x^2 + 1 = 0$

Let $u = x^2$ and $u^2 = x^4$.

$5u^2 - 7u + 1 = 0$ Substituting

$$u = \frac{-(-7) \pm \sqrt{(-7)^2 - 4 \cdot 5 \cdot 1}}{2 \cdot 5}$$

$$u = \frac{7 \pm \sqrt{29}}{10}$$

$$x^2 = \frac{7 \pm \sqrt{29}}{10} \qquad \text{Replacing } u \text{ with } x^2$$

$$x = \pm\sqrt{\frac{7 \pm \sqrt{29}}{10}}$$

All four numbers check and are the solutions.

**55.** $(x^2 - 4x - 2)^2 - 13(x^2 - 4x - 2) + 30 = 0$

Let $u = x^2 - 4x - 2$ and $u^2 = (x^2 - 4x - 2)^2$.

$u^2 - 13u + 30 = 0$ Substituting

$(u - 3)(u - 10) = 0$

$$u = 3 \quad or \quad\quad\quad u = 10$$
$$x^2 - 4x - 2 = 3 \quad or \quad x^2 - 4x - 2 = 10$$

Replacing $u$ with $x^2 - 4x - 2$

$$x^2 - 4x - 5 = 0 \quad or \quad x^2 - 4x - 12 = 0$$
$$(x - 5)(x + 1) = 0 \quad or \quad (x - 6)(x + 2) = 0$$
$$x = 5 \ or \ x = -1 \ or \ x = 6 \ or \ x = -2$$

All four numbers check and are the solutions.

**57.** $\dfrac{x}{x - 1} - 6\sqrt{\dfrac{x}{x - 1}} - 40 = 0$

Let $u = \sqrt{\dfrac{x}{x - 1}}$ and $u^2 = \dfrac{x}{x - 1}$.

$u^2 - 6u - 40 = 0$ Substituting

$(u - 10)(u + 4) = 0$

$$u = 10 \qquad\qquad or \qquad\qquad u = -4$$
$$\sqrt{\frac{x}{x - 1}} = 10 \quad or \quad \sqrt{\frac{x}{x - 1}} = -4$$
$$\frac{x}{x - 1} = 100 \qquad or \qquad \text{No solution}$$
$$x = 100x - 100 \ \text{Multiplying by } (x - 1)$$
$$100 = 99x$$
$$\frac{100}{99} = x$$

The number $\dfrac{100}{99}$ checks. It is the solution.

**59.** $a^5(a^2 - 25) + 13a^3(25 - a^2) + 36a(a^2 - 25) = 0$

$a^5(a^2 - 25) - 13a^3(a^2 - 25) + 36a(a^2 - 25) = 0$

$a(a^2 - 25)(a^4 - 13a^2 + 36) = 0$

$a(a^2 - 25)(a^2 - 4)(a^2 - 9) = 0$

$a=0 \ or \ a^2 - 25=0 \ \ or \ a^2 - 4=0 \ \ or \ a^2 - 9 = 0$

$a=0 \ or \quad\quad a^2=25 \ or \quad a^2=4 \quad or \quad\quad a^2 = 9$

$a=0 \ or \quad\quad a=\pm 5 \ or \quad a=\pm 2 \ or \quad\quad a = \pm 3$

All seven numbers check. The solutions are 0, 5, $-5$, 2, $-2$, 3, and $-3$.

**61.** $x^6 - 28x^3 + 27 = 0$

Let $u = x^3$.

$u^2 - 28u + 27 = 0$

$(u - 27)(u - 1) = 0$

$$u = 27 \quad or \quad\quad u = 1$$
$$x^3 = 27 \quad or \quad\quad x^3 = 1$$
$$x^3 - 27 = 0 \quad or \quad x^3 - 1 = 0$$

First we solve $x^3 - 27 = 0$.

$$x^3 - 27 = 0$$
$$(x - 3)(x^2 + 3x + 9) = 0$$

$x - 3 = 0$  or  $x^2 + 3x + 9 = 0$

$\quad x = 3$  or  $\qquad x = \dfrac{-3 \pm \sqrt{3^2 - 4 \cdot 1 \cdot 9}}{2 \cdot 1}$

$\quad x = 3$  or  $\qquad x = \dfrac{-3 \pm \sqrt{-27}}{2}$

$\quad x = 3$  or  $\qquad x = -\dfrac{3}{2} \pm \dfrac{3\sqrt{3}}{2}i$

Next we solve $x^3 - 1 = 0$.

$\qquad\qquad x^3 - 1 = 0$

$(x - 1)(x^2 + x + 1) = 0$

$x - 1 = 0$  or  $x^2 + x + 1 = 0$

$\quad x = 1$  or  $\qquad x = \dfrac{-1 \pm \sqrt{1^2 - 4 \cdot 1 \cdot 1}}{2 \cdot 1}$

$\quad x = 1$  or  $\qquad x = \dfrac{-1 \pm \sqrt{-3}}{2}$

$\quad x = 1$  or  $\qquad x = -\dfrac{1}{2} \pm \dfrac{\sqrt{3}}{2}i$

All six numbers check.

**63.**

## Exercise Set 8.6

**1.** The graph of $f(x) = 2(x - 1)^2 + 3$ has vertex $(1, 3)$ and opens up. Choice (h) is correct.

**3.** The graph of $f(x) = 2(x + 1)^2 + 3$ has vertex $(-1, 3)$ and opens up. Choice (f) is correct.

**5.** The graph of $f(x) = -2(x + 1)^2 + 3$ has vertex $(-1, 3)$ and opens down. Choice (b) is correct.

**7.** The graph of $f(x) = 2(x + 1)^2 - 3$ has vertex $(-1, -3)$ and opens up. Choice (e) is correct.

**9.** $f(x) = x^2$

See Example 1 in the text.

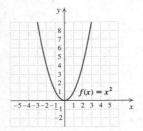

**11.** $f(x) = -2x^2$

We choose some numbers for $x$ and compute $f(x)$ for each one. Then we plot the ordered pairs $(x, f(x))$ and connect them with a smooth curve.

| $x$ | $f(x) = -4x^2$ |
|-----|----------------|
| 0   | 0              |
| 1   | $-2$           |
| 2   | $-8$           |
| $-1$| $-2$           |
| $-2$| $-8$           |

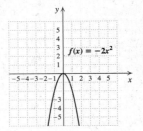

**13.** $g(x) = \dfrac{1}{3}x^2$

| $x$  | $g(x) = \dfrac{1}{3}x^2$ |
|------|--------------------------|
| 0    | 0                        |
| 1    | $\dfrac{1}{3}$           |
| 2    | $\dfrac{4}{3}$           |
| 3    | 3                        |
| $-1$ | $\dfrac{1}{3}$           |
| $-2$ | $\dfrac{4}{3}$           |
| $-3$ | 3                        |

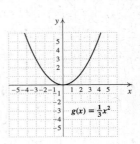

**15.** $h(x) = -\dfrac{1}{3}x^2$

Observe that the graph of $h(x) = -\dfrac{1}{3}x^2$ is the reflection of the graph of $g(x) = \dfrac{1}{3}x^2$ across the $x$-axis. We graphed $g(x)$ in Exercise 13, so we can use it to graph $h(x)$. If we did not make this observation we could find some ordered pairs, plot points, and connect them with a smooth curve.

| $x$  | $h(x) = -\dfrac{1}{3}x^2$ |
|------|---------------------------|
| 0    | 0                         |
| 1    | $-\dfrac{1}{3}$           |
| 2    | $-\dfrac{4}{3}$           |
| 3    | $-3$                      |
| $-1$ | $-\dfrac{1}{3}$           |
| $-2$ | $-\dfrac{4}{3}$           |
| $-3$ | $-3$                      |

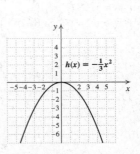

**17.** $f(x) = \frac{5}{2}x^2$

| $x$ | $f(x) = \frac{5}{2}x^2$ |
|-----|--------|
| 0 | 0 |
| 1 | $\frac{5}{2}$ |
| 2 | 10 |
| $-1$ | $\frac{5}{2}$ |
| $-2$ | 10 |

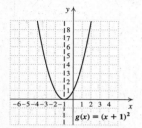

**19.** $g(x) = (x+1)^2 = [x-(-1)]^2$

We know that the graph of $g(x) = (x+1)^2$ looks like the graph of $f(x) = x^2$ (see Exercise 9) but moved to the left 1 unit.

Vertex: $(-1, 0)$, axis of symmetry: $x = -1$

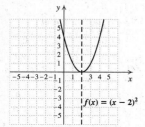

**21.** $f(x) = (x-2)^2$

The graph of $f(x) = (x-2)^2$ looks like the graph of $f(x) = x^2$ (see Exercise 9) but moved to the right 2 units.

Vertex: $(2, 0)$, axis of symmetry: $x = 2$

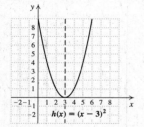

**23.** $h(x) = (x-3)^2$

The graph of $h(x) = (x-3)^2$ looks like the graph of $f(x) = x^2$ (see Exercise 9) but moved to the right 3 units.

Vertex: $(3, 0)$, axis of symmetry: $x = 3$

**25.** $g(x) = -(x+1)^2$

The graph of $g(x) = -(x+1)^2$ looks like the graph of $f(x) = x^2$ (see Exercise 9) but moved to the left 1 unit. It will also open downward because of the negative coefficient, $-1$.

Vertex: $(-1, 0)$, axis of symmetry: $x = -1$

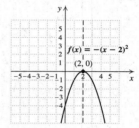

**27.** $f(x) = -(x-2)^2$

The graph of $f(x) = -(x-2)^2$ looks like the graph of $f(x) = x^2$ (see Exercise 9) but moved to the right 2 units. It will also open downward because of the negative coefficient, $-1$.

Vertex: $(2, 0)$, axis of symmetry: $x = 2$

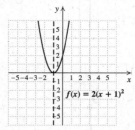

**29.** $f(x) = 2(x+1)^2$

The graph of $f(x) = 2(x+1)^2$ looks like the graph of $h(x) = 2x^2$ (see graph following Example 1) but moved to the left 1 unit.

Vertex: $(-1, 0)$, axis of symmetry: $x = -1$

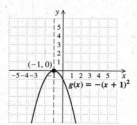

**31.** $h(x) = -\frac{1}{2}(x-4)^2$

The graph of $h(x) = -\frac{1}{2}(x-4)^2$ looks like the graph of $g(x) = \frac{1}{2}x^2$ (see graph following Example 1) but moved to the right 4 units. It will also open downward because of the negative coefficient, $-\frac{1}{2}$.

Vertex: $(4, 0)$, axis of symmetry: $x = 4$

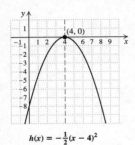

$$h(x) = -\frac{1}{2}(x-4)^2$$

**33.** $f(x) = \frac{1}{2}(x-1)^2$

The graph of $f(x) = \frac{1}{2}(x-1)^2$ looks like the graph of $g(x) = \frac{1}{2}x^2$ (see graph following Example 1) but moved to the right 1 unit.

Vertex: $(1,0)$, axis of symmetry: $x = 1$

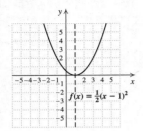

$$f(x) = \frac{1}{2}(x-1)^2$$

**35.** $f(x) = -2(x+5)^2 = -2[x-(-5)]^2$

The graph of $f(x) = -2(x+5)^2$ looks like the graph of $h(x) = 2x^2$ (see the graph following Example 1) but moved to the left 5 units. It will also open downward because of the negative coefficient, $-2$.

Vertex: $(-5,0)$, axis of symmetry: $x = -5$

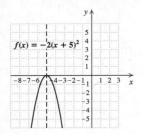

$$f(x) = -2(x+5)^2$$

**37.** $h(x) = -3\left(x - \frac{1}{2}\right)^2$

The graph of $h(x) = -3\left(x - \frac{1}{2}\right)^2$ looks like the graph of $f(x) = -3x^2$ (see Exercise 12) but moved to the right $\frac{1}{2}$ unit.

Vertex: $\left(\frac{1}{2}, 0\right)$, axis of symmetry: $x = \frac{1}{2}$

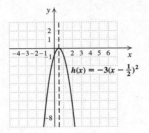

$$h(x) = -3\left(x - \frac{1}{2}\right)^2$$

**39.** $f(x) = (x-5)^2 + 2$

We know that the graph looks like the graph of $f(x) = x^2$ (see Example 1) but moved to the right 5 units and up 2 units. The vertex is $(5,2)$, and the axis of symmetry is $x = 5$. Since the coefficient of $(x-5)^2$ is positive $(1 > 0)$, there is a minimum function value, 2.

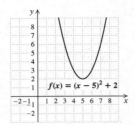

$$f(x) = (x-5)^2 + 2$$

**41.** $f(x) = (x+1)^2 - 3$

We know that the graph looks like the graph of $f(x) = x^2$ (see Example 1) but moved to the left 1 unit and down 3 units. The vertex is $(-1, -3)$, and the axis of symmetry is $x = -1$. Since the coefficient of $(x+1)^2$ is positive $(1 > 0)$, there is a minimum function value, $-3$.

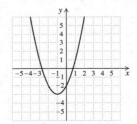

$$f(x) = (x+1)^2 - 3$$

**43.** $g(x) = (x+4)^2 + 1$

We know that the graph looks like the graph of $f(x) = x^2$ (see Example 1) but moved to the left 4 units and up 1 unit. The vertex is $(-4, 1)$, and the axis of symmetry is $x = -4$. Since the coefficient of $(x+4)^2$ is positive $(1 > 0)$, there is a minimum function value, 1.

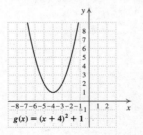

$$g(x) = (x+4)^2 + 1$$

**45.** $h(x) = -2(x - 1)^2 - 3$

We know that the graph looks like the graph of $h(x) = 2x^2$ (see graph following **Example 1**) but moved to the right 1 unit and down 3 units and turned upside down. The vertex is $(1, -3)$, and the axis of symmetry is $x = 1$. The maximum function value is $-3$.

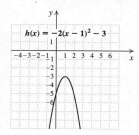

**47.** $f(x) = 2(x + 4)^2 + 1$

We know that the graph looks like the graph of $f(x) = 2x^2$ (see graph following **Example 1**) but moved to the left 4 units and up 1 unit. The vertex is $(-4, 1)$, the axis of symmetry is $x = -4$, and the minimum function value is 1.

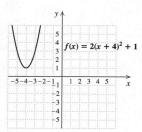

**49.** $g(x) = -\dfrac{3}{2}(x - 1)^2 + 4$

We know that the graph looks like the graph of $f(x) = \dfrac{3}{2}x^2$ (see Exercise 18) but moved to the right 1 unit and up 4 units and turned upside down. The vertex is $(1, 4)$, the axis of symmetry is $x = 1$, and the maximum function value is 4.

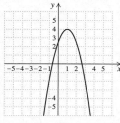

**51.** $f(x) = 6(x - 8)^2 + 7$

This function is of the form $f(x) = a(x - h)^2 + k$ with $a = 6$, $h = 8$, and $k = 7$. The vertex is $(h, k)$, or $(8, 7)$. The axis of symmetry is $x = h$, or $x = 8$. Since $a > 0$, then $k$, or 7, is the minimum function value.

**53.** $h(x) = -\dfrac{2}{7}(x + 6)^2 + 11$

This function is of the form $f(x) = a(x - h)^2 + k$ with $a = -\dfrac{2}{7}$, $h = -6$, and $k = 11$. The vertex is $(h, k)$, or $(-6, 11)$. The axis of symmetry is $x = h$, or $x = -6$. Since $a < 0$, then $k$, or 11, is the maximum function value.

**55.** $f(x) = 7\left(x + \dfrac{1}{4}\right)^2 - 13$

This function is of the form $f(x) = a(x - h)^2 + k$ with $a = 7$, $h = -\dfrac{1}{4}$, and $k = -13$. The vertex is $(h, k)$, or $\left(-\dfrac{1}{4}, -13\right)$. The axis of symmetry is $x = h$, or $x = -\dfrac{1}{4}$. Since $a > 0$, then $k$, or $-13$, is the minimum function value.

**57.** $f(x) = \sqrt{2}(x + 4.58)^2 + 65\pi$

This function is of the form $f(x) = a(x - h)^2 + k$ with $a = \sqrt{2}$, $h = -4.58$, and $k = 65\pi$. The vertex is $(h, k)$, or $(-4.58, 65\pi)$. The axis of symmetry is $x = h$, or $x = -4.58$. Since $a > 0$, then $k$, or $65\pi$, is the minimum function value.

**59.** *Writing Exercise*

**61.** Graph $2x - 7y = 28$.

Find the $x$-intercept.

$$2x - 7 \cdot 0 = 28$$
$$2x = 28$$
$$x = 14$$

The $x$-intercept is $(14, 0)$.

Find the $y$-intercept.

$$2 \cdot 0 - 7y = 28$$
$$-7y = 28$$
$$y = -4$$

The $y$-intercept is $(0, -4)$.

Plot the intercepts and draw a line through them. A third point can be plotted as a check.

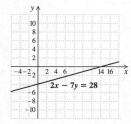

**63.** $3x + 4y = -19,$   (1)

$7x - 6y = -29$   (2)

Multiply Equation (1) by 3 and multiply Equation (2) by 2. Then add the equations to eliminate the $y$-term.

$$9x + 12y = -57$$
$$\underline{14x - 12y = -58}$$
$$23x \qquad\quad = -115$$
$$x = -5$$

Now substitute $-5$ for $x$ in one of the original equations and solve for $y$. We use Equation (1).

$$3(-5) + 4y = -19$$
$$-15 + 4y = -19$$
$$4y = -4$$
$$y = -1$$

The pair $(-5, -1)$ checks and it is the solution.

**65.** $x^2 + 5x$

We take half the coefficient of $x$ and square it.

$$\frac{1}{2} \cdot 5 = \frac{5}{2}, \left(\frac{5}{2}\right)^2 = \frac{25}{4}$$

Then we have $x^2 + 5x + \dfrac{25}{4} = \left(x + \dfrac{5}{2}\right)^2$.

**67.** *Writing Exercise*

**69.** The equation will be of the form $f(x) = \dfrac{3}{5}(x - h)^2 + k$ with $h = 4$ and $k = 1$:

$$f(x) = \frac{3}{5}(x - 4)^2 + 1$$

**71.** The equation will be of the form $f(x) = \dfrac{3}{5}(x - h)^2 + k$ with $h = 3$ and $k = -1$:

$$f(x) = \frac{3}{5}(x - 3)^2 + (-1), \text{ or}$$
$$f(x) = \frac{3}{5}(x - 3)^2 - 1$$

**73.** The equation will be of the form $f(x) = \dfrac{3}{5}(x - h)^2 + k$ with $h = -2$ and $k = -5$:

$$f(x) = \frac{3}{5}[x - (-2)]^2 + (-5), \text{ or}$$
$$f(x) = \frac{3}{5}(x + 2)^2 - 5$$

**75.** Since there is a minimum at $(2, 0)$, the parabola will have the same shape as $f(x) = 2x^2$. It will be of the form $f(x) = 2(x - h)^2 + k$ with $h = 2$ and $k = 0$: $f(x) = 2(x - 2)^2$

**77.** Since there is a maximum at $(0, 3)$, the parabola will have the same shape as $g(x) = -2x^2$. It will be of the form $g(x) = -2(x - h)^2 + k$ with $h = 0$ and $k = 3$: $g(x) = -2(x - 0)^2 + 3$, or $g(x) = -2x^2 + 3$

**79.** The maximum value of $g(x)$ is 1 and occurs at the point $(5, 1)$, so for $F(x)$ we have $h = 5$ and $k = 1$. $F(x)$ has the same shape as $f(x)$ and has a minimum, so $a = 3$. Thus, $F(x) = 3(x - 5)^2 + 1$.

**81.** The graph of $y = f(x - 1)$ looks like the graph of $y = f(x)$ moved 1 unit to the right.

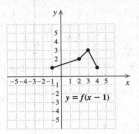

**83.** The graph of $y = f(x) + 2$ looks like the graph of $y = f(x)$ moved up 2 units.

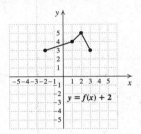

**85.** The graph of $y = f(x + 3) - 2$ looks like the graph of $y = f(x)$ moved 3 units to the left and also moved down 2 units.

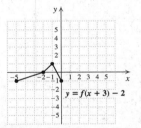

**87.**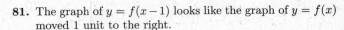

**89.** *Writing Exercise*

## Exercise Set 8.7

**1.** 9

**3.** 18

**5.** 3

**7.** $\dfrac{5}{2}$; $-4$

**9.**  $f(x) = x^2 + 4x + 5$

$\quad = (x^2 + 4x + 4 - 4) + 5$   Adding $4 - 4$

$\quad = (x^2 + 4x + 4) - 4 + 5$   Regrouping

$\quad = (x + 2)^2 + 1$

The vertex is $(-2, 1)$, the axis of symmetry is $x = -2$, and the graph opens upward since the coefficient 1 is positive. We plot a few points as a check and draw the curve.

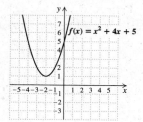

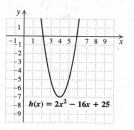

**11.**  $g(x) = x^2 - 6x + 13$

$\qquad = (x^2 - 6x + 9 - 9) + 13 \quad$ Adding $9 - 9$

$\qquad = (x^2 - 6x + 9) - 9 + 13 \quad$ Regrouping

$\qquad = (x - 3)^2 + 4$

The vertex is $(3, 4)$, the axis of symmetry is $x = 3$, and the graph opens upward since the coefficient 1 is positive. We plot a few points as a check and draw the curve.

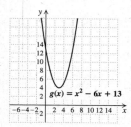

**13.**  $f(x) = x^2 + 8x + 20$

$\qquad = (x^2 + 8x + 16 - 16) + 20 \quad$ Adding $16 - 16$

$\qquad = (x^2 + 8x + 16) - 16 + 20 \quad$ Regrouping

$\qquad = (x + 4)^2 + 4$

The vertex is $(-4, 4)$, the axis of symmetry is $x = -4$, and the graph opens upward since the coefficient 1 is positive.

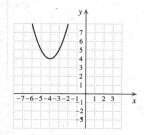

**15.**  $h(x) = 2x^2 - 16x + 25$

$\qquad = 2(x^2 - 8x) + 25 \quad$ Factoring 2 from the first two terms

$\qquad = 2(x^2 - 8x + 16 - 16) + 25 \quad$ Adding $16-16$ inside the parentheses

$\qquad = 2(x^2 - 8x + 16) + 2(-16) + 25$

$\qquad\qquad\qquad$ Distributing to obtain a trinomial square

$\qquad = 2(x - 4)^2 - 7$

The vertex is $(4, -7)$, the axis of symmetry is $x = 4$, and the graph opens upward since the coefficient 2 is positive.

**17.**  $f(x) = -x^2 + 2x + 5$

$\qquad = -(x^2 - 2x) + 5 \quad$ Factoring $-1$ from the first two terms

$\qquad = -(x^2 - 2x + 1 - 1) + 5$

$\qquad\qquad\qquad$ Adding $1 - 1$ inside the parentheses

$\qquad = -(x^2 - 2x + 1) - (-1) + 5$

$\qquad = -(x - 1)^2 + 6$

The vertex is $(1, 6)$, the axis of symmetry is $x = 1$, and the graph opens downward since the coefficient $-1$ is negative.

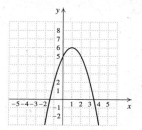

**19.**  $g(x) = x^2 + 3x - 10$

$\qquad = \left(x^2 + 3x + \dfrac{9}{4} - \dfrac{9}{4}\right) - 10$

$\qquad = \left(x^2 + 3x + \dfrac{9}{4}\right) - \dfrac{9}{4} - 10$

$\qquad = \left(x + \dfrac{3}{2}\right)^2 - \dfrac{49}{4}$

The vertex is $\left(-\dfrac{3}{2}, -\dfrac{49}{4}\right)$, the axis of symmetry is $x = -\dfrac{3}{2}$, and the graph opens upward since the coefficient 1 is positive.

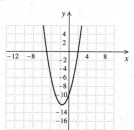

**21.**  $f(x) = 3x^2 - 24x + 50$

$\qquad = 3(x^2 - 8x) + 50 \quad$ Factoring

$\qquad = 3(x^2 - 8x + 16 - 16) + 50$

$\qquad\qquad\qquad$ Adding $16 - 16$ inside the parentheses

$\qquad = 3(x^2 - 8x + 16) - 3 \cdot 16 + 50$

$\qquad = 3(x - 4)^2 + 2$

The vertex is $(4, 2)$, the axis of symmetry is $x = 4$, and the graph opens upward since the coefficient 3 is positive.

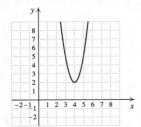

**23.**  $h(x) = x^2 + 7x$

$\qquad = \left(x^2 + 7x + \dfrac{49}{4}\right) - \dfrac{49}{4}$

$\qquad = \left(x + \dfrac{7}{2}\right)^2 - \dfrac{49}{4}$

The vertex is $\left(-\dfrac{7}{2}, -\dfrac{49}{4}\right)$, the axis of symmetry is $x = -\dfrac{7}{2}$, and the graph opens upward since the coefficient 1 is positive.

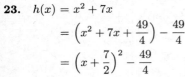
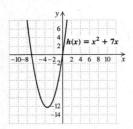

**25.**  $f(x) = -2x^2 - 4x - 6$

$\qquad = -2(x^2 + 2x) - 6 \quad \text{Factoring}$

$\qquad = -2(x^2 + 2x + 1 - 1) - 6$

$\qquad\qquad\qquad\qquad \text{Adding } 1 - 1 \text{ inside}$
$\qquad\qquad\qquad\qquad \text{the parentheses}$

$\qquad = -2(x^2 + 2x + 1) - 2(-1) - 6$

$\qquad = -2(x + 1)^2 - 4$

The vertex is $(-1, -4)$, the axis of symmetry is $x = -1$, and the graph opens downward since the coefficient $-2$ is negative.

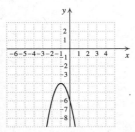

**27.**  $g(x) = 2x^2 - 8x + 3$

$\qquad = 2(x^2 - 4x) + 3 \quad \text{Factoring}$

$\qquad = 2(x^2 - 4x + 4 - 4) + 3$

$\qquad\qquad\qquad\qquad \text{Adding } 4 - 4 \text{ inside}$
$\qquad\qquad\qquad\qquad \text{the parentheses}$

$\qquad = 2(x^2 - 4x + 4) + 2(-4) + 3$

$\qquad = 2(x - 2)^2 - 5$

The vertex is $(2, -5)$, the axis of symmetry is $x = 2$, and the graph opens upward since the coefficient 2 is positive.

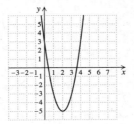

$$g(x) = 2x^2 - 8x + 3$$

**29.**  $f(x) = -3x^2 + 5x - 2$

$\qquad = -3\left(x^2 - \dfrac{5}{3}x\right) - 2 \qquad \text{Factoring}$

$\qquad = -3\left(x^2 - \dfrac{5}{3}x + \dfrac{25}{36} - \dfrac{25}{36}\right) - 2$

$\qquad\qquad\qquad \text{Adding } \dfrac{25}{36} - \dfrac{25}{36} \text{ inside}$

$\qquad\qquad\qquad \text{the parentheses}$

$\qquad = -3\left(x^2 - \dfrac{5}{3}x + \dfrac{25}{36}\right) - 3\left(-\dfrac{25}{36}\right) - 2$

$\qquad = -3\left(x - \dfrac{5}{6}\right)^2 + \dfrac{1}{12}$

The vertex is $\left(\dfrac{5}{6}, \dfrac{1}{12}\right)$, the axis of symmetry is $x = \dfrac{5}{6}$, and the graph opens downward since the coefficient $-3$ is negative.

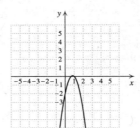

**31.**  $h(x) = \dfrac{1}{2}x^2 + 4x + \dfrac{19}{3}$

$\qquad = \dfrac{1}{2}(x^2 + 8x) + \dfrac{19}{3} \qquad \text{Factoring}$

$\qquad = \dfrac{1}{2}(x^2 + 8x + 16 - 16) + \dfrac{19}{3}$

$\qquad\qquad\qquad\qquad \text{Adding } 16 - 16 \text{ inside}$
$\qquad\qquad\qquad\qquad \text{the parentheses}$

$\qquad = \dfrac{1}{2}(x^2 + 8x + 16) + \dfrac{1}{2}(-16) + \dfrac{19}{3}$

$\qquad = \dfrac{1}{2}(x + 4)^2 - \dfrac{5}{3}$

The vertex is $\left(-4, -\dfrac{5}{3}\right)$, the axis of symmetry is $x = -4$, and the graph opens upward since the coefficient $\dfrac{1}{2}$ is positive.

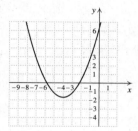

**33.** $f(x) = x^2 - 6x + 3$

To find the $x$-intercepts, solve the equation
$0 = x^2 - 6x + 3$. Use the quadratic formula.

$$x = \frac{-(-6) \pm \sqrt{(-6)^2 - 4 \cdot 1 \cdot 3}}{2 \cdot 1}$$

$$x = \frac{6 \pm \sqrt{24}}{2} = \frac{6 \pm 2\sqrt{6}}{2} = 3 \pm \sqrt{6}$$

The $x$-intercepts are $(3 - \sqrt{6}, 0)$ and $(3 + \sqrt{6}, 0)$.

The $y$-intercept is $(0, f(0))$, or $(0, 3)$.

**35.** $g(x) = -x^2 + 2x + 3$

To find the $x$-intercepts, solve the equation
$0 = -x^2 + 2x + 3$. We factor.

$0 = -x^2 + 2x + 3$

$0 = x^2 - 2x - 3$    Multiplying by $-1$

$0 = (x - 3)(x + 1)$

$x = 3 \ or \ x = -1$

The $x$-intercepts are $(-1, 0)$ and $(3, 0)$.

The $y$-intercept is $(0, g(0))$, or $(0, 3)$.

**37.** $f(x) = x^2 - 9x$

To find the $x$-intercepts, solve the equation
$0 = x^2 - 9x$. We factor.

$0 = x^2 - 9x$

$0 = x(x - 9)$

$x = 0 \ or \ x = 9$

The $x$-intercepts are $(0, 0)$ and $(9, 0)$.

Since $(0, 0)$ is an $x$-intercept, we observe that $(0, 0)$ is also the $y$-intercept.

**39.** $h(x) = -x^2 + 4x - 4$

To find the $x$-intercepts, solve the equation
$0 = -x^2 + 4x - 4$. We factor.

$0 = -x^2 + 4x - 4$

$0 = x^2 - 4x + 4$    Multiplying by $-1$

$0 = (x - 2)(x - 2)$

$x = 2 \ or \ x = 2$

The $x$-intercept is $(2, 0)$.

The $y$-intercept is $(0, h(0))$, or $(0, -4)$.

**41.** $f(x) = 2x^2 - 4x + 6$

To find the $x$-intercepts, solve the equation $0 = 2x^2 - 4x + 6$. We use the quadratic formula.

$$x = \frac{-(-4) \pm \sqrt{(-4)^2 - 4 \cdot 2 \cdot 6}}{2 \cdot 2}$$

$$x = \frac{4 \pm \sqrt{-32}}{4} = \frac{4 \pm 4i\sqrt{2}}{2} = 2 \pm 2i\sqrt{2}$$

There are no real-number solutions, so there is no $x$-intercept.

The $y$-intercept is $(0, f(0))$, or $(0, 6)$.

**43.** *Writing Exercise*

**45.** $5x - 3y = 16$,  (1)

$4x + 2y = 4$    (2)

Multiply equation (1) by 2 and equation (2) by 3 and add.

$10x - 6y = 32$

$\underline{12x + 6y = 12}$

$22x \qquad = 44$

$x = 2$

Substitute 2 for $x$ in one of the original equations and solve for $y$.

$4x + 2y = 4$    (1)

$4 \cdot 2 + 2y = 4$

$8 + 2y = 4$

$2y = -4$

$y = -2$

The solution is $(2, -2)$.

**47.** $4a - 5b + c = 3$,  (1)

$3a - 4b + 2c = 3$,  (2)

$a + b - 7c = -2$  (3)

First multiply equation (1) by $-2$ and add it to equation (2).

$-8a + 10b - 2c = -6$

$\underline{3a - 4b + 2c = 3}$

$-5a + 6b \qquad = -3$  (4)

Next multiply equation (1) by 7 and add it to equation (3).

$28a - 35b + 7c = 21$

$\underline{a + b - 7c = -2}$

$29a - 34b \qquad = 19$  (5)

Now we solve the system of equations (4) and (5). Multiply equation (4) by 29 and equation (5) by 5 and add.

$-145a + 174b = -87$

$\underline{145a - 170b = 95}$

$4b = 8$

$b = 2$

Substitute 2 for $b$ in equation (4) and solve for $a$.

$$-5a + 6 \cdot 2 = -3$$
$$-5a + 12 = -3$$
$$-5a = -15$$
$$a = 3$$

Now substitute 3 for $a$ and 2 for $b$ in equation (1) and solve for $c$.

$$4 \cdot 3 - 5 \cdot 2 + c = 3$$
$$12 - 10 + c = 3$$
$$2 + c = 3$$
$$c = 1$$

The solution is $(3, 2, 1)$.

**49.**
$$\sqrt{4x - 4} = \sqrt{x + 4} + 1$$
$$4x - 4 = x + 4 + 2\sqrt{x + 4} + 1 \quad \text{Squaring both sides}$$
$$3x - 9 = 2\sqrt{x + 4}$$
$$9x^2 - 54x + 81 = 4(x + 4) \quad \text{Squaring both sides again}$$
$$9x^2 - 54x + 81 = 4x + 16$$
$$9x^2 - 58x + 65 = 0$$
$$(9x - 13)(x - 5) = 0$$
$$x = \frac{13}{9} \quad \text{or} \quad x = 5$$

Check: For $x = \frac{13}{9}$:

$$\frac{\sqrt{4x - 4} = \sqrt{x + 4} + 1}{\sqrt{4\left(\frac{13}{9}\right) - 4} \quad \bigg| \quad \sqrt{\frac{13}{9} + 4} + 1}$$
$$\sqrt{\frac{16}{9}} \quad \bigg| \quad \sqrt{\frac{49}{9}} + 1$$
$$\frac{4}{3} \quad \bigg| \quad \frac{7}{3} + 1$$
$$\frac{4}{3} \overset{?}{=} \frac{10}{3} \qquad \text{FALSE}$$

For $x = 5$:

$$\frac{\sqrt{4x - 4} = \sqrt{x + 4} + 1}{\sqrt{4 \cdot 5 - 4} \quad \bigg| \quad \sqrt{5 + 4} + 1}$$
$$\sqrt{16} \quad \bigg| \quad \sqrt{9} + 1$$
$$4 \quad \bigg| \quad 3 + 1$$
$$4 \overset{?}{=} 4 \qquad \text{TRUE}$$

5 checks, but $\frac{13}{9}$ does not. The solution is 5.

**51.** *Writing Exercise*

**53.** a) $f(x) = 2.31x^2 - 3.135x - 5.89$
$$= 2.31(x^2 - 1.357142857x) - 5.89$$
$$= 2.31(x^2 - 1.357142857x + 0.460459183 - 0.460459183) - 5.89$$
$$= 2.31(x^2 - 1.357142857x + 0.460459183) + 2.31(-0.460459183) - 5.89$$
$$= 2.31(x - 0.678571428)^2 - 6.953660714$$

Since the coefficient 2.31 is positive, the function has a minimum value. It is $-6.953660714$.

b) To find the $x$-intercepts, solve
$$0 = 2.31x^2 - 3.135x - 5.89.$$
$$x = \frac{-(-3.135) \pm \sqrt{(-3.135)^2 - 4(2.31)(-5.89)}}{2(2.31)}$$
$$x \approx \frac{3.135 \pm 8.015723611}{4.62}$$
$$x \approx -1.056433682 \quad \text{or} \quad x \approx 2.413576539$$

The $x$-intercepts are $(-1.056433682, 0)$ and $(2.413576539, 0)$.

The $y$-intercept is $(0, f(0))$, or $(0, -5.89)$.

**55.** $f(x) = x^2 - x - 6$

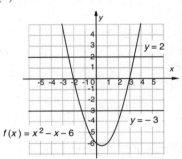

a) The solutions of $x^2 - x - 6 = 2$ are the first coordinates of the points of intersection of the graphs of $f(x) = x^2 - x - 6$ and $y = 2$. From the graph we see that the solutions are approximately $-2.4$ and $3.4$.

b) The solutions of $x^2 - x - 6 = -3$ are the first coordinates of the points of intersection of the graphs of $f(x) = x^2 - x - 6$ and $y = -3$. From the graph we see that the solutions are approximately $-1.3$ and $2.3$.

**57.** $f(x) = mx^2 - nx + p$
$$= m\left(x^2 - \frac{n}{m}x\right) + p$$
$$= m\left(x^2 - \frac{n}{m}x + \frac{n^2}{4m^2} - \frac{n^2}{4m^2}\right) + p$$
$$= m\left(x - \frac{n}{2m}\right)^2 - \frac{n^2}{4m} + p$$
$$= m\left(x - \frac{n}{2m}\right)^2 + \frac{-n^2 + 4mp}{4m}, \quad \text{or}$$
$$m\left(x - \frac{n}{2m}\right)^2 + \frac{4mp - n^2}{4m}$$

**59.** The horizontal distance from $(-1, 0)$ to $(3, -5)$ is $|3-(-1)|$, or 4, so by symmetry the other $x$-intercept is $(3+4, 0)$, or $(7, 0)$. Substituting the three ordered pairs $(-1, 0)$, $(3, -5)$, and $(7, 0)$ in the equation $f(x) = ax^2 + bx + c$ yields a system of equations:

$$0 = a - b + c,$$
$$-5 = 9a + 3b + c,$$
$$0 = 49a + 7b + c$$

The solution of this system of equations is $\left(\dfrac{5}{16}, -\dfrac{15}{8}, -\dfrac{35}{16}\right)$, so $f(x) = \dfrac{5}{16}x^2 - \dfrac{15}{8}x - \dfrac{35}{16}$.

If we complete the square we find that this function can also be expressed as $f(x) = \dfrac{5}{16}(x-3)^2 - 5$.

**61.** $f(x) = |x^2 - 1|$

We plot some points and draw the curve. Note that it will lie entirely on or above the $x$-axis since absolute value is never negative.

| $x$ | $f(x)$ |
|-----|--------|
| $-3$ | 8 |
| $-2$ | 3 |
| $-1$ | 0 |
| 0 | 1 |
| 1 | 0 |
| 2 | 3 |
| 3 | 8 |

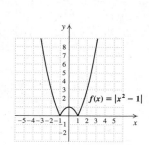

**63.** $f(x) = |2(x-3)^2 - 5|$

We plot some points and draw the curve. Note that it will lie entirely on or above the $x$-axis since absolute value is never negative.

| $x$ | $f(x)$ |
|-----|--------|
| $-1$ | 27 |
| 0 | 13 |
| 1 | 3 |
| 2 | 3 |
| 3 | 5 |
| 4 | 3 |
| 5 | 3 |
| 6 | 13 |

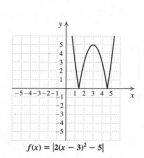

# Exercise Set 8.8

**1.** (e)

**3.** (c)

**5.** (d)

**7.** *Familiarize and Translate*. We are given the function $N(x) = -0.4x^2 + 9x + 11$.

*Carry out*. To find the value of $x$ for which $N(x)$ is a maximum, we first find $-\dfrac{b}{2a}$:

$$-\frac{b}{2a} = -\frac{9}{2(-0.4)} = 11.25$$

Now we find the maximum value of the function $N(11.25)$:

$$N(11.25) = -0.4(11.25)^2 + 9(11.25) + 11 = 61.625$$

*Check*. We can go over the calculations again. We could also solve the problem again by completing the square. The answer checks.

*State*. Daily ticket sales will peak 11 days after the concert is announced. About 62 tickets will be sold that day.

**9.** *Familiarize and Translate*. We want to find the value of $x$ for which $C(x) = 0.1x^2 - 0.7x + 2.425$ is a minimum.

*Carry out*. We complete the square.

$$C(x) = 0.1(x^2 - 7x + 12.25) + 2.425 - 1.225$$
$$C(x) = 0.1(x - 3.5)^2 + 1.2$$

The minimum function value of 1.2 occurs when $x = 3.5$.

*Check*. Check a function value for $x$ less than 3.5 and for $x$ greater than 3.5.

$$C(3) = 0.1(3)^2 - 0.7(3) + 2.425 = 1.225$$
$$C(4) = 0.1(4)^2 - 0.7(4) + 2.425 = 1.225$$

Since 1.2 is less than these numbers, it looks as though we have a minimum.

*State*. The minimum average cost is $1.2 hundred, or $120. To achieve the minimum cost, 3.5 hundred, or 350 Dobros should be built.

**11.** *Familiarize*. We make a drawing and label it.

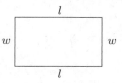

Perimeter: $2l + 2w = 128$ in.

Area: $A = l \cdot w$

*Translate*. We have a system of equations.

$$2l + 2w = 128,$$
$$A = lw$$

*Carry out*. Solving the first equation for $l$, we get $l = 64 - w$. Substituting for $l$ in the second equation we get a quadratic function $A$:

$$A = (64 - w)w$$
$$A = -w^2 + 64w$$

Completing the square, we get

$$A = -(w - 32)^2 + 1024.$$

The maximum function value is 1024. It occurs when $w$ is 32. When $w = 32$, $l = 64 - 32$, or 32.

**Check**. We check a function value for $w$ less than 32 and for $w$ greater than 32.

$$A(31) = -31^2 + 64 \cdot 31 = 1023$$
$$A(33) = -33^2 + 64 \cdot 33 = 1023$$

Since 1024 is greater than these numbers, it looks as though we have a maximum.

**State**. The maximum area occurs when the dimensions are 32 in. by 32 in.

**13. Familiarize**. We make a drawing and label it.

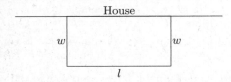

House

**Translate**. We have two equations.

$$l + 2w = 60,$$
$$A = lw$$

**Carry out**. Solve the first equation for $l$.

$$l = 60 - 2w$$

Substitute for $l$ in the second equation.

$$A = (60 - 2w)w$$
$$A = -2w^2 + 60w$$

Completing the square, we get

$$A = -2(w - 15)^2 + 450.$$

The maximum function value of 450 occurs when $w = 15$. When $w = 15$, $l = 60 - 2 \cdot 15 = 30$.

**Check**. Check a function value for $w$ less than 15 and for $w$ greater than 15.

$$A(14) = -2 \cdot 14^2 + 60 \cdot 14 = 448$$
$$A(16) = -2 \cdot 16^2 + 60 \cdot 16 = 448$$

Since 450 is greater than these numbers, it looks as though we have a maximum.

**State**. The maximum area of 450 ft$^2$ will occur when the dimensions are 15 ft by 30 ft.

**15. Familiarize**. Let $x$ represent the height of the file and $y$ represent the width. We make a drawing.

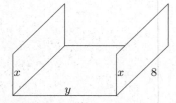

**Translate**. We have two equations.

$$2x + y = 14$$
$$V = 8xy$$

**Carry out**. Solve the first equation for $y$.

$$y = 14 - 2x$$

Substitute for $y$ in the second equation.

$$V = 8x(14 - 2x)$$
$$V = -16x^2 + 112x$$

Completing the square, we get

$$V = -16\left(x - \frac{7}{2}\right)^2 + 196.$$

The maximum function value of 196 occurs when $x = \frac{7}{2}$. When $x = \frac{7}{2}$, $y = 14 - 2 \cdot \frac{7}{2} = 7$.

**Check**. Check a function value for $x$ less than $\frac{7}{2}$ and for $x$ greater than $\frac{7}{2}$.

$$V(3) = -16 \cdot 3^2 + 112 \cdot 3 = 192$$
$$V(4) = -16 \cdot 4^2 + 112 \cdot 4 = 192$$

Since 196 is greater than these numbers, it looks as though we have a maximum.

**State**. The file should be $\frac{7}{2}$ in., or 3.5 in., tall.

**17. Familiarize**. We let $x$ and $y$ represent the numbers, and we let $P$ represent their product.

**Translate**. We have two equations.

$$x + y = 18,$$
$$P = xy$$

**Carry out**. Solving the first equation for $y$, we get $y = 18 - x$. Substituting for $y$ in the second equation we get a quadratic function $P$:

$$P = x(18 - x)$$
$$P = -x^2 + 18x$$

Completing the square, we get

$$P = -(x - 9)^2 + 81.$$

The maximum function value is 81. It occurs when $x = 9$. When $x = 9$, $y = 18 - 9$, or 9.

**Check**. We can check a function value for $x$ less than 9 and for $x$ greater than 9.

$$P(10) = -10^2 + 18 \cdot 10 = 80$$
$$P(8) = -8^2 + 18 \cdot 8 = 80$$

Since 81 is greater than these numbers, it looks as though we have a maximum.

**State**. The maximum product of 81 occurs for the numbers 9 and 9.

**19. Familiarize**. We let $x$ and $y$ represent the two numbers, and we let $P$ represent their product.

**Translate**. We have two equations.

$$x - y = 8,$$
$$P = xy$$

**Carry out**. Solve the first equation for $x$.

$$x = 8 + y$$

Substitute for $x$ in the second equation.

$$P = (8 + y)y$$
$$P = y^2 + 8y$$

Completing the square, we get
$$P = (y + 4)^2 - 16.$$

The minimum function value is $-16$. It occurs when $y = -4$. When $y = -4$, $x = 8 + (-4)$, or 4.

**Check**. Check a function value for $y$ less than $-4$ and for $y$ greater than $-4$.
$$P(-5) = (-5)^2 + 8(-5) = -15$$
$$P(-3) = (-3)^2 + 8(-3) = -15$$

Since $-16$ is less than these numbers, it looks as though we have a minimum.

**State**. The minimum product of $-16$ occurs for the numbers 4 and $-4$.

**21.** From the results of Exercises 17 and 18, we might observe that the numbers are $-5$ and $-5$ and that the maximum product is 25. We could also solve this problem as follows.

**Familiarize**. We let $x$ and $y$ represent the two numbers, and we let $P$ represent their product.

**Translate**. We have two equations.
$$x + y = -10,$$
$$P = xy$$

**Carry out**. Solve the first equation for $y$.
$$y = -10 - x$$

Substitute for $y$ in the second equation.
$$P = x(-10 - x)$$
$$P = -x^2 - 10x$$

Completing the square, we get
$$P = -(x + 5)^2 + 25$$

The maximum function value is 25. It occurs when $x = -5$. When $x = -5$, $y = -10 - (-5)$, or $-5$.

**Check**. Check a function value for $x$ less than $-5$ and for $x$ greater than $-5$.
$$P(-6) = -(-6)^2 - 10(-6) = 24$$
$$P(-4) = -(-4)^2 - 10(-4) = 24$$

Since 25 is greater than these numbers, it looks as though we have a maximum.

**State**. The maximum product of 25 occurs for the numbers $-5$ and $-5$.

**23.** The data points rise. The graph does not appear to represent a quadratic function in which the data points would rise and then fall or vice versa. Thus a linear function $f(x) = mx + b$ might be used to model the data.

**25.** The data points rise and then fall. The graph appears to represent a quadratic function that opens downward. Thus a quadratic function $f(x) = ax^2 + bx + c$, $a < 0$, might be used to model the data.

**27.** The data points fall and then rise. The graph appears to represent a quadratic function that opens upward. Thus a quadratic function $f(x) = ax^2 + bx + c$, $a > 0$, might be used to model the data.

**29.** The data points rise and then fall. The graph appears to represent a quadratic function that opens downward. Thus a quadratic function $f(x) = ax^2 + bx + c$, $a < 0$, might be used to model the data.

**31.** The data points appear to represent the right half of a quadratic function that opens upward. Thus a quadratic function $f(x) = ax^2 + bx + c$, $a > 0$, might be used to model the data.

**33.** The data points rise. The graph does not appear to represent a quadratic function in which the data points would rise and then fall or vice versa. Thus a linear function $f(x) = mx + b$ might be used to model the data.

**35.** We look for a function of the form $f(x) = ax^2 + bx + c$. Substituting the data points, we get
$$4 = a(1)^2 + b(1) + c,$$
$$-2 = a(-1)^2 + b(-1) + c,$$
$$13 = a(2)^2 + b(2) + c,$$
or
$$4 = a + b + c,$$
$$-2 = a - b + c,$$
$$13 = 4a + 2b + c.$$

Solving this system, we get
$$a = 2, \ b = 3, \text{ and } c = -1.$$

Therefore the function we are looking for is
$$f(x) = 2x^2 + 3x - 1.$$

**37.** We look for a function of the form $f(x) = ax^2 + bx + c$. Substituting the data points, we get
$$0 = a(2)^2 + b(2) + c,$$
$$3 = a(4)^2 + b(4) + c,$$
$$-5 = a(12)^2 + b(12) + c,$$
or
$$0 = 4a + 2b + c,$$
$$3 = 16a + 4b + c,$$
$$-5 = 144a + 12b + c.$$

Solving this system, we get
$$a = -\frac{1}{4}, \ b = 3, \ c = -5.$$

Therefore the function we are looking for is
$$f(x) = -\frac{1}{4}x^2 + 3x - 5.$$

**39.** a) **Familiarize**. We look for a function of the form $A(s) = as^2 + bs + c$, where $A(s)$ represents the number of nighttime accidents (for every 200 million km) and $s$ represents the travel speed (in km/h).

**Translate**. We substitute the given values of $s$ and $A(s)$.
$$400 = a(60)^2 + b(60) + c,$$
$$250 = a(80)^2 + b(80) + c,$$
$$250 = a(100)^2 + b(100) + c,$$
or

$$400 = 3600a + 60b + c,$$
$$250 = 6400a + 80b + c,$$
$$250 = 10,000a + 100b + c.$$

**Carry out**. Solving the system of equations, we get

$$a = \frac{3}{16}, \ b = -\frac{135}{4}, \ c = 1750.$$

**Check**. Recheck the calculations.

**State**. The function

$$A(s) = \frac{3}{16}s^2 - \frac{135}{4}s + 1750 \text{ fits the data.}$$

b) Find $A(50)$.

$$A(50) = \frac{3}{16}(50)^2 - \frac{135}{4}(50) + 1750 = 531.25$$

About 531 accidents occur at 50 km/h.

**41. Familiarize**. Think of a coordinate system placed on the drawing in the text with the origin at the point where the arrow is released. Then three points on the arrow's parabolic path are $(0,0)$, $(63,27)$, and $(126,0)$. We look for a function of the form $h(d) = ad^2 + bd + c$, where $h(d)$ represents the arrow's height and $d$ represents the distance the arrow has traveled horizontally.

**Translate**. We substitute the values given above for $d$ and $h(d)$.

$$0 = a \cdot 0^2 + b \cdot 0 + c,$$
$$27 = a \cdot 63^2 + b \cdot 63 + c,$$
$$0 = a \cdot 126^2 + b \cdot 126 + c$$

or

$$0 = c,$$
$$27 = 3969a + 63b + c,$$
$$0 = 15,876a + 126b + c$$

**Carry out**. Solving the system of equations, we get $a \approx -0.0068$, $b \approx 0.8571$, and $c = 0$.

**Check**. Recheck the calculations.

**State**. The function $h(d) = -0.0068d^2 + 0.8571d$ expresses the arrow's height as a function of the distance it has traveled horizontally.

**43. Writing Exercise**

**45.**
$$\frac{x}{x^2 + 17x + 72} - \frac{8}{x^2 + 15x + 56}$$
$$= \frac{x}{(x+8)(x+9)} - \frac{8}{(x+8)(x+7)}$$
$$= \frac{x}{(x+8)(x+9)} \cdot \frac{x+7}{x+7} - \frac{8}{(x+8)(x+7)} \cdot \frac{x+9}{x+9}$$
$$= \frac{x(x+7) - 8(x+9)}{(x+8)(x+9)(x+7)}$$
$$= \frac{x^2 + 7x - 8x - 72}{(x+8)(x+9)(x+7)}$$
$$= \frac{x^2 - x - 72}{(x+8)(x+9)(x+7)} = \frac{(x-9)(x+8)}{(x+8)(x+9)(x+7)}$$
$$= \frac{x-9}{(x+9)(x+7)}$$

**47.**
$$\frac{t^2 - 4}{t^2 - 7t - 8} \cdot \frac{t^2 - 64}{t^2 - 5t + 6} = \frac{(t^2-4)(t^2-64)}{(t^2-7t-8)(t^2-5t+6)}$$
$$= \frac{(t+2)(t-2)(t+8)(t-8)}{(t+1)(t-8)(t-2)(t-3)}$$
$$= \frac{(t+2)(t-2)(t+8)(t-8)}{(t+1)(t-8)(t-2)(t-3)}$$
$$= \frac{(t+2)(t+8)}{(t+1)(t-3)}$$

**49.**  $5x - 9 < 31$
$$5x < 40$$
$$x < 8$$

The solutions set is $\{x | x < 8\}$, or $(-\infty, 8)$.

**51. Writing Exercise**

**53. Familiarize**. Position the bridge on a coordinate system as shown with the vertex of the parabola at $(0, 30)$.

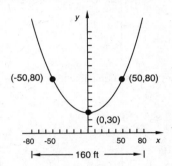

We find a function of the form $y = ax^2 + bx + c$ which represents the parabola containing the points $(0, 30)$, $(-50, 80)$, and $(50, 80)$.

**Translate**. Substitute for $x$ and $y$.

$$30 = a \cdot 0^2 + b \cdot 0 + c,$$
$$80 = a(-50)^2 + b(-50) + c,$$
$$80 = a(50)^2 + b(50) + c,$$

or

$$30 = c,$$
$$80 = 2500a - 50b + c,$$
$$80 = 2500a + 50b + c.$$

**Carry out**. Solving the system of equations, we get $a = 0.02$, $b = 0$, $c = 30$.

The function $y = 0.02x^2 + 30$ represents the parabola.

Because the cable supports are 160 ft apart, the tallest supports are positioned 160/2, or 80 ft, to the left and right of the midpoint. This means that the longest vertical cables occur at $x = -80$ and $x = 80$. For $x = \pm 80$,

$$y = 0.02(\pm 80)^2 + 30$$
$$= 128 + 30$$
$$= 158 \text{ ft}$$

**Check**. We go over the calculations.

**State**. The longest vertical cables are 158 ft long.

**55. *Familiarize*.** Let $x$ represent the number of 25¢ increases in the admission price. Then $10 + 0.25x$ represents the admission price, and $80 - x$ represents the corresponding average attendance. Let $R$ represent the total revenue.

***Translate*.** Since the total revenue is the product of the cover charge and the number attending a show, we have the following function for the amount of money the owner makes.

$$R(x) = (10 + 0.25x)(80 - x), \text{ or}$$
$$R(x) = -0.25x^2 + 10x + 800$$

***Carry out*.** Completing the square, we get

$$R(x) = -0.25(x - 20)^2 + 900$$

The maximum function value of 900 occurs when $x = 20$. The owner should charge $10 + \$0.25(20)$, or \$15.

***Check*.** We check a function value for $x$ less than 20 and for $x$ greater than 20.

$$R(19) = -0.25(19)^2 + 10 \cdot 19 + 800 = 899.75$$
$$R(21) = -0.25(21)^2 + 10 \cdot 21 + 800 = 899.75$$

Since 900 is greater than these numbers, it looks as though we have a maximum.

***State*.** The owner should charge \$15.

**57. *Familiarize*.** We add labels to the drawing in the text.

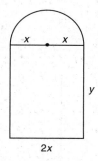

The perimeter of the semicircular portion of the window is $\frac{1}{2} \cdot 2\pi x$, or $\pi x$. The perimeter of the rectangular portion is $y + 2x + y$, or $2x + 2y$. The area of the semicircular portion of the window is $\frac{1}{2} \cdot \pi x^2$, or $\frac{\pi}{2}x^2$. The area of the rectangular portion is $2xy$.

***Translate*.** We have two equations, one giving the perimeter of the window and the other giving the area.

$$\pi x + 2x + 2y = 24,$$
$$A = \frac{\pi}{2}x^2 + 2xy$$

***Carry out*.** Solve the first equation for $y$.

$$\pi x + 2x + 2y = 24$$
$$2y = 24 - \pi x - 2x$$
$$y = 12 - \frac{\pi x}{2} - x$$

Substitute for $y$ in the second equation.

$$A = \frac{\pi}{2}x^2 + 2x\left(12 - \frac{\pi x}{2} - x\right)$$
$$A = \frac{\pi}{2}x^2 + 24x - \pi x^2 - 2x^2$$
$$A = -2x^2 - \frac{\pi}{2}x^2 + 24x$$
$$A = -\left(2x + \frac{\pi}{2}\right)x^2 + 24x$$

Completing the square, we get

$$A = -\left(2 + \frac{\pi}{2}\right)\left(x^2 + \frac{24}{-\left(2 + \frac{\pi}{2}\right)}x\right)$$
$$A = -\left(2 + \frac{\pi}{2}\right)\left(x^2 - \frac{48}{4 + \pi}x\right)$$
$$A = -\left(2 + \frac{\pi}{2}\right)\left(x - \frac{24}{4 + \pi}\right)^2 + \left(\frac{24}{4 + \pi}\right)^2$$

The maximum function value occurs when $x = \dfrac{24}{4 + \pi}$. When $x = \dfrac{24}{4 + \pi}$,

$$y = 12 - \frac{\pi}{2}\left(\frac{24}{4 + \pi}\right) - \frac{24}{4 + \pi} =$$
$$\frac{48 + 12\pi}{4 + \pi} - \frac{12\pi}{4 + \pi} - \frac{24}{4 + \pi} = \frac{24}{4 + \pi}.$$

***Check*.** Recheck the calculations.

***State*.** The radius of the circular portion of the window and the height of the rectangular portion should each be $\dfrac{24}{4 + \pi}$ ft.

**59. a)** Enter the data and use the quadratic regression operation on a graphing calculator. We get
$$c(x) = 261.875x^2 - 882.5642857x + 2134.571429,$$
where $x$ is the number of years after 1992.

**b)** In 2008, $x = 2008 - 1992 = 16$.

$$c(16) \approx 55,053 \text{ cars (rounding down)}$$

---

# Exercise Set 8.9

---

**1.** The solutions of $(x - 3)(x + 2) = 0$ are 3 and $-2$ and for a test value in $[-2, 3]$, say 0, $(x - 3)(x + 2)$ is negative so the statement is true. (Note that the endpoints must be included in the solution set because the inequality symbol is $\leq$.)

**3.** The solutions of $(x - 1)(x - 6) = 0$ are 1 and 6. For a value of $x$ less than 1, say 0, $(x - 1)(x - 6)$ is positive; for a value of $x$ greater than 6, say 7, $(x - 1)(x - 6)$ is also positive. Thus the statement is true. (Note that the endpoints of the intervals are not included because the inequality symbol is $>$.)

**5.** Since $x - 5 = 0$ when $x = 5$ and $x + 4 = 0$ when $x = -4$, the statement is true.

**7.** The only critical point is 5 and for a value of $x$ in $[5, \infty)$, say 6, $\dfrac{3}{x - 5}$ is positive, so the statement is false.

**9.** $(x+4)(x-3) < 0$

The solutions of $(x+4)(x-3) = 0$ are $-4$ and $3$. They are not solutions of the inequality, but they divide the real-number line in a natural way. The product $(x+4)(x-3)$ is positive or negative, for values other than $-4$ and $3$, depending on the signs of the factors $x+4$ and $x-3$.

$x+4 > 0$ when $x > -4$ and $x+4 < 0$ when $x < -4$.

$x-3 > 0$ when $x > 3$ and $x-3 < 0$ when $x < 3$.

We make a diagram.

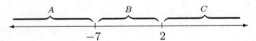

For the product $(x+4)(x-3)$ to be negative, one factor must be positive and the other negative. We see from the diagram that numbers satisfying $-4 < x < 3$ are solutions. The solution set of the inequality is $(-4, 3)$ or $\{x| -4 < x < 3\}$.

**11.** $(x+7)(x-2) \geq 0$

The solutions of $(x+7)(x-2) = 0$ are $-7$ and $2$. They divide the number line into three intervals as shown:

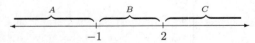

We try test numbers in each interval.

$A$: Test $-8$, $f(-8) = (-8+7)(-8-2) = 10$

$B$: Test $0$, $f(0) = (0+7)(0-2) = -14$

$C$: Test $3$, $f(3) = (3+7)(3-2) = 10$

Since $f(-8)$ and $f(3)$ are positive, the function value will be positive for all numbers in the intervals containing $-8$ and $3$. The inequality symbol is $\leq$, so we need to include the endpoints. The solution set is $(-\infty, -7] \cup [2, \infty)$, or $\{x|x \leq -7 \ or \ x \geq 2\}$.

**13.** $x^2 - x - 2 > 0$

$(x+1)(x-2) > 0$    Factoring

The solutions of $(x+1)(x-2) = 0$ are $-1$ and $2$. They divide the number line into three intervals as shown:

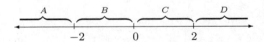

We try test numbers in each interval.

$A$: Test $-2$, $f(-2) = (-2+1)(-2-2) = 4$

$B$: Test $0$, $f(0) = (0+1)(0-2) = -2$

$C$: Test $3$, $f(3) = (3+1)(3-2) = 4$

Since $f(-2)$ and $f(3)$ are positive, the function value will be positive for all numbers in the intervals containing $-2$ and $3$. The solution set is $(-\infty, -1) \cup (2, \infty)$, or $\{x|x < -1 \ or \ x > 2\}$.

**15.** $x^2 + 4x + 4 < 0$

$(x+2)^2 < 0$

Observe that $(x+2)^2 \geq 0$ for all values of $x$. Thus, the solution set is $\emptyset$.

**17.** $\qquad x^2 - 4x < 12$

$\qquad x^2 - 4x - 12 < 0$

$\qquad (x-6)(x+2) < 0$

The solutions of $(x-6)(x+2) = 0$ are $6$ and $-2$. They are not solutions of the inequality, but they divide the real-number line in a natural way. The product $(x-6)(x+2)$ is positive or negative, for values other than $6$ and $-2$, depending on the signs of the factors $x-6$ and $x+2$.

$x-6 > 0$ when $x > 6$ and $x-6 < 0$ when $x < 6$.

$x+2 > 0$ when $x > -2$ and $x+2 < 0$ when $x < -2$.

We make a diagram.

For the product $(x-6)(x+2)$ to be negative, one factor must be positive and the other negative. The only situation in the diagram for which this happens is when $-2 < x < 6$. The solution set of the inequality is $(-2, 6)$, or $\{x| -2 < x < 6\}$.

**19.** $3x(x+2)(x-2) < 0$

The solutions of $3x(x+2)(x-2) = 0$ are $0$, $-2$, and $2$. They divide the real-number line into four intervals as shown:

We try test numbers in each interval.

$A$: Test $-3$, $f(-3) = 3(-3)(-3+2)(-3-2) = -45$

$B$: Test $-1$, $f(-1) = 3(-1)(-1+2)(-1-2) = 9$

$C$: Test $1$, $f(1) = 3(1)(1+2)(1-2) = -9$

$D$: Test $3$, $f(3) = 3(3)(3+2)(3-2) = 45$

Since $f(-3)$ and $f(1)$ are negative, the function value will be negative for all numbers in the intervals containing $-3$ and $1$. The solution set is $(-\infty, -2) \cup (0, 2)$, or $\{x|x < -2 \ or \ 0 < x < 2\}$.

**21.** $(x-1)(x+2)(x-4) \geq 0$

The solutions of $(x-1)(x+2)(x-4) = 0$ are $1$, $-2$, and $4$. They divide the real-number line in a natural way. The product $(x-1)(x+2)(x-4)$ is positive or negative depending on the signs of $x-1$, $x+2$, and $x-4$.

| Sign of $x-1$ | $-$ | $\|$ | $-$ | $\|$ | $+$ | $\|$ | $+$ |
| Sign of $x+2$ | $-$ | $\|$ | $+$ | $\|$ | $+$ | $\|$ | $+$ |
| Sign of $x-4$ | $-$ | $\|$ | $-$ | $\|$ | $-$ | $\|$ | $+$ |
| Sign of product | $-$ | $\|$ | $+$ | $\|$ | $-$ | $\|$ | $+$ |

A product of three numbers is positive when all three factors are positive or when two are negative and one is positive. Since the $\geq$ symbol allows for equality, the endpoints $-2$, 1, and 4 are solutions. From the chart we see that the solution set is $[-2,1]\cup[4,\infty)$, or $\{x|-2 \leq x \leq 1 \ or \ x \geq 4\}$.

**23.**
$$f(x) \leq 3$$
$$x^2 - 1 \leq 3$$
$$x^2 - 4 \leq 0$$
$$(x-2)(x+2) \leq 0$$

The solutions of $(x-2)(x+2) = 0$ are 2 and $-2$. They divide the real-number line as shown below.

| Sign of $x-2$ | $-$ | $\|$ | $-$ | $\|$ | $+$ |
| Sign of $x+2$ | $-$ | $\|$ | $+$ | $\|$ | $+$ |
| Sign of product | $+$ | $\|$ | $-$ | $\|$ | $+$ |

Because the inequality symbol is $\leq$, we must include the endpoints in the solution set. From the chart we see that the solution set is $[-2,2]$, or $\{x|-2 \leq x \leq 2\}$.

**25.**
$$g(x) > 0$$
$$(x-2)(x-3)(x+1) > 0$$

The solutions of $(x-2)(x-3)(x+1) = 0$ are 2, 3, and $-1$. They divide the real-number line into four intervals as shown below.

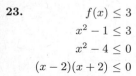

We try test numbers in each interval.

A: Test $-2$, $f(-2) = (-2-2)(-2-3)(-2+1) = -20$

B: Test 0, $f(0) = (0-2)(0-3)(0+1) = 6$

C: Test $\dfrac{5}{2}$, $f\left(\dfrac{5}{2}\right) = \left(\dfrac{5}{2}-2\right)\left(\dfrac{5}{2}-3\right)\left(\dfrac{5}{2}+1\right) = -\dfrac{7}{8}$

D: Test 4, $f(4) = (4-2)(4-3)(4+1) = 10$

The function value will be positive for all numbers in intervals B and D. The solution set is $(-1,2) \cup (3,\infty)$, or $\{x|-1 < x < 2 \ or \ x > 3\}$.

**27.**
$$F(x) \leq 0$$
$$x^3 - 7x^2 + 10x \leq 0$$
$$x(x^2 - 7x + 10) \leq 0$$
$$x(x-2)(x-5) \leq 0$$

The solutions of $x(x-2)(x-5) = 0$ are 0, 2, and 5. They divide the real-number line as shown below.

| Sign of $x$ | $-$ | $\|$ | $+$ | $\|$ | $+$ | $\|$ | $+$ |
| Sign of $x-2$ | $-$ | $\|$ | $-$ | $\|$ | $+$ | $\|$ | $+$ |
| Sign of $x-5$ | $-$ | $\|$ | $-$ | $\|$ | $-$ | $\|$ | $+$ |
| Sign of product | $-$ | $\|$ | $+$ | $\|$ | $-$ | $\|$ | $+$ |

Because the inequality symbol is $\leq$ we must include the endpoints in the solution set. From the chart we see that the solution set is $(-\infty, 0]\cup[2,5]$ or $\{x|x \leq 0 \ or \ 2 \leq x \leq 5\}$.

**29.** $\dfrac{1}{x+5} < 0$

We write the related equation by changing the $<$ symbol to $=$:
$$\frac{1}{x+5} = 0$$

We solve the related equation.
$$(x+5) \cdot \frac{1}{x+5} = (x+5) \cdot 0$$
$$1 = 0$$

The related equation has no solution.

Next we find the values that make the denominator 0 by setting the denominate equal to 0 and solving:
$$x + 5 = 0$$
$$x = -5$$

We use $-5$ to divide the number line into two intervals as shown:

We try test numbers in each interval.

A: Test $-6$, $\dfrac{1}{-6+5} = \dfrac{1}{-1} = -1 < 0$

The number $-6$ is a solution of the inequality, so the interval A is part of the solution set.

B: Test 0, $\dfrac{1}{0+5} = \dfrac{1}{5} \not< 0$

The number 0 is not a solution of the inequality, so the interval B is not part of the solution set. The solution set is $(-\infty, -5)$, or $\{x|x < -5\}$.

**31.** $\dfrac{x+1}{x-3} \geq 0$

Solve the related equation.
$$\frac{x+1}{x-3} = 0$$
$$x + 1 = 0$$
$$x = -1$$

Find the values that make the denominator 0.
$$x - 3 = 0$$
$$x = 3$$

Use the numbers $-1$ and 3 to divide the number line into intervals as shown:

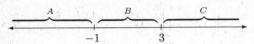

Try test numbers in each interval.

A: Test $-2$, $\dfrac{-2+1}{-2-3} = \dfrac{-1}{-5} = \dfrac{1}{5} > 0$

The number $-2$ is a solution of the inequality, so the interval $A$ is part of the solution set.

B: Test $0$, $\dfrac{0+1}{0-3} = \dfrac{1}{-3} = -\dfrac{1}{3} \not> 0$

The number $0$ is not a solution of the inequality, so the interval $B$ is not part of the solution set.

C: Test $4$, $\dfrac{4+1}{4-3} = \dfrac{5}{1} = 5 > 0$

The number $4$ is a solution of the inequality, so the interval $C$ is part of the solution set.

The solution set includes intervals $A$ and $C$. The number $-1$ is also included since the inequality symbol is $\geq$ and $-1$ is the solution of the related equation. The number $3$ is not included since $\dfrac{x+1}{x-3}$ is undefined for $x = 3$. The solution set is $(-\infty, -1] \cup (3, \infty)$, or $\{x | x \leq -1 \ or \ x > 3\}$.

**33.** $\dfrac{x+1}{x+6} \geq 1$

Solve the related equation.

$$\dfrac{x+1}{x+6} = 1$$
$$x + 1 = x + 6$$
$$1 = 6$$

The related equation has no solution.

Find the values that make the denominator 0.

$$x + 6 = 0$$
$$x = -6$$

Use the number $-6$ to divide the number line into two intervals.

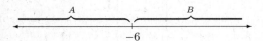

Try test numbers in each interval.

A: Test $-7$, $\dfrac{-7+1}{-7+6} = \dfrac{-6}{-1} = 6 \geq 1$.

The number $-7$ is a solution of the inequality, so the interval $A$ is part of the solution set.

B: Test $0$, $\dfrac{0+1}{0+6} = \dfrac{1}{6} \not\geq 1$

The number $0$ is not a solution of the inequality, so the interval $B$ is not part of the solution set. The number $-6$ is not included in the solution set since $\dfrac{x+1}{x+6}$ is undefined for $x = -6$.

The solution set is $(-\infty, -6)$, or $\{x | x < -6\}$.

**35.** $\dfrac{(x-2)(x+1)}{x-5} \leq 0$

Solve the related equation.

$$\dfrac{(x-2)(x+1)}{x-5} = 0$$
$$(x-2)(x+1) = 0$$
$$x = 2 \ \text{or} \ x = -1$$

Find the values that make the denominator 0.

$$x - 5 = 0$$
$$x = 5$$

Use the numbers $2$, $-1$, and $5$ to divide the number line into intervals as shown:

Try test numbers in each interval.

A: Test $-2$, $\dfrac{(-2-2)(-2+1)}{-2-5} = \dfrac{-4(-1)}{-7} =$

$-\dfrac{4}{7} \leq 0$

Interval $A$ is part of the solution set.

B: Test $0$, $\dfrac{(0-2)(0+1)}{0-5} = \dfrac{-2 \cdot 1}{-5} = \dfrac{2}{5} \not\leq 0$

Interval $B$ is not part of the solution set.

C: Test $3$, $\dfrac{(3-2)(3+1)}{3-5} = \dfrac{1 \cdot 4}{-2} = -2 \leq 0$

Interval $C$ is part of the solution set.

D: Test $6$, $\dfrac{(6-2)(6+1)}{6-5} = \dfrac{4 \cdot 7}{1} = 28 \not\leq 0$

Interval $D$ is not part of the solution set.

The solution set includes intervals $A$ and $C$. The numbers $-1$ and $2$ are also included since the inequality symbol is $\leq$ and $-1$ and $2$ are the solutions of the related equation. The number $5$ is not included since $\dfrac{(x-2)(x+1)}{x-5}$ is undefined for $x = 5$.

The solution set is $(-\infty, -1] \cup [2, 5)$, or $\{x | x \leq -1 \ or \ 2 \leq x < 5\}$.

**37.** $\dfrac{x}{x+3} \geq 0$

Solve the related equation.

$$\dfrac{x}{x+3} = 0$$
$$x = 0$$

Find the values that make the denominator 0.

$$x + 3 = 0$$
$$x = -3$$

Use the numbers $0$ and $-3$ to divide the number line into intervals as shown.

Try test numbers in each interval.

$A$: Test $-4$, $\dfrac{-4}{-4+3} = \dfrac{-4}{-1} = 4 \geq 0$

Interval $A$ is part of the solution set.

$B$: Test $-1$, $\dfrac{-1}{-1+3} = \dfrac{-1}{2} = -\dfrac{1}{2} \not\geq 0$

Interval $B$ is not part of the solution set.

$C$: Test $1$, $\dfrac{1}{1+3} = \dfrac{1}{4} \geq 0$

The interval $C$ is part of the solution set.

The solution set includes intervals $A$ and $C$. The number $0$ is also included since the inequality symbol is $\geq$ and $0$ is the solution of the related equation. The number $-3$ is not included since $\dfrac{x}{x+3}$ is undefined for $x = -3$. The solution set is $(-\infty, -3) \cup [0, \infty)$, or $\{x | x < -3 \text{ or } x \geq 0\}$.

**39.** $\dfrac{x-5}{x} < 1$

Solve the related equation.

$$\dfrac{x-5}{x} = 1$$
$$x - 5 = x$$
$$-5 = 0$$

The related equation has no solution.

Find the values that make the denominator 0.

$$x = 0$$

Use the number 0 to divide the number line into two intervals as shown.

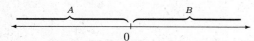

Try test numbers in each interval.

$A$: Test $-1$, $\dfrac{-1-5}{-1} = \dfrac{-6}{-1} = 6 \not< 1$

Interval $A$ is not part of the solution set.

$B$: Test $1$, $\dfrac{1-5}{1} = \dfrac{-4}{1} = -4 < 1$

Interval $B$ is part of the solution set.

The solution set is $(0, \infty)$ or $\{x | x > 0\}$.

**41.** $\dfrac{x-1}{(x-3)(x+4)} \leq 0$

Solve the related equation.

$$\dfrac{x-1}{(x-3)(x+4)} = 0$$
$$x - 1 = 0$$
$$x = 1$$

Find the values that make the denominator 0.

$$(x-3)(x+4) = 0$$
$$x = 3 \text{ or } x = -4$$

Use the numbers 1, 3, and $-4$ to divide the number line into intervals as shown:

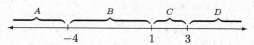

Try test numbers in each interval.

$A$: Test $-5$, $\dfrac{-5-1}{(-5-3)(-5+4)} = \dfrac{-6}{-8(-1)} =$

$-\dfrac{3}{4} < 0$

Interval $A$ is part of the solution set.

$B$: Test $0$, $\dfrac{0-1}{(0-3)(0+4)} = \dfrac{-1}{-3 \cdot 4} = \dfrac{1}{12} \not< 0$

Interval $B$ is not part of the solution set.

$C$: Test $2$, $\dfrac{2-1}{(2-3)(2+4)} = \dfrac{1}{-1 \cdot 6} = -\dfrac{1}{6} < 0$

Interval $C$ is part of the solution set.

$D$: Test $4$, $\dfrac{4-1}{(4-3)(4+4)} = \dfrac{3}{1 \cdot 8} = \dfrac{3}{8} \not< 0$

Interval $D$ is not part of the solution set.

The solution set includes intervals $A$ and $C$. The number 1 is also included since the inequality symbol is $\leq$ and 1 is the solution of the related equation. The numbers $-4$ and 3 are not included since $\dfrac{x-1}{(x-3)(x+4)}$ is undefined for $x = -4$ and for $x = 3$.

The solution set is $(-\infty, -4) \cup [1, 3)$, or $\{x | x < -4 \text{ or } 1 \leq x < 3\}$.

**43.** $f(x) \geq 0$

$$\dfrac{5-2x}{4x+3} \geq 0$$

Solve the related equation.

$$\dfrac{5-2x}{4x+3} = 0$$
$$5 - 2x = 0$$
$$5 = 2x$$
$$\dfrac{5}{2} = x$$

Find the values that make the denominator 0.

$$4x + 3 = 0$$
$$4x = -3$$
$$x = -\dfrac{3}{4}$$

Use the numbers $\dfrac{5}{2}$ and $-\dfrac{3}{4}$ to divide the number line as shown:

Try test numbers in each interval.

A: Test $-1$, $\dfrac{5-2(-1)}{4(-1)+3} = -7 \not\geq 0$

Interval A is not part of the solution set.

B: Test $0$, $\dfrac{5-2 \cdot 0}{4 \cdot 0 + 3} = \dfrac{5}{3} \geq 0$

Interval B is part of the solution set.

C: Test 3, $\dfrac{5 - 2 \cdot 3}{4 \cdot 3 + 3} = -\dfrac{1}{15} \not\geq 0$

Interval C is not part of the solution set.

The solution set includes interval B. The number $\dfrac{5}{2}$ is also included since the inequality symbol is $\geq$ and $\dfrac{5}{2}$ is the solution of the related equation. The number $-\dfrac{3}{4}$ is not included since $\dfrac{5 - 2x}{4x + 3}$ is undefined for $x = -\dfrac{3}{4}$. The solution set is $\left( -\dfrac{3}{4}, \dfrac{5}{2} \right]$, or $\left\{ x \middle| -\dfrac{3}{4} < x \leq \dfrac{5}{2} \right\}$.

**45.**   $G(x) \leq 1$

$\dfrac{1}{x - 2} \leq 1$

Solve the related equation.

$$\dfrac{1}{x - 2} = 1$$
$$1 = x - 2$$
$$3 = x$$

Find the values of $x$ that make the denominator 0.

$$x - 2 = 0$$
$$x = 2$$

Use the numbers 2 and 3 to divide the number line as shown.

Try a test number in each interval.

A: Test 0, $\dfrac{1}{0 - 2} = -\dfrac{1}{2} \leq 1$

Interval $A$ is part of the solution set.

B: Test $\dfrac{5}{2}$, $\dfrac{1}{\dfrac{5}{2} - 2} = \dfrac{1}{\dfrac{1}{2}} = 2 \not\leq 1$

Interval $B$ is not part of the solution set.

C: Test 4, $\dfrac{1}{4 - 2} = \dfrac{1}{2} \leq 1$

Interval $C$ is part of the solution set.

The solution set includes intervals $A$ and $B$. The number 3 is also included since the inequality symbol is $\leq$ and 3 is the solution of the related equation. The number 2 is not included since $\dfrac{1}{x - 2}$ is undefined for $x = 2$. The solution set is $(-\infty, 2) \cup [3, \infty)$, or $\{x | x < 2 \text{ or } x \geq 3\}$.

**47.**  *Writing Exercise*

**49.**  $(2a^3 b^2 c^4)^3 = 2^3 (a^3)^3 (b^2)^3 (c^4)^3 = 8a^{3 \cdot 3} b^{2 \cdot 3} c^{4 \cdot 3} = 8a^9 b^6 c^{12}$

**51.**  $2^{-5} = \dfrac{1}{2^5} = \dfrac{1}{32}$

**53.**  $f(x) = 3x^2$

$f(a + 1) = 3(a + 1)^2 = 3(a^2 + 2a + 1) = 3a^2 + 6a + 3$

**55.**  *Writing Exercise*

**57.**       $x^2 + 2x < 5$

$x^2 + 2x - 5 < 0$

Using the quadratic formula, we find that the solutions of the related equation are $x = -1 \pm \sqrt{6}$. These numbers divide the real-number line into three intervals as shown:

We try test numbers in each interval.

A: Test $-4$, $f(-4) = (-4)^2 + 2(-4) - 5 = 3$

B: Test 0, $f(0) = 0^2 + 2 \cdot 0 - 5 = -5$

C: Test 2, $f(2) = 2^2 + 2 \cdot 2 - 5 = 3$

The function value will be negative for all numbers in interval $B$. The solution set is $(-1 - \sqrt{6}, -1 + \sqrt{6})$, or $\{x | -1 - \sqrt{6} < x < -1 + \sqrt{6}\}$.

**59.**   $x^4 + 3x^2 \leq 0$

$x^2 (x^2 + 3) \leq 0$

$x^2 = 0$ for $x = 0$, $x^2 > 0$ for $x \neq 0$, $x^2 + 3 > 0$ for all $x$

The solution set is $\{0\}$.

**61.** a) $-3x^2 + 630x - 6000 > 0$

$x^2 - 210x + 2000 < 0$  Multiplying by $-\dfrac{1}{3}$

$(x - 200)(x - 10) < 0$

The solutions of $f(x) = (x - 200)(x - 10) = 0$ are 200 and 10. They divide the number line as shown:

A: Test 0, $f(0) = 0^2 - 210 \cdot 0 + 2000 = 2000$

B: Test 20, $f(20) = 20^2 - 210 \cdot 20 + 2000 = -1800$

C: Test 300, $f(300) = 300^2 - 210 \cdot 300 + 2000 = 29{,}000$

The company makes a profit for values of $x$ such that $10 < x < 200$, or for values of $x$ in the interval $(10, 200)$.

b) See part (a). Keep in mind that $x$ must be nonnegative since negative numbers have no meaning in this application.

The company loses money for values of $x$ such that $0 \leq x < 10$ or $x > 200$, or for values of $x$ in the interval $[0, 10) \cup (200, \infty)$.

**63.** We find values of $n$ such that $N \geq 66$ *and* $N \leq 300$.

For $N \geq 66$:

$$\dfrac{n(n - 1)}{2} \geq 66$$
$$n(n - 1) \geq 132$$
$$n^2 - n - 132 \geq 0$$
$$(n - 12)(n + 11) \geq 0$$

The solutions of $f(n) = (n - 12)(n + 11) = 0$ are 12 and $-11$. They divide the number line as shown:

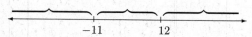

$$\begin{array}{ccc} & -11 & 12 \end{array}$$

However, only positive values of $n$ have meaning in this exercise so we need only consider the intervals shown below:

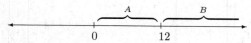

$$\begin{array}{ccc} & A & B \\ 0 & & 12 \end{array}$$

$A$: Test 1, $f(1) = 1^2 - 1 - 132 = -132$

$B$: Test 20, $f(20) = 20^2 - 20 - 132 = 248$

Thus, $N \geq 66$ for $\{n | n \geq 12\}$.

For $N \leq 300$:

$$\frac{n(n - 1)}{2} \leq 300$$

$$n(n - 1) \leq 600$$

$$n^2 - n - 600 \leq 0$$

$$(n - 25)(n + 24) \leq 0$$

The solutions of $f(n) = (n - 25)(n + 24) = 0$ are 25 and $-24$. They divide the number line as shown:

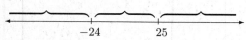

$$\begin{array}{ccc} & -24 & 25 \end{array}$$

However, only positive values of $n$ have meaning in this exercise so we need only consider the intervals shown below:

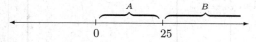

$$\begin{array}{ccc} & A & B \\ 0 & & 25 \end{array}$$

$A$: Test 1, $f(1) = 1^2 - 1 - 600 = -600$

$B$: Test 30, $f(30) = 30^2 - 30 - 600 = 270$

Thus, $N \leq 300$ (and $n > 0$) for $\{n | 0 < n \leq 25\}$.

Then $66 \leq N \leq 300$ for $\{n | n$ is an integer and $12 \leq n \leq 25\}$.

**65.** From the graph we determine the following:

The solutions of $f(x) = 0$ are $-2$, 1, and 3.

The solution of $f(x) < 0$ is $(-\infty, -2) \cup (1, 3)$, or $\{x | x < -2 \ or \ 1 < x < 3\}$.

The solution of $f(x) > 0$ is $(-2, 1) \cup (3, \infty)$, or $\{x | -2 < x < 1 \ or \ x > 3\}$.

**67.** From the graph we determine the following:

$f(x)$ has no zeros.

The solutions $f(x) < 0$ are $(-\infty, 0)$, or $\{x | x < 0\}$.

The solutions of $f(x) > 0$ are $(0, \infty)$, or $\{x | x > 0\}$.

**69.** From the graph we determine the following:

The solutions of $f(x) = 0$ are $-1$ and 0.

The solutions of $f(x) < 0$ are $(-\infty, -3) \cup (-1, 0)$, or $\{x | x < -3 \ or \ -1 < x < 0\}$.

The solutions of $f(x) > 0$ are $(-3, -1) \cup (0, 2) \cup (2, \infty)$, or $\{x | -3 < x < -1 \ or \ 0 < x < 2 \ or \ x > 2\}$.

**71.**

# Chapter 9

# Exponential and Logarithmic Functions

**1.** True; see page 585 in the text.

**3.** $(g \circ f)(x) = g(f(x)) = x^2 + 3 \neq (x+3)^2$, so the statement is false.

**5.** False; see page 588 in the text.

**7.** True; see page 589 in the text.

**9.** $(f \circ g)(1) = f(g(1)) = f(2 \cdot 1 - 3)$
$$= f(-1) = (-1)^2 + 1$$
$$= 1 + 1 = 2$$
$(g \circ f)(1) = g(f(1)) = g(1^2 + 1)$
$$= g(2) = 2 \cdot 2 - 3 = 1$$
$(f \circ g)(x) = f(g(x)) = f(2x - 3)$
$$= (2x - 3)^2 + 1$$
$$= 4x^2 - 12x + 9 + 1$$
$$= 4x^2 - 12x + 10$$
$(g \circ f)(x) = g(f(x)) = g(x^2 + 1)$
$$= 2(x^2 + 1) - 3$$
$$= 2x^2 + 2 - 3$$
$$= 2x^2 - 1$$

**11.** $(f \circ g)(1) = f(g(1)) = f(2 \cdot 1^2 - 7)$
$$= f(-5) = -5 - 3 = -8$$
$(g \circ f)(1) = g(f(1)) = g(1 - 3)$
$$= g(-2) = 2(-2)^2 - 7$$
$$= 8 - 7 = 1$$
$(f \circ g)(x) = f(g(x)) = f(2x^2 - 7)$
$$= 2x^2 - 7 - 3 = 2x^2 - 10$$
$(g \circ f)(x) = g(f(x)) = g(x - 3)$
$$= 2(x - 3)^2 - 7$$
$$= 2(x^2 - 6x + 9) - 7$$
$$= 2x^2 - 12x + 18 - 7$$
$$= 2x^2 - 12x + 11$$

**13.** $(f \circ g)(1) = f(g(1)) = f\left(\frac{1}{1^2}\right)$
$$= f(1) = 1 + 7 = 8$$
$(g \circ f)(1) = g(f(1)) = g(1 + 7)$
$$= g(8) = \frac{1}{8^2} = \frac{1}{64}$$
$(f \circ g)(x) = f(g(x))$
$$= f\left(\frac{1}{x^2}\right) = \frac{1}{x^2} + 7$$

$(g \circ f)(x) = g(f(x))$
$$= g(x + 7) = \frac{1}{(x + 7)^2}$$

**15.** $(f \circ g)(1) = f(g(1)) = f(1 + 3)$
$$= f(4) = \sqrt{4} = 2$$
$(g \circ f)(1) = g(f(1)) = g(\sqrt{1})$
$$= g(1) = 1 + 3 = 4$$
$(f \circ g)(x) = f(g(x)) = f(x + 3) = \sqrt{x + 3}$
$(g \circ f)(x) = g(f(x)) = g(\sqrt{x}) = \sqrt{x} + 3$

**17.** $(f \circ g)(1) = f(g(1)) = f\left(\frac{1}{1}\right)$
$$= f(1) = \sqrt{4 \cdot 1} = \sqrt{4} = 2$$
$(g \circ f)(1) = g(f(1)) = g(\sqrt{4 \cdot 1})$
$$= g(\sqrt{4}) = g(2) = \frac{1}{2}$$
$(f \circ g)(x) = f(g(x)) = f\left(\frac{1}{x}\right)$
$$= \sqrt{4 \cdot \frac{1}{x}} = \sqrt{\frac{4}{x}}$$
$(g \circ f)(x) = g(f(x)) = g(\sqrt{4x}) = \frac{1}{\sqrt{4x}}$

**19.** $(f \circ g)(1) = f(g(1)) = f(\sqrt{1 - 1})$
$$= f(\sqrt{0}) = f(0) = 0^2 + 4 = 4$$
$(g \circ f)(1) = g(f(1)) = g(1^2 + 4)$
$$= g(5) = \sqrt{5 - 1} = \sqrt{4} = 2$$
$(f \circ g)(x) = f(g(x)) = f(\sqrt{x - 1})$
$$= (\sqrt{x - 1})^2 + 4 = x - 1 + 4 = x + 3$$
$(g \circ f)(x) = g(f(x)) = g(x^2 + 4)$
$$= \sqrt{x^2 + 4 - 1} = \sqrt{x^2 + 3}$$

**21.** $h(x) = (7 + 5x)^2$

This is $7 + 5x$ raised to the second power, so the two most obvious functions are $f(x) = x^2$ and $g(x) = 7 + 5x$.

**23.** $h(x) = \sqrt{2x + 7}$

We have $2x + 7$ and take the square root of their expression, so the two most obvious functions are $f(x) = \sqrt{x}$ and $g(x) = 2x + 7$.

**25.** $h(x) = \dfrac{2}{x - 3}$

This is 2 divided by $x - 3$, so two functions that can be used are $f(x) = \dfrac{2}{x}$ and $g(x) = x - 3$.

**27.** The graph of $f(x) = x - 5$ is shown below.

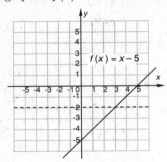

Since there is no horizontal line that crosses the graph more than once, the function is one-to-one.

**29.** $f(x) = x^2 + 1$

Observe that the graph of this function is a parabola that opens up. Thus, there are many horizontal lines that cross the graph more than once, so the function is not one-to-one. We can also draw the graph as shown below.

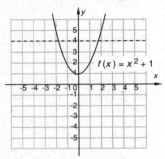

There are many horizontal lines that cross the graph more than once. In particular, the line $y = 4$ crosses the graph more than once. The function is not one-to-one.

**31.** Since there is no horizontal line that crosses the graph more than once, the function is one-to-one.

**33.** There are many horizontal lines that cross the graph more than once, so the function is not one-to-one.

**35.** a) The function $f(x) = x + 4$ is a linear function that is not constant, so it passes the horizontal-line test. Thus, $f$ is one-to-one.

b) Replace $f(x)$ by y: $y = x + 4$

Interchange $x$ and $y$: $x = y + 4$

Solve for y: $x - 4 = y$

Replace $y$ by $f^{-1}(x)$: $f^{-1}(x) = x - 4$

**37.** a) The function $f(x) = 2x$ is a linear function that is not constant, so it passes the horizontal-line test. Thus, $f$ is one-to-one.

b) Replace $f(x)$ by y: $y = 2x$

Interchange $x$ and $y$: $x = 2y$

Solve for y: $\dfrac{x}{2} = y$

Replace $y$ by $f^{-1}(x)$: $f^{-1}(x) = \dfrac{x}{2}$

**39.** a) The function $g(x) = 3x - 1$ is a linear function that is not constant, so it passes the horizontal-line test. Thus, $g$ is one-to-one.

b) Replace $g(x)$ by y: $y = 3x - 1$

Interchange variables: $x = 3y - 1$

Solve for y: $x + 1 = 3y$

$$\dfrac{x + 1}{3} = y$$

Replace $y$ by $g^{-1}(x)$: $g^{-1}(x) = \dfrac{x + 1}{3}$

**41.** a) The function $f(x) = \dfrac{1}{2}x + 1$ is a linear function that is not constant, so it passes the horizontal-line test. Thus, $f$ is one-to-one.

b) Replace $f(x)$ by y: $y = \dfrac{1}{2}x + 1$

Interchange variables: $x = \dfrac{1}{2}y + 1$

Solve for y:      $x = \dfrac{1}{2}y + 1$

$$x - 1 = \dfrac{1}{2}y$$

$$2x - 2 = y$$

Replace $y$ by $f^{-1}(x)$: $f^{-1}(x) = 2x - 2$

**43.** a) The graph of $g(x) = x^2 + 5$ is shown below. There are many horizontal lines that cross the graph more than once. For example, the line $y = 8$ crosses the graph more than once. The function is not one-to-one.

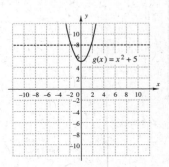

**45.** a) The function $h(x) = -2x + 4$ is a linear function that is not constant, so it passes the horizontal-line test. Thus, $h$ is one-to-one.

b) Replace $h(x)$ by y: $y = -2x + 4$

Interchange variables: $x = -2y + 4$

Solve for y:      $x = -2y + 4$

$$x - 4 = -2y$$

$$\dfrac{x - 4}{-2} = y$$

Replace $y$ by $h^{-1}(x)$: $h^{-1}(x) = \dfrac{x - 4}{-2}$

**47.** a) The graph of $f(x) = \dfrac{1}{x}$ is shown below. It passes the horizontal-line test, so the function is one-to-one.

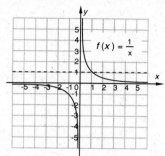

b) Replace $f(x)$ by $y$: $y = \dfrac{1}{x}$

Interchange $x$ and $y$: $x = \dfrac{1}{y}$

Solve for $y$: $xy = 1$

$$y = \dfrac{1}{x}$$

Replace $y$ by $f^{-1}(x)$: $f^{-1}(x) = \dfrac{1}{x}$

**49.** a) The graph of $G(x) = 4$ is shown below. The horizontal line $y = 4$ crosses the graph more than once, so the function is not one-to-one.

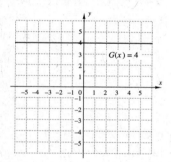

**51.** a) The function $f(x) = \dfrac{2x+1}{3} = \dfrac{2}{3}x + \dfrac{1}{3}$ is a linear function that is not constant, so it passes the horizontal-line test. Thus, $f$ is one-to-one.

b) Replace $f(x)$ by $y$: $y = \dfrac{2x+1}{3}$

Interchange $x$ and $y$: $x = \dfrac{2y+1}{3}$

Solve for $y$: $3x = 2y + 1$

$$3x - 1 = 2y$$

$$\dfrac{3x-1}{2} = y$$

Replace $y$ by $f^{-1}(x)$: $f^{-1}(x) = \dfrac{3x-1}{2}$

**53.** a) The graph of $f(x) = x^3 - 5$ is shown below. It passes the horizontal-line test, so the function is one-to-one.

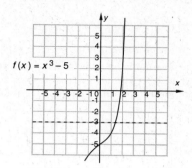

b) Replace $f(x)$ by $y$: $y = x^3 - 5$

Interchange $x$ and $y$: $x = y^3 - 5$

Solve for $y$: $x + 5 = y^3$

$$\sqrt[3]{x+5} = y$$

Replace $y$ by $f^{-1}(x)$: $f^{-1}(x) = \sqrt[3]{x+5}$

**55.** a) The graph of $g(x) = (x-2)^3$ is shown below. It passes the horizontal-line test, so the function is one-to-one.

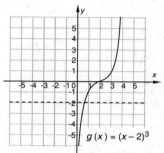

b) Replace $g(x)$ by $y$: $y = (x-2)^3$

Interchange $x$ and $y$: $x = (y-2)^3$

Solve for $y$: $\sqrt[3]{x} = y - 2$

$$\sqrt[3]{x} + 2 = y$$

Replace $y$ by $g^{-1}(x)$: $g^{-1}(x) = \sqrt[3]{x} + 2$

**57.** a) The graph of $f(x) = \sqrt{x}$ is shown below. It passes the horizontal-line test, so the function is one-to-one.

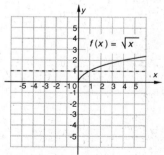

b) Replace $f(x)$ by $y$: $y = \sqrt{x}$    (Note that $f(x) \geq 0$.)

Interchange $x$ and $y$: $x = \sqrt{y}$

Solve for $y$: $x^2 = y$

Replace $y$ by $f^{-1}(x)$: $f^{-1}(x) = x^2,\ x \geq 0$

**59.** First graph $f(x) = \dfrac{2}{3}x + 4$. Then graph the inverse function by reflecting the graph of $f(x) = \dfrac{2}{3}x + 4$ across the line $y = x$. The graph of the inverse function can also be found by first finding a formula for the inverse, substituting to find function values, and then plotting points.

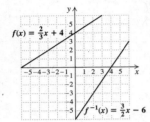

**61.** Follow the procedure described in Exercise 59 to graph the function and its inverse.

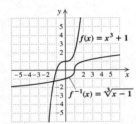

**63.** Use the procedure described in Exercise 59 to graph the function and its inverse.

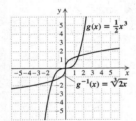

**65.** Use the procedure described in Exercise 59 to graph the function and its inverse.

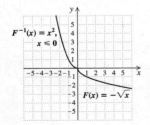

**67.** Use the procedure described in Exercise 59 to graph the function and its inverse.

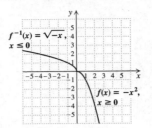

**69.** We check to see that $(f^{-1} \circ f)(x) = x$ and $(f \circ f^{-1})(x) = x$.

$(f^{-1} \circ f)(x) = f^{-1}(f(x)) = f^{-1}(\sqrt[3]{x-4}) = $
$(\sqrt[3]{x-4})^3 + 4 = x - 4 + 4 = x$
$(f \circ f^{-1})(x) = f(f^{-1}(x)) = f(x^3 + 4) = \sqrt[3]{x^3 + 4 - 4} = $
$\sqrt[3]{x^3} = x$

**71.** We check to see that $f^{-1} \circ f(x) = x$ and $f \circ f^{-1}(x) = x$.

$f^{-1} \circ f(x) = f^{-1}(f(x)) = f^{-1}\left(\dfrac{1-x}{x}\right) = \dfrac{1}{\dfrac{1-x}{x} + 1} = $

$\dfrac{1}{\dfrac{1-x}{x} + 1} \cdot \dfrac{x}{x} = \dfrac{x}{1 - x + x} = \dfrac{x}{1} = x$

$f \circ f^{-1}(x) = f(f^{-1}(x)) = f\left(\dfrac{1}{x+1}\right) = \dfrac{1 - \dfrac{1}{x+1}}{\dfrac{1}{x+1}} = $

$\dfrac{1 - \dfrac{1}{x+1}}{\dfrac{1}{x+1}} \cdot \dfrac{x+1}{x+1} = \dfrac{x+1-1}{1} = \dfrac{x}{1} = x$

**73.** a) $f(8) = 2(8 + 12) = 2 \cdot 20 = 40$

Size 40 in Italy corresponds to size 8 in the U.S.

$f(10) = 2(10 + 12) = 2 \cdot 22 = 44$

Size 44 in Italy corresponds to size 10 in the U.S.

$f(14) = 2(14 + 12) = 2 \cdot 26 = 52$

Size 52 in Italy corresponds to size 14 in the U.S.

$f(18) = 2(18 + 12) = 2 \cdot 30 = 60$

Size 60 in Italy corresponds to size 18 in the U.S.

b) The function $f(x) = 2(x + 12)$ is a linear function that is not constant, so it passes the horizontal line test and has an inverse that is a function.

Replace $f(x)$ by $y$:      $y = 2(x + 12)$

Interchange $x$ and $y$:      $x = 2(y + 12)$

Solve for $y$:      $x = 2(y + 12)$

$x = 2y + 24$

$x - 24 = 2y$

$\dfrac{x - 24}{2} = y$

Replace $y$ by $f^{-1}(x)$: $f^{-1}(x) = \dfrac{x - 24}{2}$, or

$\dfrac{x}{2} - 12$

c) $f^{-1}(40) = \dfrac{40-24}{2} = \dfrac{16}{2} = 8$

Size 8 in the U.S. corresponds to size 40 in Italy.

$f^{-1}(44) = \dfrac{44-24}{2} = \dfrac{20}{2} = 10$

Size 10 in the U.S. corresponds to size 44 in Italy.

$f^{-1}(52) = \dfrac{52-24}{2} = \dfrac{28}{2} = 14$

Size 14 in the U.S. corresponds to size 52 in Italy.

$f^{-1}(60) = \dfrac{60-24}{2} = \dfrac{36}{2} = 18$

Size 18 in the U.S. corresponds to size 60 in Italy.

**75.** *Writing Exercise*

**77.** $(a^5b^4)^2(a^3b^5) = (a^5)^2(b^4)^2(a^3b^5)$

$\qquad = a^{5\cdot2}b^{4\cdot2}a^3b^5$

$\qquad = a^{10}b^8a^3b^5$

$\qquad = a^{10+3}b^{8+5}$

$\qquad = a^{13}b^{13}$

**79.** $27^{4/3} = (3^3)^{4/3} = 3^{3\cdot\frac{4}{3}} = 3^4 = 81$

**81.**
$$x = \frac{2}{3}y - 7$$
$$x + 7 = \frac{2}{3}y$$
$$\frac{3}{2}(x+7) = y$$

**83.** *Writing Exercise*

**85.** Reflect the graph of $f$ across the line $y = x$.

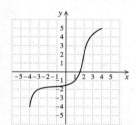

**87.** From Exercise 73(b), we know that a function that converts dress sizes in Italy to those in the United States is $g(x) = \dfrac{x-24}{2}$. From Exercise 74, we know that a function that converts dress sizes in the United States to those in France is $f(x) = x + 32$. Then a function that converts dress sizes in Italy to those in France is

$\qquad h(x) = (f \circ g)(x)$

$\qquad h(x) = f\left(\dfrac{x-24}{2}\right)$

$\qquad h(x) = \dfrac{x-24}{2} + 32$

$\qquad h(x) = \dfrac{x}{2} - 12 + 32$

$\qquad h(x) = \dfrac{x}{2} + 20.$

**89.** *Writing Exercise*

**91.** Suppose that $h(x) = (f \circ g)(x)$. First note that for $I(x) = x$, $(f \circ I)(x) = f(I(x))$ for any function $f$.

i) $\begin{aligned}[t] ((g^{-1} \circ f^{-1}) \circ h)(x) &= ((g^{-1} \circ f^{-1}) \circ (f \circ g))(x) \\ &= ((g^{-1} \circ (f^{-1} \circ f)) \circ g)(x) \\ &= ((g^{-1} \circ I) \circ g)(x) \\ &= (g^{-1} \circ g)(x) = x \end{aligned}$

ii) $\begin{aligned}[t] (h \circ (g^{-1} \circ f^{-1}))(x) &= ((f \circ g) \circ (g^{-1} \circ f^{-1}))(x) \\ &= ((f \circ (g \circ g^{-1})) \circ f^{-1})(x) \\ &= ((f \circ I) \circ f^{-1})(x) \\ &= (f \circ f^{-1})(x) = x \end{aligned}$

Therefore, $(g^{-1} \circ f^{-1})(x) = h^{-1}(x)$.

**93.** $(f \circ g)(x) = x$ and $(g \circ f)(x) = x$, so the functions are inverses.

**95.** $(f \circ g)(x) \neq x$, so the functions are not inverses. (It is also true that $(g \circ f)(x) \neq x$.)

**97.** (1) $C$; (2) $A$; (3) $B$; (4) $D$

**99.** *Writing Exercise*

## Exercise Set 9.2

**1.** True; see page 599 in the text.

**3.** True; the graph of $y = f(x-3)$ is a translation of the graph of $y = f(x)$ 3 units to the right.

**5.** False; the graph of $y = 3^x$ crosses the $y$-axis at $(0,1)$.

**7.** Graph: $y = f(x) = 3^x$

We compute some function values, thinking of $y$ as $f(x)$, and keep the results in a table.

$f(0) = 3^0 = 1$

$f(1) = 3^1 = 3$

$f(2) = 3^2 = 9$

$f(-1) = 3^{-1} = \dfrac{1}{3^1} = \dfrac{1}{3}$

$f(-2) = 3^{-2} = \dfrac{1}{3^2} = \dfrac{1}{9}$

| $x$ | $y$, or $f(x)$ |
|-----|----------------|
| 0   | 1              |
| 1   | 3              |
| 2   | 9              |
| $-1$ | $\dfrac{1}{3}$ |
| $-2$ | $\dfrac{1}{9}$ |

Next we plot these points and connect them with a smooth curve.

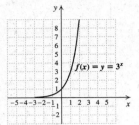

**9.** Graph:  $y = 6^x$

We compute some function values, thinking of $y$ as $f(x)$, and keep the results in a table.

$f(0) = 6^0 = 1$

$f(1) = 6^1 = 6$

$f(2) = 6^2 = 36$

$f(-1) = 6^{-1} = \dfrac{1}{6^1} = \dfrac{1}{6}$

$f(-2) = 6^{-2} = \dfrac{1}{6^2} = \dfrac{1}{36}$

| $x$ | $y$, or $f(x)$ |
|-----|----------------|
| 0   | 1              |
| 1   | 6              |
| 2   | 36             |
| $-1$ | $\dfrac{1}{6}$  |
| $-2$ | $\dfrac{1}{36}$ |

Next we plot these points and connect them with a smooth curve.

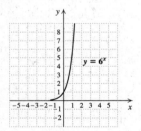

**11.** Graph:  $y = 2^x + 1$

We compute some function values, thinking of $y$ as $f(x)$, and keep the results in a table.

$f(-4) = 2^{-4} + 1 = \dfrac{1}{2^4} + 1 = \dfrac{1}{16} + 1 = 1\dfrac{1}{16}$

$f(-2) = 2^{-2} + 1 = \dfrac{1}{2^2} + 1 = \dfrac{1}{4} + 1 = 1\dfrac{1}{4}$

$f(0) = 2^0 + 1 = 1 + 1 = 2$

$f(1) = 2^1 + 1 = 2 + 1 = 3$

$f(2) = 2^2 + 1 = 4 + 1 = 5$

| $x$ | $y$, or $f(x)$ |
|-----|----------------|
| $-4$ | $1\dfrac{1}{16}$ |
| $-2$ | $1\dfrac{1}{4}$  |
| 0   | 2              |
| 1   | 3              |
| 2   | 5              |

Next we plot these points and connect them with a smooth curve.

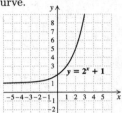

**13.** Graph:  $y = 3^x - 2$

We compute some function values, thinking of $y$ as $f(x)$, and keep the results in a table.

$f(-3) = 3^{-3} - 2 = \dfrac{1}{3^3} - 2 = \dfrac{1}{27} - 2 = -\dfrac{53}{27}$

$f(-1) = 3^{-1} - 2 = \dfrac{1}{3} - 2 = -\dfrac{5}{3}$

$f(0) = 3^0 - 2 = 1 - 2 = -1$

$f(1) = 3^1 - 2 = 3 - 2 = 1$

$f(2) = 3^2 - 2 = 9 - 2 = 7$

| $x$ | $y$, or $f(x)$ |
|-----|----------------|
| $-3$ | $-\dfrac{53}{27}$ |
| $-1$ | $-\dfrac{5}{3}$   |
| 0   | $-1$           |
| 1   | 1              |
| 2   | 7              |

Next we plot these points and connect them with a smooth curve.

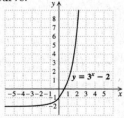

**15.** Graph:  $y = 2^x - 5$

We construct a table of values, thinking of $y$ as $f(x)$. Then we plot the points and connect them with a smooth curve.

$f(0) = 2^0 - 5 = 1 - 5 = -4$

$f(1) = 2^1 - 5 = 2 - 5 = -3$

$f(2) = 2^2 - 5 = 4 - 5 = -1$

$f(3) = 2^3 - 5 = 8 - 5 = 3$

$$f(-1) = 2^{-1} - 5 = \frac{1}{2} - 5 = -\frac{9}{2}$$

$$f(-2) = 2^{-2} - 5 = \frac{1}{4} - 5 = -\frac{19}{4}$$

$$f(-4) = 2^{-4} - 5 = \frac{1}{16} - 5 = -\frac{79}{16}$$

| $x$ | $y$, or $f(x)$ |
|-----|-----|
| 0 | $-4$ |
| 1 | $-3$ |
| 2 | $-1$ |
| 3 | $3$ |
| $-1$ | $-\dfrac{9}{2}$ |
| $-2$ | $-\dfrac{19}{4}$ |
| $-4$ | $-\dfrac{79}{16}$ |

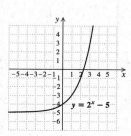

**17.** Graph: $y = 2^{x-2}$

We construct a table of values, thinking of $y$ as $f(x)$. Then we plot the points and connect them with a smooth curve.

$$f(0) = 2^{0-2} = 2^{-2} = \frac{1}{4}$$

$$f(-1) = 2^{-1-2} = 2^{-3} = \frac{1}{2^3} = \frac{1}{8}$$

$$f(-2) = 2^{-2-2} = 2^{-4} = \frac{1}{2^4} = \frac{1}{16}$$

$$f(1) = 2^{1-2} = 2^{-1} = \frac{1}{2}$$

$$f(2) = 2^{2-2} = 2^0 = 1$$

$$f(3) = 2^{3-2} = 2^1 = 2$$

$$f(4) = 2^{4-2} = 2^2 = 4$$

| $x$ | $y$, or $f(x)$ |
|-----|-----|
| 0 | $\dfrac{1}{4}$ |
| $-1$ | $\dfrac{1}{8}$ |
| $-2$ | $\dfrac{1}{16}$ |
| 1 | $\dfrac{1}{2}$ |
| 2 | $1$ |
| 3 | $2$ |
| 4 | $4$ |

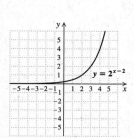

**19.** Graph: $y = 2^{x+1}$

We construct a table of values, thinking of $y$ as $f(x)$. Then we plot the points and connect them with a smooth curve.

$$f(-3) = 2^{-3+1} = 2^{-2} = \frac{1}{4}$$

$$f(-1) = 2^{-1+1} = 2^0 = 1$$

$$f(0) = 2^{0+1} = 2^1 = 2$$

$$f(1) = 2^{1+1} = 2^2 = 4$$

| $x$ | $y$, or $f(x)$ |
|-----|-----|
| $-3$ | $\dfrac{1}{4}$ |
| $-1$ | $1$ |
| $0$ | $2$ |
| $1$ | $4$ |

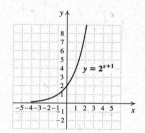

**21.** Graph: $y = \left(\frac{1}{4}\right)^x$

We construct a table of values, thinking of $y$ as $f(x)$. Then we plot the points and connect them with a smooth curve.

$$f(0) = \left(\frac{1}{4}\right)^0 = 1$$

$$f(1) = \left(\frac{1}{4}\right)^1 = \frac{1}{4}$$

$$f(2) = \left(\frac{1}{4}\right)^2 = \frac{1}{16}$$

$$f(-1) = \left(\frac{1}{4}\right)^{-1} = \frac{1}{\frac{1}{4}} = 4$$

$$f(-2) = \left(\frac{1}{4}\right)^{-2} = \frac{1}{\frac{1}{16}} = 16$$

| $x$ | $y$, or $f(x)$ |
|-----|-----|
| 0 | $1$ |
| 1 | $\dfrac{1}{4}$ |
| 2 | $\dfrac{1}{16}$ |
| $-1$ | $4$ |
| $-2$ | $16$ |

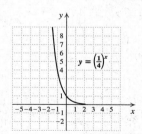

**23.** Graph: $y = \left(\frac{1}{3}\right)^x$

We construct a table of values, thinking of $y$ as $f(x)$. Then we plot the points and connect them with a smooth curve.

$$f(0) = \left(\frac{1}{3}\right)^0 = 1$$

$$f(1) = \left(\frac{1}{3}\right)^1 = \frac{1}{3}$$

$$f(2) = \left(\frac{1}{3}\right)^2 = \frac{1}{9}$$

$$f(3) = \left(\frac{1}{3}\right)^3 = \frac{1}{27}$$

$$f(-1) = \left(\frac{1}{3}\right)^{-1} = \frac{1}{\left(\frac{1}{3}\right)^1} = \frac{1}{\frac{1}{3}} = 3$$

$$f(-2) = \left(\frac{1}{3}\right)^{-2} = \frac{1}{\left(\frac{1}{3}\right)^2} = \frac{1}{\frac{1}{9}} = 9$$

$$f(-3) = \left(\frac{1}{3}\right)^{-3} = \frac{1}{\left(\frac{1}{3}\right)^3} = \frac{1}{\frac{1}{27}} = 27$$

| $x$ | $y$, or $f(x)$ |
|---|---|
| 0 | 1 |
| 1 | $\frac{1}{3}$ |
| 2 | $\frac{1}{9}$ |
| 3 | $\frac{1}{27}$ |
| $-1$ | 3 |
| $-2$ | 9 |
| $-3$ | 27 |

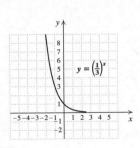

**25.** Graph: $y = 2^{x+1} - 3$

We construct a table of values, thinking of $y$ as $f(x)$. Then we plot the points and connect them with a smooth curve.

$$f(0) = 2^{0+1} - 3 = 2 - 3 = -1$$
$$f(1) = 2^{1+1} - 3 = 4 - 3 = 1$$
$$f(2) = 2^{2+1} - 3 = 8 - 3 = 5$$
$$f(-1) = 2^{-1+1} - 3 = 1 - 3 = -2$$
$$f(-2) = 2^{-2+1} - 3 = \frac{1}{2} - 3 = -\frac{5}{2}$$
$$f(-3) = 2^{-3+1} - 3 = \frac{1}{4} - 3 = -\frac{11}{4}$$

| $x$ | $y$, or $f(x)$ |
|---|---|
| 0 | $-1$ |
| 1 | 1 |
| 2 | 5 |
| $-1$ | $-2$ |
| $-2$ | $-\frac{5}{2}$ |
| $-3$ | $-\frac{11}{4}$ |

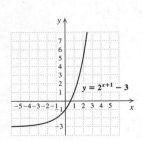

**27.** Graph: $x = 6^y$

We can find ordered pairs by choosing values for $y$ and then computing values for $x$.

For $y = 0$, $x = 6^0 = 1$.

For $y = 1$, $x = 6^1 = 6$.

For $y = -1$, $x = 6^{-1} = \frac{1}{6^1} = \frac{1}{6}$.

For $y = -2$, $x = 6^{-2} = \frac{1}{6^2} = \frac{1}{36}$.

| $x$ | $y$ |
|---|---|
| 1 | 0 |
| 6 | 1 |
| $\frac{1}{6}$ | $-1$ |
| $\frac{1}{36}$ | $-2$ |

    (1) Choose values for $y$.

    (2) Compute values for $x$.

We plot the points and connect them with a smooth curve.

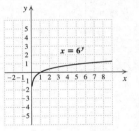

**29.** Graph: $x = 3^{-y} = \left(\frac{1}{3}\right)^y$

We can find ordered pairs by choosing values for $y$ and then computing values for $x$. Then we plot these points and connect them with a smooth curve.

For $y = 0$, $x = \left(\frac{1}{3}\right)^0 = 1$.

For $y = 1$, $x = \left(\frac{1}{3}\right)^1 = \frac{1}{3}$.

For $y = 2$, $x = \left(\frac{1}{3}\right)^2 = \frac{1}{9}$.

For $y = -1$, $x = \left(\frac{1}{3}\right)^{-1} = \frac{1}{\frac{1}{3}} = 3$.

For $y = -2$, $x = \left(\frac{1}{3}\right)^{-2} = \frac{1}{\frac{1}{9}} = 9$.

| $x$ | $y$ |
|---|---|
| 1 | 0 |
| $\frac{1}{3}$ | 1 |
| $\frac{1}{9}$ | 2 |
| 3 | $-1$ |
| 9 | $-2$ |

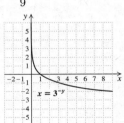

**31.** Graph: $x = 4^y$

We can find ordered pairs by choosing values for $y$ and then computing values for $x$. Then we plot these points and connect them with a smooth curve.

For $y = 0$, $x = 4^0 = 1$.

For $y = 1$, $x = 4^1 = 4$.

For $y = 2$, $x = 4^2 = 16$.

For $y = -1$, $x = 4^{-1} = \frac{1}{4}$.

For $y = -2$, $x = 4^{-2} = \frac{1}{16}$.

| $x$ | $y$ |
|---|---|
| 1 | 0 |
| 4 | 1 |
| 16 | 2 |
| $\frac{1}{4}$ | $-1$ |
| $\frac{1}{16}$ | $-2$ |

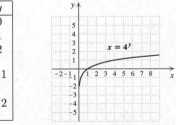

**33.** Graph: $x = \left(\dfrac{4}{3}\right)^y$

We can find ordered pairs by choosing values for $y$ and then computing values for $x$. Then we plot these points and connect them with a smooth curve.

For $y = 0$, $x = \left(\dfrac{4}{3}\right)^0 = 1$.

For $y = 1$, $x = \left(\dfrac{4}{3}\right)^1 = \dfrac{4}{3}$.

For $y = 2$, $x = \left(\dfrac{4}{3}\right)^2 = \dfrac{16}{9}$.

For $y = 3$, $x = \left(\dfrac{4}{3}\right)^3 = \dfrac{64}{27}$.

For $y = -1$, $x = \left(\dfrac{4}{3}\right)^{-1} = \dfrac{3}{4}$.

For $y = -2$, $x = \left(\dfrac{4}{3}\right)^{-2} = \left(\dfrac{3}{4}\right)^2 = \dfrac{9}{16}$.

For $y = -3$, $x = \left(\dfrac{4}{3}\right)^{-3} = \left(\dfrac{3}{4}\right)^3 = \dfrac{27}{64}$.

| $x$ | $y$ |
|-----|-----|
| 1 | 0 |
| $\dfrac{4}{3}$ | 1 |
| $\dfrac{16}{9}$ | 2 |
| $\dfrac{64}{27}$ | 3 |
| $\dfrac{3}{4}$ | $-1$ |
| $\dfrac{9}{16}$ | $-2$ |
| $\dfrac{27}{64}$ | $-3$ |

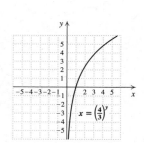

**35.** Graph $y = 3^x$ (see Exercise 8) and $x = 3^y$ (see Exercise 28) using the same set of axes.

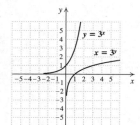

**37.** Graph $y = \left(\dfrac{1}{2}\right)^x$ (see Exercise 24) and $x = \left(\dfrac{1}{2}\right)^y$ (see Exercise 30) using the same set of axes.

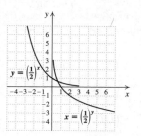

**39.** a) In 2008, $t = 2008 - 1980 = 28$.

$$P(28) = 4.495(1.015)^{28} \approx 6.8 \text{ billion}$$

In 2012, $t = 2012 - 1980 = 32$.

$$P(32) = 4.495(1.015)^{32} \approx 7.2 \text{ billion}$$

In 2016, $t = 2016 - 1980 = 36$.

$$P(36) = 4.495(1.015)^{36} \approx 7.7 \text{ billion}$$

b) Use the function values computed in part (a) and others, if desired, and draw the graph.

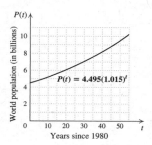

**41.** a) $P(1) = 21.4(0.914)^1 \approx 19.6\%$

$P(3) = 21.4(0.914)^3 \approx 16.3\%$

1 yr = 12 months; $P(12) = 21.4(0.914)^{12} \approx 7.3\%$

b)

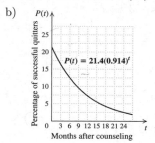

**43.** a) In 1930, $t = 1930 - 1900 = 30$.

$$P(t) = 150(0.960)^t$$
$$P(30) = 150(0.960)^{30}$$
$$\approx 44.079$$

In 1930, about 44.079 thousand, or 44,079, humpback whales were alive.

In 1960, $t = 1960 - 1900 = 60$.

$$P(t) = 150(0.960)^t$$
$$P(60) = 150(0.960)^{60}$$
$$\approx 12.953$$

In 1960, about 12.953 thousand, or 12,953, humpback whales were alive.

b) Plot the points found in part (a), $(30, \ 44,079)$ and $(60, \ 12,953)$ and additional points as needed and graph the function.

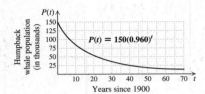

45. a) In 1992, $t = 1992 - 1982 = 10$.

$$P(10) = 5.5(1.047)^{10} \approx 8.706$$

In 1992, about 8.706 thousand, or 8706, humpback whales were alive.

In 2004, $t = 2004 - 1982 = 22$.

$$P(22) = 5.5(1.047)^{22} \approx 15.107$$

In 2004, about 15.107 thousand, or 15,107, humpback whales were alive.

b) Use the function values computed in part (a) and others, if desired, and draw the graph.

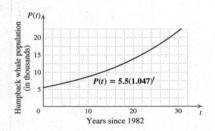

47. a)  $A(5) = 10 \cdot 34^5 = 454,354,240 \text{ cm}^2$

$A(7) = 10 \cdot 34^7 = 525,233,501,400 \text{ cm}^2$

b) Use the function values computed in part (a) and others, if desired, and draw the graph.

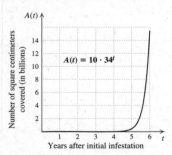

49. *Writing Exercise*

51. $5^{-2} = \dfrac{1}{5^2} = \dfrac{1}{25}$

53. $1000^{2/3} = (10^3)^{2/3} = 10^{3 \cdot \frac{2}{3}} = 10^2 = 100$

55. $\dfrac{10a^8b^7}{2a^2b^4} = \dfrac{10}{2}a^{8-2}b^{7-4} = 5a^6b^3$

57. *Writing Exercise*

59. Since the bases are the same, the one with the larger exponent is the larger number. Thus $\pi^{2.4}$ is larger.

61. Graph:  $f(x) = 3.8^x$

Use a calculator with a power key to construct a table of values. (We will round values of $f(x)$ to the nearest hundredth.) Then plot these points and connect them with a smooth curve.

| $x$ | $y$ |
|-----|--------|
| 0 | 1 |
| 1 | 3.8 |
| 2 | 14.44 |
| 3 | 54.872 |
| $-1$ | 0.26 |
| $-2$ | 0.7 |

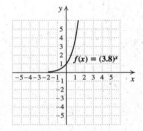

63. Graph:  $y = 2^x + 2^{-x}$

Construct a table of values, thinking of $y$ as $f(x)$. Then plot these points and connect them with a curve.

$$f(0) = 2^0 + 2^{-0} = 1 + 1 = 2$$

$$f(1) = 2^1 + 2^{-1} = 2 + \frac{1}{2} = 2\frac{1}{2}$$

$$f(2) = 2^2 + 2^{-2} = 4 + \frac{1}{4} = 4\frac{1}{4}$$

$$f(3) = 2^3 + 2^{-3} = 8 + \frac{1}{8} = 8\frac{1}{8}$$

$$f(-1) = 2^{-1} + 2^{-(-1)} = \frac{1}{2} + 2 = 2\frac{1}{2}$$

$$f(-2) = 2^{-2} + 2^{-(-2)} = \frac{1}{4} + 4 = 4\frac{1}{4}$$

$$f(-3) = 2^{-3} + 2^{-(-3)} = \frac{1}{8} + 8 = 8\frac{1}{8}$$

| $x$ | $y$, or $f(x)$ |
|-----|------------------|
| 0 | 2 |
| 1 | $2\frac{1}{2}$ |
| 2 | $4\frac{1}{4}$ |
| 3 | $8\frac{1}{8}$ |
| $-1$ | $2\frac{1}{2}$ |
| $-2$ | $4\frac{1}{4}$ |
| $-3$ | $8\frac{1}{8}$ |

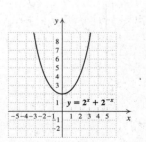

**65.** Graph: $y = |2^x - 2|$

We construct a table of values, thinking of $y$ as $f(x)$. Then plot these points and connect them with a curve.

$f(0) = |2^0 - 2| = |1 - 2| = |-1| = 1$

$f(1) = |2^1 - 2| = |2 - 2| = |0| = 0$

$f(2) = |2^2 - 2| = |4 - 2| = |2| = 2$

$f(3) = |2^3 - 2| = |8 - 2| = |6| = 6$

$f(-1) = |2^{-1} - 2| = \left|\dfrac{1}{2} - 2\right| = \left|-\dfrac{3}{2}\right| = \dfrac{3}{2}$

$f(-3) = |2^{-3} - 2| = \left|\dfrac{1}{8} - 2\right| = \left|-\dfrac{15}{8}\right| = \dfrac{15}{8}$

$f(-5) = |2^{-5} - 2| = \left|\dfrac{1}{32} - 2\right| = \left|-\dfrac{63}{32}\right| = \dfrac{63}{32}$

| $x$ | $y$, or $f(x)$ |
|-----|------|
| 0 | 1 |
| 1 | 0 |
| 2 | 2 |
| 3 | 6 |
| $-1$ | $\dfrac{3}{2}$ |
| $-3$ | $\dfrac{15}{8}$ |
| $-5$ | $\dfrac{63}{32}$ |

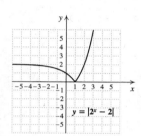

**67.** Graph: $y = |2x^2 - 1|$

We construct a table of values, thinking of $y$ as $f(x)$. Then we plot these points and connect them with a curve.

$f(0) = |2 \cdot 0^2 - 1| = |1 - 1| = 0$

$f(1) = |2 \cdot 1^2 - 1| = |2 - 1| = 1$

$f(2) = |2 \cdot 2^2 - 1| = |16 - 1| = 15$

$f(-1) = |2(-1)^2 - 1| = |2 - 1| = 1$

$f(-2) = |2(-2)^2 - 1| = |16 - 1| = 15$

| $x$ | $y$, or $f(x)$ |
|-----|------|
| 0 | 0 |
| 1 | 1 |
| 2 | 15 |
| $-1$ | 1 |
| $-2$ | 15 |

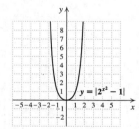

**69.** $y = 3^{-(x-1)}$ $\qquad$ $x = 3^{-(y-1)}$

| $x$ | $y$ |
|-----|-----|
| 0 | 3 |
| 1 | 1 |
| 2 | $\dfrac{1}{3}$ |
| 3 | $\dfrac{1}{9}$ |
| $-1$ | 9 |

| $x$ | $y$ |
|-----|-----|
| 3 | 0 |
| 1 | 1 |
| $\dfrac{1}{3}$ | 2 |
| $\dfrac{1}{9}$ | 3 |
| 9 | $-1$ |

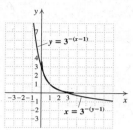

**71.** Enter the data points $(0, 171)$, $(2, 1099)$, and $(4, 2697)$ and then use the ExpReg option from the STAT CALC menu of a graphing calculator to find an exponential function that models the data:

$A(t) = 200.7624553(1.992834389)^t$, where $A(t)$ is total sales, in millions of dollars, $t$ years after 1997.

In 2008, $t = 2008 - 1997 = 11$.

$A(11) \approx \$395,244.4657$ million, or $\$395,244,465,700$

**73.** *Writing Exercise*

**75.** ▮

## Exercise Set 9.3

**1.** $5^2 = 25$, so choice (g) is correct.

**3.** $5^1 = 5$, so choice (a) is correct.

**5.** The exponent to which we raise 5 to get $5^x$ is $x$, so choice (b) is correct.

**7.** $8 = 2^x$ is equivalent to $\log_2 8 = x$, so choice (e) is correct.

**9.** $\log_{10} 1000$ is the exponent to which we raise 10 to get 1000. Since $10^3 = 1000$, $\log_{10} 1000 = 3$.

**11.** $\log_2 16$ is the exponent to which we raise 2 to get 16. Since $2^4 = 16$, $\log_2 16 = 4$.

**13.** $\log_3 81$ is the exponent to which we raise 3 to get 81. Since $3^4 = 81$, $\log_3 81 = 4$.

**15.** $\log_4 \dfrac{1}{16}$ is the exponent to which we raise 4 to get $\dfrac{1}{16}$. Since $4^{-2} = \dfrac{1}{16}$, $\log_4 \dfrac{1}{16} = -2$.

**17.** Since $7^{-1} = \dfrac{1}{7}$, $\log_7 \dfrac{1}{7} = -1$.

**19.** Since $5^4 = 625$, $\log_5 625 = 4$.

**21.** Since $8^1 = 8$, $\log_8 8 = 1$.

**23.** Since $8^0 = 1$, $\log_8 1 = 0$.

**25.** $\log_9 9^5$ is the exponent to which we raise 9 to get $9^5$. Clearly, this power is 5, so $\log_9 9^5 = 5$.

**27.** Since $10^{-2} = \dfrac{1}{100} = 0.01$, $\log_{10} 0.01 = -2$.

**29.** Since $9^{1/2} = 3$, $\log_9 3 = \dfrac{1}{2}$.

**31.** Since $9 = 3^2$ and $(3^2)^{3/2} = 3^3 = 27$, $\log_9 27 = \dfrac{3}{2}$.

**33.** Since $1000 = 10^3$ and $(10^3)^{2/3} = 10^2 = 100$,
$\log_{1000} 100 = \dfrac{2}{3}$.

**35.** Since $\log_5 7$ is the power to which we raise 5 to get 7, then 5 raised to this power is 7. That is, $5^{\log_5 7} = 7$.

**37.** Graph: $y = \log_{10} x$

The equation $y = \log_{10} x$ is equivalent to $10^y = x$. We can find ordered pairs by choosing values for $y$ and computing the corresponding $x$-values.

For $y = 0$, $x = 10^0 = 1$.

For $y = 1$, $x = 10^1 = 10$.

For $y = 2$, $x = 10^2 = 100$.

For $y = -1$, $x = 10^{-1} = \dfrac{1}{10}$.

For $y = -2$, $x = 10^{-2} = \dfrac{1}{100}$.

| $x$, or $10^y$ | $y$ |
|---|---|
| 1 | 0 |
| 10 | 1 |
| 100 | 2 |
| $\dfrac{1}{10}$ | $-1$ |
| $\dfrac{1}{100}$ | $-2$ |

↑ └─ (1) Select $y$.

└─ (2) Compute $x$.

We plot the set of ordered pairs and connect the points with a smooth curve.

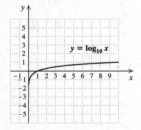

**39.** Graph: $y = \log_3 x$

The equation $y = \log_3 x$ is equivalent to $3^y = x$. We can find ordered pairs by choosing values for $y$ and computing the corresponding $x$-values.

For $y = 0$, $x = 3^0 = 1$.

For $y = 1$, $x = 3^1 = 3$.

For $y = 2$, $x = 3^2 = 9$.

For $y = -1$, $x = 3^{-1} = \dfrac{1}{3}$.

For $y = -2$, $x = 3^{-2} = \dfrac{1}{9}$.

| $x$, or $3^y$ | $y$ |
|---|---|
| 1 | 0 |
| 3 | 1 |
| 9 | 2 |
| $\dfrac{1}{3}$ | $-1$ |
| $\dfrac{1}{9}$ | $-2$ |

We plot the set of ordered pairs and connect the points with a smooth curve.

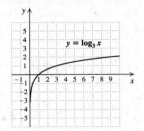

**41.** Graph: $f(x) = \log_6 x$

Think of $f(x)$ as $y$. Then $y = \log_6 x$ is equivalent to $6^y = x$. We find ordered pairs by choosing values for $y$ and computing the corresponding $x$-values. Then we plot the points and connect them with a smooth curve.

For $y = 0$, $x = 6^0 = 1$.

For $y = 1$, $x = 6^1 = 6$.

For $y = 2$, $x = 6^2 = 36$.

For $y = -1$, $x = 6^{-1} = \dfrac{1}{6}$.

For $y = -2$, $x = 6^{-2} = \dfrac{1}{36}$.

| $x$, or $6^y$ | $y$ |
|---|---|
| 1 | 0 |
| 6 | 1 |
| 36 | 2 |
| $\dfrac{1}{6}$ | $-1$ |
| $\dfrac{1}{36}$ | $-2$ |

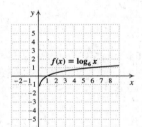

**43.** Graph: $f(x) = \log_{2.5} x$

Think of $f(x)$ as $y$. Then $y = \log_{2.5} x$ is equivalent to $2.5^y = x$. We construct a table of values, plot these points and connect them with a smooth curve.

For $y = 0$, $x = 2.5^0 = 1$.

For $y = 1$, $x = 2.5^1 = 2.5$.

For $y = 2$, $x = 2.5^2 = 6.25$.

For $y = 3$, $x = 2.5^3 = 15.625$.

For $y = -1$, $x = 2.5^{-1} = 0.4$.

For $y = -2$, $x = 2.5^{-2} = 0.16$.

| $x$, or $2.5^y$ | $y$ |
|---|---|
| 1 | 0 |
| 2.5 | 1 |
| 6.25 | 2 |
| 15.625 | 3 |
| 0.4 | $-1$ |
| 0.16 | $-2$ |

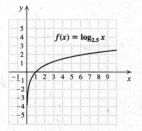

**45.** Graph $f(x) = 3^x$ (see Exercise Set 9.2, Exercise 7) and $f^{-1}(x) = \log_3 x$ (see Exercise 39 above) on the same set of axes.

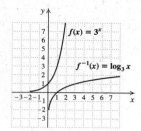

**47.**

The base remains the same.

$t = \log_5 9 \Rightarrow 5^t = 9$

The logarithm is the exponent.

**49.**

The logarithm is the exponent.

$\log_5 25 = 2 \Rightarrow 5^2 = 25$

The base remains the same.

**51.** $\log_{10} 0.1 = -1$ is equivalent to $10^{-1} = 0.1$.

**53.** $\log_{10} 7 = 0.845$ is equivalent to $10^{0.845} = 7$.

**55.** $\log_c m = 8$ is equivalent to $c^8 = m$.

**57.** $\log_t Q = r$ is equivalent to $t^r = Q$.

**59.** $\log_e 0.25 = -1.3863$ is equivalent to $e^{-1.3863} = 0.25$.

**61.** $\log_r T = -x$ is equivalent to $r^{-x} = T$.

**63.**

The exponent is the logarithm.

$10^2 = 100 \Rightarrow 2 = \log_{10} 100$

The base remains the same.

**65.**

The exponent is the logarithm.

$4^{-5} = \dfrac{1}{1024} \Rightarrow -5 = \log_4 \dfrac{1}{1024}$

The base remains the same.

**67.** $16^{3/4} = 8$ is equivalent to $\dfrac{3}{4} = \log_{16} 8$.

**69.** $10^{0.4771} = 3$ is equivalent to $0.4771 = \log_{10} 3$.

**71.** $z^m = 6$ is equivalent to $m = \log_z 6$.

**73.** $p^m = V$ is equivalent to $m = \log_p V$.

**75.** $e^3 = 20.0855$ is equivalent to $3 = \log_e 20.0855$.

**77.** $e^{-4} = 0.0183$ is equivalent to $-4 = \log_e 0.0183$.

**79.** $\log_3 x = 2$

$\quad 3^2 = x \quad$ Converting to an exponential equation

$\quad 9 = x \quad$ Computing $3^2$

**81.** $\log_5 125 = x$

$\quad 5^x = 125 \quad$ Converting to an exponential equation

$\quad 5^x = 5^3$

$\quad x = 3 \quad$ The exponents must be the same.

**83.** $\log_2 16 = x$

$\quad 2^x = 16 \quad$ Converting to an exponential equation

$\quad 2^x = 2^4$

$\quad x = 4 \quad$ The exponents must be the same.

**85.** $\log_x 7 = 1$

$\quad x^1 = 7 \quad$ Converting to an exponential equation

$\quad x = 7 \quad$ Siimplifying $x^1$

**87.** $\log_3 x = -2$

$\quad 3^{-2} = x \quad$ Converting to an exponential equation

$\quad \dfrac{1}{9} = x \quad$ Simplifying

**89.** $\log_{32} x = \dfrac{2}{5}$

$\quad 32^{2/5} = x \quad$ Converting to an exponential equation

$\quad (2^5)^{2/5} = x$

$\quad 4 = x$

**91.** *Writing Exercise*

**93.** $\dfrac{x^{12}}{x^4} = x^{12-4} = x^8$

**95.** $(a^4 b^6)(a^3 b^2) = a^{4+3} b^{6+2} = a^7 b^8$

**97.** $\dfrac{\dfrac{3}{x} - \dfrac{2}{xy}}{\dfrac{2}{x^2} + \dfrac{1}{xy}}$

The LCD of all the denominators is $x^2 y$. We multiply numerator and denominator by the LCD.

$$\dfrac{\dfrac{3}{x} - \dfrac{2}{xy}}{\dfrac{2}{x^2} + \dfrac{1}{xy}} \cdot \dfrac{x^2 y}{x^2 y} = \dfrac{\left(\dfrac{3}{x} - \dfrac{2}{xy}\right) x^2 y}{\left(\dfrac{2}{x^2} + \dfrac{1}{xy}\right) x^2 y}$$

$$= \dfrac{\dfrac{3}{x} \cdot x^2 y - \dfrac{2}{xy} \cdot x^2 y}{\dfrac{2}{x^2} \cdot x^2 y + \dfrac{1}{xy} \cdot x^2 y}$$

$$= \dfrac{3xy - 2x}{2y + x}, \text{ or}$$

$$\dfrac{x(3y - 2)}{2y + x}$$

**99.** *Writing Exercise*

**101.** Graph: $y = \left(\frac{3}{2}\right)^x$          Graph: $y = \log_{3/2} x$, or

$$x = \left(\frac{3}{2}\right)^y$$

| $x$ | $y$, or $\left(\frac{3}{2}\right)^x$ |
|-----|------|
| 0 | 1 |
| 1 | $\frac{3}{2}$ |
| 2 | $\frac{9}{4}$ |
| 3 | $\frac{27}{8}$ |
| −1 | $\frac{2}{3}$ |
| −2 | $\frac{4}{9}$ |

| $x$, or $\left(\frac{3}{2}\right)^y$ | $y$ |
|------|-----|
| 1 | 0 |
| $\frac{3}{2}$ | 1 |
| $\frac{9}{4}$ | 2 |
| $\frac{27}{8}$ | 3 |
| $\frac{2}{3}$ | −1 |
| $\frac{4}{9}$ | −2 |

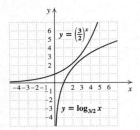

**103.** Graph: $y = \log_3 |x + 1|$

| $x$ | $y$ |
|-----|-----|
| 0 | 0 |
| 2 | 1 |
| 8 | 2 |
| −2 | 0 |
| −4 | 1 |
| −9 | 2 |

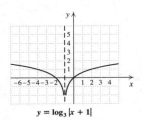

$y = \log_3 |x + 1|$

**105.**  $\log_4(3x - 2) = 2$

$$4^2 = 3x - 2$$
$$16 = 3x - 2$$
$$18 = 3x$$
$$6 = x$$

**107.**  $\log_{10}(x^2 + 21x) = 2$

$$10^2 = x^2 + 21x$$
$$0 = x^2 + 21x - 100$$
$$0 = (x + 25)(x - 4)$$

$$x = -25 \ or \ x = 4$$

**109.** Let $\log_{1/5} 25 = x$. Then

$$\left(\frac{1}{5}\right)^x = 25$$
$$(5^{-1})^x = 25$$
$$5^{-x} = 5^2$$
$$-x = 2$$
$$x = -2.$$

Thus, $\log_{1/5} 25 = -2$.

**111.**     $\log_{10} (\log_4 (\log_3 81))$

$= \log_{10} (\log_4 4)$          $(\log_3 81 = 4)$

$= \log_{10} 1$          $(\log_4 4 = 1)$

$= 0$

**113.** Let $b = 0$, $x = 1$, and $y = 2$. Then $0^1 = 0^2$, but $1 \neq 2$. Let $b = 1$, $x = 1$, and $y = 2$. Then $1^1 = 1^2$, but $1 \neq 2$.

## Exercise Set 9.4

**1.** Use the product rule for logarithms.

$\log_7 20 = \log_7(5 \cdot 4) = \log_7 5 + \log_7 4$; choice (e) is correct.

**3.** Use the quotient rule for logarithms.

$\log_7 \dfrac{5}{4} = \log_7 5 - \log_7 4$; choice (a) is correct.

**5.** The exponent to which we raise 7 to get 1 is 0, so choice (c) is correct.

**7.** $\log_3 (81 \cdot 27) = \log_3 81 + \log_3 27$  Using the product rule

**9.** $\log_4 (64 \cdot 16) = \log_4 64 + \log_4 16$  Using the product rule

**11.**     $\log_c rst$

$= \log_c r + \log_c s + \log_c t$  Using the product rule

**13.** $\log_a 5 + \log_a 14 = \log_a (5 \cdot 14)$  Using the product rule

The result can also be expressed as $\log_a 70$.

**15.** $\log_c t + \log_c y = \log_c (t \cdot y)$  Using the product rule

**17.** $\log_a r^8 = 8 \log_a r$  Using the power rule

**19.** $\log_c y^6 = 6 \log_c y$  Using the power rule

**21.** $\log_b C^{-3} = -3 \log_b C$  Using the power rule

**23.** $\log_2 \dfrac{25}{13} = \log_2 25 - \log_2 13$  Using the quotient rule

**25.** $\log_b \dfrac{m}{n} = \log_b m - \log_b n$  Using the quotient rule

**27.**     $\log_a 17 - \log_a 6$

$= \log_a \dfrac{17}{6}$

**29.** $\quad \log_b 36 - \log_b 4$

$= \log_b \dfrac{36}{4}, \qquad$ Using the quotient rule

or $\log_b 9$

**31.** $\log_a 7 - \log_z 18 = \log_a \dfrac{7}{18}\quad$ Using the quotient rule

**33.** $\quad \log_a(xyz)$

$= \log_a x + \log_a y + \log_a z \quad$ Using the product rule

**35.** $\quad \log_a(x^3 z^4)$

$= \log_a x^3 + \log_a z^4 \quad$ Using the product rule

$= 3\log_a x + 4\log_a z \quad$ Using the power rule

**37.** $\quad \log_a(x^2 y^{-2} z)$

$= \log_a x^2 + \log_a y^{-2} + \log_a z \quad$ Using the product rule

$= 2\log_a x - 2\log_a y + \log_a z \quad$ Using the power rule

**39.** $\quad \log_a \dfrac{x^4}{y^3 z}$

$= \log_a x^4 - \log_a y^3 z \quad$ Using the quotient rule

$= \log_a x^4 - (\log_a y^3 + \log_a z) \quad$ Using the product rule

$= \log_a x^4 - \log_a y^3 - \log_a z \quad$ Removing parentheses

$= 4\log_a x - 3\log_a y - \log_a z \quad$ Using the power rule

**41.** $\quad \log_b \dfrac{xy^2}{wz^3}$

$= \log_b xy^2 - \log_b wz^3 \quad$ Using the quotient rule

$= \log_b x + \log_b y^2 - (\log_b w + \log_b z^3)$
$\qquad\qquad$ Using the product rule

$= \log_b x + \log_b y^2 - \log_b w - \log_b z^3$
$\qquad\qquad$ Removing parentheses

$= \log_b x + 2\log_b y - \log_b w - 3\log_b z$
$\qquad\qquad$ Using the power rule

**43.** $\quad \log_a \sqrt{\dfrac{x^7}{y^5 z^8}}$

$= \log_a \left(\dfrac{x^7}{y^5 z^8}\right)^{1/2}$

$= \dfrac{1}{2}\log_a \dfrac{x^7}{y^5 z^8} \quad$ Using the power rule

$= \dfrac{1}{2}(\log_a x^7 - \log_a y^5 z^8) \quad$ Using the quotient rule

$= \dfrac{1}{2}\left[\log_a x^7 - (\log_a y^5 + \log_a z^8)\right]$
$\qquad\qquad$ Using the product rule

$= \dfrac{1}{2}(\log_a x^7 - \log_a y^5 - \log_a z^8)$
$\qquad\qquad$ Removing parentheses

$= \dfrac{1}{2}(7\log_a x - 5\log_a y - 8\log_a z)$
$\qquad\qquad$ Using the power rule

**45.** $\quad \log_a \sqrt[3]{\dfrac{x^6 y^3}{a^2 z^7}}$

$= \log_a \left(\dfrac{x^6 y^3}{a^2 z^7}\right)^{1/3}$

$= \dfrac{1}{3}\log_a \dfrac{x^6 y^3}{a^2 z^7} \quad$ Using the power rule

$= \dfrac{1}{3}(\log_a x^6 y^3 - \log_a a^2 z^7) \quad$ Using the quotient rule

$= \dfrac{1}{3}[\log_a x^6 + \log_a y^3 - (\log_a a^2 + \log_a z^7)]$
$\qquad\qquad$ Using the product rule

$= \dfrac{1}{3}(\log_a x^6 + \log_a y^3 - \log_a a^2 - \log_a z^7)$
$\qquad\qquad$ Removing parentheses

$= \dfrac{1}{3}(\log_a x^6 + \log_a y^3 - 2 - \log_a z^7)$
$\qquad\qquad$ 2 is the number to which we raise $a$ to get $a^2$.

$= \dfrac{1}{3}(6\log_a x + 3\log_a y - 2 - 7\log_a z)$
$\qquad\qquad$ Using the power rule

**47.** $\quad 8\log_a x + 3\log_a z$

$= \log_a x^8 + \log_a z^3 \quad$ Using the power rule

$= \log_a (x^8 z^3) \quad$ Using the product rule

**49.** $\quad \log_a x^2 - 2\log_a \sqrt{x}$

$= \log_a x^2 - \log_a (\sqrt{x})^2 \quad$ Using the power rule

$= \log_a x^2 - \log_a x \qquad (\sqrt{x})^2 = x$

$= \log_a \dfrac{x^2}{x} \quad$ Using the quotient rule

$= \log_a x \quad$ Simplifying

**51.** $\quad \dfrac{1}{2}\log_a x + 5\log_a y - 2\log_a x$

$= \log_a x^{1/2} + \log_a y^5 - \log_a x^2 \quad$ Using the power rule

$= \log_a x^{1/2} y^5 - \log_a x^2 \quad$ Using the product rule

$= \log_a \dfrac{x^{1/2} y^5}{x^2} \quad$ Using the quotient rule

The result can also be expressed as $\log_a \dfrac{\sqrt{x}\,y^5}{x^2}$ or as $\log_a \dfrac{y^5}{x^{3/2}}$.

**53.** $\quad \log_a(x^2 - 4) - \log_a(x + 2)$

$= \log_a \dfrac{x^2 - 4}{x + 2} \quad$ Using the quotient rule

$= \log_a \dfrac{(x+2)(x-2)}{x+2}$

$= \log_a \dfrac{(x\!\!\not{+}\!\!2)(x-2)}{x\!\!\not{+}\!\!2} \quad$ Simplifying

$= \log_a(x - 2)$

**55.** $\log_b 15 = \log_b (3 \cdot 5)$

$= \log_b 3 + \log_b 5 \quad$ Using the product rule

$= 0.792 + 1.161$

$= 1.953$

**57.** $\log_b \dfrac{3}{5} = \log_b 3 - \log_b 5$   Using the quotient rule

$\hspace{3.2em} = 0.792 - 1.161$

$\hspace{3.2em} = -0.369$

**59.** $\log_b \dfrac{1}{5} = \log_b 1 - \log_b 5$   Using the quotient rule

$\hspace{3.2em} = 0 - 1.161 \hspace{2em} (\log_b 1 = 0)$

$\hspace{3.2em} = -1.161$

**61.** $\log_b \sqrt{b^3} = \log_b b^{3/2} = \dfrac{3}{2}$   3/2 is the number to which we raise $b$ to get $b^{3/2}$.

**63.** $\log_b 8$

Since 8 cannot be expressed using the numbers 1, 3, and 5, we cannot find $\log_b 8$ using the given information.

**65.** $\log_t t^7 = 7$   7 is the exponent to which we raise $t$ to get $t^7$.

**67.** $\log_e e^m = m$   $m$ is the exponent to which we raise $e$ to get $e^m$.

**69.** *Writing Exercise*

**71.** Graph $f(x) = \sqrt{x} - 3$.

We construct a table of values, plot points, and connect them with a smooth curve. Note that we must choose nonnegative values of $x$ in order for $\sqrt{x}$ to be a real number.

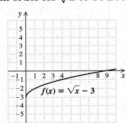

| $x$ | $f(x)$ |
|-----|--------|
| 0   | $-3$   |
| 1   | $-2$   |
| 4   | $-1$   |
| 9   | $0$    |

**73.** Graph $g(x) = \sqrt[3]{x} + 1$.

We construct a table of values, plot points, and connect them with a smooth curve.

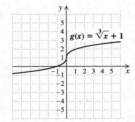

| $x$ | $g(x)$ |
|-----|--------|
| $-8$ | $-1$  |
| $-1$ | $0$   |
| $0$  | $1$   |
| $1$  | $2$   |
| $8$  | $3$   |

**75.** $(a^3 b^2)^5 (a^2 b^7) = (a^{3 \cdot 5} b^{2 \cdot 5})(a^2 b^7) = a^{15} b^{10} a^2 b^7 = a^{15+2} b^{10+7} = a^{17} b^{17}$

**77.** *Writing Exercise*

---

**79.** $\log_a (x^8 - y^8) - \log_a (x^2 + y^2)$

$= \log_a \dfrac{x^8 - y^8}{x^2 + y^2}$

$= \log_a \dfrac{(x^4 + y^4)(x^2 + y^2)(x + y)(x - y)}{x^2 + y^2}$

$= \log_a [(x^4 + y^4)(x^2 - y^2)]$   Simplifying

$= \log_a (x^6 - x^4 y^2 + x^2 y^4 - y^6)$

**81.** $\log_a \sqrt{1 - s^2}$

$= \log_a (1 - s^2)^{1/2}$

$= \dfrac{1}{2} \log_a (1 - s^2)$

$= \dfrac{1}{2} \log_a [(1 - s)(1 + s)]$

$= \dfrac{1}{2} \log_a (1 - s) + \dfrac{1}{2} \log_a (1 + s)$

**83.** $\log_a \dfrac{\sqrt[3]{x^2 z}}{\sqrt[3]{y^2 z^{-2}}}$

$= \log_a \left( \dfrac{x^2 z^3}{y^2} \right)^{1/3}$

$= \dfrac{1}{3} (\log_a x^2 z^3 - \log_a y^2)$

$= \dfrac{1}{3} (2 \log_a x + 3 \log_a z - 2 \log_a y)$

$= \dfrac{1}{3} [2 \cdot 2 + 3 \cdot 4 - 2 \cdot 3]$

$= \dfrac{1}{3} (10)$

$= \dfrac{10}{3}$

**85.** $\log_a x = 2$, so $a^2 = x$.

Let $\log_{1/a} x = n$ and solve for $n$.

$\log_{1/a} a^2 = n$   Substituting $a^2$ for $x$

$\left( \dfrac{1}{a} \right)^n = a^2$

$(a^{-1})^n = a^2$

$a^{-n} = a^2$

$-n = 2$

$n = -2$

Thus, $\log_{1/a} x = -2$ when $\log_a x = 2$.

**87.** True; $\log_a(Q + Q^2) = \log_a[Q(1 + Q)] = \log_a Q + \log_a(1 + Q) = \log_a Q + \log_a(Q + 1)$.

---

## Exercise Set 9.5

**1.** True; see page 619 in the text.

**3.** True; see page 621 in the text.

**5.** Using the change-of-base formula with $a = e$, $b = 2$, and $M = 9$, we see that the statement is true.

**7.** True; see Example 7.

**9.** 0.7782

**11.** 1.8621

**13.** Since $10^3 = 1000$, $\log 1000 = 3$.

**15.** $-0.2782$

**17.** 1.7986

**19.** 199.5262

**21.** 1.4894

**23.** 0.0011

**25.** 1.6094

**27.** 4.0431

**29.** $-5.0832$

**31.** 96.7583

**33.** 15.0293

**35.** 0.0305

**37.** 109.9472

**39.** We will use common logarithms for the conversion. Let $a = 10$, $b = 6$, and $M = 92$ and substitute in the change-of-base formula.

$$\log_b M = \frac{\log_a M}{\log_a b}$$

$$\log_6 92 = \frac{\log_{10} 92}{\log_{10} 6}$$

$$\approx \frac{1.963787827}{0.7781512504}$$

$$\approx 2.5237$$

**41.** We will use common logarithms for the conversion. Let $a = 10$, $b = 2$, and $M = 100$ and substitute in the change-of-base formula.

$$\log_2 100 = \frac{\log_{10} 100}{\log_{10} 2}$$

$$\approx \frac{2}{0.3010}$$

$$\approx 6.6439$$

**43.** We will use natural logarithms for the conversion. Let $a = e$, $b = 7$, and $M = 65$ and substitute in the change-of-base formula.

$$\log_7 65 = \frac{\ln 65}{\ln 7}$$

$$\approx \frac{4.1744}{1.9459}$$

$$\approx 2.1452$$

**45.** We will use natural logarithms for the conversion. Let $a = e$, $b = 0.5$, and $M = 5$ and substitute in the change-of-base formula.

$$\log_{0.5} 5 = \frac{\ln 5}{\ln 0.5}$$

$$\approx \frac{1.6094}{-0.6931}$$

$$\approx -2.3219$$

**47.** We will use common logarithms for the conversion. Let $a = 10$, $b = 2$, and $M = 0.2$ and substitute in the change-of-base formula.

$$\log_2 0.2 = \frac{\log_{10} 0.2}{\log_{10} 2}$$

$$\approx \frac{-0.6990}{0.3010}$$

$$\approx -2.3219$$

**49.** We will use natural logarithms for the conversion. Let $a = e$, $b = \pi$, and $M = 58$ and substitute in the change-of-base formula.

$$\log_\pi 58 = \frac{\ln 58}{\ln \pi}$$

$$\approx \frac{4.0604}{1.1447}$$

$$\approx 3.5471$$

**51.** Graph: $f(x) = e^x$

We find some function values with a calculator. We use these values to plot points and draw the graph.

| $x$ | $e^x$ |
|-----|-------|
| 0 | 1 |
| 1 | 2.7 |
| 2 | 7.4 |
| 3 | 20.1 |
| $-1$ | 0.4 |
| $-2$ | 0.1 |

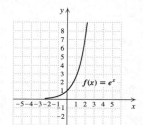

The domain is the set of real numbers and the range is $(0, \infty)$.

**53.** Graph: $f(x) = e^x + 3$

We find some function values, plot points, and draw the graph.

| $x$ | $e^x + 3$ |
|-----|-----------|
| 0 | 4 |
| 1 | 5.72 |
| 2 | 10.39 |
| $-1$ | 3.37 |
| $-2$ | 3.14 |

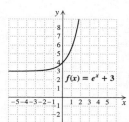

The domain is the set of real numbers and the range is $(3, \infty)$.

**55.** Graph: $f(x) = e^x - 2$

We find some function values, plot points, and draw the graph.

| $x$ | $e^x - 2$ |
|-----|-----------|
| 0 | $-1$ |
| 1 | 0.72 |
| 2 | 5.4 |
| $-1$ | $-1.6$ |
| $-2$ | $-1.9$ |

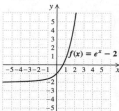

The domain is the set of real numbers and the range is $(-2, \infty)$.

**57.** Graph: $f(x) = 0.5e^x$

We find some function values, plot points, and draw the graph.

| $x$ | $0.5e^x$ |
|-----|----------|
| 0 | 0.5 |
| 1 | 1.36 |
| 2 | 3.69 |
| $-1$ | 0.18 |
| $-2$ | 0.07 |

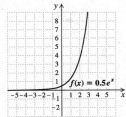

The domain is the set of real numbers and the range is $(0, \infty)$.

**59.** Graph: $f(x) = 0.5e^{2x}$

We find some function values, plot points, and draw the graph.

| $x$ | $0.5e^{2x}$ |
|-----|-------------|
| 0 | 0.5 |
| 1 | 3.69 |
| 2 | 27.30 |
| $-1$ | 0.07 |
| $-2$ | 0.01 |

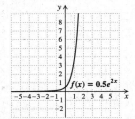

The domain is the set of real numbers and the range is $(0, \infty)$.

**61.** Graph: $f(x) = e^{x-3}$

We find some function values, plot points, and draw the graph.

| $x$ | $e^{x-3}$ |
|-----|-----------|
| 0 | 0.05 |
| 2 | 0.37 |
| 3 | 1 |
| 4 | 2.72 |
| $-2$ | 0.01 |

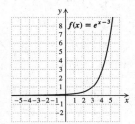

The domain is the set of real numbers and the range is $(0, \infty)$.

**63.** Graph: $f(x) = e^{x+2}$

We find some function values, plot points, and draw the graph.

| $x$ | $e^{x+2}$ |
|-----|-----------|
| 0 | 7.39 |
| $-1$ | 2.72 |
| $-2$ | 1 |
| $-3$ | 0.37 |
| $-4$ | 0.14 |

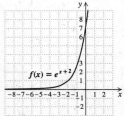

The domain is the set of real numbers and the range is $(0, \infty)$.

**65.** Graph: $f(x) = -e^x$

We find some function values, plot points, and draw the graph.

| $x$ | $-e^x$ |
|-----|--------|
| 0 | $-1$ |
| 1 | $-2.72$ |
| 2 | $-7.39$ |
| $-1$ | $-0.37$ |
| $-3$ | $-0.05$ |

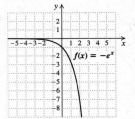

The domain is the set of real numbers and the range is $(-\infty, 0)$.

**67.** Graph: $g(x) = \ln x + 1$

We find some function values, plot points, and draw the graph.

| $x$ | $\ln x + 1$ |
|-----|-------------|
| 0.5 | 0.31 |
| 1 | 1 |
| 3 | 2.10 |
| 5 | 2.61 |
| 7 | 2.95 |

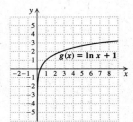

The domain is $(0, \infty)$ and the range is the set of real numbers.

**69.** Graph: $g(x) = \ln x - 2$

| $x$ | $\ln x - 2$ |
|-----|-------------|
| 1 | $-2$ |
| 2 | $-1.31$ |
| 3 | $-0.90$ |
| 4 | $-0.61$ |
| 5 | $-0.39$ |

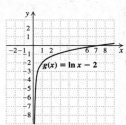

The domain is $(0, \infty)$ and the range is the set of real numbers.

**71.** Graph: $f(x) = 2 \ln x$

| $x$ | $2 \ln x$ |
|-----|-----------|
| 0.5 | $-1.4$ |
| 1 | 0 |
| 2 | 1.4 |
| 3 | 2.2 |
| 4 | 2.8 |
| 5 | 3.2 |
| 6 | 3.6 |

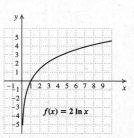

The domain is $(0, \infty)$ and the range is the set of real numbers.

**73.** Graph: $g(x) = -2 \ln x$

| $x$ | $-2 \ln x$ |
|-----|-----------|
| 0.5 | 1.4 |
| 1 | 0 |
| 2 | -1.4 |
| 3 | -2.2 |
| 4 | -2.8 |
| 5 | -3.2 |
| 6 | -3.6 |

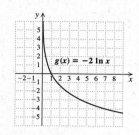

$g(x) = -2 \ln x$

The domain is $(0, \infty)$ and the range is the set of real numbers.

**75.** Graph: $f(x) = \ln(x + 2)$

We find some function values, plot points, and draw the graph.

| $x$ | $\ln(x + 2)$ |
|-----|-------------|
| 0 | 0.69 |
| 1 | 1.10 |
| 3 | 1.61 |
| 5 | 1.95 |
| -1 | 0 |
| -2 | Undefined |

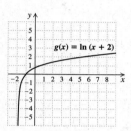

$g(x) = \ln (x + 2)$

The domain is $(-2, \infty)$ and the range is the set of real numbers.

**77.** Graph: $g(x) = \ln(x - 1)$

We find some function values, plot points, and draw the graph.

| $x$ | $\ln(x - 1)$ |
|-----|-------------|
| 1.1 | -2.30 |
| 2 | 0 |
| 3 | 0.69 |
| 4 | 1.10 |
| 6 | 1.61 |

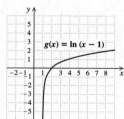

$g(x) = \ln (x - 1)$

The domain is $(1, \infty)$ and the range is the set of real numbers.

**79.** *Writing Exercise*

**81.**
$$4x^2 - 25 = 0$$
$$(2x + 5)(2x - 5) = 0$$
$$2x + 5 = 0 \quad \text{or} \quad 2x - 5 = 0$$
$$2x = -5 \quad \text{or} \quad 2x = 5$$
$$x = -\frac{5}{2} \quad \text{or} \quad x = \frac{5}{2}$$

The solutions are $-\dfrac{5}{2}$ and $\dfrac{5}{2}$.

**83.**
$$17x - 15 = 0$$
$$17x = 15$$
$$x = \frac{15}{17}$$

The solution is $\dfrac{15}{17}$.

**85.** $x^{1/2} - 6x^{1/4} + 8 = 0$

Let $u = x^{1/4}$.

$$u^2 - 6u + 8 = 0 \qquad \text{Substituting}$$
$$(u - 4)(u - 2) = 0$$
$$u = 4 \quad \text{or} \quad u = 2$$
$$x^{1/4} = 4 \quad \text{or} \quad x^{1/4} = 2$$
$$x = 256 \quad \text{or} \quad x = 16 \quad \begin{array}{l}\text{Raising both sides to}\\\text{the fourth power}\end{array}$$

Both numbers check. The solutions are 256 and 16.

**87.** *Writing Exercise*

**89.** We use the change-of-base formula.

$$\log_6 81 = \frac{\log 81}{\log 6}$$
$$= \frac{\log 3^4}{\log(2 \cdot 3)}$$
$$= \frac{4 \log 3}{\log 2 + \log 3}$$
$$\approx \frac{4(0.477)}{0.301 + 0.477}$$
$$\approx 2.452$$

**91.** We use the change-of-base formula.

$$\log_{12} 36 = \frac{\log 36}{\log 12}$$
$$= \frac{\log(2 \cdot 3)^2}{\log(2^2 \cdot 3)}$$
$$= \frac{2 \log(2 \cdot 3)}{\log 2^2 + \log 3}$$
$$= \frac{2(\log 2 + \log 3)}{2 \log 2 + \log 3}$$
$$\approx \frac{2(0.301 + 0.477)}{2(0.301) + 0.477}$$
$$\approx 1.442$$

**93.** Use the change-of-base formula with $a = e$ and $b = 10$. We obtain

$$\log M = \frac{\ln M}{\ln 10}.$$

**95.**
$$\log(492x) = 5.728$$
$$10^{5.728} = 492x$$
$$\frac{10^{5.728}}{492} = x$$
$$1086.5129 \approx x$$

**97.**
$$\log 692 + \log x = \log 3450$$
$$\log x = \log 3450 - \log 692$$
$$\log x = \log \frac{3450}{692}$$
$$x = \frac{3450}{692}$$
$$x \approx 4.9855$$

**99.** (a) Domain: $\{x | x > 0\}$, or $(0, \infty)$;

range: $\{y | y < 0.5135\}$, or $(-\infty, 0.5135)$;

(b) $[-1, 5, -10, 5]$;

(c)

$y = 3.4 \ln x - 0.25 e^x$

**101.** (a) Domain $\{x | x > 0\}$, or $(0, \infty)$;

range: $\{y | y > -0.2453\}$, or $(-0.2453, \infty)$

(b) $[-1, 5, -1, 10]$;

(c)

$y = 2x^3 \ln x$

**103.**

## Exercise Set 9.6

**1.** If we take the common logarithm on both sides, we see that choice (e) is correct.

**3.** $\ln x = 3$ means that 3 is the exponent to which we raise $e$ to get $x$, so choice (f) is correct.

**5.** By the product rule for logarithms, $\log_5 x + \log_5(x - 2) = \log_5[x(x - 2)] = \log_5(x^2 - 2x)$, so choice (b) is correct.

**7.** By the quotient rule for logarithms,

$\ln x - \ln(x - 2) = \ln \dfrac{x}{x - 2}$, so choice (g) is correct.

**9.**
$$2^x = 19$$
$$\log 2^x = \log 19$$
$$x \log 2 = \log 19$$
$$x = \frac{\log 19}{\log 2}$$
$$x \approx 4.248$$

The solution is $\dfrac{\log 19}{\log 2}$, or approximately 4.248.

**11.**
$$8^{x-1} = 17$$
$$\log 8^{x-1} = \log 17$$
$$(x - 1) \log 8 = \log 17$$
$$x - 1 = \frac{\log 17}{\log 8}$$
$$x = \frac{\log 17}{\log 8} + 1$$
$$x \approx 2.362$$

The solution is $\dfrac{\log 17}{\log 8} + 1$, or approximately 2.362.

**13.**

| | | |
|---|---|---|
| $e^t = 1000$ | | |
| $\ln e^t = \ln 1000$ | Taking ln on both sides |
| $t = \ln 1000$ | Finding the logarithm of the base to a power |
| $t \approx 6.908$ | Using a calculator |

**15.**
$$e^{0.03t} + 2 = 7$$
$$e^{0.03t} = 5$$

$\ln e^{0.03t} = \ln 5$    Taking ln on both sides

$0.03t = \ln 5$    Finding the logarithm of the base to a power

$$t = \frac{\ln 5}{0.03}$$
$$t \approx 53.648$$

**17.**
$$5 = 3^{x+1}$$
$$\log 5 = \log 3^{x+1}$$
$$\log 5 = (x + 1) \log 3$$
$$\log 5 = x \log 3 + \log 3$$
$$\log 5 - \log 3 = x \log 3$$
$$\frac{\log 5 - \log 3}{\log 3} = x, \text{ or}$$
$$\frac{\log 5}{\log 3} - 1 = x$$
$$0.465 \approx x$$

**19.** $2^{x+3} = 16$

Observe that $16 = 2^4$. Equating exponents, we have $x + 3 = 4$, or $x = 1$.

**21.**
$$4.9^x - 87 = 0$$
$$4.9^x = 87$$
$$\log 4.9^x = \log 87$$
$$x \log 4.9 = \log 87$$
$$x = \frac{\log 87}{\log 4.9}$$
$$x \approx 2.810$$

**23.**
$$19 = 2e^{4x}$$
$$\frac{19}{2} = e^{4x}$$
$$\ln\left(\frac{19}{2}\right) = \ln e^{4x}$$
$$\ln\left(\frac{19}{2}\right) = 4x$$
$$\frac{\ln\left(\frac{19}{2}\right)}{4} = x$$
$$0.563 \approx x$$

**25.** $7 + 3e^{5x} = 13$

$3e^{5x} = 6$

$e^{5x} = 2$

$\ln e^{5x} = \ln 2$

$5x = \ln 2$

$x = \dfrac{\ln 2}{5}$

$x \approx 0.139$

**27.** $\log_3 x = 4$

$x = 3^4$    Writing an equivalent exponential equation

$x = 81$

**29.** $\log_2 x = -3$

$x = 2^{-3}$    Writing an equivalent exponential equation

$x = \dfrac{1}{2^3}$, or $\dfrac{1}{8}$

**31.** $\ln x = 5$

$x = e^5$    Writing an equivalent exponential equation

$x \approx 148.413$

**33.** $\log_8 x = \dfrac{1}{3}$

$x = 8^{1/3} = \sqrt[3]{8}$

$x = 2$

**35.** $\ln 4x = 3$

$4x = e^3$

$x = \dfrac{e^3}{4} \approx 5.021$

**37.** $\log x = 2.5$    The base is 10.

$x = 10^{2.5}$

$x \approx 316.228$

**39.** $\ln(2x + 1) = 4$

$2x + 1 = e^4$

$2x = e^4 - 1$

$x = \dfrac{e^4 - 1}{2} \approx 26.799$

**41.** $\ln x = 1$

$x = e \approx 2.718$

**43.** $5 \ln x = -15$

$\ln x = -3$

$x = e^{-3} \approx 0.050$

**45.** $\log_2(8 - 6x) = 5$

$8 - 6x = 2^5$

$8 - 6x = 32$

$-6x = 24$

$x = -4$

The answer checks. The solution is $-4$.

**47.** $\log(x - 9) + \log x = 1$    The base is 10.

$\log_{10}[(x - 9)(x)] = 1$    Using the product rule

$x(x - 9) = 10^1$

$x^2 - 9x = 10$

$x^2 - 9x - 10 = 0$

$(x + 1)(x - 10) = 0$

$x = -1$ or $x = 10$

Check: For $-1$:

$$\log(x - 9) + \log x = 1$$

$$\log(-1 + 9) + \log(-1) \overset{?}{=} 1 \quad \text{FALSE}$$

For 10:

$$\log(x - 9) + \log x = 1$$

$$
\begin{array}{c|c}
\log(10 - 9) + \log (10) & 1 \\
\log 1 + \log 10 & \\
0 + 1 & \\
1 \overset{?}{=} 1 & \text{TRUE}
\end{array}
$$

The number $-1$ does not check, because negative numbers do not have logarithms. The solution is 10.

**49.** $\log x - \log(x + 3) = 1$    The base is 10.

$\log_{10} \dfrac{x}{x + 3} = 1$    Using the quotient rule

$\dfrac{x}{x + 3} = 10^1$

$x = 10(x + 3)$

$x = 10x + 30$

$-9x = 30$

$x = -\dfrac{10}{3}$

The number $-\dfrac{10}{3}$ does not check. The equation has no solution.

**51.** $\log_4(x + 3) = 2 + \log_4(x - 5)$

$\log_4(x + 3) - \log_4(x - 5) = 2$

$\log_4 \dfrac{x + 3}{x - 5} = 2$    Using the quotient rule

$\dfrac{x + 3}{x - 5} = 4^2$

$\dfrac{x + 3}{x - 5} = 16$

$x + 3 = 16(x - 5)$

$x + 3 = 16x - 80$

$83 = 15x$

$\dfrac{83}{15} = x$

The number $\dfrac{83}{15}$ checks. It is the solution.

**53.** $\log_7(x+1) + \log_7(x+2) = \log_7 6$

$\quad\log_7[(x+1)(x+2)] = \log_7 6$  Using the
$\qquad\qquad\qquad\qquad\qquad\qquad$ product rule

$\quad\log_7(x^2 + 3x + 2) = \log_7 6$

$\qquad\quad x^2 + 3x + 2 = 6$  $\quad$ Using the
$\qquad\qquad$ property of logarithmic equality

$\qquad\qquad x^2 + 3x - 4 = 0$

$\qquad\qquad (x+4)(x-1) = 0$

$x = -4$ $\;or\;$ $x = 1$

The number 1 checks, but $-4$ does not. The solution is 1.

**55.** $\log_5(x+4) + \log_5(x-4) = \log_5 20$

$\quad\log_5[(x+4)(x-4)] = \log_5 20$  Using the
$\qquad\qquad\qquad\qquad\qquad\qquad$ product rule

$\quad\quad\log_5(x^2 - 16) = \log_5 20$

$\qquad\qquad x^2 - 16 = 20$  $\quad$ Using the
$\qquad\qquad$ property of logarithmic equality

$\qquad\qquad\qquad x^2 = 36$

$\qquad\qquad\qquad x = \pm 6$

The number 6 checks, but $-6$ does not. The solution is 6.

**57.** $\ln(x+5) + \ln(x+1) = \ln 12$

$\quad\ln[(x+5)(x+1)] = \ln 12$

$\quad\ln(x^2 + 6x + 5) = \ln 12$

$\qquad\quad x^2 + 6x + 5 = 12$

$\qquad\quad x^2 + 6x - 7 = 0$

$\qquad\quad (x+7)(x-1) = 0$

$x = -7$ $\;or\;$ $x = 1$

The number $-7$ does not check, but 1 does. The solution is 1.

**59.** $\log_2(x+3) + \log_2(x-3) = 4$

$\quad\log_2[(x+3)(x-3)] = 4$

$\qquad\quad (x+3)(x-3) = 2^4$

$\qquad\qquad\qquad x^2 - 9 = 16$

$\qquad\qquad\qquad x^2 = 25$

$\qquad\qquad\qquad x = \pm 5$

The number 5 checks, but $-5$ does not. The solution is 5.

**61.** $\log_{12}(x+5) - \log_{12}(x-4) = \log_{12} 3$

$\quad\log_{12}\dfrac{x+5}{x-4} = \log_{12} 3$

$\qquad\quad\dfrac{x+5}{x-4} = 3$  Using the prop-
$\qquad\qquad\qquad$ erty of logarithmic
$\qquad\qquad\qquad$ equality

$\qquad\quad x + 5 = 3(x-4)$

$\qquad\quad x + 5 = 3x - 12$

$\qquad\qquad 17 = 2x$

$\qquad\qquad \dfrac{17}{2} = x$

The number $\dfrac{17}{2}$ checks and is the solution.

**63.** $\log_2(x-2) + \log_2 x = 3$

$\quad\log_2[(x-2)(x)] = 3$

$\qquad\quad x(x-2) = 2^3$

$\qquad\qquad x^2 - 2x = 8$

$\qquad\quad x^2 - 2x - 8 = 0$

$\quad (x-4)(x+2) = 0$

$x = 4$ $\;or\;$ $x = -2$

The number 4 checks, but $-2$ does not. The solution is 4.

**65.** *Writing Exercise*

**67.** $\quad y = kx$

$\quad 7.2 = k(0.8)$  $\quad$ Substituting

$\qquad 9 = k$  $\qquad\qquad$ Variation constant

$\qquad y = 9x$  $\qquad\quad$ Equation of variation

**69.**
$$T = 2\pi\sqrt{\dfrac{L}{32}}$$

$$\dfrac{T}{2\pi} = \sqrt{\dfrac{L}{32}}$$

$$\left(\dfrac{T}{2\pi}\right)^2 = \left(\sqrt{\dfrac{L}{32}}\right)^2$$

$$\dfrac{T^2}{4\pi^2} = \dfrac{L}{32}$$

$$32 \cdot \dfrac{T^2}{4\pi^2} = L$$

$$\dfrac{8T^2}{\pi^2} = L$$

**71. Familiarize.** Let $t =$ the time, in hours, it takes Joni and Miles to key in the score, working together. Then in $t$ hours Joni does $\dfrac{t}{2}$ of the job, Miles does $\dfrac{t}{3}$, and together they do 1 entire job.

**Translate.**

$$\dfrac{t}{2} + \dfrac{t}{3} = 1$$

**Carry out.** We solve the equation. First we multiply by the LCD, 6.

$$6\left(\dfrac{t}{2} + \dfrac{t}{3}\right) = 6 \cdot 1$$

$$6 \cdot \dfrac{t}{2} + 6 \cdot \dfrac{t}{3} = 6$$

$$3t + 2t = 6$$

$$5t = 6$$

$$t = \dfrac{6}{5}$$

**Check.** In $\dfrac{6}{5}$ hr Joni does $\dfrac{6/5}{2}$, or $\dfrac{3}{5}$ of the job, and Miles does $\dfrac{6/5}{3}$, or $\dfrac{2}{5}$ of the job. Together they do $\dfrac{3}{5} + \dfrac{2}{5}$ or 1 entire job. The answer checks.

**State.** It takes Joni and Miles $\dfrac{6}{5}$ hr, or $1\dfrac{1}{5}$ hr, to do the job, working together.

**73.** *Writing Exercise*

**75.**
$$27^x = 81^{2x-3}$$
$$(3^3)^x = (3^4)^{2x-3}$$
$$3^{3x} = 3^{8x-12}$$
$$3x = 8x - 12$$
$$12 = 5x$$
$$\frac{12}{5} = x$$

The solution is $\frac{12}{5}$.

**77.** $\log_x (\log_3 27) = 3$
$$\log_3 27 = x^3$$
$$3 = x^3 \qquad (\log_3 27 = 3)$$
$$\sqrt[3]{3} = x$$

The solution is $\sqrt[3]{3}$.

**79.**
$$x \cdot \log \frac{1}{8} = \log 8$$
$$x \cdot \log 8^{-1} = \log 8$$
$$x(-\log 8) = \log 8 \quad \text{Using the power rule}$$
$$x = -1$$

The solution is $-1$.

**81.**
$$2^{x^2+4x} = \frac{1}{8}$$
$$2^{x^2+4x} = \frac{1}{2^3}$$
$$2^{x^2+4x} = 2^{-3}$$
$$x^2 + 4x = -3$$
$$x^2 + 4x + 3 = 0$$
$$(x+3)(x+1) = 0$$
$$x = -3 \text{ or } x = -1$$

The solutions are $-3$ and $-1$.

**83.** $\log_5 |x| = 4$
$$|x| = 5^4$$
$$|x| = 625$$
$$x = 625 \text{ or } x = -625$$

The solutions are $625$ and $-625$.

**85.**
$$\log \sqrt{2x} = \sqrt{\log 2x}$$
$$\log (2x)^{1/2} = \sqrt{\log 2x}$$
$$\frac{1}{2} \log 2x = \sqrt{\log 2x}$$
$$\frac{1}{4} (\log 2x)^2 = \log 2x \quad \text{Squaring both sides}$$
$$\frac{1}{4}(\log 2x)^2 - \log 2x = 0$$

Let $u = \log 2x$.
$$\frac{1}{4}u^2 - u = 0$$
$$u\left(\frac{1}{4}u - 1\right) = 0$$

$$u = 0 \quad \text{or} \quad \frac{1}{4}u - 1 = 0$$
$$u = 0 \quad \text{or} \quad \frac{1}{4}u = 1$$
$$u = 0 \quad \text{or} \quad u = 4$$
$$\log 2x = 0 \quad \text{or} \quad \log 2x = 4 \quad \begin{array}{l}\text{Replacing } u \\ \text{with } \log 2x\end{array}$$
$$2x = 10^0 \quad \text{or} \quad 2x = 10^4$$
$$2x = 1 \quad \text{or} \quad 2x = 10,000$$
$$x = \frac{1}{2} \quad \text{or} \quad x = 5000$$

Both numbers check. The solutions are $\frac{1}{2}$ and $5000$.

**87.**
$$3^{x^2} \cdot 3^{4x} = \frac{1}{27}$$
$$3^{x^2+4x} = 3^{-3}$$
$$x^2 + 4x = -3 \quad \text{The exponents must be equal.}$$
$$x^2 + 4x + 3 = 0$$
$$(x+1)(x+3) = 0$$
$$x = -1 \text{ or } x = -3$$

Both numbers check. The solutions are $-1$ and $-3$.

**89.**
$$\log x^{\log x} = 25$$
$$\log x (\log x) = 25 \quad \text{Using the power rule}$$
$$(\log x)^2 = 25$$
$$\log x = \pm 5$$
$$x = 10^5 \quad \text{or} \quad x = 10^{-5}$$
$$x = 100,000 \quad \text{or} \quad x = \frac{1}{100,000}$$

Both numbers check. The solutions are $100,000$ and $\frac{1}{100,000}$.

**91.**
$$(81^{x-2})(27^{x+1}) = 9^{2x-3}$$
$$[(3^4)^{x-2}][(3^3)^{x+1}] = (3^2)^{2x-3}$$
$$(3^{4x-8})(3^{3x+3}) = 3^{4x-6}$$
$$3^{7x-5} = 3^{4x-6}$$
$$7x - 5 = 4x - 6$$
$$3x = -1$$
$$x = -\frac{1}{3}$$

The solution is $-\frac{1}{3}$.

**93.** $2^y = 16^{x-3}$ and $3^{y+2} = 27^x$
$$2^y = (2^4)^{x-3} \text{ and } 3^{y+2} = (3^3)^x$$
$$y = 4x - 12 \text{ and } y + 2 = 3x$$
$$12 = 4x - y \text{ and } 2 = 3x - y$$

Solving this system of equations we get $x = 10$ and $y = 28$. Then $x + y = 10 + 28 = 38$.

**95.** Find the first coordinate of the point of intersection of $y_1 = \ln x$ and $y_2 = \log x$. The value of $x$ for which the natural logarithm of $x$ is the same as the common logarithm of $x$ is $1$.

## Exercise Set 9.7

**1. a)** Replace $S(t)$ with 2800 and solve for $t$.

$$S(t) = 200 \cdot 2^t$$
$$2800 = 200 \cdot 2^t$$
$$14 = 2^t$$
$$\ln 14 = \ln 2^t$$
$$\ln 14 = t \ln 2$$
$$\frac{\ln 14}{\ln 2} = t$$
$$4 \approx t$$

Sales of DVD players first reached \$2800 million about 4 yr after 1997, or in 2001.

**b)** $S(0) = 200 \cdot 2^0 = 200 \cdot 1 = 200$, so to find the doubling time we replace $S(t)$ with 400 and solve for $t$.

$$400 = 200 \cdot 2^t$$
$$2 = 2^t$$
$$1 = t \quad \text{The exponents must be the same.}$$

The doubling time is about 1 year.

**3. a)** Find $N(21)$.

$$N(x) = 1337(0.9)^x$$
$$N(21) = 1337(0.9)^{21}$$
$$N(21) \approx 146.293$$

We estimate that there are about 146.293 thousand, or 146,293,000 21-year-old skateboarders.

**b)** $6300 = 6.3$ thousand; substitute 6.3 for $N(x)$ and solve for $x$.

$$6.3 = 1337(0.9)^x$$
$$0.0047 \approx 0.9^x$$
$$\log 0.0047 \approx \log 0.9^x$$
$$\log 0.0047 \approx x \log 0.9$$
$$\frac{\log 0.0047}{\log 0.9} \approx x$$
$$51 \approx x$$

At about age 51 there are only 6300 skateboarders.

**5. a)** Replace $A(t)$ with 35,000 and solve for $t$.

$$A(t) = 29,000(1.03)^t$$
$$35,000 = 29,000(1.03)^t$$
$$1.207 \approx (1.03)^t$$
$$\log 1.207 \approx \log(1.03)^t$$
$$\log 1.207 \approx t \log 1.03$$
$$\frac{\log 1.207}{\log 1.03} \approx t$$
$$6.4 \approx t$$

The amount due will reach \$35,000 after about 6.4 years.

**b)** Replace $A(t)$ with 2(29,000), or 58,000, and solve for $t$.

$$58,000 = 29,000(1.03)^t$$
$$2 = (1.03)^t$$
$$\log 2 = \log(1.03)^t$$
$$\log 2 = t \log 1.03$$
$$\frac{\log 2}{\log 1.03} = t$$
$$23.4 \approx t$$

The doubling time is about 23.4 years.

**7. a)** Substitute 50 for $P(t)$ and solve for $t$.

$$P(t) = 63.03(0.95)^t$$
$$50 = 63.03(0.95)^t$$
$$0.7933 \approx 0.95^t$$
$$\ln 0.7933 \approx \ln 0.95^t$$
$$\ln 0.7933 \approx t \ln 0.95$$
$$\frac{\ln 0.7933}{\ln 0.95} \approx t$$
$$4.5 \approx t$$

The percentage of phones that are land lines will be 50% about 4.5 yr after 2000, so the percentage will drop below 50% in about 2005.

**b)** Substitute 25 for $P(t)$ and solve for $t$.

$$25 = 63.03(0.95)^t$$
$$0.3966 \approx 0.95^t$$
$$\ln 0.3966 \approx \ln 0.95^t$$
$$\ln 0.3966 \approx t \ln 0.95$$
$$\frac{\ln 0.3966}{\ln 0.95} \approx t$$
$$18 \approx t$$

25% of phones will be land lines about 18 yr after 2000, so the percentage will drop below 25% in about 2018.

**9. a)** $P(t)$ is given in thousands, so we substitute 30 for $P(t)$ and solve for $t$.

$$P(t) = 5.5(1.047)^t$$
$$30 = 5.5(1.047)^t$$
$$5.455 \approx 1.047^t$$
$$\log 5.455 \approx \log 1.047^t$$
$$\log 5.455 \approx t \log 1.047$$
$$\frac{\log 5.455}{\log 1.047} \approx t$$
$$37 \approx t$$

The humpback whale population will reach 30,000 about 37 yr after 1982, or in 2019.

b) $P(0) = 5.5(1.047)^0 = 5.5(1) = 5.5$ and $2(5.5) = 11$, so we substitute 11 for $P(t)$ and solve for $t$.

$$11 = 5.5(1.047)^t$$
$$2 = 1.047^t$$
$$\log 2 = \log 1.047^t$$
$$\log 2 = t \log 1.047$$
$$\frac{\log 2}{\log 1.047} = t$$
$$15.1 \approx t$$

The doubling time is about 15.1 yr.

**11.** $pH = -\log[H^+]$
$$= -\log[1.3 \times 10^{-5}]$$
$$\approx -(-4.886057) \quad \text{Using a calculator}$$
$$\approx 4.9$$

The pH of fresh-brewed coffee is about 4.9.

**13.** $pH = -\log[H^+]$
$$7.0 = -\log[H^+]$$
$$-7.0 = \log[H^+]$$
$$10^{-7.0} = [H^+] \quad \text{Converting to an exponential equation}$$

The hydrogen ion concentration is $10^{-7}$ moles per liter.

**15.** $L = 10 \cdot \log \dfrac{I}{I_0}$
$$= 10 \cdot \log \frac{3.2 \times 10^{-6}}{10^{-12}}$$
$$= 10 \cdot \log(3.2 \times 10^6)$$
$$\approx 10(6.5)$$
$$\approx 65$$

The intensity of sound in normal conversation is about 65 decibels.

**17.**
$$L = 10 \cdot \log \frac{I}{I_0}$$
$$105 = 10 \cdot \log \frac{I}{10^{-12}}$$
$$10.5 = \log \frac{I}{10^{-12}}$$
$$10.5 = \log I - \log 10^{-12} \quad \text{Using the quotient rule}$$
$$10.5 = \log I - (-12) \quad (\log 10^a = a)$$
$$10.5 = \log I + 12$$
$$-1.5 = \log I$$
$$10^{-1.5} = I \quad \text{Converting to an exponential equation}$$
$$3.2 \times 10^{-2} \approx I$$

The intensity of the sound is $10^{-1.5}$ W/m$^2$, or about $3.2 \times 10^{-2}$ W/m$^2$.

**19.** a) Substitute 0.025 for $k$:
$$P(t) = P_0 e^{0.025t}$$

b) To find the balance after one year, replace $P_0$ with 5000 and $t$ with 1. We find $P(1)$:
$$P(1) = 5000 \, e^{0.025(1)} = 5000 \, e^{0.025} \approx \$5126.58$$

To find the balance after 2 years, replace $P_0$ with 5000 and $t$ with 2. We find $P(2)$:
$$P(2) = 5000 \, e^{0.025(2)} = 5000 \, e^{0.05} \approx \$5256.36$$

c) To find the doubling time, replace $P_0$ with 5000 and $P(t)$ with 10,000 and solve for $t$.
$$10,000 = 5000 \, e^{0.025t}$$
$$2 = e^{0.025t}$$
$$\ln 2 = \ln e^{0.025t} \quad \text{Taking the natural logarithm on both sides}$$
$$\ln 2 = 0.025t \quad \text{Finding the logarithm of the base to a power}$$
$$\frac{\ln 2}{0.025} = t$$
$$27.7 \approx t$$

The investment will double in about 27.7 years.

**21.** a) $P(t) = 292.80e^{0.009t}$, where $P(t)$ is in millions and $t$ is the number of years after 2004.

b) In 2005, $t = 2005 - 2004 = 1$. Find $P(1)$.
$$P(1) = 292.80e^{0.009(1)} = 292.80e^{0.009} \approx 295.45$$

The U.S. population will be about 295.45 million in 2005.

c) Substitute 325 for $P(t)$ and solve for $t$.
$$325 = 292.80e^{0.009t}$$
$$1.1100 \approx e^{0.009t}$$
$$\ln 1.1100 \approx \ln e^{0.009t}$$
$$\ln 1.1100 \approx 0.009t$$
$$\frac{\ln 1.1100}{0.009} \approx t$$
$$12 \approx t$$

The U.S. population will reach 325 million about 12 yr after 2004, or in 2016.

**23.** The exponential growth function is $S(t) = S_0 e^{0.103t}$. We replace $S(t)$ with $2S_0$ and solve for $t$.
$$2S_0 = S_0 e^{0.103t}$$
$$2 = e^{0.103t}$$
$$\ln 2 = \ln e^{0.103t}$$
$$\ln 2 = 0.103t$$
$$\frac{\ln 2}{0.103} = t$$
$$6.7 \approx t$$

The doubling time for iPod sales is about 6.7 months.

**25.** $Y(x) = 67.17 \ln \dfrac{x}{4.5}$

a) $Y(7) = 67.17 \ln \dfrac{7}{4.5} \approx 30$

The world population will reach 7 billion about 30 yr after 1980, or in 2010.

b) $Y(8) = 67.17 \ln \dfrac{8}{4.5} \approx 39$

The world population will reach 8 billion about 39 yr after 1980, or in 2019.

c) Plot the points found in parts (a) and (b) and others as necessary and draw the graph.

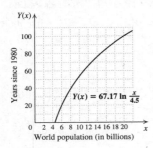

Y(x) = 67.17 ln $\frac{x}{4.5}$

27. a) $S(0) = 68 - 20 \log (0 + 1) = 68 - 20 \ \log \ 1 =$
$68 - 20(0) = 68\%$

b) $S(4) = 68 - 20 \log (4 + 1) = 68 - 20 \ \log \ 5 \approx$
$68 - 20(0.69897) \approx 54\%$

$S(24) = 68 - 20 \log (24 + 1) =$
$68 - 20 \ \log \ 25 \approx 68 - 20 \ (1.39794) \approx$
$40\%$

c) Using the values we computed in parts (a) and (b) and any others we wish to calculate, we sketch the graph:

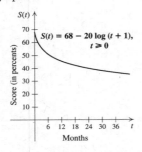

S(t) = 68 − 20 log (t + 1), t ≥ 0

d) $50 = 68 - 20 \ \log \ (t + 1)$
$-18 = -20 \ \log \ (t + 1)$
$0.9 = \log \ (t + 1)$
$10^{0.9} = t + 1$
$7.9 \approx t + 1$
$6.9 \approx t$

After about 6.9 months, the average score was 50.

29. a) We use the growth equation $N(t) = N_0 e^{kt}$, where $t$ is the number of years since 2000. In 2000, at $t = 0$, 17 people were infected. We substitute 17 for $N_0$:
$N(t) = 17e^{kt}$.

To find the exponential growth rate $k$, observe that 1 year later 29 people were infected.

$N(1) = 17e^{k \cdot 1}$   Substituting 1 for $t$
$29 = 17e^k$   Substituting 29 for $N(1)$
$1.706 \approx e^k$
$\ln 1.706 \approx \ln e^k$
$\ln 1.706 \approx k$
$0.534 \approx k$

The exponential function is $N(t) = 17e^{0.534t}$, where $t$ is the number of years since 2000.

b) In 2006, $t = 2006 - 2000$, or 6. Find $N(6)$.
$N(6) = 17e^{0.534(6)}$
$= 17e^{3.204}$
$\approx 418.7$

Approximately 419 people will be infected in 2006.

31. a) Let $P(t)$ represent farmland, in millions of acres, and let $t$ represent the number of years after 1990.
$P(t) = 987e^{-kt}$
$P(12) = 987e^{-k \cdot 12}$
$941 = 987e^{-12k}$
$0.9534 \approx e^{-12k}$
$\ln 0.9534 \approx \ln e^{-12k}$
$\ln 0.9534 \approx -12k$
$\dfrac{\ln 0.9534}{-12} \approx k$
$0.004 \approx k$

The exponential function is $P(t) = 987e^{-0.004t}$, where $P(t)$ and $t$ are as described above.

b) In 2008, $t = 2008 - 1990 = 18$.
$P(18) = 987e^{-0.004(18)}$
$= 987e^{-0.072}$
$\approx 918$

In 2008 there will be about 918 million acres of U.S. farmland.

c) Substitute 800 for $P(t)$ and solve for $t$.
$800 = 987e^{-0.004t}$
$0.8105 \approx e^{-0.004t}$
$\ln 0.8105 \approx \ln e^{-0.004t}$
$\ln 0.8105 \approx -0.004t$
$\dfrac{\ln 0.8105}{-0.004} \approx t$
$53 \approx t$

There will be 800 million acres of U.S. farmland about 53 yr after 1990, or in 2043.

33. We will use the function derived in Example 7:
$P(t) = P_0 e^{-0.00012t}$

If the scrolls had lost 22.3% of their carbon-14 from an initial amount $P_0$, then 77.7%($P_0$) is the amount present. To find the age $t$ of the scrolls, we substitute 77.7%($P_0$), or $0.777P_0$, for $P(t)$ in the function above and solve for $t$.

$$0.777 P_0 = P_0 e^{-0.00012t}$$

$$0.777 = e^{-0.00012t}$$

$$\ln 0.777 = \ln e^{-0.00012t}$$

$$-0.2523 \approx -0.00012t$$

$$t \approx \frac{-0.2523}{-0.00012} \approx 2103$$

The scrolls are about 2103 years old.

**35.** The function $P(t) = P_0 e^{-kt}$, $k > 0$, can be used to model decay. For iodine-131, $k = 9.6\%$, or 0.096. To find the half-life we substitute 0.096 for $k$ and $\frac{1}{2} P_0$ for $P(t)$, and solve for $t$.

$$\frac{1}{2} P_0 = P_0 e^{-0.096t}, \text{ or } \frac{1}{2} = e^{-0.096t}$$

$$\ln \frac{1}{2} = \ln e^{-0.096t} = -0.096t$$

$$t = \frac{\ln 0.5}{-0.096} \approx \frac{-0.6931}{-0.096} \approx 7.2 \text{ days}$$

**37.** The function $P(t) = P_0 e^{-kt}$, $k > 0$, can be used to model decay. We substitute $\frac{1}{2} P_0$ for $P(t)$ and 1 for $t$ and solve for the decay rate $k$.

$$\frac{1}{2} P_0 = P_0 e^{-k \cdot 1}$$

$$\frac{1}{2} = e^{-k}$$

$$\ln \frac{1}{2} = \ln e^{-k}$$

$$-0.693 \approx -k$$

$$0.693 \approx k$$

The decay rate is 0.693, or 69.3% per year.

**39.** a) We start with the exponential growth equation

$$V(t) = V_0 e^{kt}, \text{ where } t \text{ is the number}$$
of years after 1991.

Substituting 451,000 for $V_0$, we have

$$V(t) = 451,000 e^{kt}.$$

To find the exponential growth rate $k$, observe that the card sold for \$1.1 million, or \$1,100,000 in 2000, or 9 years after 1991. We substitute and solve for $k$.

$$V(9) = 451,000 e^{k \cdot 9}$$

$$1,100,000 = 451,000 e^{9k}$$

$$2.4390 \approx e^{9k}$$

$$\ln 2.4390 \approx \ln e^{9k}$$

$$\ln 2.4390 \approx 9k$$

$$\frac{\ln 2.4390}{9} \approx k$$

$$0.099 \approx k$$

Thus, the exponential growth function is $V(t) = 451,000 e^{0.099t}$, where $t$ is the number of years after 1991.

b) In 2006, $t = 2006 - 1991 = 15$

$$V(15) = 451,000 e^{0.099(15)} \approx 1,991,149$$

The card's value in 2006 will be about \$1.99 million

c) Substitute 2(\$451,000), or \$902,000 for $V(t)$ and solve for $t$.

$$902,000 = 451,000 e^{0.099t}$$

$$2 = e^{0.099t}$$

$$\ln 2 = \ln e^{0.099t}$$

$$\ln 2 = 0.099t$$

$$\frac{\ln 2}{0.099} = t$$

$$7.0 \approx t$$

The doubling time is about 7.0 years.

d) Substitute \$3,000,000 for $V(t)$ and solve for $t$.

$$3,000,000 = 451,000 e^{0.099t}$$

$$6.6519 \approx e^{0.099t}$$

$$\ln 6.6519 \approx \ln e^{0.099t}$$

$$\ln 6.6519 \approx 0.099t$$

$$\frac{\ln 6.6519}{0.099} \approx t$$

$$19 \approx t$$

The value of the card will first exceed \$3,000,000 about 19 years after 1991, or in 2010.

**41.** *Writing Exercise*

**43.** Graph $y = x^2 - 8x$.

First we find the vertex.

$$-\frac{b}{2a} = -\frac{-8}{2 \cdot 1} = 4$$

When $x = 4$, $y = 4^2 - 8 \cdot 4 = 16 - 32 = -16$.

The vertex is $(4, -16)$ and the axis of symmetry is $x = 4$. We plot a few points on either side of the vertex and graph the parabola.

| $x$ | $y$ |
|---|---|
| 4 | -16 |
| 0 | 0 |
| 2 | -12 |
| 5 | -15 |
| 6 | -12 |

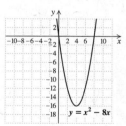

**45.** Graph $f(x) = 3x^2 - 5x - 1$.

First we find the vertex.

$$-\frac{b}{2a} = -\frac{-5}{2 \cdot 3} = \frac{5}{6}$$

$$f\left(\frac{5}{6}\right) = 3\left(\frac{5}{6}\right)^2 - 5 \cdot \frac{5}{6} - 1 = -\frac{37}{12}$$

The vertex is $\left(\frac{5}{6}, -\frac{37}{12}\right)$ and the axis of symmetry is

$x = \frac{5}{6}$. We plot a few points on either side of the vertex and graph the parabola.

| $x$ | $f(x)$ |
|-----|--------|
| $\frac{5}{6}$ | $-\frac{37}{12}$ |
| 0 | $-1$ |
| $-1$ | 7 |
| 2 | 1 |
| 3 | 11 |

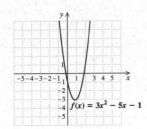

$f(x) = 3x^2 - 5x - 1$

**47.**
$$x^2 - 8x = 7$$

$$x^2 - 8x + 16 = 7 + 16 \qquad \text{Adding } \left[\frac{1}{2}(-8)\right]^2$$

$$(x-4)^2 = 23$$

$$x - 4 = \pm\sqrt{23}$$

$$x = 4 \pm \sqrt{23}$$

The solutions are $4 \pm \sqrt{23}$.

**49.** *Writing Exercise*

**51.** We will use the exponential growth function $V(t) = V_0 e^{kt}$, where $t$ is the number of years after 2004 and $V(t)$ is in millions of dollars. Substitute 24 for $V(t)$, 0.04 for $k$, and 6 for $t$ and solve for $V_0$.

$$24 = V_0 e^{0.04(6)}$$

$$24 = V_0 e^{0.24}$$

$$\frac{24}{e^{0.24}} = V_0$$

$$18.9 \approx V_0$$

About \$18.9 million would need to be invested.

**53.** From Exercise 7 we know that the percentage of U.S. phone lines that are land lines is given by $P(t) = 63.03(0.95)^t$, where $t$ is the number of years after 2000. Then the percentage of U.S. phone lines that are cellular $t$ years after 2000 is given by the function $P(t) = 100 - 63.03(0.95)^t$.

**55.** First we find $k$. When $t = 24{,}360$, $P(t) = 0.5P_0$.

$$0.5P_0 = P_0 e^{-k \cdot 24{,}360}$$

$$0.5 = e^{-24{,}360k}$$

$$\ln 0.5 = \ln e^{-24{,}360k}$$

$$\ln 0.5 = -24{,}360k$$

$$\frac{\ln 0.5}{-24{,}360} = k$$

$$0.0000285 \approx k$$

Now we have a function for the decay of plutonium-239.

$$P(t) = P_0 e^{-0.0000285t}$$

If a fuel rod has lost 90% of its plutonium, then 10% of the initial amount is still present. We substitute and solve for $t$.

$$0.1P_0 = P_0 e^{-0.0000285t}$$

$$0.1 = e^{-0.0000285t}$$

$$\ln 0.1 = \ln e^{-0.0000285t}$$

$$\ln 0.1 = -0.0000285t$$

$$\frac{\ln 0.1}{-0.0000285} = t$$

$$80{,}792 \approx t$$

It will take about 80,792 yr for the fuel rod of plutonium -239 to lose 90% of its radioactivity.

**57.** Consider an exponential growth function $P(t) = P_0 e^{kt}$. Suppose that at time $T$, $P(T) = 2P_0$.

Solve for $T$:

$$2P_0 = P_0 e^{kT}$$

$$2 = e^{kT}$$

$$\ln 2 = \ln e^{kT}$$

$$\ln 2 = kT$$

$$\frac{\ln 2}{k} = T$$

**59.** *Writing Exercise*

# Chapter 10

# Conic Sections

**1.** $(x-2)^2 + (y+5)^2 = 9$, or $(x-2)^2 + [y-(-5)]^2 = 3^2$, is the equation of a circle with center $(2,-5)$ and radius 3, so choice $(f)$ is correct.

**3.** $(x-5)^2 + (y+2)^2 = 9$, or $(x-5)^2 + [y-(-2)]^2 = 3^2$, is the equation of a circle with center $(5,-2)$ and radius 3, so choice (g) is correct.

**5.** $y = (x-2)^2 - 5$ is the equation of a parabola with vertex $(2,-5)$ that opens upward, so choice (c) is correct.

**7.** $x = (y-2)^2 - 5$ is the equation of a parabola with vertex $(-5,2)$ that opens to the right, so choice (d) is correct.

**9.** $y = -x^2$

This is equivalent to $y = -(x-0)^2 + 0$. The vertex is $(0,0)$.

We choose some $x$-values on both sides of the vertex and compute the corresponding values of $y$. The graph opens down, because the coefficient of $x^2$, $-1$, is negative.

| $x$ | $y$ |
|---|---|
| 0 | 0 |
| 1 | $-1$ |
| 2 | $-4$ |
| $-1$ | $-1$ |
| $-2$ | $-4$ |

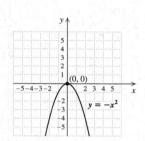

**11.** $y = -x^2 + 4x - 5$

We can find the vertex by computing the first coordinate, $x = -b/2a$, and then substituting to find the second coordinate:

$$x = -\frac{b}{2a} = -\frac{4}{2(-1)} = 2$$
$$y = -x^2 + 4x - 5 = -(2)^2 + 4(2) - 5 = -1$$

The vertex is $(2,-1)$.

We choose some $x$-values and compute the corresponding values for $y$. The graph opens downward because the coefficient of $x^2$, $-1$, is negative.

| $x$ | $y$ |
|---|---|
| 2 | $-1$ |
| 3 | $-2$ |
| 4 | $-5$ |
| 1 | $-2$ |
| 0 | $-5$ |

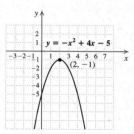

**13.** $x = y^2 - 4y + 2$

We find the vertex by completing the square.

$$x = (y^2 - 4y + 4) + 2 - 4$$
$$x = (y-2)^2 - 2$$

The vertex is $(-2,2)$.

To find ordered pairs, we choose values for $y$ and compute the corresponding values of $x$. The graph opens to the right, because the coefficient of $y^2$, 1, is positive.

| $x$ | $y$ |
|---|---|
| 7 | $-1$ |
| 2 | 0 |
| $-1$ | 1 |
| $-2$ | 2 |
| $-1$ | 3 |

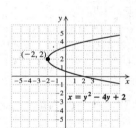

**15.** $x = y^2 + 3$

$$x = (y-0)^2 + 3$$

The vertex is $(3,0)$.

To find the ordered pairs, we choose $y$-values and compute the corresponding values for $x$. The graph opens to the right, because the coefficient of $y^2$, 1, is positive.

| $x$ | $y$ |
|---|---|
| 3 | 0 |
| 4 | 1 |
| 7 | 2 |
| 4 | $-1$ |
| 7 | $-2$ |

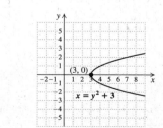

**17.** $x = -\dfrac{1}{2}y^2$

$x = -\dfrac{1}{2}(y-0)^2 + 0$

The vertex is $(0,0)$.

We choose $y$-values and compute the corresponding values for $x$. The graph opens to the left, because the coefficient of $y^2$, $-\dfrac{1}{2}$, is negative.

| $x$ | $y$ |
|-----|-----|
| 0   | 0   |
| -2  | 2   |
| -8  | 4   |
| -2  | -2  |
| -8  | -4  |

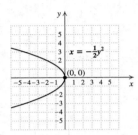

**19.** $x = -y^2 - 4y$

We find the vertex by computing the second coordinate, $y = -b/2a$, and then substituting to find the first coordinate:

$y = -\dfrac{b}{2a} = -\dfrac{-4}{2(-1)} = -2$

$x = -y^2 - 4y = (-2)^2 - 4(-2) = 4$

The vertex is $(4, -2)$.

We choose $y$-values and compute the corresponding values for $x$. The graph opens to the left, because the coefficient of $y^2$, $-1$, is negative.

| $x$ | $y$ |
|-----|-----|
| 4   | -2  |
| -5  | 1   |
| 0   | 0   |
| 3   | -1  |
| 3   | -3  |

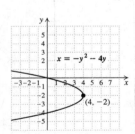

**21.** $x = 4 - y - y^2$

We find the vertex by completing the square.

$x = -(y^2 + y) + 4$

$x = -\left(y^2 + y + \dfrac{1}{4}\right) + 4 + \dfrac{1}{4}$

$x = -\left(y + \dfrac{1}{2}\right)^2 + \dfrac{17}{4}$

The vertex is $\left(\dfrac{17}{4}, -\dfrac{1}{2}\right)$.

We choose $y$-values and compute the corresponding values for $x$. The graph opens to the left, because the coefficient of $y^2$, $-1$, is negative.

| $x$ | $y$ |
|----------------|----------------|
| $\dfrac{17}{4}$ | $-\dfrac{1}{2}$ |
| 4              | 0              |
| 2              | 1              |
| -2             | 2              |
| 4              | -1             |
| 2              | -2             |
| -2             | -3             |

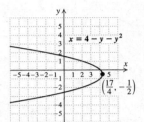

**23.** $y = x^2 - 2x + 1$

$y = (x-1)^2 + 0$

The vertex is $(1, 0)$.

We choose $x$-values and compute the corresponding values for $y$. The graph opens upward, because the coefficient of $x^2$, $1$, is positive.

| $x$ | $y$ |
|-----|-----|
| 1   | 0   |
| 0   | 1   |
| -1  | 4   |
| 2   | 1   |
| 3   | 4   |

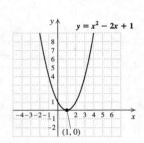

**25.** $x = -y^2 + 2y - 1$

We find the vertex by computing the second coordinate, $y = -b/2a$, and then substituting to find the first coordinate.

$y = -\dfrac{b}{2a} = -\dfrac{2}{2(-1)} = 1$

$x = -y^2 + 2y - 1 = -(1)^2 + 2(1) - 1 = 0$

The vertex is $(0, 1)$.

We choose $y$-values and compute the corresponding values for $x$. The graph opens to the left, because the coefficient of $y^2$, $-1$, is negative.

| $x$ | $y$ |
|-----|-----|
| -4  | 3   |
| -1  | 2   |
| -1  | 0   |
| -4  | -1  |
| -4  | 3   |

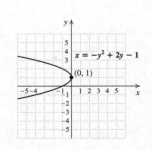

**27.** $x = -2y^2 - 4y + 1$

We find the vertex by completing the square.

$x = -2(y^2 + 2y) + 1$

$x = -2(y^2 + 2y + 1) + 1 + 2$

$x = -2(y + 1)^2 + 3$

The vertex is $(-3, -1)$.

We choose $y$-values and compute the corresponding values for $x$. The graph opens to the left, because the coefficient of $y^2$, $-2$, is negative.

| $x$ | $y$ |
|-----|-----|
| 3 | $-1$ |
| 1 | $-2$ |
| $-5$ | $-3$ |
| 1 | 0 |
| $-5$ | 1 |

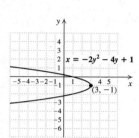

**29.** $d = \sqrt{(x_2 - x_1)^2 + (y_2 + y_1)^2}$  Distance formula

$= \sqrt{(5 - 1)^2 + (9 - 6)^2}$  Substituting

$= \sqrt{4^2 + 3^2}$

$= \sqrt{25} = 5$

**31.** $d = \sqrt{(x_2 - x_1)^2 + (y_2 - y_1)^2}$  Distance formula

$= \sqrt{(3 - 0)^2 + [-4 - (-7)]^2}$  Substituting

$= \sqrt{3^2 + 3^2}$

$= \sqrt{18} \approx 4.243$  Simplifying and approximating

**33.** $d = \sqrt{(x_2 - x_1)^2 + (y_2 - y_1)^2}$

$= \sqrt{[6 - (-4)]^2 + (-6 - 4)^2}$

$= \sqrt{200} \approx 14.142$

**35.** $d = \sqrt{(x_2 - x_1)^2 + (y_2 - y_1)^2}$

$= \sqrt{(-9.2 - 8.6)^2 + [-3.4 - (-3.4)]^2}$

$= \sqrt{(-17.8)^2 + 0^2}$

$= \sqrt{316.84} = 17.8$

(Since these points are on a horizontal line, we could have found the distance between them by finding $|x_2 - x_1| = |-9.2 - 8.6| = |-17.8| = 17.8$.)

**37.** $d = \sqrt{(x_2 - x_1)^2 + (y_2 - y_1)^2}$

$d = \sqrt{\left(\frac{5}{7} - \frac{1}{7}\right)^2 + \left(\frac{1}{14} - \frac{11}{14}\right)^2}$

$= \sqrt{\left(\frac{4}{7}\right)^2 + \left(-\frac{5}{7}\right)^2}$

$= \sqrt{\frac{16}{49} + \frac{25}{49}}$

$= \sqrt{\frac{41}{49}} = \frac{\sqrt{41}}{7} \approx 0.915$

**39.** $d = \sqrt{(x_2 - x_1)^2 + (y_2 - y_1)^2}$

$d = \sqrt{[0 - (-\sqrt{6})]^2 + (0 - \sqrt{2})^2}$

$= \sqrt{6 + 2}$

$= \sqrt{8} \approx 2.828$

**41.** $d = \sqrt{(x_2 - x_1)^2 + (y_2 - y_1)^2}$

$d = \sqrt{[-7 - (-4)]^2 + [-11 - (-2)]^2}$

$= \sqrt{(-3)^2 + (-9)^2}$

$= \sqrt{9 + 81}$

$= \sqrt{90} \approx 9.487$

**43.** We use the midpoint formula:

$\left(\frac{x_1 + x_2}{2}, \frac{y_1 + y_2}{2}\right) = \left(\frac{-7 + 9}{2}, \frac{6 + 2}{2}\right)$, or

$\left(\frac{2}{2}, \frac{8}{2}\right)$, or $(1, 4)$

**45.** We use the midpoint formula:

$\left(\frac{x_1 + x_2}{2}, \frac{y_1 + y_2}{2}\right) = \left(\frac{2 + 5}{2}, \frac{-1 + 8}{2}\right)$, or

$\left(\frac{7}{2}, \frac{7}{2}\right)$

**47.** We use the midpoint formula:

$\left(\frac{x_1 + x_2}{2}, \frac{y_1 + y_2}{2}\right) = \left(\frac{-8 + 6}{2}, \frac{-5 + (-1)}{2}\right)$, or

$\left(\frac{-2}{2}, \frac{-6}{2}\right)$, or $(-1, -3)$

**49.** We use the midpoint formula:

$\left(\frac{x_1 + x_2}{2}, \frac{y_1 + y_2}{2}\right) = \left(\frac{-3.4 + 2.9}{2}, \frac{8.1 + (-8.7)}{2}\right)$,

or $\left(\frac{-0.5}{2}, \frac{-0.6}{2}\right)$, or $(-0.25, -0.3)$

**51.** We use the midpoint formula:

$\left(\frac{x_1 + x_2}{2}, \frac{y_1 + y_2}{2}\right) = \left(\frac{\frac{1}{6} + \left(-\frac{1}{3}\right)}{2}, \frac{-\frac{3}{4} + \frac{5}{6}}{2}\right)$,

or $\left(\frac{-\frac{1}{6}}{2}, \frac{\frac{1}{12}}{2}\right)$, or $\left(-\frac{1}{12}, \frac{1}{24}\right)$

**53.** We use the midpoint formula:

$\left(\frac{x_1 + x_2}{2}, \frac{y_1 + y_2}{2}\right) = \left(\frac{\sqrt{2} + \sqrt{3}}{2}, \frac{-1 + 4}{2}\right)$, or

$\left(\frac{\sqrt{2} + \sqrt{3}}{2}, \frac{3}{2}\right)$

**55.** $(x - h)^2 + (y - k)^2 = r^2$  Standard form

$(x - 0)^2 + (y - 0)^2 = 6^2$  Substituting

$x^2 + y^2 = 36$  Simplifying

**57.** $(x - h)^2 + (y - k)^2 = r^2$  Standard form

$(x - 7)^2 + (y - 3)^2 = (\sqrt{5})^2$  Substituting

$(x - 7)^2 + (y - 3)^2 = 5$

**59.**
$$(x-h)^2 + (y-k)^2 = r^2 \qquad \text{Standard form}$$
$$[x-(-4)]^2 + (y-3)^2 = (4\sqrt{3})^2 \quad \text{Substituting}$$
$$(x+4)^2 + (y-3)^2 = 48$$
$$[(4\sqrt{3})^2 = 16\cdot 3 = 48]$$

**61.**
$$(x-h)^2 + (y-k)^2 = r^2$$
$$[x-(-7)]^2 + [y-(-2)]^2 = (5\sqrt{2})^2$$
$$(x+7)^2 + (y+2)^2 = 50$$

**63.** Since the center is $(0,0)$, we have
$$(x-0)^2 + (y-0)^2 = r^2 \text{ or } x^2 + y^2 = r^2$$
The circle passes through $(-3,4)$. We find $r^2$ by substituting $-3$ for $x$ and $4$ for $y$.
$$(-3)^2 + 4^2 = r^2$$
$$9 + 16 = r^2$$
$$25 = r^2$$
Then $x^2 + y^2 = 25$ is an equation of the circle.

**65.** Since the center is $(-4,1)$, we have
$$[x-(-4)]^2 + (y-1)^2 = r^2, \text{ or}$$
$$(x+4)^2 + (y-1)^2 = r^2.$$
The circle passes through $(-2,5)$. We find $r^2$ by substituting $-2$ for $x$ and $5$ for $y$.
$$(-2+4)^2 + (5-1)^2 = r^2$$
$$4 + 16 = r^2$$
$$20 = r^2$$
Then $(x+4)^2 + (y-1)^2 = 20$ is an equation of the circle.

**67.** We write standard form.
$$(x-0)^2 + (y-0)^2 = 8^2$$
The center is $(0,0)$, and the radius is 8.

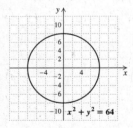

**69.**
$$(x+1)^2 + (y+3)^2 = 36$$
$$[x-(-1)]^2 + [y-(-3)]^2 = 6^2 \quad \text{Standard form}$$

The center is $(-1,-3)$, and the radius is 6.

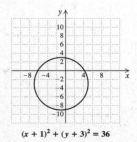

$(x+1)^2 + (y+3)^2 = 36$

**71.**
$$(x-4)^2 + (y+3)^2 = 10$$
$$(x-4)^2 + [y-(-3)]^2 = (\sqrt{10})^2$$
The center is $(4,-3)$, and the radius is $\sqrt{10}$.

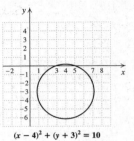

$(x-4)^2 + (y+3)^2 = 10$

**73.**
$$x^2 + y^2 = 10$$
$$(x-0)^2 + (y-0)^2 = (\sqrt{10})^2 \quad \text{Standard form}$$
The center is $(0,0)$, and the radius is $\sqrt{10}$.

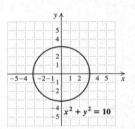

$x^2 + y^2 = 10$

**75.**
$$(x-5)^2 + y^2 = \frac{1}{4}$$
$$(x-5)^2 + (y-0)^2 = \left(\frac{1}{2}\right)^2 \quad \text{Standard form}$$
The center is $(5,0)$, and the radius is $\frac{1}{2}$.

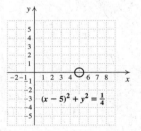

$(x-5)^2 + y^2 = \frac{1}{4}$

**77.**
$$x^2 + y^2 + 8x - 6y - 15 = 0$$
$$x^2 + 8x + y^2 - 6y = 15$$
$$(x^2+8x+16) + (y^2-6y+9) = 15 + 16 + 9$$
$$\text{Completing the square twice}$$
$$(x+4)^2 + (y-3)^2 = 40$$
$$[x-(-4)]^2 + (y-3)^2 = (\sqrt{40})^2$$
$$\text{Standard form}$$
The center is $(-4,3)$, and the radius is $\sqrt{40}$, or $2\sqrt{10}$.

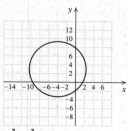

$$x^2 + y^2 + 8x - 6y - 15 = 0$$

**79.**
$$x^2 + y^2 - 8x + 2y + 13 = 0$$
$$x^2 - 8x + y^2 + 2y = -13$$
$$(x^2 - 8x + 16) + (y^2 + 2y + 1) = -13 + 16 + 1$$
Completing the square twice
$$(x - 4)^2 + (y + 1)^2 = 4$$
$$(x - 4)^2 + [y - (-1)]^2 = 2^2$$
Standard form

The center is $(4, -1)$, and the radius is 2.

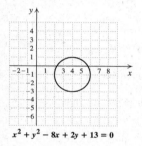

$$x^2 + y^2 - 8x + 2y + 13 = 0$$

**81.**
$$x^2 + y^2 + 10y - 75 = 0$$
$$x^2 + y^2 + 10y = 75$$
$$x^2 + (y^2 + 10y + 25) = 75 + 25$$
$$(x - 0)^2 + (y + 5)^2 = 100$$
$$(x - 0)^2 + [y - (-5)]^2 = 10^2$$

The center is $(0, -5)$, and the radius is 10.

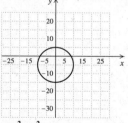

$$x^2 + y^2 + 10y - 75 = 0$$

**83.**
$$x^2 + y^2 + 7x - 3y - 10 = 0$$
$$x^2 + 7x + y^2 - 3y = 10$$
$$\left(x^2 + 7x + \frac{49}{4}\right) + \left(y^2 - 3y + \frac{9}{4}\right) = 10 + \frac{49}{4} + \frac{9}{4}$$
$$\left(x + \frac{7}{2}\right)^2 + \left(y - \frac{3}{2}\right)^2 = \frac{98}{4}$$
$$\left[x - \left(-\frac{7}{2}\right)\right]^2 + \left(y - \frac{3}{2}\right)^2 = \left(\sqrt{\frac{98}{4}}\right)^2$$

The center is $\left(-\frac{7}{2}, \frac{3}{2}\right)$, and the radius is $\sqrt{\frac{98}{4}}$, or $\frac{\sqrt{98}}{2}$, or $\frac{7\sqrt{2}}{2}$.

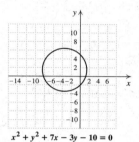

$$x^2 + y^2 + 7x - 3y - 10 = 0$$

**85.**
$$36x^2 + 36y^2 = 1$$
$$x^2 + y^2 = \frac{1}{36} \quad \text{Multiplying by } \frac{1}{36} \text{ on both sides}$$
$$(x - 0)^2 + (y - 0)^2 = \left(\frac{1}{6}\right)^2$$

The center is $(0, 0)$, and the radius is $\frac{1}{6}$.

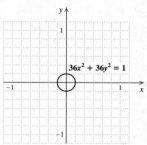

$$36x^2 + 36y^2 = 1$$

**87.** *Writing Exercise*

**89.**
$$\frac{x}{4} + \frac{5}{6} = \frac{2}{3}, \quad \text{LCD is } 12$$
$$12\left(\frac{x}{4} + \frac{5}{6}\right) = 12 \cdot \frac{2}{3}$$
$$12 \cdot \frac{x}{4} + 12 \cdot \frac{5}{6} = 8$$
$$3x + 10 = 8$$
$$3x = -2$$
$$x = -\frac{2}{3}$$

The solution is $-\frac{2}{3}$.

**91.** *Familiarize.* We make a drawing and label it. Let $x$ represent the width of the border.

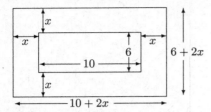

The perimeter of the larger rectangle is

$$2(10 + 2x) + 2(6 + 2x), \text{ or } 8x + 32.$$

The perimeter of the smaller rectangle is

$$2(10) + 2(6), \text{ or } 32.$$

*Translate.* The perimeter of the larger rectangle is twice the perimeter of the smaller rectangle.

$$8x + 32 = 2 \cdot 32$$

*Carry out.* We solve the equation.

$$8x + 32 = 64$$
$$8x = 32$$
$$x = 4$$

*Check.* If the width of the border is 4 in., then the length and width of the larger rectangle are 18 in. and 14 in. Thus its perimeter is $2(18) + 2(14)$, or 64 in. The perimeter of the smaller rectangle is 32 in. The perimeter of the larger rectangle is twice the perimeter of the smaller rectangle.

*State.* The width of the border is 4 in.

**93.**
$$3x - 8y = 5, \quad (1)$$
$$2x + 6y = 5 \quad (2)$$

Multiply Equation (1) by 3, multiply Equation (2) by 4, and add.

$$9x - 24y = 15$$
$$\underline{8x + 24y = 20}$$
$$17x \qquad = 35$$
$$x = \frac{35}{17}$$

Now substitute $\frac{35}{17}$ for $x$ in one of the original equations and solve for $y$. We use Equation (2).

$$2x + 6y = 5$$
$$2\left(\frac{35}{17}\right) + 6y = 5$$
$$\frac{70}{17} + 6y = 5$$
$$6y = \frac{15}{17}$$
$$y = \frac{5}{34}$$

The solution is $\left(\dfrac{35}{17}, \dfrac{5}{34}\right)$.

**95.** *Writing Exercise*

**97.** We make a drawing of the circle with center $(3, -5)$ and tangent to the $y$-axis.

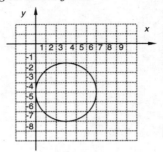

We see that the circle touches the $y$-axis at $(0, -5)$. Hence the radius is the distance between $(0, -5)$ and $(3, -5)$, or $\sqrt{(3 - 0)^2 + [-5 - (-5)]^2}$, or 3. Now we write the equation of the circle.

$$(x - h)^2 + (y - k)^2 = r^2$$
$$(x - 3)^2 + [y - (-5)]^2 = 3^2$$
$$(x - 3)^2 + (y + 5)^2 = 9$$

**99.** First we use the midpoint formula to find the center:

$$\left(\frac{7 + (-1)}{2}, \frac{3 + (-3)}{2}\right), \text{ or } \left(\frac{6}{2}, \frac{0}{2}\right), \text{ or } (3, 0)$$

The length of the radius is the distance between the center $(3, 0)$ and either endpoint of a diameter. We will use endpoint $(7, 3)$ in the distance formula:

$$r = \sqrt{(7 - 3)^2 + (3 - 0)^2} = \sqrt{25} = 5$$

Now we write the equation of the circle:

$$(x - h)^2 + (y - k)^2 = r^2$$
$$(x - 3)^2 + (y - 0)^2 = 5^2$$
$$(x - 3)^2 + y^2 = 25$$

**101.** Let $(0, y)$ be the point on the $y$-axis that is equidistant from $(2, 10)$ and $(6, 2)$. Then the distance between $(2, 10)$ and $(0, y)$ is the same as the distance between $(6, 2)$ and $(0, y)$.

$$\sqrt{(0 - 2)^2 + (y - 10)^2} = \sqrt{(0 - 6)^2 + (y - 2)^2}$$
$$(-2)^2 + (y - 10)^2 = (-6)^2 + (y - 2)^2$$

$$\text{Squaring both sides}$$

$$4 + y^2 - 20y + 100 = 36 + y^2 - 4y + 4$$
$$64 = 16y$$
$$4 = y$$

This number checks. The point is $(0, 4)$.

**103.** For the outer circle, $r^2 = \dfrac{81}{4}$. For the inner circle, $r^2 = 16$. The area of the red zone is the difference between the areas of the outer and inner circles. Recall that the area A of a circle with radius $r$ is given by the formula $A = \pi r^2$.

$$\pi \cdot \frac{81}{4} - \pi \cdot 16 = \frac{81}{4}\pi - \frac{64}{4}\pi = \frac{17}{4}\pi$$

The area of the red zone is $\dfrac{17}{4}\pi$ m$^2$, or about 13.4 m$^2$.

**105.** Superimposing a coordinate system on the snowboard as in Exercise 104, and observing that $1170/2 = 585$, we know that three points on the circle are $(-585, 0)$, $(0, 23)$, and $(585, 0)$. Let $(0, k)$ represent the center of the circle. Use the fact that $(0, k)$ is equidistant from $(-585, 0)$ and $(0, 23)$.

$$\sqrt{(-585-0)^2 + (0-k)^2} = \sqrt{(0-0)^2 + (23-k)^2}$$
$$\sqrt{342{,}225 + k^2} = \sqrt{529 - 46k + k^2}$$
$$342{,}225 + k^2 = 529 - 46k + k^2 \quad \text{Squaring both sides}$$
$$341{,}696 = -46k$$
$$-7428.2 \approx k$$

Then to find the radius, find the distance from the center, $(0, -7428.2)$, to any one of the three known points on the circle. We use $(0, 23)$.

$$r = \sqrt{(0-0)^2 + (-7428.2 - 23)^2} \approx 7451.2 \text{ mm}$$

**107.** a) When the circle is positioned on a coordinate system as shown in the text, the center lies on the $y$-axis. To find the center, we will find the point on the $y$-axis that is equidistant from $(-4, 0)$ and $(0, 2)$. Let $(0, y)$ be this point.

$$\sqrt{[0-(-4)]^2 + (y-0)^2} = \sqrt{(0-0)^2 + (y-2)^2}$$
$$4^2 + y^2 = 0^2 + (y-2)^2$$
$$\text{Squaring both sides}$$
$$16 + y^2 = y^2 - 4y + 4$$
$$12 = -4y$$
$$-3 = y$$

The center of the circle is $(0, -3)$.

b) We find the radius of the circle.

$$(x-0)^2 + [y-(-3)]^2 = r^2 \quad \text{Standard form}$$
$$x^2 + (y+3)^2 = r^2$$
$$(-4)^2 + (0+3)^2 = r^2 \quad \text{Substituting}$$
$$\qquad\qquad\qquad\qquad (-4, 0) \text{ for } (x, y)$$
$$16 + 9 = r^2$$
$$25 = r^2$$
$$5 = r$$

The radius is 5 ft.

**109.** We write the equation of a circle with center $(0, 30.6)$ and radius 24.3:

$$x^2 + (y - 30.6)^2 = 590.49$$

**111.** Substitute 6 for $N$.

$$H = \frac{D^2 N}{2.5} = \frac{D^2 \cdot 6}{2.5} = 2.4D^2$$

Find some ordered pairs for $2.5 \le D \le 8$ and draw the graph.

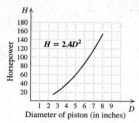

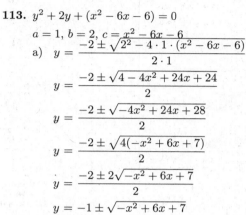

**113.** $y^2 + 2y + (x^2 - 6x - 6) = 0$

$a = 1, \ b = 2, \ c = x^2 - 6x - 6$

a) $\quad y = \dfrac{-2 \pm \sqrt{2^2 - 4 \cdot 1 \cdot (x^2 - 6x - 6)}}{2 \cdot 1}$

$\qquad y = \dfrac{-2 \pm \sqrt{4 - 4x^2 + 24x + 24}}{2}$

$\qquad y = \dfrac{-2 \pm \sqrt{-4x^2 + 24x + 28}}{2}$

$\qquad y = \dfrac{-2 \pm \sqrt{4(-x^2 + 6x + 7)}}{2}$

$\qquad y = \dfrac{-2 \pm 2\sqrt{-x^2 + 6x + 7}}{2}$

$\qquad y = -1 \pm \sqrt{-x^2 + 6x + 7}$

b)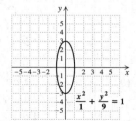

**115.** *Writing Exercise*

---

## Exercise Set 10.2

**1.** True; see page 671 in the text.

**3.** True; see page 671 in the text.

**5.** False; see page 671 in the text.

**7.** True; see page 673 in the text.

**9.** $\dfrac{x^2}{1} + \dfrac{y^2}{9} = 1$

$\dfrac{x^2}{1^2} + \dfrac{y^2}{3^2} = 1$

The $x$-intercepts are $(1, 0)$ and $(-1, 0)$, and the $y$-intercepts are $(0, 3)$ and $(0, -3)$. We plot these points and connect them with an oval-shaped curve.

**11.**  $\dfrac{x^2}{25} + \dfrac{y^2}{9} = 1$

$\dfrac{x^2}{5^2} + \dfrac{y^2}{3^2} = 1$

The $x$-intercepts are $(5, 0)$ and $(-5, 0)$, and the $y$-intercepts are $(0, 3)$ and $(0, -3)$. We plot these points and connect them with an oval-shaped curve.

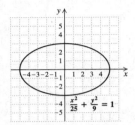

**13.**  $4x^2 + 9y^2 = 36$

$\dfrac{1}{36}(4x^2 + 9y^2) = \dfrac{1}{36}(36)$      Multiplying by $\dfrac{1}{36}$

$\dfrac{x^2}{9} + \dfrac{y^2}{4} = 1$

$\dfrac{x^2}{3^2} + \dfrac{y^2}{2^2} = 1$

The $x$-intercepts are $(-3, 0)$ and $(3, 0)$, and the $y$-intercepts are $(0, -2)$ and $(0, 2)$. We plot these points and connect them with an oval-shaped curve.

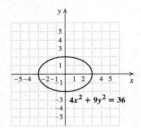

**15.**  $16x^2 + 9y^2 = 144$

$\dfrac{x^2}{9} + \dfrac{y^2}{16} = 1$   Multiplying by $\dfrac{1}{144}$

$\dfrac{x^2}{3^2} + \dfrac{y^2}{4^2} = 1$

The $x$-intercepts are $(3, 0)$ and $(-3, 0)$, and the $y$-intercepts are $(0, 4)$ and $(0, -4)$. We plot these points and connect them with an oval-shaped curve.

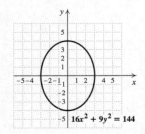

**17.**  $2x^2 + 3y^2 = 6$

$\dfrac{x^2}{3} + \dfrac{y^2}{2} = 1$   Multiplying by $\dfrac{1}{6}$

$\dfrac{x^2}{(\sqrt{3})^2} + \dfrac{y^2}{(\sqrt{2})^2} = 1$

The $x$-intercepts are $(\sqrt{3}, 0)$ and $(-\sqrt{3}, 0)$, and the $y$-intercepts are $(0, \sqrt{2})$ and $(0, -\sqrt{2})$. We plot these points and connect them with an oval-shaped curve.

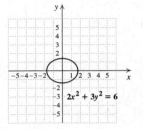

**19.**  $5x^2 + 5y^2 = 125$

Observe that the $x^2$- and $y^2$-terms have the same coefficient. We divide both sides of the equation by 5 to obtain $x^2 + y^2 = 25$. This is the equation of a circle with center $(0, 0)$ and radius 5.

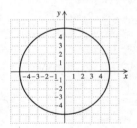

**21.**  $3x^2 + 7y^2 - 63 = 0$

$3x^2 + 7y^2 = 63$

$\dfrac{x^2}{21} + \dfrac{y^2}{9} = 1$   Multiplying by $\dfrac{1}{63}$

$\dfrac{x^2}{(\sqrt{21})^2} + \dfrac{y^2}{3^2} = 1$

The $x$-intercepts are $(\sqrt{21}, 0)$ and $(-\sqrt{21}, 0)$, or about $(4.583, 0)$ and $(-4.583, 0)$. The $y$-intercepts are $(0, 3)$ and $(0, -3)$. We plot these points and connect them with an oval-shaped curve.

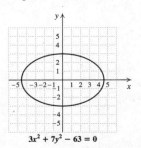

**23.**
$$8x^2 = 96 - 3y^2$$
$$8x^2 + 3y^2 = 96$$
$$\frac{x^2}{12} + \frac{y^2}{32} = 1$$
$$\frac{x^2}{(\sqrt{12})^2} + \frac{y^2}{(\sqrt{32})^2} = 1$$

The $x$-intercepts are $(\sqrt{12}, 0)$ and $(-\sqrt{12}, 0)$, or about $(3.464, 0)$ and $(-3.464, 0)$. The $y$-intercepts are $(0, \sqrt{32})$ and $(0, -\sqrt{32})$, or about $(0, 5.657)$ and $(0, -5.657)$. We plot these points and connect them with an oval-shaped curve.

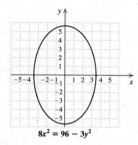

$8x^2 = 96 - 3y^2$

**25.** $16x^2 + 25y^2 = 1$

Note that $16 = \dfrac{1}{\dfrac{1}{16}}$ and $25 = \dfrac{1}{\dfrac{1}{25}}$. Thus, we can rewrite the equation:

$$\frac{x^2}{\dfrac{1}{16}} + \frac{y^2}{\dfrac{1}{25}} = 1$$

$$\frac{x^2}{\left(\dfrac{1}{4}\right)^2} + \frac{y^2}{\left(\dfrac{1}{5}\right)^2} = 1$$

The $x$-intercepts are $\left(\dfrac{1}{4}, 0\right)$ and $\left(-\dfrac{1}{4}, 0\right)$, and the $y$-intercepts are $\left(0, \dfrac{1}{5}\right)$ and $\left(0, -\dfrac{1}{5}\right)$. We plot these points and connect them with an oval-shaped curve.

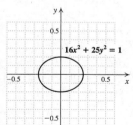

$16x^2 + 25y^2 = 1$

**27.** $\dfrac{(x-3)^2}{9} + \dfrac{(y-2)^2}{25} = 1$

$$\frac{(x-3)^2}{3^2} + \frac{(y-2)^2}{5^2} = 1$$

The center of the ellipse is $(3, 2)$. Note that $a = 3$ and $b = 5$. We locate the center and then plot the points $(3+3, 2)$ $(3-3, 2)$, $(3, 2+5)$, and $(3, 2-5)$, or $(6, 2)$, $(0, 2)$, $(3, 7)$, and $(3, -3)$. Connect these points with an oval-shaped curve.

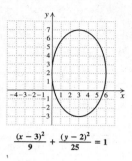

$\dfrac{(x-3)^2}{9} + \dfrac{(y-2)^2}{25} = 1$

**29.** $\dfrac{(x+4)^2}{16} + \dfrac{(y-3)^2}{49} = 1$

$$\frac{(x-(-4))^2}{4^2} + \frac{(y-3)^2}{7^2} = 1$$

The center of the ellipse is $(-4, 3)$. Note that $a = 4$ and $b = 7$. We locate the center and then plot the points $(-4+4, 3)$, $(-4-4, 3)$, $(-4, 3+7)$, and $(-4, 3-7)$, or $(0, 3)$, $(-8, 3)$, $(-4, 10)$, and $(-4, -4)$. Connect these points with an oval-shaped curve.

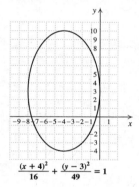

$\dfrac{(x+4)^2}{16} + \dfrac{(y-3)^2}{49} = 1$

**31.** $12(x-1)^2 + 3(y+4)^2 = 48$

$$\frac{(x-1)^2}{4} + \frac{(y+4)^2}{16} = 1$$

$$\frac{(x-1)^2}{2^2} + \frac{(y-(-4))^2}{4^2} = 1$$

The center of the ellipse is $(1, -4)$. Note that $a = 2$ and $b = 4$. We locate the center and then plot the points $(1+2, -4)$, $(1-2, -4)$, $(1, -4+4)$, and $(1, -4-4)$, or $(3, -4)$, $(-1, -4)$, $(1, 0)$, and $(1, -8)$. Connect these points with an oval-shaped curve.

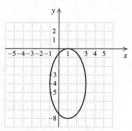

$12(x-1)^2 + 3(y+4)^2 = 48$

**33.**  $4(x+3)^2 + 4(y+1)^2 - 10 = 90$

$$4(x+3)^2 + 4(y+1)^2 = 100$$

Observe that the $x^2$- and $y^2$-terms have the some coefficient. Dividing both sides by 4, we have

$$(x+3)^2 + (y+1)^2 = 25.$$

This is the equation of a circle with center $(-3, -1)$ and radius 5.

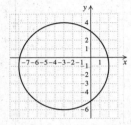

$4(x+3)^2 + 4(y+1)^2 - 10 = 90$

**35.** *Writing Exercise*

**37.**  $\dfrac{3}{x-2} - \dfrac{5}{x-2} = 9$

Note that the denominators are 0 when $x = 2$, so 2 cannot be a solution. We multiply by the LCD, $x - 2$.

$$x - 2\left(\frac{3}{x-2} - \frac{5}{x-2}\right) = (x-2)9$$

$$(x-2)\cdot\frac{3}{x-2} - (x-2)\cdot\frac{5}{x-2} = 9x - 18$$

$$3 - 5 = 9x - 18$$

$$-2 = 9x - 18$$

$$16 = 9x$$

$$\frac{16}{9} = x$$

The number $\dfrac{16}{9}$ checks and is the solution.

**39.**  $\dfrac{x}{x-4} - \dfrac{3}{x-5} = \dfrac{2}{x-4}$

Note that $x - 4$ is 0 when $x = 4$ and $x - 5$ is 0 when $x$ is 5, so 4 and 5 cannot be solutions. We multiply by the LCD, $(x-4)(x-5)$.

$$(x-4)(x-5)\left(\frac{x}{x-4} - \frac{3}{x-5}\right) =$$
$$(x-4)(x-5)\cdot\frac{2}{x-4}$$

$$(x-4)(x-5)\cdot\frac{x}{x-4} - (x-4)(x-5)\frac{3}{x-5} = 2(x-5)$$

$$x(x-5) - 3(x-4) = 2(x-5)$$

$$x^2 - 5x - 3x + 12 = 2x - 10$$

$$x^2 - 8x + 12 = 2x - 10$$

$$x^2 - 10x + 22 = 0$$

We use the quadratic formula with $a = 1$, $b = -10$, and $c = 22$.

$$x = \frac{-b \pm \sqrt{b^2 - 4ac}}{2a}$$

$$x = \frac{-(-10) \pm \sqrt{(-10)^2 - 4\cdot 1\cdot 22}}{2\cdot 1}$$

$$x = \frac{10 \pm \sqrt{12}}{2} = \frac{10 \pm 2\sqrt{3}}{2}$$

$$x = \frac{\cancel{2}(5 \pm \sqrt{3})}{\cancel{2}\cdot 1} = 5 \pm \sqrt{3}$$

Both numbers check. The solutions are $5 \pm \sqrt{3}$.

**41.**  $9 - \sqrt{2x+1} = 7$

$$-\sqrt{2x+1} = -2 \qquad \text{Isolating the radical}$$

$$(-\sqrt{2x+1})^2 = (-2)^2$$

$$2x + 1 = 4$$

$$2x = 3$$

$$x = \frac{3}{2}$$

The number $\dfrac{3}{2}$ checks and is the solution.

**43.** *Writing Exercise*

**45.** Plot the given points.

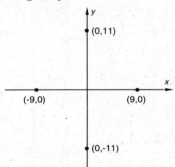

From the location of these points, we see that the ellipse that contains them is centered at the origin with $a = 9$ and $b = 11$. We write the equation of the ellipse:

$$\frac{x^2}{9^2} + \frac{y^2}{11^2} = 1$$

$$\frac{x^2}{81} + \frac{y^2}{121} = 1$$

**47.** Plot the given points.

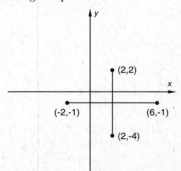

The midpoint of the segment from $(-2,-1)$ to $(6,-1)$ is $\left(\dfrac{-2+6}{2}, \dfrac{-1-1}{2}\right)$, or $(2,-1)$. The midpoint of the segment from $(2,-4)$ to $(2,2)$ is $\left(\dfrac{2+2}{2}, \dfrac{-4+2}{2}\right)$, or $(2,-1)$. Thus, we can conclude that $(2,-1)$ is the center of the ellipse. The distance from $(-2,-1)$ to $(2,-1)$ is $\sqrt{[2-(-2)]^2 + [-1-(-1)]^2} = \sqrt{16} = 4$, so $a = 4$. The distance from $(2,2)$ to $(2,-1)$ is $\sqrt{(2-2)^2 + (-1-2)^2} = \sqrt{9} = 3$, so $b = 3$. We write the equation of the ellipse.

$$\frac{(x-2)^2}{4^2} + \frac{(y-(-1))^2}{3^2} = 1$$

$$\frac{(x-2)^2}{16} + \frac{(y+1)^2}{9} = 1$$

**49.** We have a vertical ellipse centered at the origin with $a = 6/2$, or $3$, and $b = 10/2$, or $5$. Then the equation is $\dfrac{x^2}{3^2} + \dfrac{y^2}{5^2} = 1$, or $\dfrac{x^2}{9} + \dfrac{y^2}{25} = 1$.

**51.** a) Let $F_1 = (-c, 0)$ and $F_2 = (c, 0)$. Then the sum of the distances from the foci to $P$ is $2a$. By the distance formula,

$$\sqrt{(x+c)^2 + y^2} + \sqrt{(x-c)^2 + y^2} = 2a, \text{ or}$$

$$\sqrt{(x+c)^2 + y^2} = 2a - \sqrt{(x-c)^2 + y^2}.$$

Squaring, we get

$$(x+c)^2 + y^2 = 4a^2 - 4a\sqrt{(x-c)^2 + y^2} + (x-c)^2 + y^2,$$

or $x^2 + 2cx + c^2 + y^2$

$$= 4a^2 - 4a\sqrt{(x-c)^2 + y^2} + x^2 - 2cx + c^2 + y^2.$$

Thus

$$-4a^2 + 4cx = -4a\sqrt{(x-c)^2 + y^2}$$

$$a^2 - cx = a\sqrt{(x-c)^2 + y^2}.$$

Squaring again, we get

$$a^4 - 2a^2cx + c^2x^2 = a^2(x^2 - 2cx + c^2 + y^2)$$

$$a^4 - 2a^2cx + c^2x^2 = a^2x^2 - 2a^2cx + a^2c^2 + a^2y^2,$$

or

$$x^2(a^2 - c^2) + a^2y^2 = a^2(a^2 - c^2)$$

$$\frac{x^2}{a^2} + \frac{y^2}{a^2 - c^2} = 1.$$

b) When $P$ is at $(0, b)$, it follows that $b^2 = a^2 - c^2$. Substituting, we have

$$\frac{x^2}{a^2} + \frac{y^2}{b^2} = 1.$$

**53.** For the given ellipse, $a = 6/2$, or $3$, and $b = 2/2$, or $1$. The patient's mouth should be at a distance $2c$ from the light source, where the coordinates of the foci of the ellipse are $(-c, 0)$ and $(c, 0)$. From Exercise 38(b), we know $b^2 = a^2 - c^2$. We use this to find $c$.

$$b^2 = a^2 - c^2$$

$$1^2 = 3^2 - c^2 \quad \text{Substituting}$$

$$c^2 = 8$$

$$c = \sqrt{8}$$

Then $2c = 2\sqrt{8} \approx 5.66$. The patient's mouth should be about 5.66 ft from the light source.

**55.**

$$x^2 - 4x + 4y^2 + 8y - 8 = 0$$

$$x^2 - 4x + 4y^2 + 8y = 8$$

$$x^2 - 4x + 4(y^2 + 2y) = 8$$

$$(x^2 - 4x + 4 - 4) + 4(y^2 + 2y + 1 - 1) = 8$$

$$(x^2 - 4x + 4) + 4(y^2 + 2y + 1) = 8 + 4 + 4 \cdot 1$$

$$(x-2)^2 + 4(y+1)^2 = 16$$

$$\frac{(x-2)^2}{16} + \frac{(y+1)^2}{4} = 1$$

$$\frac{(x-2)^2}{4^2} + \frac{(y-(-1))^2}{2^2} = 1$$

The center of the ellipse is $(2, -1)$. Note that $a = 4$ and $b = 2$. We locate the center and then plot the points $(2+4, -1)$, $(2-4, -1)$, $(2, -1+2)$, $(2, -1-2)$, or $(6, -1)$, $(-2, -1)$, $(2, 1)$, and $(2, -3)$. Connect these points with an oval-shaped curve.

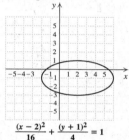

$$\frac{(x-2)^2}{16} + \frac{(y+1)^2}{4} = 1$$

**57.**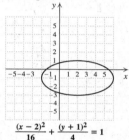

---

## Exercise Set 10.3

**1.** (d); see page 678 in the text.

**3.** (h); see page 685 in the text.

**5.** (g); see page 683 in the text.

**7.** (c); see page 683 in the text.

**9.** $\dfrac{x^2}{16} - \dfrac{y^2}{16} = 1$

$$\frac{x^2}{4^2} - \frac{y^2}{4^2} = 1$$

$a = 4$ and $b = 4$, so the asymptotes are $y = \dfrac{4}{4}x$ and $y = -\dfrac{4}{4}x$, or $y = x$ and $y = -x$. We sketch them.

Replacing $y$ with $0$ and solving for $x$, we get $x = \pm 4$, so the intercepts are $(4, 0)$ and $(-4, 0)$.

We plot the intercepts and draw smooth curves through them that approach the asymptotes.

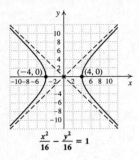

$$\frac{x^2}{16} - \frac{y^2}{16} = 1$$

**11.**  $\dfrac{x^2}{4} - \dfrac{y^2}{25} = 1$

$$\frac{x^2}{2^2} - \frac{y^2}{5^2} = 1$$

$a = 2$ and $b = 5$, so the asymptotes are $y = \dfrac{5}{2}x$ and

$y = -\dfrac{5}{2}x$. We sketch them.

Replacing $y$ with 0 and solving for $x$, we get $x = \pm 2$, so the intercepts are $(2, 0)$ and $(-2, 0)$.

We plot the intercepts and draw smooth curves through them that approach the asymptotes.

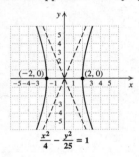

$$\frac{x^2}{4} - \frac{y^2}{25} = 1$$

**13.**  $\dfrac{y^2}{36} - \dfrac{x^2}{9} = 1$

$$\frac{y^2}{6^2} - \frac{x^2}{3^2} = 1$$

$a = 3$ and $b = 6$, so the asymptotes are $y = \dfrac{6}{3}x$ and

$y = -\dfrac{6}{3}x$, or $y = 2x$ and $y = -2x$. We sketch them.

Replacing $x$ with 0 and solving for $y$, we get $y = \pm 6$, so the intercepts are $(0, 6)$ and $(0, -6)$.

We plot the intercepts and draw smooth curves through them that approach the asymptotes.

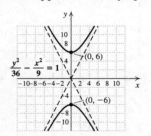

$$\frac{y^2}{36} - \frac{x^2}{9} = 1$$

**15.**  $y^2 - x^2 = 25$

$$\frac{y^2}{25} - \frac{x^2}{25} = 1$$

$$\frac{y^2}{5^2} - \frac{x^2}{5^2} = 1$$

$a = 5$ and $b = 5$, so the asymptotes are $y = \dfrac{5}{5}x$ and

$y = -\dfrac{5}{5}x$, or $y = x$ and $y = -x$. We sketch them.

Replacing $x$ with 0 and solving for $y$, we get $y = \pm 5$, so the intercepts are $(0, 5)$ and $(0, -5)$.

We plot the intercepts and draw smooth curves through them that approach the asymptotes.

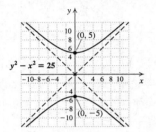

$$y^2 - x^2 = 25$$

**17.**  $25x^2 - 16y^2 = 400$

$$\frac{x^2}{16} - \frac{y^2}{25} = 1 \qquad \text{Multiplying by } \frac{1}{400}$$

$$\frac{x^2}{4^2} - \frac{y^2}{5^2} = 1$$

$a = 4$ and $b = 5$, so the asymptotes are $y = \dfrac{5}{4}x$ and

$y = -\dfrac{5}{4}x$. We sketch them.

Replacing $y$ with 0 and solving for $x$, we get $x = \pm 4$, so the intercepts are $(4, 0)$ and $(-4, 0)$.

We plot the intercepts and draw smooth curves through them that approach the asymptotes.

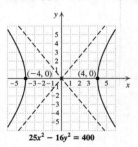

$$25x^2 - 16y^2 = 400$$

**19.** $xy = -5$

$\quad y = -\dfrac{5}{x}$    Solving for $y$

We find some solutions, keeping the results in a table.

| $x$ | $y$ |
|---|---|
| $\dfrac{1}{5}$ | $-25$ |
| $1$ | $-5$ |
| $2$ | $-\dfrac{5}{2}$ |
| $5$ | $-1$ |
| $10$ | $-\dfrac{1}{2}$ |
| $-\dfrac{1}{5}$ | $25$ |
| $-1$ | $5$ |
| $-5$ | $1$ |
| $-10$ | $\dfrac{1}{2}$ |

Note that we cannot use 0 for $x$. The $x$-axis and the $y$-axis are the asymptotes.

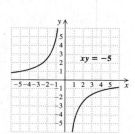

**21.** $xy = 4$

$\quad y = \dfrac{4}{x}$    Solving for $y$

We find some solutions, keeping the results in a table.

| $x$ | $y$ |
|---|---|
| $\dfrac{1}{2}$ | $8$ |
| $1$ | $4$ |
| $4$ | $1$ |
| $8$ | $\dfrac{1}{2}$ |
| $-\dfrac{1}{2}$ | $-8$ |
| $-1$ | $-4$ |
| $-2$ | $-2$ |
| $-4$ | $-1$ |

Note that we cannot use 0 for $x$. The $x$-axis and the $y$-axis are the asymptotes.

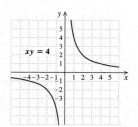

**23.** $xy = -2$

$\quad y = -\dfrac{2}{x}$    Solving for $y$

| $x$ | $y$ |
|---|---|
| $\dfrac{1}{2}$ | $-4$ |
| $1$ | $-2$ |
| $2$ | $-1$ |
| $4$ | $-\dfrac{1}{2}$ |
| $-\dfrac{1}{2}$ | $4$ |
| $-1$ | $2$ |
| $-2$ | $1$ |
| $-4$ | $\dfrac{1}{2}$ |

Note that we cannot use 0 for $x$. The $x$-axis and the $y$-axis are the asymptotes.

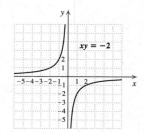

**25.** $xy = 1$

$\quad y = \dfrac{1}{x}$    Solving for $y$

| $x$ | $y$ |
|---|---|
| $\dfrac{1}{4}$ | $4$ |
| $\dfrac{1}{2}$ | $2$ |
| $1$ | $1$ |
| $2$ | $\dfrac{1}{2}$ |
| $4$ | $\dfrac{1}{4}$ |
| $-\dfrac{1}{4}$ | $-4$ |
| $-\dfrac{1}{2}$ | $-2$ |
| $-1$ | $-1$ |
| $-2$ | $-\dfrac{1}{2}$ |
| $-4$ | $-\dfrac{1}{4}$ |

Note that we cannot use 0 for $x$. The $x$-axis and the $y$-axis are the asymptotes.

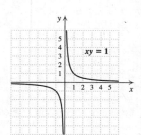

**27.** $x^2 + y^2 - 6x + 4y - 30 = 0$

Completing the square twice, we obtain an equivalent equation:

$$(x^2 - 6x) + (y^2 + 4y) = 30$$
$$(x^2 - 6x + 9) + (y^2 + 4y + 4) = 30 + 9 + 4$$
$$(x - 3)^2 + (y + 2)^2 = 43$$

The graph is a circle.

**29.** $9x^2 + 4y^2 - 36 = 0$

$$9x^2 + 4y^2 = 36$$

$$\frac{x^2}{4} + \frac{y^2}{9} = 1$$

The graph is an ellipse.

**31.** $4x^2 - 9y^2 - 72 = 0$

$$4x^2 - 9y^2 = 72$$

$$\frac{x^2}{18} - \frac{y^2}{8} = 1$$

The graph is a hyperbola.

**33.**
$$x^2 + y^2 = 2x + 4y + 4$$
$$x^2 - 2x + y^2 - 4y = 4$$
$$(x^2 - 2x + 1) + (y^2 - 4y + 4) = 4 + 1 + 4$$
$$(x - 1)^2 + (y - 2)^2 = 9$$

The graph is a circle.

**35.**
$$4x^2 = 64 - y^2$$
$$4x^2 + y^2 = 64$$
$$\frac{x^2}{16} + \frac{y^2}{64} = 1$$

The graph is an ellipse.

**37.** $x - \dfrac{8}{y} = 0$

$$x = \frac{8}{y}$$

$$xy = 8$$

The graph is a hyperbola.

**39.** $y + 6x = x^2 + 5$

$$y = x^2 - 6x + 5$$

The graph is a parabola.

**41.**
$$9y^2 = 36 + 4x^2$$
$$9y^2 - 4x^2 = 36$$
$$\frac{y^2}{4} - \frac{x^2}{9} = 1$$

The graph is a hyperbola.

**43.**
$$3x^2 + y^2 - x = 2x^2 - 9x + 10y + 40$$
$$x^2 + y^2 + 8x - 10y = 40$$

Both variables are squared, so the graph is not a parabola. The plus sign between $x^2$ and $y^2$ indicates that we have either a circle or an ellipse. Since the coefficients of $x^2$ and $y^2$ are the same, the graph is a circle.

**45.** $16x^2 + 5y^2 - 12x^2 + 8y^2 - 3x + 4y = 568$

$$4x^2 + 13y^2 - 3x + 4y = 568$$

Both variables are squared, so the graph is not a parabola. The plus sign between $x^2$ and $y^2$ indicates that we have either a circle or an ellipse. Since the coefficients of $x^2$ and $y^2$ are different, the graph is an ellipse.

**47.** *Writing Exercise*

**49.** $5x + 6y = -12,$    (1)

$3x + 9y = 15$      (2)

We will use the elimination method. First multiply equation (1) by 3 and equation (2) by $-2$ and add.

$$\begin{array}{r} 15x + 18y = -36 \\ -6x - 18y = -30 \\ \hline 9x \qquad\quad = -66 \end{array}$$

$$x = -\frac{22}{3}$$

Now substitute $-\dfrac{22}{3}$ for $x$ in one of the original equations and solve for $y$.

$$5x + 6y = -12 \quad (1)$$

$$5\left(-\frac{22}{3}\right) + 6y = -12$$

$$-\frac{110}{3} + 6y = -12$$

$$6y = \frac{74}{3}$$

$$y = \frac{37}{9}$$

The solution is $\left(-\dfrac{22}{3}, \dfrac{37}{9}\right)$.

**51.** $y^2 - 3 = 6$

$$y^2 = 9$$

$y = 3 \;\; or \;\; y = -3$   Principle of square roots

The solutions are 3 and $-3$.

**53.** *Familiarize*. Let $p =$ the price of the lawn chair before the tax was added. Then the total price is $p + 5\%p$, or $p + 0.05p$, or $1.05p$.

*Translate*.

$\underbrace{\text{The total price}}$  is  \$36.75.

$\qquad\quad \downarrow \qquad\qquad \downarrow \quad\;\; \downarrow$

$\qquad 1.05p \qquad = \;\; 36.75$

*Carry out*. We solve the equation.

$$1.05p = 36.75$$

$$p = \frac{36.75}{1.05}$$

$$p = 35$$

*Check*. 5% of \$35 is \$1.75 and \$35 + \$1.75 = \$36.75. The answer checks.

*State*. The price before tax was \$35.

**55.** *Writing Exercise*

**57.** Since the intercepts are $(0, 6)$ and $(0, -6)$, we know that the hyperbola is of the form $\dfrac{y^2}{b^2} - \dfrac{x^2}{a^2} = 1$ and that $b = 6$. The equations of the asymptotes tell us that $b/a = 3$, so

$$\frac{6}{a} = 3$$

$$a = 2.$$

The equation is $\dfrac{y^2}{6^2} - \dfrac{x^2}{2^2} = 1$, or $\dfrac{y^2}{36} - \dfrac{x^2}{4} = 1$.

**59.** $\dfrac{(x-5)^2}{36} - \dfrac{(y-2)^2}{25} = 1$

$\dfrac{(x-5)^2}{6^2} - \dfrac{(y-2)^2}{5^2} = 1$

$h = 5, k = 2, a = 6, b = 5$

Center: $(5, 2)$

Vertices: $(5-6, 2)$ and $(5+6, 2)$, or $(-1, 2)$ and $(11, 2)$

Asymptotes: $y - 2 = \dfrac{5}{6}(x-5)$ and $y - 2 = -\dfrac{5}{6}(x-5)$

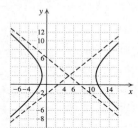

**61.** $8(y+3)^2 - 2(x-4)^2 = 32$

$\dfrac{(y+3)^2}{4} - \dfrac{(x-4)^2}{16} = 1$

$\dfrac{(y-(-3))^2}{2^2} - \dfrac{(x-4)^2}{4^2} = 1$

$h = 4, k = -3, a = 4, b = 2$

Center: $(4, -3)$

Vertices: $(4, -3+2)$ and $(4, -3-2)$, or $(4, -1)$ and $(4, -5)$

Asymptotes: $y - (-3) = \dfrac{2}{4}(x-4)$ and

$y - (-3) = -\dfrac{2}{4}(x-4)$, or $y + 3 = \dfrac{1}{2}(x-4)$ and

$y + 3 = -\dfrac{1}{2}(x-4)$

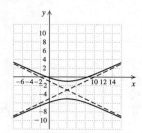

**63.** $4x^2 - y^2 + 24x + 4y + 28 = 0$

$4(x^2 + 6x) - (y^2 - 4y) = -28$

$4(x^2+6x+9-9) - (y^2-4y+4-4) = -28$

$4(x^2 + 6x + 9) - (y^2 - 4y + 4) = -28 + 4\cdot 9 - 4$

$4(x+3)^2 - (y-2)^2 = 4$

$\dfrac{(x+3)^2}{1} - \dfrac{(y-2)^2}{4} = 1$

$\dfrac{(x-(-3))^2}{1^2} - \dfrac{(y-2)^2}{2^2} = 1$

$h = -3, k = 2, a = 1, b = 2$

Center: $(-3, 2)$

Vertices: $(-3-1, 2)$, and $(-3+1, 2)$, or $(-4, 2)$ and $(-2, 2)$

Asymptotes: $y - 2 = \dfrac{2}{1}(x-(-3))$ and

$y - 2 = -\dfrac{2}{1}(x-(-3))$, or $y - 2 = 2(x+3)$ and

$y - 2 = -2(x+3)$

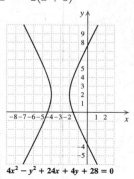

$4x^2 - y^2 + 24x + 4y + 28 = 0$

**65.** ▮▮▮

## Exercise Set 10.4

**1.** True

**3.** False; see page 692 in the text.

**5.** True; see page 689 in the text.

**7.** $x^2 + y^2 = 25$,  (1)

$y - x = 1$  (2)

First solve Eq. (2) for $y$.

$y = x + 1$  (3)

Then substitute $x + 1$ for $y$ in Eq. (1) and solve for $x$.

$x^2 + y^2 = 25$

$x^2 + (x+1)^2 = 25$

$x^2 + x^2 + 2x + 1 = 25$

$2x^2 + 2x - 24 = 0$

$x^2 + x - 12 = 0$

$(x+4)(x-3) = 0$

$x + 4 = 0 \quad or \quad x - 3 = 0$  Principle of zero products

$x = -4 \quad or \qquad x = 3$

Now substitute these numbers in Eq. (3) and solve for $y$.

$y = -4 + 1 = -3$

$y = 3 + 1 = 4$

The pairs $(-4, -3)$ and $(3, 4)$ check, so they are the solutions.

**9.** $4x^2 + 9y^2 = 36$,  (1)

$3y + 2x = 6$  (2)

First solve Eq. (2) for $y$.

$3y = -2x + 6$

$y = -\dfrac{2}{3}x + 2$  (3)

Then substitute $-\frac{2}{3}x + 2$ for $y$ in Eq. (1) and solve for $x$.

$$4x^2 + 9y^2 = 36$$

$$4x^2 + 9\left(-\frac{2}{3}x + 2\right)^2 = 36$$

$$4x^2 + 9\left(\frac{4}{9}x^2 - \frac{8}{3}x + 4\right) = 36$$

$$4x^2 + 4x^2 - 24x + 36 = 36$$

$$8x^2 - 24x = 0$$

$$x^2 - 3x = 0$$

$$x(x - 3) = 0$$

$x = 0 \ or \ x = 3$

Now substitute these numbers in Eq. (3) and solve for $y$.

$$y = -\frac{2}{3} \cdot 0 + 2 = 2$$

$$y = -\frac{2}{3} \cdot 3 + 2 = 0$$

The pairs $(0, 2)$ and $(3, 0)$ check, so they are the solutions.

**11.** $y^2 = x + 3,$      (1)

    $2y = x + 4$      (2)

First solve Eq. (2) for $x$.

    $2y - 4 = x$     (3)

Then substitute $2y - 4$ for $x$ in Eq. (1) and solve for $y$.

$$y^2 = x + 3$$

$$y^2 = (2y - 4) + 3$$

$$y^2 = 2y - 1$$

$$y^2 - 2y + 1 = 0$$

$$(y - 1)(y - 1) = 0$$

$$y - 1 = 0 \ \ or \ \ y - 1 = 0$$

$$y = 1 \ \ or \ \ \ \ \ \ y = 1$$

Now substitute 1 for $y$ in Eq. (3) and solve for $x$.

    $2 \cdot 1 - 4 = x$

      $-2 = x$

The pair $(-2, 1)$ checks. It is the solution.

**13.** $x^2 - xy + 3y^2 = 27,$     (1)

    $x - y = 2$           (2)

First solve Eq. (2) for $y$.

    $x - 2 = y$          (3)

Then substitute $x - 2$ for $y$ in Eq. (1) and solve for $x$.

$$x^2 - xy + 3y^2 = 27$$

$$x^2 - x(x - 2) + 3(x - 2)^2 = 27$$

$$x^2 - x^2 + 2x + 3x^2 - 12x + 12 = 27$$

$$3x^2 - 10x - 15 = 0$$

$$x = \frac{-(-10) \pm \sqrt{(-10)^2 - 4(3)(-15)}}{2 \cdot 3}$$

$$x = \frac{10 \pm \sqrt{100 + 180}}{6} = \frac{10 \pm \sqrt{280}}{6}$$

$$x = \frac{10 \pm 2\sqrt{70}}{6} = \frac{5 \pm \sqrt{70}}{3}$$

Now substitute these numbers in Eq. (3) and solve for $y$.

$$y = \frac{5 + \sqrt{70}}{3} - 2 = \frac{-1 + \sqrt{70}}{3}$$

$$y = \frac{5 - \sqrt{70}}{3} - 2 = \frac{-1 - \sqrt{70}}{3}$$

The pairs $\left(\dfrac{5 + \sqrt{70}}{3}, \dfrac{-1 + \sqrt{70}}{3}\right)$ and

$\left(\dfrac{5 - \sqrt{70}}{3}, \dfrac{-1 - \sqrt{70}}{3}\right)$ check, so they are the solutions.

**15.** $x^2 + 4y^2 = 25,$   (1)

    $x + 2y = 7$      (2)

First solve Eq. (2) for $x$.

    $x = -2y + 7$     (3)

Then substitute $-2y + 7$ for $x$ in Eq. (1) and solve for $y$.

$$x^2 + 4y^2 = 25$$

$$(-2y + 7)^2 + 4y^2 = 25$$

$$4y^2 - 28y + 49 + 4y^2 = 25$$

$$8y^2 - 28y + 24 = 0$$

$$2y^2 - 7y + 6 = 0$$

$$(2y - 3)(y - 2) = 0$$

$$y = \frac{3}{2} \ or \ y = 2$$

Now substitute these numbers in Eq. (3) and solve for $x$.

$$x = -2 \cdot \frac{3}{2} + 7 = 4$$

$$x = -2 \cdot 2 + 7 = 3$$

The pairs $\left(4, \dfrac{3}{2}\right)$ and $(3, 2)$ check, so they are the solutions.

**17.** $x^2 - xy + 3y^2 = 5,$     (1)

    $x - y = 2$           (2)

First solve Eq. (2) for $y$.

    $x - 2 = y$          (3)

Then substitute $x - 2$ for $y$ in Eq. (1) and solve for $x$.

$$x^2 - xy + 3y^2 = 5$$

$$x^2 - x(x - 2) + 3(x - 2)^2 = 5$$

$$x^2 - x^2 + 2x + 3x^2 - 12x + 12 = 5$$

$$3x^2 - 10x + 7 = 0$$

$$(3x - 7)(x - 1) = 0$$

$$x = \frac{7}{3} \ \ or \ \ x = 1$$

Now substitute these numbers in Eq. (3) and solve for $y$.

$$y = \frac{7}{3} - 2 = \frac{1}{3}$$

$$y = 1 - 2 = -1$$

The pairs $\left(\dfrac{7}{3}, \dfrac{1}{3}\right)$ and $(1, -1)$ check, so they are the solutions.

**19.** $3x + y = 7$,    (1)

$4x^2 + 5y = 24$  (2)

First solve Eq. (1) for $y$.

$y = 7 - 3x$    (3)

Then substitute $7 - 3x$ for $y$ in Eq. (2) and solve for $x$.

$$4x^2 + 5y = 24$$
$$4x^2 + 5(7 - 3x) = 24$$
$$4x^2 + 35 - 15x = 24$$
$$4x^2 - 15x + 11 = 0$$
$$(4x - 11)(x - 1) = 0$$

$x = \dfrac{11}{4}$ or $x = 1$

Now substitute these numbers into Eq. (3) and solve for $y$.

$y = 7 - 3 \cdot \dfrac{11}{4} = -\dfrac{5}{4}$

$y = 7 - 3 \cdot 1 = 4$

The pairs $\left(\dfrac{11}{4}, -\dfrac{5}{4}\right)$ and $(1, 4)$ check, so they are the solutions.

**21.**  $a + b = 7$,    (1)

$ab = 4$      (2)

First solve Eq. (1) for $a$.

$a = -b + 7$   (3)

Then substitute $-b + 7$ for $a$ in Eq. (2) and solve for $b$.

$$(-b + 7)b = 4$$
$$-b^2 + 7b = 4$$
$$0 = b^2 - 7b + 4$$

$b = \dfrac{-(-7) \pm \sqrt{(-7)^2 - 4 \cdot 1 \cdot 4}}{2 \cdot 1}$

$b = \dfrac{7 \pm \sqrt{33}}{2}$

Now substitute these numbers in Eq. (3) and solve for $a$.

$a = -\left(\dfrac{7 + \sqrt{33}}{2}\right) + 7 = \dfrac{7 - \sqrt{33}}{2}$

$a = -\left(\dfrac{7 - \sqrt{33}}{2}\right) + 7 = \dfrac{7 + \sqrt{33}}{2}$

The pairs $\left(\dfrac{7 - \sqrt{33}}{2}, \dfrac{7 + \sqrt{33}}{2}\right)$ and

$\left(\dfrac{7 + \sqrt{33}}{2}, \dfrac{7 - \sqrt{33}}{2}\right)$ check, so they are the solutions.

**23.**  $2a + b = 1$,      (1)

$b = 4 - a^2$       (2)

Eq. (2) is already solved for $b$. Substitute $4 - a^2$ for $b$ in Eq. (1) and solve for $a$.

$2a + 4 - a^2 = 1$

$0 = a^2 - 2a - 3$

$0 = (a - 3)(a + 1)$

$a = 3$  or  $a = -1$

Substitute these numbers in Eq. (2) and solve for $b$.

$b = 4 - 3^2 = -5$

$b = 4 - (-1)^2 = 3$

The pairs $(3, -5)$ and $(-1, 3)$ check, so they are the solutions.

**25.**   $a^2 + b^2 = 89$,   (1)

$a - b = 3$       (2)

First solve Eq. (2) for $a$.

$a = b + 3$      (3)

Then substitute $b + 3$ for $a$ in Eq. (1) and solve for $b$.

$$(b + 3)^2 + b^2 = 89$$
$$b^2 + 6b + 9 + b^2 = 89$$
$$2b^2 + 6b - 80 = 0$$
$$b^2 + 3b - 40 = 0$$
$$(b + 8)(b - 5) = 0$$

$b = -8$ or $b = 5$

Substitute these numbers in Eq. (3) and solve for $a$.

$a = -8 + 3 = -5$

$a = 5 + 3 = 8$

The pairs $(-5, -8)$ and $(8, 5)$ check, so they are the solutions.

**27.**  $y = x^2$,   (1)

$x = y^2$   (2)

Eq. (1) is already solved for $y$. Substitute $x^2$ for $y$ in Eq. (2) and solve for $x$.

$x = y^2$

$x = (x^2)^2$

$x = x^4$

$0 = x^4 - x$

$0 = x(x^3 - 1)$

$0 = x(x - 1)(x^2 + x + 1)$

$x = 0$  or  $x = 1$  or  $x = \dfrac{-1 \pm \sqrt{1^2 - 4 \cdot 1 \cdot 1}}{2}$

$x = 0$  or  $x = 1$  or  $x = -\dfrac{1}{2} \pm \dfrac{\sqrt{3}}{2}i$

Substitute these numbers in Eq. (1) and solve for $y$.

$y = 0^2 = 0$

$y = 1^2 = 1$

$y = \left(-\dfrac{1}{2} + \dfrac{\sqrt{3}}{2}i\right)^2 = -\dfrac{1}{2} - \dfrac{\sqrt{3}}{2}i$

$y = \left(-\dfrac{1}{2} - \dfrac{\sqrt{3}}{2}i\right)^2 = -\dfrac{1}{2} + \dfrac{\sqrt{3}}{2}i$

The pairs $(0, 0)$, $(1, 1)$, $\left(-\dfrac{1}{2} + \dfrac{\sqrt{3}}{2}i, -\dfrac{1}{2} - \dfrac{\sqrt{3}}{2}i\right)$,

and $\left(-\dfrac{1}{2} - \dfrac{\sqrt{3}}{2}i, -\dfrac{1}{2} + \dfrac{\sqrt{3}}{2}i\right)$ check, so they are the solutions.

**29.** $x^2 + y^2 = 9,$     (1)

     $x^2 - y^2 = 9$     (2)

Here we use the elimination method.

$$x^2 + y^2 = \phantom{0}9 \quad (1)$$
$$\underline{x^2 - y^2 = \phantom{0}9} \quad (2)$$
$$2x^2 \phantom{xxx} = 18 \quad \text{Adding}$$
$$x^2 = \phantom{0}9$$
$$x = \pm 3$$

If $x = 3$, $x^2 = 9$, and if $x = -3$, $x^2 = 9$, so substituting 3 or $-3$ in Eq. (1) gives us

$$x^2 + y^2 = 9$$
$$9 + y^2 = 9$$
$$y^2 = 0$$
$$y = 0.$$

The pairs $(3, 0)$ and $(-3, 0)$ check. They are the solutions.

**31.** $x^2 + y^2 = 25,$    (1)

     $xy = 12$       (2)

First we solve Eq. (2) for $y$.

$$xy = 12$$
$$y = \frac{12}{x}$$

Then we substitute $\dfrac{12}{x}$ for $y$ in Eq. (1) and solve for $x$.

$$x^2 + y^2 = 25$$
$$x^2 + \left(\frac{12}{x}\right)^2 = 25$$
$$x^2 + \frac{144}{x^2} = 25$$
$$x^4 + 144 = 25x^2 \quad \text{Multiplying by } x^2$$
$$x^4 - 25x^2 + 144 = 0$$
$$u^2 - 25u + 144 = 0 \quad \text{Letting } u = x^2$$
$$(u - 9)(u - 16) = 0$$
$$u = 9 \quad \text{or} \quad u = 16$$

We now substitute $x^2$ for $u$ and solve for $x$.

$$x^2 = 9 \quad or \quad x^2 = 16$$
$$x = \pm 3 \quad or \quad x = \pm 4$$

Since $y = 12/x$, if $x = 3$, $y = 4$; if $x = -3$, $y = -4$; if $x = 4$, $y = 3$; and if $x = -4$, $y = -3$. The pairs $(3, 4)$, $(-3, -4)$, $(4, 3)$, and $(-4, -3)$ check. They are the solutions.

**33.** $x^2 + y^2 = 9,$      (1)

     $25x^2 + 16y^2 = 400$    (2)

$$\phantom{2}-16x^2 - 16y^2 = -144 \quad \text{Multiplying (1) by } -16$$
$$\underline{\phantom{-}25x^2 + 16y^2 = \phantom{-}400}$$
$$9x^2 \phantom{xxxx} = \phantom{-}256 \quad \text{Adding}$$
$$x = \pm\frac{16}{3}$$

$$\frac{256}{9} + y^2 = 9 \quad \text{Substituting in (1)}$$
$$y^2 = 9 - \frac{256}{9}$$
$$y^2 = -\frac{175}{9}$$
$$y = \pm\sqrt{-\frac{175}{9}} = \pm\frac{5\sqrt{7}}{3}i$$

The pairs $\left(\dfrac{16}{3}, \dfrac{5\sqrt{7}}{3}i\right)$, $\left(\dfrac{16}{3}, -\dfrac{5\sqrt{7}}{3}i\right)$,

$\left(-\dfrac{16}{3}, \dfrac{5\sqrt{7}}{3}i\right)$, and $\left(-\dfrac{16}{3}, -\dfrac{5\sqrt{7}}{3}i\right)$ check. They are the solutions.

**35.** $x^2 + y^2 = 14,$    (1)

$$\underline{x^2 - y^2 = \phantom{0}4} \quad (2)$$
$$2x^2 \phantom{xxx} = 18 \quad \text{Adding}$$
$$x^2 = \phantom{0}9$$
$$x = \pm 3$$

$$9 + y^2 = 14 \quad \text{Substituting in Eq. (1)}$$
$$y^2 = 5$$
$$y = \pm\sqrt{5}$$

The pairs $(-3, -\sqrt{5})$, $(-3, \sqrt{5})$, $(3, -\sqrt{5})$, and $(3, \sqrt{5})$ check. They are the solutions.

**37.** $x^2 + y^2 = 20,$    (1)

     $xy = 8$       (2)

First we solve Eq. (2) for $y$.

$$y = \frac{8}{x}$$

Then we substitute $\dfrac{8}{x}$ for $y$ in Eq. (1) and solve for $x$.

$$x^2 + \left(\frac{8}{x}\right)^2 = 20$$
$$x^2 + \frac{64}{x^2} = 20$$
$$x^4 + 64 = 20x^2$$
$$x^4 - 20x^2 + 64 = 0$$
$$u^2 - 20u + 64 = 0 \quad \text{Letting } u = x^2$$
$$(u - 16)(u - 4) = 0$$
$$u = 16 \quad or \quad u = 4$$
$$x^2 = 16 \quad or \quad x^2 = 4$$
$$x = \pm 4 \quad or \quad x = \pm 2$$

$y = 8/x$, so if $x = 4$, $y = 2$; if $x = -4$, $y = -2$; if $x = 2$, $y = 4$; if $x = -2$, $y = -4$. The pairs $(4, 2)$, $(-4, -2)$, $(2, 4)$, and $(-2, -4)$ check. They are the solutions.

**39.** $x^2 + 4y^2 = 20,$    (1)

     $xy = 4$       (2)

First we solve Eq. (2) for $y$.

$$y = \frac{4}{x}$$

Then we substitute $\frac{4}{x}$ for $y$ in Eq. (1) and solve for $x$.

$$x^2 + 4\left(\frac{4}{x}\right)^2 = 20$$

$$x^2 + \frac{64}{x^2} = 20$$

$$x^4 + 64 = 20x^2$$

$$x^4 - 20x^2 + 64 = 0$$

$$u^2 - 20u + 64 = 0 \quad \text{Letting } u = x^2$$

$$(u - 16)(u - 4) = 0$$

$$u = 16 \quad or \quad u = 4$$

$$x^2 = 16 \quad or \quad x^2 = 4$$

$$x = \pm 4 \quad or \quad x = \pm 2$$

$y = 4/x$, so if $x = 4$, $y = 1$; if $x = -4$, $y = -1$; if $x = 2$, $y = 2$; and if $x = -2$, $y = -2$. The pairs $(4,1)$, $(-4,-1)$, $(2,2)$, and $(-2,-2)$ check. They are the solutions.

**41.** $2xy + 3y^2 = 7,$ (1)

$3xy - 2y^2 = 4$ (2)

$6xy + 9y^2 = 21$ Multiplying (1) by 3

$\underline{-6xy + 4y^2 = -8}$ Multiplying (2) by $-2$

$13y^2 = 13$

$y^2 = 1$

$y = \pm 1$

Substitute for $y$ in Eq. (1) and solve for $x$.

When $y = 1$: $2 \cdot x \cdot 1 + 3 \cdot 1^2 = 7$

$2x = 4$

$x = 2$

When $y = -1$: $2 \cdot x \cdot (-1) + 3(-1)^2 = 7$

$-2x = 4$

$x = -2$

The pairs $(2,1)$ and $(-2,-1)$ check. They are the solutions.

**43.** $4a^2 - 25b^2 = 0,$ (1)

$2a^2 - 10b^2 = 3b + 4$ (2)

$4a^2 - 25b^2 = 0$

$\underline{-4a^2 + 20b^2 = -6b - 8}$ Multiplying (2) by $-2$

$-5b^2 = -6b - 8$

$0 = 5b^2 - 6b - 8$

$0 = (5b + 4)(b - 2)$

$b = -\frac{4}{5} \quad or \quad b = 2$

Substitute for $b$ in Eq. (1) and solve for $a$.

When $b = -\frac{4}{5}$: $4a^2 - 25\left(-\frac{4}{5}\right)^2 = 0$

$4a^2 = 16$

$a^2 = 4$

$a = \pm 2$

When $b = 2$: $4a^2 - 25(2)^2 = 0$

$4a^2 = 100$

$a^2 = 25$

$a = \pm 5$

The pairs $\left(2, -\frac{4}{5}\right)$, $\left(-2, -\frac{4}{5}\right)$, $(5,2)$ and $(-5,2)$ check. They are the solutions.

**45.** $ab - b^2 = -4,$ (1)

$ab - 2b^2 = -6$ (2)

$ab - b^2 = -4$

$\underline{-ab + 2b^2 = 6}$ Multiplying (2) by $-1$

$b^2 = 2$

$b = \pm\sqrt{2}$

Substitute for $b$ in Eq. (1) and solve for $a$.

When $b = \sqrt{2}$: $a(\sqrt{2}) - (\sqrt{2})^2 = -4$

$a\sqrt{2} = -2$

$a = -\frac{2}{\sqrt{2}} = -\sqrt{2}$

When $b = -\sqrt{2}$: $a(-\sqrt{2}) - (-\sqrt{2})^2 = -4$

$-a\sqrt{2} = -2$

$a = \frac{-2}{-\sqrt{2}} = \sqrt{2}$

The pairs $(-\sqrt{2}, \sqrt{2})$ and $(\sqrt{2}, -\sqrt{2})$ check. They are the solutions.

**47. *Familiarize.*** We first make a drawing. We let $l$ and $w$ represent the length and width, respectively.

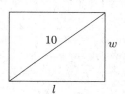

***Translate.*** The perimeter is 28 cm.

$2l + 2w = 28$, or $l + w = 14$

Using the Pythagorean theorem we have another equation.

$l^2 + w^2 = 10^2$, or $l^2 + w^2 = 100$

***Carry out.*** We solve the system:

$l + w = 14,$ (1)

$l^2 + w^2 = 100$ (2)

First solve Eq. (1) for $w$.

$w = 14 - l$ (3)

Then substitute $14 - l$ for $w$ in Eq. (2) and solve for $l$.

$$l^2 + w^2 = 100$$

$$l^2 + (14 - l)^2 = 100$$

$$l^2 + 196 - 28l + l^2 = 100$$

$$2l^2 - 28l + 96 = 0$$

$$l^2 - 14l + 48 = 0$$

$$(l - 8)(l - 6) = 0$$

$$l = 8 \; or \; l = 6$$

If $l = 8$, then $w = 14 - 8$, or 6. If $l = 6$, then $w = 14 - 6$, or 8. Since the length is usually considered to be longer than the width, we have the solution $l = 8$ and $w = 6$, or $(8, 6)$.

**Check**. If $l = 8$ and $w = 6$, then the perimeter is $2 \cdot 8 + 2 \cdot 6$, or 28. The length of a diagonal is $\sqrt{8^2 + 6^2}$, or $\sqrt{100}$, or 10. The numbers check.

**State**. The length is 8 cm, and the width is 6 cm.

**49. Familiarize**. We first make a drawing. Let $l =$ the length and $w =$ the width of the rectangle.

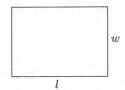

**Translate**.

Area: $lw = 20$

Perimeter: $2l + 2w = 18$, or $l + w = 9$

**Carry out**. We solve the system:

Solve the second equation for $l$:  $l = 9 - w$

Substitute $9 - w$ for $l$ in the first equation and solve for $w$.

$(9 - w)w = 20$

$9w - w^2 = 20$

$0 = w^2 - 9w + 20$

$0 = (w - 5)(w - 4)$

$w = 5 \ or \ w = 4$

If $w = 5$, then $l = 9 - w$, or 4. If $w = 4$, then $l = 9 - 4$, or 5. Since length is usually considered to be longer than width, we have the solution $l = 5$ and $w = 4$, or $(5, 4)$.

**Check**. If $l = 5$ and $w = 4$, the area is $5 \cdot 4$, or 20. The perimeter is $2 \cdot 5 + 2 \cdot 4$, or 18. The numbers check.

**State**. The length is 5 in. and the width is 4 in.

**51. Familiarize**. We first make a drawing. Let $l =$ the length and $w =$ the width of the cargo area, in feet.

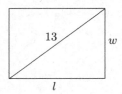

**Translate**. The cargo area must be 60 ft$^2$, so we have one equation:

$lw = 60$

The Pythagorean equation gives us another equation:

$l^2 + w^2 = 13^2$, or $l^2 + w^2 = 169$

**Carry out**. We solve the system of equations.

$lw = 60$,          (1)

$l^2 + w^2 = 169$   (2)

First solve Eq. (1) for $w$:

$lw = 60$

$w = \dfrac{60}{l}$   (3)

Then substitute $60/l$ for $w$ in Eq. (2) and solve for $l$.

$$l^2 + w^2 = 169$$

$$l^2 + \left(\frac{60}{l}\right)^2 = 169$$

$$l^2 + \frac{3600}{l^2} = 169$$

$$l^4 + 3600 = 169l^2$$

$$l^4 - 169l^2 + 3600 = 0$$

Let $u = l^2$ and $u^2 = l^4$ and substitute.

$u^2 - 169u + 3600 = 0$

$(u - 144)(u - 25) = 0$

$u = 144 \ \ or \ \ u = 25$

$l^2 = 144 \ \ or \ \ l^2 = 25$   Replacing $u$ with $l^2$

$l = \pm 12 \ \ or \ \ l = \pm 5$

Since the length cannot be negative, we consider only 12 and 5. We substitute in Eq. (3) to find $w$. When $l = 12$, $w = 60/12 = 5$; when $l = 5$, $w = 60/5 = 12$. Since we usually consider length to be longer than width, we check the pair $(12.5)$.

**Check**. If the length is 12 ft and the width is 5 ft, then the area is $12 \cdot 5$, or 60 ft$^2$. Also $12^2 + 5^2 = 144 + 25 = 169 = 13^2$. The answer checks.

**State**. The length is 12 ft and the width is 5 ft.

**53. Familiarize**. Let $x$ and $y$ represent the numbers.

**Translate**. The product of the numbers is 60, so we have

$xy = 60.$     (1)

The sum of the squares of the numbers is 136, so we have

$x^2 + y^2 = 136.$     (2)

**Carry out**. We solve the system of equations. First solve Eq. (1) for $y$.

$xy = 60$

$y = \dfrac{60}{x}$   (3)

Now substitute $\dfrac{60}{x}$ for $y$ in Eq. (2) and solve for $x$.

$$x^2 + \left(\frac{60}{x}\right)^2 = 136$$

$$x^2 + \frac{3600}{x^2} = 136$$

$x^4 + 3600 = 136x^2$   Multiplying by $x^2$

$x^4 - 136x^2 + 3600 = 0$

$u^2 - 136u + 3600 = 0$        Letting $u = x^2$

$(u - 36)(u - 100) = 0$

$u = 36 \ or \ u = 100$

Now substitute $x^2$ for $u$ and solve for $x$.

$x^2 = 36 \ \ or \ \ x^2 = 100$

$x = \pm 6 \ \ or \ \ x = \pm 10$

We use Eq. (3) to find $y$. When $x = 6$, $y = 60/6 = 10$; when $x = -6$, $y = 60/-6 = -10$; when $x = 10$, $y = 60/10 = 6$; when $x = -10$, $y = 60/-10 = -6$. We see that the numbers can be 6 and 10 or $-6$ and $-10$.

**Check.** $6 \cdot 10 = 60$ and $6^2 + 10^2 = 136$; also $-6(-10) = 60$ and $(-6)^2 + (-10)^2 = 136$. The solutions check.

**State.** The numbers are 6 and 10 or $-6$ and $-10$.

**55. Familiarize.** We let $x =$ the length of a side of one peanut bed, in feet, and $y =$ the length of a side of the other peanut bed. Make a drawing.

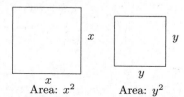

Area: $x^2$      Area: $y^2$

**Translate.** The sum of the areas is 832 ft², so we have

$$x^2 + y^2 = 832. \qquad (1)$$

The difference of the areas is 320 ft², so we have

$$x^2 - y^2 = 320. \qquad (2)$$

**Carry out.** We solve the system of equations.

$$
\begin{array}{ll}
x^2 + y^2 = 832 & (1) \\
x^2 - y^2 = 320 & (2) \\
\hline
2x^2 \quad\quad = 1152 & \text{Adding} \\
x^2 = 576 \\
x = \pm 24
\end{array}
$$

Since the length cannot be negative we consider only 24. We substitute 24 for $x$ in Eq. (1) and solve for $y$.

$$24^2 + y^2 = 832$$
$$576 + y^2 = 832$$
$$y^2 = 256$$
$$y = \pm 16$$

Again we consider only the positive number.

**Check.** If the lengths of the sides of the beds are 24 ft and 16 ft, the areas of the beds are $24^2$, or 576 ft², and $16^2$, or 256 ft², respectively. Then 576 ft² + 256 ft² = 832 ft², and 576 ft² − 256 ft² = 320 ft², so the answer checks.

**State.** The lengths of the sides of the beds are 24 ft and 16 ft.

**57. Familiarize.** We make a drawing and label it. Let $x$ and $y$ represent the lengths of the legs of the triangle.

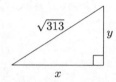

**Translate.** The product of the lengths of the legs is 156, so we have:

$$xy = 156$$

We use the Pythagorean theorem to get a second equation:

$$x^2 + y^2 = (\sqrt{313})^2, \text{ or } x^2 + y^2 = 313$$

**Carry out.** We solve the system of equations.

$$
\begin{array}{ll}
xy = 156, & (1) \\
x^2 + y^2 = 313 & (2)
\end{array}
$$

First solve Equation (1) for $y$.

$$xy = 156$$
$$y = \frac{156}{x}$$

Then we substitute $\dfrac{156}{x}$ for $y$ in Eq. (2) and solve for $x$.

$$x^2 + y^2 = 313 \qquad (2)$$
$$x^2 + \left(\frac{156}{x}\right)^2 = 313$$
$$x^2 + \frac{24{,}336}{x^2} = 313$$
$$x^4 + 24{,}336 = 313x^2$$
$$x^4 - 313x^2 + 24{,}336 = 0$$
$$u^2 - 313u + 24{,}336 = 0 \qquad \text{Letting } u = x^2$$
$$(u - 169)(u - 144) = 0$$
$$u - 169 = 0 \quad or \quad u - 144 = 0$$
$$u = 169 \quad or \quad\quad\quad u = 144$$

We now substitute $x^2$ for $u$ and solve for $x$.

$$x = \pm 13 \quad or \quad x = \pm 12$$

Since $y = 156/x$, if $x = 13$, $y = 12$; if $x = -13$, $y = -12$; if $x = 12$, $y = 13$; and if $x = -12$, $y = -13$. The possible solutions are $(13, 12)$, $(-13, -12)$, $(12, 13)$, and $(-12, -13)$.

**Check.** Since measurements cannot be negative, we consider only $(13, 12)$ and $(12, 13)$. Since both possible solutions give the same pair of legs, we only need to check $(13, 12)$. If $x = 13$ and $y = 12$, their product is 156. Also, $\sqrt{13^2 + 12^2} = \sqrt{313}$. The numbers check.

**State.** The lengths of the legs are 13 and 12.

**59.** *Writing Exercise*

**61.** $(-1)^9 (-2)^4 = -1 \cdot 16 = -16$

**63.** $\dfrac{(-1)^k}{k - 5} = \dfrac{(-1)^6}{6 - 5} = \dfrac{1}{1} = 1$

**65.** $\dfrac{n}{2}(3 + n) = \dfrac{8}{2}(3 + 8) = 4 \cdot 11 = 44$

**67.** *Writing Exercise*

**69.** Let $(h, k)$ represent the point on the line $5x + 8y = -2$ which is the center of a circle that passes through the points $(-2, 3)$ and $(-4, 1)$. The distance between $(h, k)$ and $(-2, 3)$ is the same as the distance between $(h, k)$ and $(-4, 1)$. This gives us one equation:

$$\sqrt{[h-(-2)]^2+(k-3)^2} = \sqrt{[h-(-4)]^2+(k-1)^2}$$
$$(h+2)^2 + (k-3)^2 = (h+4)^2 + (k-1)^2$$
$$h^2+4h+4+k^2-6k+9 = h^2+8h+16+k^2-2k+1$$
$$4h-6k+13 = 8h-2k+17$$
$$-4h-4k = 4$$
$$h+k = -1$$

We get a second equation by substituting $(h,k)$ in $5x+8y=-2$.

$$5h+8k = -2$$

We now solve the following system:

$$h+\phantom{5}k = -1,$$
$$5h+8k = -2$$

The solution, which is the center of the circle, is $(-2,1)$.

Next we find the length of the radius. We can find the distance between either $(-2,3)$ or $(-4,1)$ and the center $(-2,1)$. We use $(-2,3)$.

$$r = \sqrt{[-2-(-2)]^2+(1-3)^2}$$
$$r = \sqrt{0^2+(-2)^2}$$
$$r = \sqrt{4} = 2$$

We can write the equation of the circle with center $(-2,1)$ and radius 2.

$$(x-h)^2+(y-k)^2 = r^2$$
$$[x-(-2)]^2+(y-1)^2 = 2^2$$
$$(x+2)^2+(y-1)^2 = 4$$

**71.** $p^2+q^2 = 13,$     (1)

$$\frac{1}{pq} = -\frac{1}{6} \qquad (2)$$

Solve Eq. (2) for $p$.

$$\frac{1}{q} = -\frac{p}{6}$$
$$-\frac{6}{q} = p$$

Substitute $-6/q$ for $p$ in Eq. (1) and solve for $q$.

$$\left(-\frac{6}{q}\right)^2+q^2 = 13$$
$$\frac{36}{q^2}+q^2 = 13$$
$$36+q^4 = 13q^2$$
$$q^4-13q^2+36 = 0$$
$$u^2-13u+36 = 0 \qquad \text{Letting } u = q^2$$
$$(u-9)(u-4) = 0$$
$$u = 9 \quad \text{or} \quad u = 4$$
$$x^2 = 9 \quad \text{or} \quad x^2 = 4$$
$$x = \pm 3 \quad \text{or} \quad x = \pm 2$$

Since $p = -6/q$, if $q = 3$, $p = -2$; if $q = -3$, $p = 2$; if $q = 2$, $p = -3$; and if $q = -2$, $p = 3$. The pairs $(-2,3)$, $(2,-3)$, $(-3,2)$, and $(3,-2)$ check. They are the solutions.

**73. Familiarize.** Let $l = $ the length of the rectangle, in feet, and let $w = $ the width.

**Translate.** 100 ft of fencing is used, so we have

$$l+w = 100. \qquad (1)$$

The area is 2475 ft$^2$, so we have

$$lw = 2475. \qquad (2)$$

**Carry out.** Solving the system of equations, we get $(55,45)$ and $(45,55)$. Since length is usually considered to be longer than width, we have $l = 55$ and $w = 45$.

**Check.** If the length is 55 ft and the width is 45 ft, then $55+45$, or 100 ft, of fencing is used. The area is $55 \cdot 45$, or 2475 ft$^2$. The answer checks.

**State.** The length of the rectangle is 55 ft, and the width is 45 ft.

**75. Familiarize.** We let $x$ and $y$ represent the length and width of the base of the box, in inches, respectively. Make a drawing.

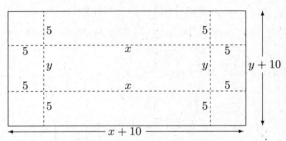

The dimensions of the metal sheet are $x+10$ and $y+10$.

**Translate.** The area of the sheet of metal is 340 in$^2$, so we have

$$(x+10)(y+10) = 340. \qquad (1)$$

The volume of the box is 350 in$^3$, so we have

$$x \cdot y \cdot 5 = 350. \qquad (2)$$

**Carry out.** Solving the system of equations, we get $(10,7)$ and $(7,10)$. Since length is usually considered to be longer than width, we have $l = 10$ and $w = 7$.

**Check.** The dimensions of the metal sheet are $10+10$, or 20, and $7+10$, or 17, so the area is $20 \cdot 17$, or 340 in$^2$. The volume of the box is $7 \cdot 10 \cdot 5$, or 350 in$^3$. The answer checks.

**State.** The dimensions of the box are 10 in. by 7 in. by 5 in.

**77. Familiarize.** Let $l = $ the length and $h = $ the height, in inches.

**Translate.** Since the ratio of the length to the height is 16 to 9, we have one equation:

$$\frac{l}{h} = \frac{16}{9}$$

The Pythagorean equation gives us a second equation:

$$l^2+h^2 = (\sqrt{4901})^2, \text{ or } l^2+h^2 = 4901$$

We have a system of equations.

$$\frac{l}{h} = \frac{16}{9}, \qquad (1)$$
$$l^2+h^2 = 4901 \qquad (2)$$

*Carry out*. Solving the system of equations, we get $(34.32, 61.02)$ and $(-34.32, -61.02)$. Since the dimensions cannot be negative, we consider only $(34.32, 61.02)$.

*Check*. The ratio of 61.02 to 34.32 is $\dfrac{61.02}{34.32} \approx 1.778 \approx \dfrac{16}{9}$. Also $(61.02)^2 + (34.32)^2 = 4901.3028 \approx 4901$. The answer checks.

*State*. The length is about 61.02 in., and the height is about 34.32 in.

79.

# Chapter 11

# Sequences, Series, and the Binomial Theorem

**1.** (f)

**3.** (d)

**5.** (c)

**7.** $a_n = 2n + 3$

$a_1 = 2 \cdot 1 + 3 = 5$,

$a_2 = 2 \cdot 2 + 3 = 7$,

$a_3 = 2 \cdot 3 + 3 = 9$,

$a_4 = 2 \cdot 4 + 3 = 11$;

$a_{10} = 2 \cdot 10 + 3 = 23$;

$a_{15} = 2 \cdot 15 + 3 = 33$

**9.** $a_n = n^2 + 2$

$a_1 = 1^2 + 2 = 3$,

$a_2 = 2^2 + 2 = 6$,

$a_3 = 3^2 + 2 = 11$,

$a_4 = 4^2 + 2 = 18$;

$a_{10} = 10^2 + 2 = 102$;

$a_{15} = 15^2 + 2 = 227$

**11.** $a_n = \dfrac{n^2 - 1}{n^2 + 1}$

$a_1 = \dfrac{1^2 - 1}{1^2 + 1} = 0$,

$a_2 = \dfrac{2^2 - 1}{2^2 + 1} = \dfrac{3}{5}$,

$a_3 = \dfrac{3^2 - 1}{3^2 + 1} = \dfrac{8}{10} = \dfrac{4}{5}$,

$a_4 = \dfrac{4^2 - 1}{4^2 + 1} = \dfrac{15}{17}$;

$a_{10} = \dfrac{10^2 - 1}{10^2 + 1} = \dfrac{99}{101}$;

$a_{15} = \dfrac{15^2 - 1}{15^2 + 1} = \dfrac{224}{226} = \dfrac{112}{113}$

**13.** $a_n = \left( -\dfrac{1}{2} \right)^{n-1}$

$a_1 = \left( -\dfrac{1}{2} \right)^{1-1} = 1$,

$a_2 = \left( -\dfrac{1}{2} \right)^{2-1} = -\dfrac{1}{2}$,

$a_3 = \left( -\dfrac{1}{2} \right)^{3-1} = \dfrac{1}{4}$,

$a_4 = \left( -\dfrac{1}{2} \right)^{4-1} = -\dfrac{1}{8}$;

$a_{10} = \left( -\dfrac{1}{2} \right)^{10-1} = -\dfrac{1}{512}$;

$a_{15} = \left( -\dfrac{1}{2} \right)^{15-1} = \dfrac{1}{16,384}$

**15.** $a_n = (-1)^n(n + 3)$

$a_1 = (-1)^1(1 + 3) = -4$,

$a_2 = (-1)^2(2 + 3) = 5$,

$a_3 = (-1)^3(3 + 3) = -6$,

$a_4 = (-1)^4(4 + 3) = 7$;

$a_{10} = (-1)^{10}(10 + 3) = 13$;

$a_{15} = (-1)^{15}(15 + 3) = -18$

**17.** $a_n = (-1)^n(n^3 - 1)$

$a_1 = (-1)^1(1^3 - 1) = 0$,

$a_2 = (-1)^2(2^3 - 1) = 7$,

$a_3 = (-1)^3(3^3 - 1) = -26$,

$a_4 = (-1)^4(4^3 - 1) = 63$;

$a_{10} = (-1)^{10}(10^3 - 1) = 999$;

$a_{15} = (-1)^{15}(15^3 - 1) = -3374$

**19.** $a_n = 2n - 3$

$a_8 = 2 \cdot 8 - 3 = 16 - 3 = 13$

**21.** $a_n = (3n + 1)(2n - 5)$

$a_9 = (3 \cdot 9 + 1)(2 \cdot 9 - 5) = 28 \cdot 13 = 364$

**23.** $a_n = (-1)^{n-1}(3.4n - 17.3)$

$a_{12} = (-1)^{12-1}[3.4(12) - 17.3] = -23.5$

**25.** $a_n = 3n^2(9n - 100)$

$a_{11} = 3 \cdot 11^2(9 \cdot 11 - 100) = 3 \cdot 121(-1) = -363$

**27.** $a_n = \left( 1 + \dfrac{1}{n} \right)^2$

$a_{20} = \left( 1 + \dfrac{1}{20} \right)^2 = \left( \dfrac{21}{20} \right)^2 = \dfrac{441}{400}$

**29.** 2, 4, 6, 8, 10, . . .

These are even integers beginning with 2, so the general term could be $2n$.

**31.** 1, −1, 1, −1, . . .

1 and −1 alternate, beginning with 1, so the general term could be $(-1)^{n+1}$.

**33.** −1, 2, −3, 4, . . .

These are the first four natural numbers, but with alternating signs, beginning with a negative number. The general term could be $(-1)^n \cdot n$.

**35.** 3, 5, 7, 9, . . .

These are odd integers beginning with 3, so the general term could be $2n + 1$.

**37.** $-2, 6, -18, 54, \ldots$

We can see a pattern if we write the sequence as

$-1 \cdot 2 \cdot 1, 1 \cdot 2 \cdot 3, -1 \cdot 2 \cdot 9, 1 \cdot 2 \cdot 27, \ldots$

The general term could be $(-1)^n \cdot 2 \cdot (3)^{n-1}$.

**39.** $\dfrac{1}{2}, \dfrac{2}{3}, \dfrac{3}{4}, \dfrac{4}{5}, \dfrac{5}{6}, \ldots$

These are fractions in which the denominator is 1 greater than the numerator. Also, each numerator is 1 greater than the preceding numerator. The general term could be $\dfrac{n}{n+1}$.

**41.** 5, 25, 125, 625, . . .

This is powers of 5, so the general term could be $5^n$.

**43.** $-1, 4, -9, 16, \ldots$

This is the squares of the first four natural numbers, but with alternating signs, beginning with a negative number. The general term could be $(-1)^n \cdot n^2$.

**45.** $1, -2, 3, -4, 5, -6, \ldots$

$S_7 = 1 - 2 + 3 - 4 + 5 - 6 + 7 = 4$

**47.** 2, 4, 6, 8, . . .

$S_5 = 2 + 4 + 6 + 8 + 10 = 30$

**49.** $\displaystyle\sum_{k=1}^{5} \dfrac{1}{2k} = \dfrac{1}{2 \cdot 1} + \dfrac{1}{2 \cdot 2} + \dfrac{1}{2 \cdot 3} + \dfrac{1}{2 \cdot 4} + \dfrac{1}{2 \cdot 5}$

$\qquad = \dfrac{1}{2} + \dfrac{1}{4} + \dfrac{1}{6} + \dfrac{1}{8} + \dfrac{1}{10}$

$\qquad = \dfrac{60}{120} + \dfrac{30}{120} + \dfrac{20}{120} + \dfrac{15}{120} + \dfrac{12}{120}$

$\qquad = \dfrac{137}{120}$

**51.** $\displaystyle\sum_{k=0}^{4} 3^k = 3^0 + 3^1 + 3^2 + 3^3 + 3^4$

$\qquad = 1 + 3 + 9 + 27 + 81$

$\qquad = 121$

**53.** $\displaystyle\sum_{k=2}^{8} \dfrac{k}{k-1} = \dfrac{2}{2-1} + \dfrac{3}{3-1} + \dfrac{4}{4-1} +$

$\qquad \dfrac{5}{5-1} + \dfrac{6}{6-1} + \dfrac{7}{7-1} + \dfrac{8}{8-1}$

$\qquad = \dfrac{2}{1} + \dfrac{3}{2} + \dfrac{4}{3} + \dfrac{5}{4} + \dfrac{6}{5} + \dfrac{7}{6} + \dfrac{8}{7}$

$\qquad = \dfrac{1343}{140}$

**55.** $\displaystyle\sum_{k=1}^{8} (-1)^{k+1} 2^k = (-1)^2 2^1 + (-1)^3 2^2 + (-1)^4 2^3 +$

$\qquad (-1)^5 2^4 + (-1)^6 2^5 + (-1)^7 2^6 +$

$\qquad (-1)^8 2^7 + (-1)^9 2^8$

$\qquad = 2 - 4 + 8 - 16 + 32 - 64 +$

$\qquad 128 - 256$

$\qquad = -170$

**57.** $\displaystyle\sum_{k=0}^{5} (k^2 - 2k + 3)$

$= (0^2 - 2 \cdot 0 + 3) + (1^2 - 2 \cdot 1 + 3) +$

$\quad (2^2 - 2 \cdot 2 + 3) + (3^2 - 2 \cdot 3 + 3) +$

$\quad (4^2 - 2 \cdot 4 + 3) + (5^2 - 2 \cdot 5 + 3)$

$= 3 + 2 + 3 + 6 + 11 + 18$

$= 43$

**59.** $\displaystyle\sum_{k=3}^{5} \dfrac{(-1)^k}{k(k+1)} = \dfrac{(-1)^3}{3(3+1)} + \dfrac{(-1)^4}{4(4+1)} + \dfrac{(-1)^5}{5(5+1)}$

$\qquad = \dfrac{-1}{3 \cdot 4} + \dfrac{1}{4 \cdot 5} + \dfrac{-1}{5 \cdot 6}$

$\qquad = -\dfrac{1}{12} + \dfrac{1}{20} - \dfrac{1}{30}$

$\qquad = -\dfrac{4}{60} = -\dfrac{1}{15}$

**61.** $\dfrac{2}{3} + \dfrac{3}{4} + \dfrac{4}{5} + \dfrac{5}{6} + \dfrac{6}{7}$

This is a sum of fractions in which the denominator is one greater than the numerator. Also, each numerator is 1 greater than the preceding numerator. Sigma notation is

$\displaystyle\sum_{k=1}^{5} \dfrac{k+1}{k+2}.$

**63.** $1 + 4 + 9 + 16 + 25 + 36$

This is the sum of the squares of the first six natural numbers. Sigma notation is

$\displaystyle\sum_{k=1}^{6} k^2.$

**65.** $4 - 9 + 16 - 25 + \ldots + (-1)^n n^2$

This is a sum of terms of the form $(-1)^k k^2$, beginning with $k = 2$ and continuing through $k = n$. Sigma notation is

$\displaystyle\sum_{k=2}^{n} (-1)^k k^2.$

**67.** $5 + 10 + 15 + 20 + 25 + \ldots$

This is a sum of multiples of 5, and it is an infinite series. Sigma notation is

$\displaystyle\sum_{k=1}^{\infty} 5k.$

**69.** $\dfrac{1}{1 \cdot 2} + \dfrac{1}{2 \cdot 3} + \dfrac{1}{3 \cdot 4} + \dfrac{1}{4 \cdot 5} + \ldots$

This is a sum of fractions in which the numerator is 1 and the denominator is a product of two consecutive integers. The larger integer in each product is the smaller integer in the succeeding product. It is an infinite series. Sigma notation is

$\displaystyle\sum_{k=1}^{\infty} \dfrac{1}{k(k+1)}.$

**71.** *Writing Exercise*

**73.** $\dfrac{7}{2}(a_1 + a_7) = \dfrac{7}{2}(8 + 14) = \dfrac{7}{2} \cdot 22 = 77$

**75.** $(x + y)^3$
$$= (x + y)(x + y)^2$$
$$= (x + y)(x^2 + 2xy + y^2)$$
$$= x(x^2 + 2xy + y^2) + y(x^2 + 2xy + y^2)$$
$$= x^3 + 2x^2y + xy^2 + x^2y + 2xy^2 + y^3$$
$$= x^3 + 3x^2y + 3xy^2 + y^3$$

**77.** $(2a - b)^3$
$$= (2a - b)(2a - b)^2$$
$$= (2a - b)(4a^2 - 4ab + b^2)$$
$$= 2a(4a^2 - 4ab + b^2) - b(4a^2 - 4ab + b^2)$$
$$= 8a^3 - 8a^2b + 2ab^2 - 4a^2b + 4ab^2 - b^3$$
$$= 8a^3 - 12a^2b + 6ab^2 - b^3$$

**79.** *Writing Exercise*

**81.** $a_1 = 1$, $a_{n+1} = 5a_n - 2$
$$a_1 = 1$$
$$a_2 = 5 \cdot 1 - 2 = 3$$
$$a_3 = 5 \cdot 3 - 2 = 13$$
$$a_4 = 5 \cdot 13 - 2 = 63$$
$$a_5 = 5 \cdot 63 - 2 = 313$$
$$a_6 = 5 \cdot 313 - 2 = 1563$$

**83.** Find each term by multiplying the preceding term by 0.75:

$5200, $3900, $2925, $2193.75, $1645.31,

$1233.98, $925.49, $694.12, $520.59, $390.44

**85.** $a_n = (-1)^n$

This sequence is of the form $-1, 1, -1, 1, \ldots$. Each pair of terms adds to 0. $S_{100}$ has 50 such pairs, so $S_{100} = 0$. $S_{101}$ consists of the 50 pairs in $S_{100}$ that add to 0 as well as $a_{101}$, or $-1$, so $S_{101} = -1$.

**87.** $a_n = i^n$
$$a_1 = i^1 = i$$
$$a_2 = i^2 = -1$$
$$a_3 = i^3 = i^2 \cdot i = -1 \cdot i = -i$$
$$a_4 = i^4 = (i^2)^2 = (-1)^2 = 1$$
$$a_5 = i^5 = (i^2)^2 \cdot i = (-1)^2 \cdot i = 1 \cdot i = i$$
$$S_5 = i - 1 - i + 1 + i = i$$

**89.** Enter $y_1 = 14x^4 + 6x^3 + 416x^2 - 655x - 1050$. Then scroll through a table of values. We see that $y_1 = 6144$ when $x = 11$, so the 11th term of the sequence is 6144.

## Exercise Set 11.2

**1.** True; see page 711 in the text.

**3.** False; see page 711 in the text.

**5.** True; see page 714 in the text.

**7.** False; $S_5 = a_1 + a_2 + a_3 + a_4 + a_5$.

**9.** 2, 6, 10, 14, ...
$$a_1 = 2$$
$$d = 4 \qquad (6 - 2 = 4, \ 10 - 6 = 4, \ 14 - 10 = 4)$$

**11.** 7, 3, −1, −5, ...
$$a_1 = 7$$
$$d = -4 \qquad (3 - 7 = -4, -1 - 3 = -4,$$
$$-5 - (-1) = -4)$$

**13.** $\dfrac{3}{2}, \dfrac{9}{4}, 3, \dfrac{15}{4}, \ldots$
$$a_1 = \frac{3}{2}$$
$$d = \frac{3}{4} \qquad \left(\frac{9}{4} - \frac{3}{2} = \frac{3}{4}, \ 3 - \frac{9}{4} = \frac{3}{4}\right)$$

**15.** $5.12, $5.24, $5.36, $5.48, ...
$$a_1 = \$5.12$$
$$d = \$0.12 \quad (\$5.24 - \$5.12 = \$0.12, \ \$5.36 -$$
$$\$5.24 = \$0.12, \ \$5.48 - \$5.36 =$$
$$\$0.12)$$

**17.** 7, 10, 13, ...
$$a_1 = 7, \ d = 3, \text{ and } n = 15$$
$$a_n = a_1 + (n - 1)d$$
$$a_{15} = 7 + (15 - 1)3 = 7 + 14 \cdot 3 = 7 + 42 = 49$$

**19.** 8, 2, −4, ...
$$a_1 = 8, \ d = -6, \text{ and } n = 18$$
$$a_n = a_1 + (n - 1)d$$
$$a_{18} = 8 + (18 - 1)(-6) = 8 + 17(-6) =$$
$$8 - 102 = -94$$

**21.** $1200, $964.32, $728.64, ...
$$a_1 = \$1200, \ d = \$964.32 - \$1200 = -\$235.68,$$
$$\text{and } n = 13$$
$$a_n = a_1 + (n - 1)d$$
$$a_{13} = \$1200 + (13 - 1)(-\$235.68) =$$
$$\$1200 + 12(-\$235.68) = \$1200 - \$2828.16 =$$
$$-\$1628.16$$

**23.** $a_1 = 7$, $d = 3$
$$a_n = a_1 + (n - 1)d$$
Let $a_n = 82$, and solve for $n$.
$$82 = 7 + (n - 1)(3)$$
$$82 = 7 + 3n - 3$$
$$82 = 4 + 3n$$
$$78 = 3n$$
$$26 = n$$

The 26th term is 82.

**25.**     $a_1 = 8,\ d = -6$

$\qquad a_n = a_1 + (n-1)d$

$\quad -328 = 8 + (n-1)(-6)$

$\quad -328 = 8 - 6n + 6$

$\quad -328 = 14 - 6n$

$\quad -342 = -6n$

$\qquad 57 = n$

The 57th term is $-328$.

**27.**     $a_n = a_1 + (n-1)d$

$\quad a_{17} = 2 + (17-1)5 \qquad$ Substituting 17 for $n$,

$\qquad\qquad\qquad\qquad\qquad$ 2 for $a_1$, and 5 for $d$

$\qquad = 2 + 16\cdot 5$

$\qquad = 2 + 80$

$\qquad = 82$

**29.**     $a_n = a_1 + (n-1)d$

$\quad 33 = a_1 + (8-1)4 \qquad$ Substituting 33 for $a_8$,

$\qquad\qquad\qquad\qquad\qquad$ 8 for $n$, and 4 for $d$

$\quad 33 = a_1 + 28$

$\quad 5 = a_1$

(Note that this procedure is equivalent to subtracting $d$ from $a_8$ seven times to get $a_1$: $33 - 7(4) = 33 - 28 = 5$)

**31.**     $a_n = a_1 + (n-1)d$

$\quad -76 = 5 + (n-1)(-3) \qquad$ Substituting $-76$ for

$\qquad\qquad\qquad\qquad\qquad$ $a_n$, 5 for $a_1$, and $-3$

$\qquad\qquad\qquad\qquad\qquad$ for $d$

$\quad -76 = 5 - 3n + 3$

$\quad -76 = 8 - 3n$

$\quad -84 = -3n$

$\quad 28 = n$

**33.** We know that $a_{17} = -40$ and $a_{28} = -73$. We would have to add $d$ eleven times to get from $a_{17}$ to $a_{28}$. That is,

$\quad -40 + 11d = -73$

$\qquad\quad 11d = -33$

$\qquad\qquad d = -3.$

Since $a_{17} = -40$, we subtract $d$ sixteen times to get to $a_1$.

$\quad a_1 = -40 - 16(-3) = -40 + 48 = 8$

We write the first five terms of the sequence:

$\quad 8,\ 5,\ 2,\ -1,\ -4$

**35.** $a_{13} = 13$ and $a_{54} = 54$

Observe that for this to be true, $a_1 = 1$ and $d = 1$.

**37.** $1 + 5 + 9 + 13 + \ldots$

Note that $a_1 = 1$, $d = 4$, and $n = 20$. Before using the formula for $S_n$, we find $a_{20}$:

$\quad a_{20} = 1 + (20-1)4 \qquad$ Substituting into

$\qquad\qquad\qquad\qquad\qquad$ the formula for $a_n$

$\qquad\quad = 1 + 19\cdot 4$

$\qquad\quad = 77$

Then

$\quad S_{20} = \dfrac{20}{2}(1 + 77) \qquad$ Using the formula for $S_n$

$\qquad\quad = 10(78)$

$\qquad\quad = 780.$

**39.** The sum is $1 + 2 + 3 + \ldots + 249 + 250$. This is the sum of the arithmetic sequence for which $a_1 = 1$, $a_n = 250$, and $n = 250$. We use the formula for $S_n$.

$$S_n = \frac{n}{2}(a_1 + a_n)$$

$$S_{300} = \frac{250}{2}(1 + 250) = 125(251) = 31,375$$

**41.** The sum is $2 + 4 + 6 + \ldots + 98 + 100$. This is the sum of the arithmetic sequence for which $a_1 = 2$, $a_n = 100$, and $n = 50$. We use the formula for $S_n$.

$$S_n = \frac{n}{2}(a_1 + a_n)$$

$$S_{50} = \frac{50}{2}(2 + 100) = 25(102) = 2550$$

**43.** The sum is $6 + 12 + 18 + \ldots + 96 + 102$. This is the sum of the arithmetic sequence for which $a_1 = 6$, $a_n = 102$, and $n = 17$. We use the formula for $S_n$.

$$S_n = \frac{n}{2}(a_1 + a_n)$$

$$S_{17} = \frac{17}{2}(6 + 102) = \frac{17}{2}(108) = 918$$

**45.** Before using the formula for $S_n$, we find $a_{20}$:

$\quad a_{20} = 4 + (20-1)5 \qquad$ Substituting into

$\qquad\qquad\qquad\qquad\qquad$ the formula for $a_n$

$\qquad\quad = 4 + 19\cdot 5 = 99$

Then

$\quad S_{20} = \dfrac{20}{2}(4 + 99) \qquad$ Using the formula

$\qquad\qquad\qquad\qquad\qquad$ for $S_n$

$\qquad\quad = 10(103) = 1030.$

**47. _Familiarize._** We want to find the fifteenth term and the sum of an arithmetic sequence with $a_1 = 7$, $d = 2$, and $n = 15$. We will first use the formula for $a_n$ to find $a_{15}$. This result is the number of marchers in the last row. Then we will use the formula for $S_n$ to find $S_{15}$. This is the total number of marchers.

**_Translate._** Substituting into the formula for $a_n$, we have

$\quad a_{15} = 7 + (15-1)2.$

**_Carry out._** We first find $a_{15}$.

$\quad a_{15} = 7 + 14\cdot 2 = 35$

Then use the formula for $S_n$ to find $S_{15}$.

$\quad S_{15} = \dfrac{15}{2}(7 + 35) = \dfrac{15}{2}(42) = 315$

**_Check._** We can do the calculations again. We can also do the entire addition.

$\quad 7 + 9 + 11 + \ldots + 35.$

**_State._** There are 35 marchers in the last row, and there are 315 marchers altogether.

**49.** *Familiarize.* We want to find the sum of the arithmetic sequence $36 + 32 + \ldots + 4$. Note that $a_1 = 36$ and $d = -4$. We will first use the formula for $a_n$ to find $n$. Then we will use the formula for $S_n$.

*Translate.* Substituting into the formula for $a_n$, we have

$$4 = 36 + (n-1)(-4).$$

*Carry out.* We solve for $n$.

$$4 = 36 + (n-1)(-4)$$
$$4 = 36 - 4n + 4$$
$$4 = 40 - 4n$$
$$-36 = -4n$$
$$9 = n$$

Now we find $S_9$.

$$S_9 = \frac{9}{2}(36 + 4) = \frac{9}{2} \cdot 40 = 180$$

*Check.* We can do the calculations again. We can also do the entire addition.

$$36 + 32 + \ldots + 4.$$

*State.* There are 180 stones in the pyramid.

**51.** *Familiarize.* We want to find the sum of the arithmetic sequence with $a_1 = 10\cent$, $d = 10\cent$, and $n = 31$. First we will find $a_{31}$ and then we will find $S_{31}$.

*Translate.* Substituting in the formula for $a_n$, we have

$$a_{31} = 10 + (31-1)(10).$$

*Carry out.* First we find $a_{31}$.

$$a_{31} = 10 + 30 \cdot 10 = 10 + 300 = 310$$

Then we use the formula for $S_n$ to find $S_{31}$.

$$S_{31} = \frac{31}{2}(10 + 310) = \frac{31}{2} \cdot 320 = 4960$$

*Check.* We can do the calculations again.

*State.* The amount saved is $4960\cent$, or \$49.60.

**53.** *Familiarize.* We want to find the sum of an arithmetic sequence with $a_1 = 20$, $d = 2$, and $n = 19$. We will use the formula for $a_n$ to find $a_{19}$, and then we will use the formula for $S_n$ to find $S_{19}$.

*Translate.* Substituting into the formula for $a_n$, we have

$$a_{19} = 20 + (19-1)(2).$$

*Carry out.* We find $a_{19}$.

$$a_{19} = 20 + 18 \cdot 2 = 56$$

Then we use the formula for $S_n$ to find $S_{19}$.

$$S_{19} = \frac{19}{2}(20 + 56) = 722$$

*Check.* We can do the calculation again.

*State.* There are 722 seats.

**55.** *Writing Exercise*

**57.**

$$\frac{3}{10x} + \frac{2}{15x}, \text{ LCD is } 30x$$
$$= \frac{3}{10x} \cdot \frac{3}{3} + \frac{2}{15x} \cdot \frac{2}{2}$$
$$= \frac{9}{30x} + \frac{4}{30x}$$
$$= \frac{13}{30x}$$

**59.**

$\log_a P = k \qquad a^k = P$ — The logarithm is the exponent. The base does not change.

**61.** Standard form for the equation of a circle with center $(h, k)$ and radius $r$ is

$$(x - h)^2 + (y - k)^2 = r^2.$$

We substitute 0 for $h$, 0 for $k$, and 9 for $r$:

$$(x - 0)^2 + (y - 0)^2 = 9^2$$
$$x^2 + y^2 = 81$$

**63.** *Writing Exercise*

**65.** The frog climbs $4 - 1$, or 3 ft, with each jump. Then the total distance the frog has jumped with each successive jump is given by the arithmetic sequence $3, 6, 9, \ldots, 96$. When the frog has climbed 96 ft, it will reach the top of the hole on the next jump because it will have climbed $96 + 4$, or 100 ft with that jump. Then the total number of jumps is the number of terms of the sequence above plus the final jump. We find $n$ for the sequence with $a_1 = 3$, $d = 3$, and $a_n = 96$:

$$a_n = a_1 + (n-1)d$$
$$96 = 3 + (n-1)3$$
$$96 = 3 + 3n - 3$$
$$96 = 3n$$
$$32 = n$$

Then the total number of jumps is $32 + 1$, or 33 jumps.

**67.**

$$a_1 = \$8760$$
$$a_2 = \$8760 + (-\$798.23) = \$7961.77$$
$$a_3 = \$8760 + 2(-\$798.23) = \$7163.54$$
$$a_4 = \$8760 + 3(-\$798.23) = \$6365.31$$
$$a_5 = \$8760 + 4(-\$798.23) = \$5567.08$$
$$a_6 = \$8760 + 5(-\$798.23) = \$4768.85$$
$$a_7 = \$8760 + 6(-\$798.23) = \$3970.62$$
$$a_8 = \$8760 + 7(-\$798.23) = \$3172.39$$
$$a_9 = \$8760 + 8(-\$798.23) = \$2374.16$$
$$a_{10} = \$8760 + 9(-\$798.23) = \$1575.93$$

**69.** See the answer section in the text.

**71.** Each integer from 501 through 750 is 500 more than the corresponding integer from 1 through 250. There are 250 integers from 501 through 750, so their sum is the sum of the integers from 1 to 250 plus $250 \cdot 500$. From Exercise 39, we know that the sum of the integers from 1 through 250 is 31,375. Thus, we have

$$31,375 + 250 \cdot 500, \text{ or } 156,375.$$

## Exercise Set 11.3

**1.** $\dfrac{a_{n+1}}{a_n} = 3$, so this is a geometric sequence.

**3.** $a_{n+1} = a_n + 5$, so this is an arithmetic sequence.

**5.** $\dfrac{a_{n+1}}{a_n} = 5$, so this is a geometric series.

**7.** $\dfrac{a_{n+1}}{a_n} = -\dfrac{1}{2}$, so this is a geometric series.

**9.** $7, 14, 28, 56, \ldots$

$\dfrac{14}{7} = 2, \quad \dfrac{28}{14} = 2, \quad \dfrac{56}{28} = 2$

$r = 2$

**11.** $6, -0.6, 0.06, -0.006, \ldots$

$-\dfrac{0.6}{6} = -0.1, \quad \dfrac{0.06}{-0.6} = -0.1, \quad \dfrac{-0.006}{0.06} = -0.1$

$r = -0.1$

**13.** $\dfrac{1}{2}, -\dfrac{1}{4}, \dfrac{1}{8}, -\dfrac{1}{16}, \ldots$

$\dfrac{-\dfrac{1}{4}}{\dfrac{1}{2}} = -\dfrac{1}{4} \cdot \dfrac{2}{1} = -\dfrac{2}{4} = -\dfrac{1}{2}$

$\dfrac{\dfrac{1}{8}}{-\dfrac{1}{4}} = \dfrac{1}{8} \cdot \left(-\dfrac{4}{1}\right) = -\dfrac{4}{8} = -\dfrac{1}{2}$

$\dfrac{-\dfrac{1}{16}}{\dfrac{1}{8}} = -\dfrac{1}{16} \cdot \dfrac{8}{1} = -\dfrac{8}{16} = -\dfrac{1}{2}$

$r = -\dfrac{1}{2}$

**15.** $75, 15, 3, \dfrac{3}{5}, \ldots$

$\dfrac{15}{75} = \dfrac{1}{5}, \quad \dfrac{3}{15} = \dfrac{1}{5}, \quad \dfrac{\dfrac{3}{5}}{3} = \dfrac{3}{5} \cdot \dfrac{1}{3} = \dfrac{1}{5}$

$r = \dfrac{1}{5}$

**17.** $\dfrac{1}{m}, \dfrac{6}{m^2}, \dfrac{36}{m^3}, \dfrac{216}{m^4}, \ldots$

$\dfrac{\dfrac{6}{m^2}}{\dfrac{1}{m}} = \dfrac{6}{m^2} \cdot \dfrac{m}{1} = \dfrac{6}{m}$

$\dfrac{\dfrac{36}{m^3}}{\dfrac{6}{m^2}} = \dfrac{36}{m^3} \cdot \dfrac{m^2}{6} = \dfrac{6}{m}$

$\dfrac{\dfrac{216}{m^4}}{\dfrac{36}{m^3}} = \dfrac{216}{m^4} \cdot \dfrac{m^3}{36} = \dfrac{6}{m}$

$r = \dfrac{6}{m}$

**19.** $3, 6, 12, \ldots$

$a_1 = 3, n = 7, \text{ and } r = \dfrac{6}{3} = 2$

We use the formula $a_n = a_1 r^{n-1}$.

$a_7 = 3 \cdot 2^{7-1} = 3 \cdot 2^6 = 3 \cdot 64 = 192$

**21.** $7, 7\sqrt{2}, 14, \ldots$

$a_1 = 7, n = 10, \text{ and } r = \dfrac{7\sqrt{2}}{7} = \sqrt{2}$

$a_n = a_1 r^{n-1}$

$a_{10} = 7(\sqrt{2})^{10-1} = 7(\sqrt{2})^9 = 7 \cdot 16\sqrt{2} = 112\sqrt{2}$

**23.** $-\dfrac{8}{243}, \dfrac{8}{81}, -\dfrac{8}{27}, \ldots$

$a_1 = -\dfrac{8}{243}, n = 14, \text{ and } r = \dfrac{\dfrac{8}{81}}{-\dfrac{8}{243}} =$

$\dfrac{8}{81}\left(-\dfrac{243}{8}\right) = -3$

$a_n = a_1 r^{n-1}$

$a_{14} = -\dfrac{8}{243}(-3)^{14-1} = -\dfrac{8}{243}(-3)^{13} =$

$-\dfrac{8}{243}(-1,594,323) = 52,488$

**25.** $\$1000, \$1080, \$1166.40, \ldots$

$a_1 = \$1000, n = 12, \text{ and } r = \dfrac{\$1080}{\$1000} = 1.08$

$a_n = a_1 r^{n-1}$

$a_{12} = \$1000(1.08)^{12-1} \approx \$1000(2.331638997) \approx$
$\$2331.64$

**27.** $1, 5, 25, 125, \ldots$

$a_1 = 1 \text{ and } r = \dfrac{5}{1} = 5$

$a_n = a_1 r^{n-1}$

$a_n = 1 \cdot 5^{n-1} = 5^{n-1}$

**29.** $1, -1, 1, -1, \ldots$

$a_1 = 1 \text{ and } r = \dfrac{-1}{1} = -1$

$a_n = a_1 r^{n-1}$

$a_n = 1(-1)^{n-1} = (-1)^{n-1}$

**31.** $\dfrac{1}{x}, \dfrac{1}{x^2}, \dfrac{1}{x^2}, \ldots$

$a_1 = \dfrac{1}{x}$ and $r = \dfrac{\frac{1}{x^2}}{\frac{1}{x}} = \dfrac{1}{x^2} \cdot \dfrac{x}{1} = \dfrac{1}{x}$

$a_n = a_1 r^{n-1}$

$a_n = \dfrac{1}{x}\left(\dfrac{1}{x}\right)^{n-1} = \dfrac{1}{x} \cdot \dfrac{1}{x^{n-1}} = \dfrac{1}{x^{1+n-1}} = \dfrac{1}{x^n}$, or $x^{-n}$

**33.** $6 + 12 + 24 + \ldots$

$a_1 = 6$, $n = 9$, and $r = \dfrac{12}{6} = 2$

$S_n = \dfrac{a_1(1 - r^n)}{1 - r}$

$S_9 = \dfrac{6(1 - 2^9)}{1 - 2} = \dfrac{6(1 - 512)}{-1} = \dfrac{6(-511)}{-1} = 3066$

**35.** $\dfrac{1}{18} - \dfrac{1}{6} + \dfrac{1}{2} - \ldots$

$a_1 = \dfrac{1}{18}$, $n = 7$, and $r = \dfrac{-\frac{1}{6}}{\frac{1}{18}} = -\dfrac{1}{6} \cdot \dfrac{18}{1} = -3$

$S_n = \dfrac{a_1(1 - r^n)}{1 - r}$

$S_7 = \dfrac{\frac{1}{18}\left[1 - (-3)^7\right]}{1 - (-3)} = \dfrac{\frac{1}{18}(1 + 2187)}{4} = \dfrac{\frac{1}{18}(2188)}{4} =$

$\dfrac{1}{18}(2188)\left(\dfrac{1}{4}\right) = \dfrac{547}{18}$

**37.** $1 + x + x^2 + x^3 + \ldots$

$a_1 = 1$, $n = 8$, and $r = \dfrac{x}{1}$, or $x$

$S_n = \dfrac{a_1(1 - r^n)}{1 - r}$

$S_8 = \dfrac{1(x - x^8)}{1 - x} = \dfrac{(1 + x^4)(1 - x^4)}{1 - x} =$

$\dfrac{(1 + x^4)(1 + x^2)(1 - x^2)}{1 - x} =$

$\dfrac{(1 + x^4)(1 + x^2)(1 + x)(1 - x)}{1 - x} =$

$(1 + x^4)(1 + x^2)(1 + x)$

**39.** $\$200$, $\$200(1.06)$, $\$200(1.06)^2, \ldots$

$a_1 = \$200$, $n = 16$, and $r = \dfrac{\$200(1.06)}{\$200} = 1.06$

$S_n = \dfrac{a_1(1 - r^n)}{1 - r}$

$S_{16} = \dfrac{\$200[1 - (1.06)^{16}]}{1 - 1.06} \approx$

$\dfrac{\$200(1 - 2.540351685)}{-0.06} \approx \$5134.51$

**41.** $16 + 4 + 1 + \ldots$

$|r| = \left|\dfrac{4}{16}\right| = \left|\dfrac{1}{4}\right| = \dfrac{1}{4}$, and since $|r| < 1$, the series does have a sum.

$S_\infty = \dfrac{a_1}{1 - r} = \dfrac{16}{1 - \frac{1}{4}} = \dfrac{16}{\frac{3}{4}} = 16 \cdot \dfrac{4}{3} = \dfrac{64}{3}$

**43.** $7 + 3 + \dfrac{9}{7} + \ldots$

$|r| = \left|\dfrac{3}{7}\right| = \dfrac{3}{7}$, and since $|r| < 1$, the series does have a sum.

$S_\infty = \dfrac{a_1}{1 - r} = \dfrac{7}{1 - \frac{3}{7}} = \dfrac{7}{\frac{4}{7}} = 7 \cdot \dfrac{7}{4} = \dfrac{49}{4}$

**45.** $3 + 15 + 75 + \ldots$

$|r| = \left|\dfrac{15}{3}\right| = |5| = 5$, and since $|r| \not< 1$ the series does not have a sum.

**47.** $4 - 6 + 9 - \dfrac{27}{2} + \ldots$

$|r| = \left|\dfrac{-6}{4}\right| = \left|-\dfrac{3}{2}\right| = \dfrac{3}{2}$, and since $|r| \not< 1$ the series does not have a sum.

**49.** $0.43 + 0.0043 + 0.000043 + \ldots$

$|r| = \left|\dfrac{0.0043}{0.43}\right| = |0.01| = 0.01$, and since $|r| < 1$, the series does have a sum.

$S_\infty = \dfrac{a_1}{1 - r} = \dfrac{0.43}{1 - 0.01} = \dfrac{0.43}{0.99} = \dfrac{43}{99}$

**51.** $\$500(1.02)^{-1} + \$500(1.02)^{-2} + \$500(1.02)^{-3} + \ldots$

$|r| = \left|\dfrac{\$500(1.02)^{-2}}{\$500(1.02)^{-1}}\right| = |(1.02)^{-1}| = (1.02)^{-1}$, or

$\dfrac{1}{1.02}$, and since $|r| < 1$, the series does have a sum.

$S_\infty = \dfrac{a_1}{1 - r} = \dfrac{\$500(1.02)^{-1}}{1 - \left(\frac{1}{1.02}\right)} = \dfrac{\frac{\$500}{1.02}}{\frac{0.02}{1.02}} =$

$\dfrac{\$500}{1.02} \cdot \dfrac{1.02}{0.02} = \$25,000$

**53.** $0.7777\ldots = 0.7 + 0.07 + 0.007 + 0.0007 + \ldots$

This is an infinite geometric series with $a_1 = 0.7$.

$|r| = \left|\dfrac{0.07}{0.7}\right| = |0.1| = 0.1 < 1$, so the series has a sum.

$S_\infty = \dfrac{a_1}{1 - r} = \dfrac{0.7}{1 - 0.1} = \dfrac{0.7}{0.9} = \dfrac{7}{9}$

Fractional notation for $0.7777\ldots$ is $\dfrac{7}{9}$.

**55.** $8.3838\ldots = 8.3 + 0.083 + 0.00083 + \ldots$

This is an infinite geometric series with $a_1 = 8.3$.

$|r| = \left|\dfrac{0.083}{8.3}\right| = |0.01| = 0.01 < 1$, so the series has a sum.

$S_\infty = \dfrac{a_1}{1 - r} = \dfrac{8.3}{1 - 0.01} = \dfrac{8.3}{0.99} = \dfrac{830}{99}$

Fractional notation for $8.3838\ldots$ is $\dfrac{830}{99}$.

**57.** $0.15151515\ldots = 0.15 + 0.0015 + 0.000015 + \ldots$

This is an infinite geometric series with $a_1 = 0.15$.

$|r| = \left|\dfrac{0.0015}{0.15}\right| = |0.01| = 0.01 < 1$, so the series has a sum.

$$S_\infty = \frac{a_1}{1-r} = \frac{0.15}{1-0.01} = \frac{0.15}{0.99} = \frac{15}{99} = \frac{5}{33}$$

Fractional notation for $0.15151515\ldots$ is $\dfrac{5}{33}$.

**59.** **Familiarize.** The rebound distances form a geometric sequence:

$$\frac{1}{4} \times 20, \quad \left(\frac{1}{4}\right)^2 \times 20, \quad \left(\frac{1}{4}\right)^3 \times 20, \ldots,$$

or $\quad 5, \quad \dfrac{1}{4} \times 5, \quad \left(\dfrac{1}{4}\right)^2 \times 5, \ldots$

The height of the 6th rebound is the 6th term of the sequence.

**Translate.** We will use the formula $a_n = a_1 r^{n-1}$, with $a_1 = 5$, $r = \dfrac{1}{4}$, and $n = 6$:

$$a_6 = 5\left(\frac{1}{4}\right)^{6-1}$$

**Carry out.** We calculate to obtain $a_6 = \dfrac{5}{1024}$.

**Check.** We can do the calculation again.

**State.** It rebounds $\dfrac{5}{1024}$ ft the 6th time.

**61.** **Familiarize.** In one year, the population will be $100,000 + 0.03(100,000)$, or $(1.03)100,000$. In two years, the population will be $(1.03)100,000 + 0.03(1.03)100,000$, or $(1.03)^2 100,000$. Thus the populations form a geometric sequence:

$$100,000, \quad (1.03)100,000, \quad (1.03)^2 100,000, \ldots$$

The population in 15 years will be the 16th term of the sequence.

**Translate.** We will use the formula $a_n = a_1 r^{n-1}$ with $a_1 = 100,000$, $r = 1.03$, and $n = 16$:

$$a_{16} = 100,000(1.03)^{16-1}$$

**Carry out.** We calculate to obtain $a_{16} \approx 155,797$.

**Check.** We can do the calculation again.

**State.** In 15 years the population will be about $155,797$.

**63.** **Familiarize.** At the end of each minute the population is 96% of the previous population.

We have a geometric sequence:

$$5000, \ 5000(0.96), \ 5000(0.96)^2, \ldots$$

The number of fruit flies remaining alive after 15 minutes is given by the 16th term of the sequence.

**Translate.** We use the formula $a_n = a_1 r^{n-1}$ with $a_1 = 5000$, $r = 0.96$, and $n = 16$:

$$a_{16} = 5000(0.96)^{16-1}$$

**Carry out.** We calculate to obtain $a_{16} \approx 2710$.

**Check.** We can do the calculation again.

**State.** About 2710 flies will be alive after 15 min.

**65.** **Familiarize.** Each year the number of apartments and houses built in the U.S. is 105.35% of the number built the previous year. These numbers form a geometric sequence:

$$534,000, \ 534,000(1.0535), \ 534,000(1.0535)^2, \ldots$$

The number of apartments and houses built from 1991 through 2004 is the sum of the first 14 terms of this sequence.

**Translate.** We use the formula $S_n = \dfrac{a_1(1-r^n)}{1-r}$ with $a_1 = 534,000$, $r = 1.0535$, and $n = 14$.

$$S_{14} = \frac{534,000(1-1.0535^{14})}{1-1.0535}$$

**Carry out.** We use a calculator to obtain

$$S_{14} \approx 10,723.419.$$

**Check.** We can do the calculation again.

**State.** About 10,723,491 apartments and houses were built in the U.S. from 1991 through 2004.

**67.** **Familiarize.** The lengths of the falls form a geometric sequence:

$$556, \ \left(\frac{3}{4}\right)556, \ \left(\frac{3}{4}\right)^2 556, \ \left(\frac{3}{4}\right)^3 556, \ldots$$

The total length of the first 6 falls is the sum of the first six terms of this sequence. The heights of the rebounds also form a geometric sequence:

$$\left(\frac{3}{4}\right)556, \ \left(\frac{3}{4}\right)^2 556, \ \left(\frac{3}{4}\right)^3 556, \ldots, \quad \text{or}$$

$$417, \ \left(\frac{3}{4}\right)417, \ \left(\frac{3}{4}\right)^2 417, \ldots$$

When the ball hits the ground for the 6th time, it will have rebounded 5 times. Thus the total length of the rebounds is the sum of the first five terms of this sequence.

**Translate.** We use the formula $S_n = \dfrac{a_1(1-r^n)}{1-r}$ twice, once with $a_1 = 556$, $r = \dfrac{3}{4}$, and $n = 6$ and a second time with $a_1 = 417$, $r = \dfrac{3}{4}$, and $n = 5$.

$D = $ Length of falls + length of rebounds

$$= \frac{556\left[1-\left(\frac{3}{4}\right)^6\right]}{1-\frac{3}{4}} + \frac{417\left[1-\left(\frac{3}{4}\right)^5\right]}{1-\frac{3}{4}}.$$

**Carry out.** We use a calculator to obtain $D \approx 3100.35$.

**Check.** We can do the calculations again.

**State.** The ball will have traveled about 3100.35 ft.

**69.** **Familiarize.** The heights of the stack form a geometric sequence:

$$0.02, 0.02(2), 0.02(2^2), \ldots$$

The height of the stack after it is doubled 10 times is given by the 11th term of this sequence.

**Translate.** We have a geometric sequence with $a_1 = 0.02$, $r = 2$, and $n = 11$. We use the formula

$$a_n = a_1 r^{n-1}.$$

**Carry out.** We substitute and calculate.

$$a_{11} = 0.02(2^{11-1})$$
$$a_{11} = 0.02(1024) = 20.48$$

**Check**. We can do the calculation again.

**State**. The final stack will be 20.48 in. high.

**71.** *Writing Exercise*

**73.**
$$(x + y)(x^2 + 2xy + y^2)$$
$$= x(x^2 + 2xy + y^2) + y(x^2 + 2xy + y^2)$$
$$= x^3 + 2x^2y + xy^2 + x^2y + 2xy^2 + y^3$$
$$= x^3 + 3x^2y + 3xy^2 + y^3$$

**75.** $5x - 2y = -3, \quad (1)$

$2x + 5y = -24 \quad (2)$

Multiply Eq. (1) by 5 and Eq. (2) by 2 and add.

$$25x - 10y = -15$$
$$\underline{4x + 10y = -48}$$
$$29x \qquad = -63$$
$$x = -\frac{63}{29}$$

Substitute $-\dfrac{63}{29}$ for $x$ in the second equation and solve for $y$.

$$2\left(-\frac{63}{29}\right) + 5y = -24$$
$$-\frac{126}{29} + 5y = -24$$
$$5y = -\frac{570}{29}$$
$$y = -\frac{114}{29}$$

The solution is $\left(-\dfrac{63}{29}, -\dfrac{114}{29}\right)$.

**77.** *Writing Exercise*

**79.** $\displaystyle\sum_{k=1}^{\infty} 6(0.9)^k = 6(0.9) + 6(0.9)^2 + 6(0.9)^3 + \ldots$

$|r| = \left|\dfrac{6(0.9)^2}{6(0.9)}\right| = |0.9| = 0.9 < 1$ so the series has a sum.

$$S_\infty = \frac{6(0.9)}{1 - 0.9} = \frac{6(0.9)}{0.1} = 54$$

**81.** $x^2 - x^3 + x^4 + x^5 + \ldots$

This is a geometric series with $a_1 = x^2$ and $r = -x$.

$$S_n = \frac{a_1(1 - r^n)}{1 - r} = \frac{x^2[1 - (-x)^n]}{1 - (-x)} = \frac{x^2[1 - (-x)^n]}{1 + x}$$

**83.** The length of a side of the first square is 16 cm. The length of a side of the next square is the length of the hypotenuse of a right triangle with legs 8 cm and 8 cm, or $8\sqrt{2}$ cm. The length of a side of the next square is the length of the hypotenuse of a right triangle with legs $4\sqrt{2}$ cm and $4\sqrt{2}$ cm, or 8 cm. The areas of the squares form a sequence:

$$(16)^2, \ (8\sqrt{2})^2, \ (8)^2, \ldots, \ \text{or}$$
$$256, \ 128, \ 64, \ldots.$$

This is a geometric sequence with $a_1 = 256$ and $r = \dfrac{1}{2}$.

We find the sum of the infinite geometric series $256 + 128 + 64 + \ldots$.

$$S_\infty = \frac{256}{1 - \dfrac{1}{2}} = \frac{256}{\dfrac{1}{2}} = 512 \text{ cm}^2$$

**85.** *Writing Exercise*

## Exercise Set 11.4

**1.** $2^5$, or 32

**3.** 9

**5.** $\begin{pmatrix} 8 \\ 5 \end{pmatrix}$

**7.** $x^7 y^2$

**9.** $9! = 9 \cdot 8 \cdot 7 \cdot 6 \cdot 5 \cdot 4 \cdot 3 \cdot 2 \cdot 1 = 362,880$

**11.** $11! = 11 \cdot 10 \cdot 9 \cdot 8 \cdot 7 \cdot 6 \cdot 5 \cdot 4 \cdot 3 \cdot 2 \cdot 1 = 39,916,800$

**13.** $\dfrac{8!}{6!} = \dfrac{8 \cdot 7 \cdot 6!}{6!} = 8 \cdot 7 = 56$

**15.** $\dfrac{9!}{5!} = \dfrac{9 \cdot 8 \cdot 7 \cdot 6 \cdot 5!}{5!} = 9 \cdot 8 \cdot 7 \cdot 6 = 3024$

**17.** $\begin{pmatrix} 7 \\ 4 \end{pmatrix} = \dfrac{7!}{3!4!} = \dfrac{7 \cdot 6 \cdot 5 \cdot 4!}{3 \cdot 2 \cdot 1 \cdot 4!} = \dfrac{7 \cdot 6 \cdot 5}{3 \cdot 2} = 35$

**19.** $\begin{pmatrix} 9 \\ 5 \end{pmatrix} = \dfrac{9!}{4!5!} = \dfrac{9 \cdot 8 \cdot 7 \cdot 6 \cdot 5!}{4 \cdot 3 \cdot 2 \cdot 5!} =$

$\dfrac{9 \cdot 8 \cdot 7 \cdot 6}{4 \cdot 3 \cdot 2} = 3 \cdot 7 \cdot 6 = 126$

**21.** $\begin{pmatrix} 30 \\ 3 \end{pmatrix} = \dfrac{30!}{27!3!} = \dfrac{30 \cdot 29 \cdot 28 \cdot 27!}{27! \cdot 3 \cdot 2 \cdot 1} = \dfrac{30 \cdot 29 \cdot 28}{3 \cdot 2} =$
4060

**23.** $\begin{pmatrix} 40 \\ 38 \end{pmatrix} = \dfrac{40!}{2!38!} = \dfrac{40 \cdot 39 \cdot 38!}{2 \cdot 1 \cdot 38!} = \dfrac{40 \cdot 39}{2} = 780$

**25.** Expand $(a - b)^4$.

We have $a = a$, $b = -b$, and $n = 4$.

Form 1: We use the fifth row of Pascal's triangle:

$$1 \qquad 4 \qquad 6 \qquad 4 \qquad 1$$
$$(a - b)^4$$
$$= 1 \cdot a^4 + 4a^3(-b) + 6a^2(-b)^2 + 4a(-b)^3 + (-b)^4$$
$$= a^4 - 4a^3b + 6a^2b^2 - 4ab^3 + b^4$$

Form 2:

$$(a - b)^4 = \begin{pmatrix} 4 \\ 0 \end{pmatrix} a^4 + \begin{pmatrix} 4 \\ 1 \end{pmatrix} a^3(-b) + \begin{pmatrix} 4 \\ 2 \end{pmatrix} a^2(-b)^2 +$$
$$\begin{pmatrix} 4 \\ 3 \end{pmatrix} a(-b)^3 + \begin{pmatrix} 4 \\ 4 \end{pmatrix} (-b)^4$$
$$= \frac{4!}{4!0!} a^4 + \frac{4!}{3!1!} a^3(-b) + \frac{4!}{2!2!} a^2(-b)^2 +$$
$$\frac{4!}{1!3!} a(-b)^3 + \frac{4!}{0!4!} (-b)^4$$
$$= a^4 - 4a^3b + 6a^2b^2 - 4ab^3 + b^4$$

**27.** Expand $(p + q)^7$.

We have $a = p$, $b = q$, and $n = 7$.

Form 1: We use the eighth row of Pascal's triangle:

$$1 \quad 7 \quad 21 \quad 35 \quad 35 \quad 21 \quad 7 \quad 1$$

$(p + q)^7 = p^7 + 7p^6q + 21p^5q^2 + 35p^4q^3 + 35p^3q^4 +$
$\qquad 21p^2q^5 + 7pq^6 + q^7$

Form 2:

$(p + q)^7 = \binom{7}{0} p^7 + \binom{7}{1} p^6q + \binom{7}{2} p^5q^2 +$

$\qquad \binom{7}{3} p^4q^3 + \binom{7}{4} p^3q^4 + \binom{7}{5} p^2q^5 +$

$\qquad \binom{7}{6} pq^6 + \binom{7}{7} q^7$

$\qquad = \dfrac{7!}{7!0!}p^7 + \dfrac{7!}{6!1!}p^6q + \dfrac{7!}{5!2!}p^5q^2 +$

$\qquad \dfrac{7!}{4!3!}p^4q^3 \dfrac{7!}{3!4!}p^3q^4 + \dfrac{7!}{2!5!}p^2q^5 +$

$\qquad \dfrac{7!}{1!6!}pq^6 + \dfrac{7!}{0!7!}q^7$

$\qquad = p^7 + 7p^6q + 21p^5q^2 + 35p^4q^3 + 35p^3q^4 +$
$\qquad 21p^2q^5 + 7pq^6 + q^7$

**29.** Expand $(3c - d)^7$.

We have $a = 3c$, $b = -d$, and $n = 7$.

Form 1: We use the eighth row of Pascal's triangle:

$$1 \quad 7 \quad 21 \quad 35 \quad 35 \quad 21 \quad 7 \quad 1$$

$(3c - d)^7 = 1 \cdot (3c)^7 + 7(3c)^6(-d) + 21(3c)^5(-d)^2 +$
$\qquad 35(3c)^4(-d)^3 + 35(3c)^3(-d)^4 +$
$\qquad 21(3c)^2(-d)^5 + 7(3c)(-d)^6 + 1 \cdot (-d)^7$

$\qquad = 2187c^7 - 5103c^6d + 5103c^5d^2 -$
$\qquad 2835c^4d^3 + 945c^3d^4 - 189c^2d^5 +$
$\qquad 21cd^6 - d^7$

Form 2:

$(3c - d)^7 = \binom{7}{0}(3c)^7 + \binom{7}{1}(3c)^6(-d) +$

$\qquad \binom{7}{2}(3c)^5(-d)^2 + \binom{7}{3}(3c)^4(-d)^3 +$

$\qquad \binom{7}{4}(3c)^3(-d)^4 + \binom{7}{5}(3c)^2(-d)^5 +$

$\qquad \binom{7}{6}(3c)(-d)^6 + \binom{7}{7}(-d)^7$

$\qquad = \dfrac{7!}{7!0!}(3c)^7 + \dfrac{7!}{6!1!}(3c)^6(-d) +$

$\qquad \dfrac{7!}{5!2!}(3c)^5(-d)^2 + \dfrac{7!}{4!3!}(3c)^4(-d)^3 +$

$\qquad \dfrac{7!}{3!4!}(3c)^3(-d)^4 + \dfrac{7!}{2!5!}(3c)^2(-d)^5 +$

$\qquad \dfrac{7!}{1!6!}(3c)(-d)^6 + \dfrac{7!}{0!7!}(-d)^7$

$\qquad = 2187c^7 - 5103c^6d + 5103c^5d^2 -$
$\qquad 2835c^4d^3 + 945c^3d^4 - 189c^2d^5 +$
$\qquad 21cd^6 - d^7$

**31.** Expand $(t^{-2} + 2)^6$.

We have $a = t^{-2}$, $b = 2$, and $n = 6$.

Form 1: We use the 7th row of Pascal's triangle:

$$1 \quad 6 \quad 15 \quad 20 \quad 15 \quad 6 \quad 1$$

$(t^{-2} + 2)^6 = 1 \cdot (t^{-2})^6 + 6(t^{-2})^5(2) + 15(t^{-2})^4(2^2) +$
$\qquad 20(t^{-2})^3(2^3) + 15(t^{-2})^2(2^4) +$
$\qquad 6t^{-2}(2^5) + 1 \cdot 2^6$

$\qquad = t^{-12} + 12t^{-10} + 60t^{-8} + 160t^{-6} +$
$\qquad 240t^{-4} + 192t^{-2} + 64$

Form 2:

$(t^{-2} + 2)^6 = \binom{6}{0}(t^{-2})^6 + \binom{6}{1}(t^{-2})^5(2) +$

$\qquad \binom{6}{2}(t^{-2})^4(2^2) + \binom{6}{3}(t^{-2})^3(2^3) +$

$\qquad \binom{6}{4}(t^{-2})^2(2^4) + \binom{6}{5}(t^{-2})^3(2^5) +$

$\qquad \binom{6}{6}2^6$

$\qquad = \dfrac{6!}{6!0!}(t^{-2})^6 + \dfrac{6!}{5!1!}(t^{-2})^5(2) +$

$\qquad \dfrac{6!}{4!2!}(t^{-2})^4(2^2) + \dfrac{6!}{3!3!}(t^{-2})^3(2^3) +$

$\qquad \dfrac{6!}{2!4!}(t^{-2})^2(2^4) + \dfrac{6!}{1!5!}(t^{-2})(2^5) +$

$\qquad \dfrac{6!}{0!6!}(2^6)$

$\qquad = t^{-12} + 12t^{-10} + 60t^{-8} + 160t^{-6} +$
$\qquad 240t^{-4} + 192t^{-2} + 64$

**33.** Expand $(x - y)^5$.

We have $a = x$, $b = -y$, and $n = 5$.

Form 1: We use the sixth row of Pascal's triangle.

$$1 \quad 5 \quad 10 \quad 10 \quad 5 \quad 1$$

$(x - y)^5$
$= 1 \cdot x^5 + 5x^4(-y) + 10x^3(-y)^2 + 10x^2(-y)^3 +$
$\qquad 5x(-y)^4 + 1 \cdot (-y)^5$
$= x^5 - 5x^4y + 10x^3y^2 - 10x^2y^3 + 5xy^4 - y^5$

Form 2:

$(x - y)^5$

$= \binom{5}{0} x^5 + \binom{5}{1} x^4(-y) + \binom{5}{2} x^3(-y)^2 +$

$\qquad \binom{5}{3} x^2(-y)^3 + \binom{5}{4} x(-y)^4 + \binom{5}{5} (-y)^5$

$= \dfrac{5!}{5!0!}x^5 + \dfrac{5!}{4!1!}x^4(-y) + \dfrac{5!}{3!2!}x^3(-y)^2 +$

$\qquad \dfrac{5!}{2!3!}x^2(-y)^3 + \dfrac{5!}{1!4!}x(-y)^4 + \dfrac{5!}{0!5!}(-y)^5$

$= x^5 - 5x^4y + 10x^3y^2 - 10x^2y^3 + 5xy^4 - y^5$

**35.** Expand $\left(3s + \dfrac{1}{t}\right)^9$.

We have $a = 3s$, $b = \dfrac{1}{t}$, and $n = 9$.

Form 1: We use the tenth row of Pascal's triangle:

1   9   36   84   126   126   84   36   9   1

$\left(3s + \dfrac{1}{t}\right)^9$

$= 1 \cdot (3s)^9 + 9(3s)^8\left(\dfrac{1}{t}\right) + 36(3s)^7\left(\dfrac{1}{t}\right)^2 +$

$84(3s)^6\left(\dfrac{1}{t}\right)^3 + 126(3s)^5\left(\dfrac{1}{t}\right)^4 +$

$126(3s)^4\left(\dfrac{1}{t}\right)^5 + 84(3s)^3\left(\dfrac{1}{t}\right)^6 +$

$36(3s)^2\left(\dfrac{1}{t}\right)^7 + 9(3s)\left(\dfrac{1}{t}\right)^8 +$

$1 \cdot \left(\dfrac{1}{t}\right)^9$

$= 19,683s^9 + \dfrac{59,049s^8}{t} + \dfrac{78,732s^7}{t^2} +$

$\dfrac{61,236s^6}{t^3} + \dfrac{30,618s^5}{t^4} + \dfrac{10,206s^4}{t^5} +$

$\dfrac{2268s^3}{t^6} + \dfrac{324s^2}{t^7} + \dfrac{27s}{t^8} + \dfrac{1}{t^9}$

Form 2:

$\left(3s + \dfrac{1}{t}\right)^9$

$= \binom{9}{0}(3s)^9 + \binom{9}{1}(3s)^8\left(\dfrac{1}{t}\right) +$

$\binom{9}{2}(3s)^7\left(\dfrac{1}{t}\right)^2 + \binom{9}{3}(3s)^6\left(\dfrac{1}{t}\right)^3 +$

$\binom{9}{4}(3s)^5\left(\dfrac{1}{t}\right)^4 + \binom{9}{5}(3s)^4\left(\dfrac{1}{t}\right)^5 +$

$\binom{9}{6}(3s)^3\left(\dfrac{1}{t}\right)^6 + \binom{9}{7}(3s)^2\left(\dfrac{1}{t}\right)^7 +$

$\binom{9}{8}(3s)\left(\dfrac{1}{t}\right)^8 + \binom{9}{9}\left(\dfrac{1}{t}\right)^9$

$= \dfrac{9!}{9!0!}(3s)^9 + \dfrac{9!}{8!1!}(3s)^8\left(\dfrac{1}{t}\right) +$

$\dfrac{9!}{7!2!}(3s)^7\left(\dfrac{1}{t}\right)^2 + \dfrac{9!}{6!3!}(3s)^6\left(\dfrac{1}{t}\right)^3 +$

$\dfrac{9!}{5!4!}(3s)^5\left(\dfrac{1}{t}\right)^4 + \dfrac{9!}{4!5!}(3s)^4\left(\dfrac{1}{t}\right)^5 +$

$\dfrac{9!}{3!6!}(3s)^3\left(\dfrac{1}{t}\right)^6 + \dfrac{9!}{2!7!}(3s)^2\left(\dfrac{1}{t}\right)^7 +$

$\dfrac{9!}{1!8!}(3s)\left(\dfrac{1}{t}\right)^8 + \dfrac{9!}{0!9!}\left(\dfrac{1}{t}\right)^9$

$= 19,683s^9 + \dfrac{59,049s^8}{t} + \dfrac{78,732s^7}{t^2} +$

$\dfrac{61,236s^6}{t^3} + \dfrac{30,618s^5}{t^4} + \dfrac{10,206s^4}{t^5} +$

$\dfrac{2268s^3}{t^6} + \dfrac{324s^2}{t^7} + \dfrac{27s}{t^8} + \dfrac{1}{t^9}$

**37.** $(x^3 - 2y)^5$

We have $a = x^3$, $b = -2y$, and $n = 5$.

Form 1: We use the 6th row of Pascal's triangle.

1     5     10     10     5     1

$(x^3 - 2y)^5$

$= 1 \cdot (x^3)^5 + 5(x^3)^4(-2y) + 10(x^3)^3(-2y)^2 +$

$10(x^3)^2(-2y)^3 + 5(x^3)(-2y)^4 + 1 \cdot (-2y)^5$

$= x^{15} - 10x^{12}y + 40x^9y^2 - 80x^6y^3 +$

$80x^3y^4 - 32y^5$

Form 2:

$(x^3 - 2y)^5$

$= \binom{5}{0}(x^3)^5 + \binom{5}{1}(x^3)^4(-2y) +$

$\binom{5}{2}(x^3)^3(-2y)^2 + \binom{5}{3}(x^3)^2(-2y)^3 +$

$\binom{5}{4}(x^3)(-2y)^4 + \binom{5}{5}(-2y)^5$

$= \dfrac{5!}{5!0!}(x^3)^5 + \dfrac{5!}{4!1!}(x^3)^4(-2y) +$

$\dfrac{5!}{3!2!}(x^3)^3(-2y)^2 + \dfrac{5!}{2!3!}(x^3)^2(-2y)^3 +$

$\dfrac{5!}{1!4!}(x^3)(-2y)^4 + \dfrac{5!}{0!5!}(-2y)^5$

$= x^{15} - 10x^{12}y + 40x^9y^2 - 80x^6y^3 +$

$80x^3y^4 - 32y^5$

**39.** Expand $(\sqrt{5} + t)^6$.

We have $a = \sqrt{5}$, $b = t$, and $n = 6$.

Form 1: We use the seventh row of Pascal's triangle:

1     6     15     20     15     6     1

$(\sqrt{5} + t)^6 = 1 \cdot (\sqrt{5})^6 + 6(\sqrt{5})^5(t) +$

$15(\sqrt{5})^4(t^2) + 20(\sqrt{5})^3(t^3) +$

$15(\sqrt{5})^2(t^4) + 6\sqrt{5}t^5 + 1 \cdot t^6$

$= 125 + 150\sqrt{5}\,t + 375t^2 + 100\sqrt{5}\,t^3 +$

$75t^4 + 6\sqrt{5}\,t^5 + t^6$

Form 2:

$$(\sqrt{5}+t)^6 = \binom{6}{0}(\sqrt{5})^6 + \binom{6}{1}(\sqrt{5})^5(t) +$$

$$\binom{6}{2}(\sqrt{5})^4(t^2) + \binom{6}{3}(\sqrt{5})^3(t^3) +$$

$$\binom{6}{4}(\sqrt{5})^2(t^4) + \binom{6}{5}(\sqrt{5})(t^5) +$$

$$\binom{6}{6}(t^6)$$

$$= \frac{6!}{6!0!}(\sqrt{5})^6 + \frac{6!}{5!1!}(\sqrt{5})^5(t) +$$

$$\frac{6!}{4!2!}(\sqrt{5})^4(t^2) + \frac{6!}{3!3!}(\sqrt{5})^3(t^3) +$$

$$\frac{6!}{2!4!}(\sqrt{5})^2(t^4) + \frac{6!}{1!5!}(\sqrt{5})(t^5) +$$

$$\frac{6!}{0!6!}(t^6)$$

$$= 125 + 150\sqrt{5}\,t + 375t^2 + 100\sqrt{5}\,t^3 +$$

$$75t^4 + 6\sqrt{5}\,t^5 + t^6$$

**41.** Expand $\left(\dfrac{1}{\sqrt{x}} - \sqrt{x}\right)^6$.

We have $a = \dfrac{1}{\sqrt{x}}$, $b = -\sqrt{x}$, and $n = 6$.

Form 1: We use the seventh row of Pascal's triangle:

$$1 \quad 6 \quad 15 \quad 20 \quad 15 \quad 6 \quad 1$$

$$\left(\frac{1}{\sqrt{x}} - \sqrt{x}\right)^6$$

$$= 1 \cdot \left(\frac{1}{\sqrt{x}}\right)^6 + 6\left(\frac{1}{\sqrt{x}}\right)^5(-\sqrt{x}) +$$

$$15\left(\frac{1}{\sqrt{x}}\right)^4(-\sqrt{x})^2 + 20\left(\frac{1}{\sqrt{x}}\right)^3(-\sqrt{x})^3 +$$

$$15\left(\frac{1}{\sqrt{x}}\right)^2(-\sqrt{x})^4 + 6\left(\frac{1}{\sqrt{x}}\right)(-\sqrt{x})^5 + 1 \cdot (-\sqrt{x})^6$$

$$= x^{-3} - 6x^{-2} + 15x^{-1} - 20 + 15x - 6x^2 + x^3$$

Form 2:

$$\left(\frac{1}{\sqrt{x}} - \sqrt{x}\right)^6$$

$$= \binom{6}{0}\left(\frac{1}{\sqrt{x}}\right)^6 + \binom{6}{1}\left(\frac{1}{\sqrt{x}}\right)^5(-\sqrt{x}) +$$

$$\binom{6}{2}\left(\frac{1}{\sqrt{x}}\right)^4(-\sqrt{x})^2 + \binom{6}{3}\left(\frac{1}{\sqrt{x}}\right)^3(-\sqrt{x})^3 +$$

$$\binom{6}{4}\left(\frac{1}{\sqrt{x}}\right)^2(-\sqrt{x})^4 + \binom{6}{5}\left(\frac{1}{\sqrt{x}}\right)(-\sqrt{x})^5 +$$

$$\binom{6}{6}(-\sqrt{x})^6$$

$$= \frac{6!}{6!0!}\left(\frac{1}{\sqrt{x}}\right)^6 + \frac{6!}{5!1!}\left(\frac{1}{\sqrt{x}}\right)^5(-\sqrt{x}) +$$

$$\frac{6!}{4!2!}\left(\frac{1}{\sqrt{x}}\right)^4(-\sqrt{x})^2 + \frac{6!}{3!3!}\left(\frac{1}{\sqrt{x}}\right)^3(-\sqrt{x})^3 +$$

$$\frac{6!}{2!4!}\left(\frac{1}{\sqrt{x}}\right)^2(-\sqrt{x})^4 + \frac{6!}{1!5!}\left(\frac{1}{\sqrt{x}}\right)(-\sqrt{x})^5 +$$

$$\frac{6!}{0!6!}(-\sqrt{x})^6$$

$$= x^{-3} - 6x^{-2} + 15x^{-1} - 20 + 15x - 6x^2 + x^3$$

**43.** Find the 3rd term of $(a+b)^6$.

First, we note that $3 = 2+1$, $a = a$, $b = b$, and $n = 6$. Then the 3rd term of the expansion of $(a+b)^6$ is

$$\binom{6}{2}a^{6-2}b^2, \text{ or } \frac{6!}{4!2!}a^4b^2, \text{ or } 15a^4b^2.$$

**45.** Find the 12th term of $(a-3)^{14}$.

First, we note that $12 = 11+1$, $a = a$, $b = -3$, and $n = 14$. Then the 12th term of the expansion of $(a-3)^{14}$ is

$$\binom{14}{11}a^{14-11} \cdot (-3)^{11} = \frac{14!}{3!11!}a^3(-177,147)$$

$$= 364a^3(-177,147)$$

$$= -64,481,508a^3$$

**47.** Find the 5th term of $(2x^3 - \sqrt{y})^8$.

First, we note that $5 = 4+1$, $a = 2x^3$, $b = -\sqrt{y}$, and $n = 8$. Then the 5th term of the expansion of $(2x^3 - \sqrt{y})^8$ is

$$\binom{8}{4}(2x^3)^{8-4}(-\sqrt{y})^4$$

$$= \frac{8!}{4!4!}(2x^3)^4(-\sqrt{y})^4$$

$$= 70(16x^{12})(y^2)$$

$$= 1120x^{12}y^2$$

**49.** The expansion of $(2u - 3v^2)^{10}$ has 11 terms so the 6th term is the middle term. Note that $6 = 5+1$, $a = 2u$, $b = -3v^2$, and $n = 10$. Then the 6th term of the expansion of $(2u - 3v^2)^{10}$ is

$$\binom{10}{5}(2u)^{10-5}(-3v^2)^5$$

$$= \frac{10!}{5!5!}(2u)^5(-3v^2)^5$$

$$= 252(32u^5)(-243v^{10})$$

$$= -1,959,552u^5v^{10}$$

**51.** The 9th term of $(x-y)^8$ is the last term, $y^8$.

**53.** *Writing Exercise*

**55.** $\log_2 x + \log_2(x-2) = 3$

$\qquad \log_2 x(x-2) = 3$

$\qquad\qquad x(x-2) = 2^3$

$\qquad\qquad x^2 - 2x = 8$

$\qquad\quad x^2 - 2x - 8 = 0$

$\qquad (x-4)(x+2) = 0$

$x = 4 \ or \ x = -2$

Only 4 checks. It is the solution.

**57.** $\quad e^t = 280$

$\ln e^t = \ln 280$

$\quad\ t = \ln 280$

$\quad\ t \approx 5.6348$

**59.** *Writing Exercise*

**61.** Consider the set of 5 elements $\{a, b, c, d, e\}$. List all the subsets of size 3:

$\{a,b,c\}, \{a,b,d\}, \{a,b,e\}, \{a,c,d\},$
$\{a,c,e\}, \{a,d,e\}, \{b,c,d\}, \{b,c,e\},$
$\{b,d,e\}, \{c,d,e\}.$

There are exactly 10 subsets of size 3 and $\begin{pmatrix} 5 \\ 3 \end{pmatrix} = 10$, so there are exactly $\begin{pmatrix} 5 \\ 3 \end{pmatrix}$ ways of forming a subset of size 3 from a set of 5 elements.

**63.** Find the sixth term of $(0.15 + 0.85)^8$:

$$\begin{pmatrix} 8 \\ 5 \end{pmatrix}(0.15)^{8-5}(0.85)^5 = \frac{8!}{3!5!}(0.15)^3(0.85)^5 \approx 0.084$$

**65.** Find and add the 7th through the 9th terms of $(0.15 + 0.85)^9$:

$$\begin{pmatrix} 8 \\ 6 \end{pmatrix}(0.15)^2(0.85)^6 + \begin{pmatrix} 8 \\ 7 \end{pmatrix}(0.15)(0.85)^7 +$$

$$\begin{pmatrix} 8 \\ 8 \end{pmatrix}(0.85)^8 \approx 0.89$$

**67.** $\begin{pmatrix} n \\ n-r \end{pmatrix} = \dfrac{n!}{[n-(n-r)!](n-r)!} = \dfrac{n!}{r!(n-r)!} =$

$\begin{pmatrix} n \\ r \end{pmatrix}$

**69.** $\dfrac{\begin{pmatrix} 5 \\ 3 \end{pmatrix}(p^2)^2\left(-\dfrac{1}{2}p\sqrt[3]{q}\right)^3}{\begin{pmatrix} 5 \\ 2 \end{pmatrix}(p^2)^3\left(-\dfrac{1}{2}p\sqrt[3]{q}\right)^2} = \dfrac{-\dfrac{1}{8}p^7 q}{\dfrac{1}{4}p^8\sqrt[3]{q^2}} =$

$-\dfrac{\dfrac{1}{8}p^7 q}{\dfrac{1}{4}p^8 q^{2/3}} = -\dfrac{1}{8} \cdot \dfrac{4}{1} \cdot p^{7-8} \cdot q^{1-2/3} =$

$-\dfrac{1}{2}p^{-1}q^{1/3} = -\dfrac{\sqrt[3]{q}}{2p}$

**71.** $(x^2 + 2xy + y^2)(x^2 + 2xy + y^2)^2(x+y) =$

$(x+y)^2[(x+y)^2]^2(x+y) = (x+y)^7$

We can find the given product by finding the binomial expansion of $(x+y)^7$. It is $x^7 + 7x^6y + 21x^5y^2 + 35x^4y^3 + 35x^3y^4 + 21x^2y^5 + 7xy^6 + y^7$. (See Exercise 27.)